Buildings: 10th Anniversary-Volume II

Buildings: 10th Anniversary-Volume II

Editor

David Arditi

Basel • Beijing • Wuhan • Barcelona • Belgrade • Novi Sad • Cluj • Manchester

Editor
David Arditi
Illinois Institute of Technology
Chicago
USA

Editorial Office
MDPI
St. Alban-Anlage 66
4052 Basel, Switzerland

This is a reprint of articles from the Special Issue published online in the open access journal *Buildings* (ISSN 2075-5309)   (available at: https://www.mdpi.com/journal/buildings/special_issues/Buildings_10th_Anniversary).

For citation purposes, cite each article independently as indicated on the article page online and as indicated below:

Lastname, A.A.; Lastname, B.B. Article Title. *Journal Name* **Year**, *Volume Number*, Page Range.

Volume II
ISBN 978-3-7258-1163-2 (Hbk)
ISBN 978-3-7258-1164-9 (PDF)
doi.org/10.3390/books978-3-7258-1164-9

Set
ISBN 978-3-7258-1159-5 (Hbk)
ISBN 978-3-7258-1160-1 (PDF)

Cover image courtesy of S. R. Crown Hall, Illinois Institute of Technology, Chicago IL 60616

Contents

About the Editor

David Arditi

Dr. David Arditi is currently Professor Emeritus of Civil and Architectural Engineering at Illinois Institute of Technology. He is the founder and former Director of the Construction Engineering and Management Program. He established the Burton and Erma Lewis Construction Engineering and Management Laboratory in 1995 and the Trimble Technology Laboratory in 2022. Moreover, the sum of the funds that he raised in his academic career for research and lab development amounts to around USD 20 million. During his 50-year academic career, Professor Arditi conducted several funded research projects sponsored by federal and state agencies. He supervised the research of many PhD and MS students with diverse nationalities who occupy respectable academic and industry positions in over 20 countries. Dr. Arditi and his research associates have published close to 400 research papers in refereed national and international journals, edited books, and peer-reviewed conference proceedings. According to Google Scholar, by March 2024, Dr. Arditi's research papers received 15,814 citations and had an h-index of 73. He is currently serving as the Editor-in-Chief of *Buildings* (MDPI) and is Associate Editor of the *International Journal of Project Organization and Management*, in addition to serving on the editorial board of 12 journals in the field of construction project management. He regularly reviews research proposals, research papers, and textbooks for several research agencies and publishers located in the U.S., Canada, Saudi Arabia, Hong Kong, Israel, and Georgia. Dr. Arditi is an active member of several professional societies including CMAA, ASCE, AACE, and PMI. Furthermore, he is the recipient of multiple awards: he was elected to the College of Fellows of CMAA in 2013, named Professor of the Year by CAEE students in 2016, and elected a member of the National Academy of Construction (NAC) in 2022.

Preface

The scope of *Buildings* includes subjects related to building science, building engineering, building construction, and architecture. *Buildings* is organized into five sections: (1) Building Structures; (2) Building Materials and Repair and Renovation; (3) Building Energy, Physics, Environment, and Systems; (4) Architectural Design, Urban Science, and Real Estate; and (5) Construction Management, Computers, and Digitization. The "10th Anniversary" Special Issue comprises 39 research papers covering historical trends, the state of the art, current challenges, creative and innovative solutions, and future developments in each of these five sections (with 8 papers in Section 1, 6 in Section 2, 7 in Section 3, 6 in Section 4, and 12 in Section 5). Twenty-one papers focusing on the first three sections are in Book 1, whereas the remaining eighteen papers are in Book 2. I would like to thank the researchers who contributed to these two books of the Special Issue, as well as the reviewers, editors, and managerial staff (especially our Managing Editor at the time, Ms. May Zheng) who were involved in the production of this Special Issue. The multi-disciplinary nature of the 39 high-quality papers included in the "10th Anniversary" Special Issue reflects the high level of maturity that has been reached by *Buildings* over the last 10 years, as well as the wider impact of *Buildings* on building-related research and practices.

David Arditi
Editor

 *buildings*

Article

Energy Rating of Buildings to Promote Energy-Conscious Design in Israel

Abraham Yezioro and Isaac Guedi Capeluto *

Climate and Energy Lab in Architecture, Faculty of Architecture and Town Planning, Technion-Israel Institute of Technology, Haifa 3200003, Israel; ayez@ar.technion.ac.il
* Correspondence: arrguedi@technion.ac.il

Abstract: Improving the energy efficiency of existing and new buildings is an important step towards achieving more sustainable environments. There are various methods for grading buildings that are required according to regulations in different places for green building certification. However, in new buildings, these rating systems are usually implemented at late design stages due to their complexity and lack of integration in the architectural design process, thus limiting the available options for improving their performance. In this paper, the model ENERGYui used for design and rating buildings in Israel is presented. One of its main advantages is that it can be used at any design stage, including the early ones. It requires information that is available at each stage only, as the additional necessary information is supplemented by the model. In this way, architects can design buildings in a way where they are aware of each design decision and its impact on their energy performance, while testing different design directions. ENERGYui rates the energy performance of each basic unit, as well as the entire building. The use of the model is demonstrated in two different scenarios: an office building in which basic architectural features such as form and orientation are tested from the very beginning, and a residential building in which the intervention focuses on its envelope, highlighting the possibilities of improving their design during the whole design process.

Keywords: energy rating; green buildings; design tools; energy-conscious design

Citation: Yezioro, A.; Capeluto, I.G. Energy Rating of Buildings to Promote Energy-Conscious Design in Israel. *Buildings* **2021**, *11*, 59. https://doi.org/10.3390/buildings11020059

Academic Editor: David Arditi
Received: 30 December 2020
Accepted: 2 February 2021
Published: 8 February 2021

1. Introduction

In recent years, there has been a significant increase in interest in subjects concerning sustainable design and construction of buildings that save energy and emissions, while ensuring comfortable conditions inside and outside of them.

In order to evaluate the energetic performance of buildings, different methods and rating systems have been developed in various places in the world, such as LEED [1] in North America and EPBD [2] in Europe. In this work, we introduce the model ENERGYui as a tool for design and rating buildings in Israel. The paper significantly expands upon two previously published conference papers, where an early limited version of the model was introduced, and demonstrates its use not only for rating buildings but also as a design tool [3,4]. These methods can help enable consumers and businesses to make more informed choices and decisions to save energy and money. Despite the development of these methods, there is still uncertainty about their relationship to property value and the understanding of the meaning of the energy performance certificates by the general public [5]. As part of these directives, various methods were developed, which can be used by designers and advisors for evaluating the green performance of buildings in general, and their energy performance in particular [6]. The complexity involved in these evaluations and the special requirements of each method has led to the development of a large variety of tools with different levels of difficulty. A comprehensive list has been featured and evaluated according to various criteria by the United States Department of Energy [7,8]. Although there are tools to evaluate the implications of design changes with an emphasis on architectural parameters, there is not always clear what are the assumptions that these tools take regarding the rest of the parameters. This is a critical point for a design tool that rates buildings according to the strict requirements of a standard. Different approaches for rating buildings are used; some of them are based on actual performance while others rely on design data. The former reflects the performance of the building after its construction and occupancy, while the latter rate the proposed building before its construction, which poses a significant challenge.

To evaluate the energy consumption and performance of the building, sophisticated dynamic hourly energy simulation models, which require a high degree of detail, are generally used. Expertise is required to define the input needed for simulation as well as for understanding the output produced by the models. Moreover, tedious work is required for defining all the building parameters and details needed to perform the simulation. For these reasons, these simulation tools are generally used late in the design process [9], mainly by external expert consultants and not by architects, and therefore their impact during the design process is limited. At late design stages, it is very expensive and sometimes impossible to propose and implement major design changes in the building, even if they may bring a significant improvement in its performance [10]. Furthermore, using energy simulation tools usually is not aligned with the design process and requires capabilities beyond those commonly available to designers. The tools and knowledge that required setting the proper conditions of the simulation usually deprive architects of using this kind of tool during the design process and prevent the possibility of asking important what-if questions that can encourage them to examine different design directions. As a result, these are generally inappropriate as practical design aids for architects, especially early in the process since they share the following characteristics:

- Demand expertise and specific knowledge;
- Require the definition of multiple variables related to mechanical systems, load, schedules, etc.;
- Produce complicated outputs that are difficult to understand and translate to architectural changes or use for answering "what to do next?" questions.

The Israeli Standard IS5281 "Buildings with Reduced Environmental Impact (Green Buildings)" was approved in November 2005 [11]; it was recently updated [12] after a comprehensive revision and has an important impact on the architectural practice in Israel. It provides a multi-disciplinary approach for the assessment of new and thoroughly renovated buildings, by scoring points and compliance thresholds. The standard was adopted in 2013 by the forum of the 15 main cities in Israel and deemed as compulsory for building permits in their jurisdiction. Following this initiative, the planning authorities have decided that the standard will be mandatory for all construction in Israel starting in 2021. It is worth emphasizing that so far in Israel, the only mandatory requirement for obtaining a building permit has been in compliance with IS1045 "Thermal Insulation of Buildings" [13,14], which determines the minimum required levels of thermal insulation of envelope elements according to the building type and climatic zone in which the building is located. Standard IS5281 is divided into nine main chapters that focus on the different aspects of sustainable design and green architecture: (1) energy, (2) site, (3) water, (4) materials, (5) health and wellbeing, (6) waste, (7) transport, (8) management, and (9) innovation. Among them, the energy chapter is the most significant in terms of its relative weight, and its verification and compliance rely on Standard IS5282 "Energy Rating of Buildings" [15].

Standard IS5282 uses two basic approaches to demonstrate compliance: (1) the prescriptive-descriptive approach [15,16] which defines various pre-set solutions (prescriptions) to achieve energy conservation according to the desired ranking, and (2) the performance approach, which defines the energy performance of the building that should be met, considering site energy. In this case, the energy consumption of the proposed building is compared with a theoretical reference building, which determines the energy budget. The rating of the building is determined according to the ratio of energy savings in relation to the reference building, between level F (worst) and level A+ (best). For the implementation of this approach, the use of a dynamic energy hourly simulation model is required. For residential buildings, Table 1 shows the required improvement percentages for each level in accordance to the climate zone the project is located on. Depending on the level obtained, a grade value (*GradeValue$_u$*) is assigned for each evaluated unit (apartment, office, etc.). The improvement percentage for each unit is calculated according to Equation (1), while the rating of the whole building is calculated according to Equation (2).

$$IP = 100 \times \frac{EC_{ref} - EC_{des}}{EC_{ref}} \tag{1}$$

where:

IP—Improvement percentage (%) of energy consumption for floor area
EC_{ref}—Reference energy consumption (kWh/m^2 year)
EC_{des}—Unit energy consumption (kWh/m^2 year)

$$Bld_{rate} = \frac{\sum_{u=1}^{m} Area_u \times GradeValue_u}{\sum_{u=1}^{m} Area_u} \tag{2}$$

where:

Bld_{rate}—Energy rating of building
$GradeValue_u$—Energy rating of unit (apartment/office)
$Area_u$—Area of unit (m^2)
u—Unit
m—Number of units

Table 1. Unit energy efficiency rating (residential) with *GradeValue$_u$*.

Rating of Unit	Grade Value	Energy Efficiency Improvement Percentage by Climatic Zone (%)			
		Climate Zone A	Climate Zone B	Climate Zone C	Climate Zone D
A+	5	$\geq$35	$\geq$35	$\geq$40	$\geq$29
A	4	$\geq$30	$\geq$30	$\geq$34	$\geq$26
B	3	$\geq$25	$\geq$25	$\geq$27	$\geq$23
C	2	$\geq$20	$\geq$20	$\geq$20	$\geq$20
D	1	$\geq$10	$\geq$10	$\geq$10	$\geq$10
E	0	<10	<10	<10	<10
F	-1	<0	<0	<0	<0

In the following sections, we present the conceptual idea and development of ENERGYui, a model that allows designers to understand, evaluate, rate, and improve the design and energy performance of buildings during all the design stages including the early ones, by easily using sophisticated and reliable energy simulation models. The simulation engine of the model is the robust hourly dynamic model EnergyPlus developed by the US Department of Energy [17]. The proposed model provides a graphic user interface (GUI) that includes information that helps with fulfilling the requirements of Standard IS5282 for the energy rating of buildings. It also includes a materials library certified by the Standards Institution of Israel, which provides the definition of the properties of the building's materials easily and efficiently. Hence, the user is required only to provide or choose simple data related to the architectural characteristics of the project, such as location, building type, building geometry (envelope, internal walls, and openings), materials, etc., and during the early stages of design they can rely on pre-set smart default values for non-architectural data such as mechanical systems, schedules, etc., to evaluate the proposed design alternative. The idea behind the model is allowing designers to improve the understanding of the influence of design decisions on the energy performance of the building to improve the decision-making process. In this way, a simple easy-to-use model from the user's point of view is provided, while a reliable and robust one creates and simulates a full-detailed building description.

2. Description of the ENERGYui Model

A scheme describing the workflow of ENERGYui is shown in Figure 1. One of the advantages of the proposed model is that it requires users to only provide the available information related to architectural characteristics and features of the project. Among them, variables such as project location, building type, geometry, windows, shading elements, materials, and ventilation, are selected or defined by the user. Non-architectural parameters are defined by the model behind the scenes (see Table 2 for a list of architectural and non-architectural parameters for residential building type). Hence, this avoids confusion for the users regarding the information they are supposed to provide for the simulation to be performed on the one hand, and it avoids errors or manipulation of different simulation settings in obtaining reliable results on the other hand. In this way, the proposed model adapts to the way architects work and allows for performing sophisticated simulations without the need of dealing with complex definitions. Accordingly, this allows them to correct and improve the design to meet the architectural objectives on the one hand and obtain the desired ranking on the other. Moreover, it provides authorities a way of controlling the correctness of the input data and the reliability of obtained results, which result in the rating of the building.

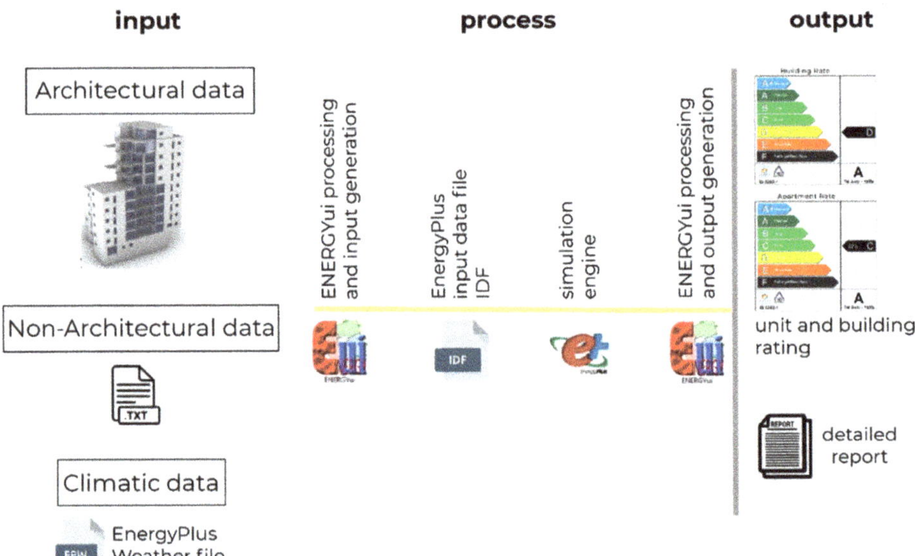

Figure 1. ENERGYui workflow.

ENERGYui is controlled and organized by a command tool chest that guides and advises the user concerning the information and input required or missing to perform the simulation (Figure 2), as will be demonstrated in the following sections.

Table 2. Architectural and non-architectural parameters for residential building type.

Info	Parameters		Variation		
Architectural (user-defined)	Location		Set by user *		
	Geometry	Opaque	Set by user * (See Figures 3 and 4)		
		Windows	Set by user * (See Figures 3 and 5)		
		Blinds		Shading Coefficient Winter	Shading Coefficient Summer
			No Blinds	1.0	1.0
			2/3 opening	0.6	0.4
			1/2 opening	0.5	0.4
			Full opening	0.8	0.4
		Sunshades	Set by user * (See Figure 5)		
	Materials		Set by user * (See Figures 4 and 7)		
	Ventilation		Night Crossed, Comfort		
Non-Architectural (Tool defined)	Loads	People	From 4 to 8—According to apartment size		
		Constant	From 1 to 0.5 W/m^2—According to apartment size		
		Non-constant	From 8 to 4 W/m^2—According to apartment size		
		Lighting	5 W/m^2		
	Mechanical system		Ideal system—Heating/Cooling Loads Calculation		
	Setpoint	Heating	20 °C		
		Cooling	24 °C		
	Infiltration		1 ach		
	Seasons		According to location climatic zone		

Set by user *—User is not constrained by requirements for each parameter.

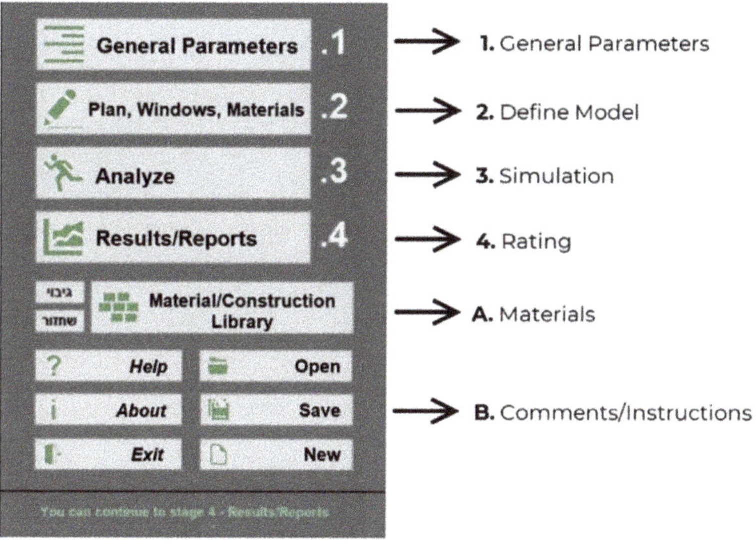

Figure 2. Command tool chest.

2.1. Step 1—General Parameters

The first step relates to the definition of the general information on the project at hand, with the definition of the location (i.e., setting weather conditions for the project) being the most important. Additional information includes designer and developer details and contact, terrain data, etc. After these general parameters are set, the user is allowed to continue to step 2. If information is missing or incorrect in a certain step, the tool chest prompts a notification with details for designers and does not allow the user to continue to the next step, guaranteeing in this way the completeness of the data needed in order to perform the full simulation.

2.2. Steps 2 and 3—Building Model Definition and Simulation

In step 2, the user defines the project geometry: plan, external envelope, openings, materials, number of floors, thermal zoning (offices, apartments, cores, corridors, etc.). The user can start this stage from scratch or can use one of the templates provided by ENERGYui as a starting point. The templates define some of the typical building layouts for various building types in Israel (see Figure 3 for some examples of residential buildings).

Figure 3. ENERGYui residential templates.

As mentioned above, the definition of the architectural design parameters is done by the designer according to the information available at each design stage, while all non-architectural parameters are defaulted by ENERGYui. The idea behind this setting is to encourage designers to improve the performance of the buildings from the very beginning, based on their basic architectural characteristics, rather than relying on mechanical systems exclusively. Figure 4 shows the range of information needed, aside from the geometry itself, i.e., walls, windows, and thermal zones (see A in Figure 4). The different envelope elements need to be assigned a composite construction, which can be selected from a provided library (Figures 5 and 8) or can be newly created by users according to their needs. These composite constructions can be applied to a specific wall or all walls in the floor or building. In the same manner, for any window, the user needs to choose its frame and glazing material and internal or external shading type, i.e., blinds and/or sunshades (Figure 6). While working on this step, ENERGYui provides graphic feedback on the completeness of the information provided. For instance, a mustard-yellow color element means that a construction definition is still missing and acts to guide the designer to complete it, while a green one indicates that it is fully defined. Additional information that needs to be defined at this step relates to the determination of the north direction, the number of floors, thermal zones, assigning composite constructions for the floor, roofs, internal floors, as well as first-floor type, i.e., on the ground, on columns, or over an unheated space (all seen in B, C, and D in Figure 4). Since some of this information can be unknown in the early stages of design, the user can choose from one of the smart defaults offered by the model (insulation, window size, etc.), which are based on requirements from IS5282.

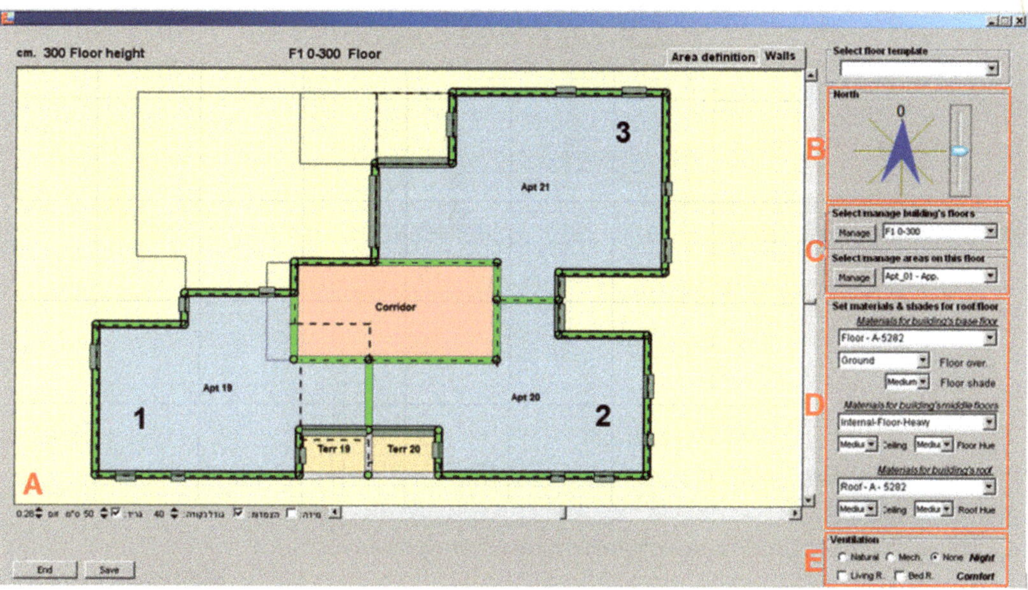

Figure 4. ENERGYui modeling GUI.

Once step 2 is completed and the model contains all the required information for the simulation, the user is allowed to proceed with step 3, which involves running the simulation itself. ENERGYui automatically creates and processes the full input file used by the simulation engine, i.e., EnergyPlus [17].

2.3. Step 4—Rating Individual Units and the Whole Building

After completion of the simulation, ENERGYui rates both the whole building and each of the zones (apartments, offices, etc., according to the building type chosen for evaluation), as seen in Figure 7. Instead of having to deal with a large and complicated number and types of outputs, the building rating and the energy certificate, together with a detailed report for further analysis, are provided. IS5281 requires the whole building to be rated, while IS5282 allows for the rating of both the entire building and individual units. In this sense, the individual rating can help owners or potential buyers to know the energy performance of the specific property and conduct informed decision-making. This information can also be used by planning authorities to stimulate green buildings by providing economic incentives, such as low-interest rates mortgages.

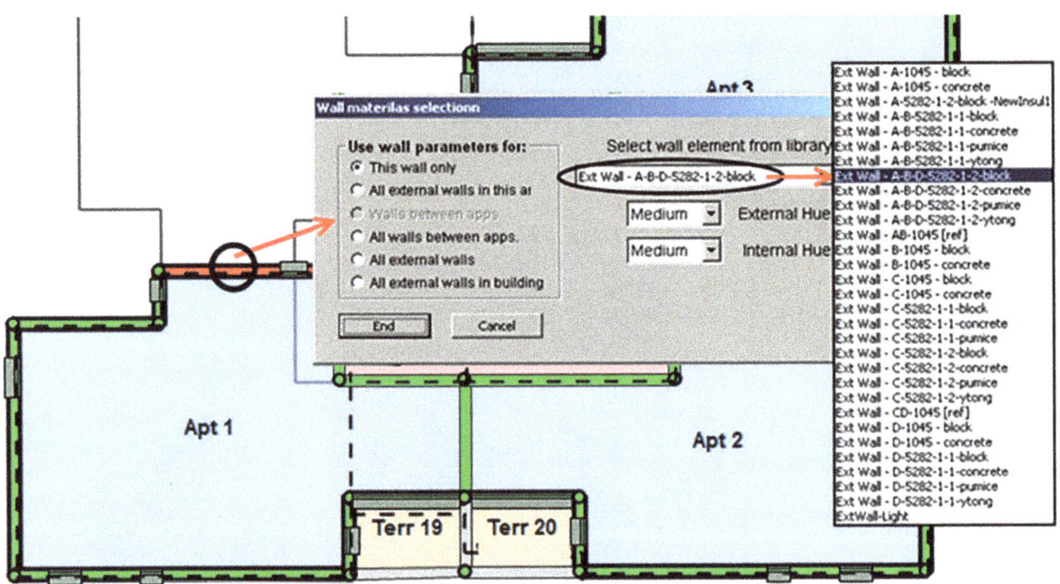

Figure 5. Opaque envelope: Assignment of composite elements.

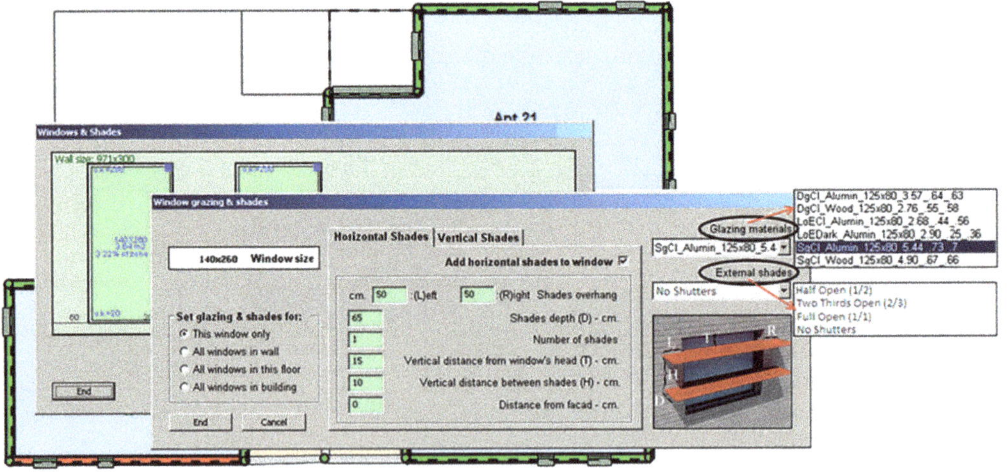

Figure 6. Openings: Assignment of window and shading elements.

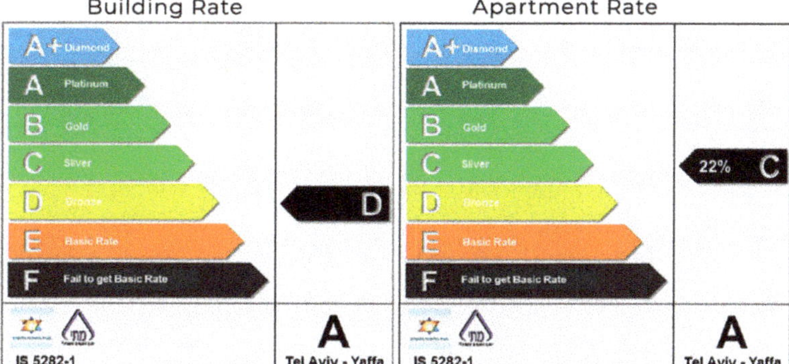

Figure 7. Energy rating for a building (**left**) and an individual unit (**right**).

2.4. Material/Construction Library

ENERGYui contains a library of basic materials and combinations that allows for easy user choice and ensures compliance with requirements as well as the quality and consistency of the data. The materials are categorized according to different types (concrete, wood, glazing, etc.) in a way where users can quickly find the most relevant one for their needs (see Figure 8 part A). Individual materials are used for the creation of composite elements for the building envelope. Those assemblies can be assigned to different geometry elements in the building, as in floors, windows, walls, roofs, etc. (see Figure 8 part B). It is possible to choose from default predefined composite elements or create new ones as shown in Figure 8 part C. New composite constructions can be created based on existing and certified basic materials included in the library, or based on new materials on the market as defined by the user. In the last case, a notice will be printed in the detailed report, meaning that the designer will be requested to provide approved documentation certifying the properties of the new material. The library manager differentiates between certified materials/composite elements and new ones defined by the users by different colors, while the first category is protected and cannot be altered by users.

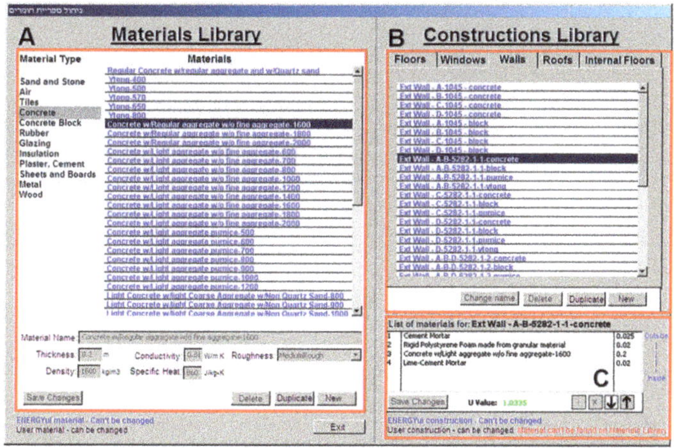

Figure 8. Material/construction library. List of materials (**left**); list of constructions (**right**).

3. Case Studies

In this section, we demonstrate the use of the ENERGYui model through the analysis and exploration of several design alternatives in two different examples. The first case study demonstrates the use of the model for the analysis of alternatives and decision-making in the design of an office building from the first stages of the design process to the detailed design. The second one shows the analysis of a residential building in a more advanced state of design, in which several basic decisions have already been made and therefore the freedom of action is more limited.

3.1. Office Building

This case study demonstrate the design of a theoretical office building in the city of Tel Aviv in the coastal plan zone of Israel. As stated above, this case allows the designer to explore various basic design alternatives for the building from the very beginning, where the criterion for choosing which alternative(s) to continue to develop is the achieved energy rating of the building. The use of the tool at early stages allows for exploring fundamental architectural decisions such as massing and main orientation of the building. Three basic options for the office building proportions were explored based on their depth, going from a deep office space of 14 m, a more typical depth of 8.2 m, and a shallow deep of 5 m. The expected total area is about 4500 m^2 for all buildings, resulting in two 4-story and one 5-story buildings, respectively as shown in Figure 9.

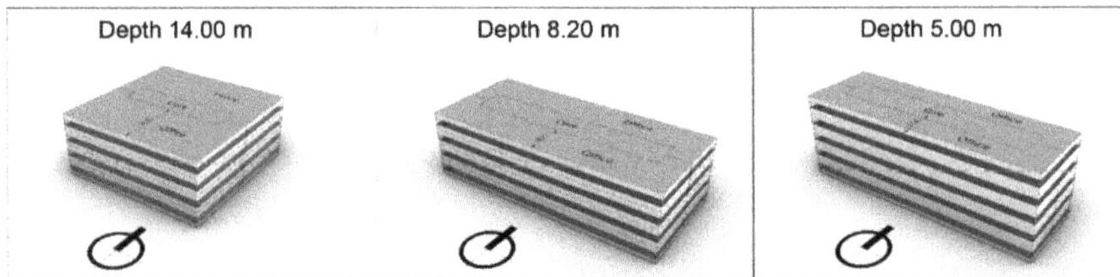

Figure 9. Case study building 1: Basic massing alternatives. **Left**: Deep offices (14 m); **Middle**: Standard depth (8.2 m); **Right**: Shallow depth (5 m).

To allow for the use of the simulation model at this early stage of design, basic properties for this case study were set as smart-default values based on requirements and common practices: A window size is predetermined for all buildings as a strip of 1.40 m on all facades, and double-glazing clear glazing type is used for openings. The opaque part of the façade uses as a starting point a basic level of insulation, meeting the minimum requirements required for heavy construction as set out in IS1045. Although some improvement in insulation levels beyond this minimum value may be beneficial in terms of overall energy consumption and rating (as we demonstrate later), it should be emphasized that in Israel's coastal climate zone there is a balance point between winter and summer. In winter, adding insulation may help lower heating requirements, although during the dominant warm period, adding insulation beyond the recommended level may make it difficult to cool the building and may require night ventilation [18]. Therefore, the insulation requirements in this climate zone are less stringent than in cold regions. No internal or external shading devices and no ventilation were implemented, and the light control was set to one step on/off (LC1S) for the basic set of alternatives. The full set of characteristics of the buildings (basic set and design alternatives) are presented in Table 3.

Table 3. Office building case with design parameters.

Parameter	Acronym		Description	Value
Proportion			Office's depth	5.0 m, 8.2 m, 14.0 m
Orientation	Des_Alt	0 Deg, 45 Deg, 315 Deg	Main façade orientation	-
Insulation	Bsc	StIns	Standard Insulation	$U = 1.25\ \text{W/m}^2\text{K}$
	Des_Alt	ImIns	Improved insulation	$U = 0.5\ \text{W/m}^2\text{K}$
Glazing	Bsc	DgCl	Double Glazing Clear	$U = 3.95\ \text{W/m}^2\text{K}$, SHGC = 0.65, Vt = 0.63
	Des_Alt	LowE	Low Emissivity glazing	$U = 3.14\ \text{W/m}^2\text{K}$, SHGC = 0.52, Vt = 0.6
Shading	Bsc	noShd	No shading	-
	Des_Alt	IntShd	Internal dynamic shading	SC = 0.55
	Des_Alt	ExtShd	External dynamic shading	-
	Des_Alt	BrSol	External fixed shading	-
Window Size	Bsc	MedWin	Medium window size	H = 1.4 m
	Des_Alt	MaxWin	Maximal window size	H = 2.7 m
	Des_Alt	MinWin	Reduced window size	H = 1.0 m
	Des_Alt	MinWinSouth	Reduced window size South	H = 1.0 m
Light Control	Bsc	LC1S	Light control 1 step	on/off (500 lx)
	Des_Alt	LC2S	Light control 2 step	on/off (500 lx)
	Des_Alt	LCdim	Light control dimmer	dimmer (500 lx)
Ventilation	Bsc	noV	No Ventilation	-
	Des_Alt	nNV	Natural night ventilation	3 ach
	Des_Alt	mNV	Mechanical night ventilation	10 ach
	Des_Alt	ComfV	Comfort ventilation	Ceiling fan (Allows to raise summer temperature by 0.5 °C)

Basic alternative (Bsc)					
Insulation	Glazing	Shading	Window Size	Light Control	Ventilation
StIns	DgCl	noShd	MedWin	LC1S	noV

Design alternative (Des_Alt): Alternative get value from a set of given parameters or from user defined values.

Table 4 and Figure 10 present the description and results for the alternatives of the theoretical office building. As previously stated for this case study, the exploratory process starts from the most basic design decisions at the first stage, i.e., massing and orientation of the building according to site constraints (0 deg, 45 deg, and 315 deg). The results were assessed at two levels: the whole building and at two typical middle floor offices (north and south). The best results (higher rating and lower energy consumption) for each building depth were kept for the next stages. For the 5 m depth building, all orientations were rated E or above, and hence it was decided to keep them all. For the 8.2 m depth case, 0 deg and 45 deg were kept, and for the 14 m depth, only the 0 deg option was suitable to be kept for further analysis.

Table 4. Office building case with design path showing improvements of design alternatives.

Category	Description (EC$_{ref}$ Offices: 42.38 kWh/m² year)	D5.0 ON Improv %	D5.0 ON Rating	D5.0 ON EC$_{des}$	D5.0 OS Improv %	D5.0 OS Rating	D5.0 OS EC$_{des}$	D5.0 Whole Building Rating	D8.2 ON Improv %	D8.2 ON Rating	D8.2 ON EC$_{des}$	D8.2 OS Improv %	D8.2 OS Rating	D8.2 OS EC$_{des}$	D8.2 Whole Building Rating	D14.0 ON Improv %	D14.0 ON Rating	D14.0 ON EC$_{des}$	D14.0 OS Improv %	D14.0 OS Rating	D14.0 OS EC$_{des}$	D14.0 Whole Building Rating
Orientation	0 Deg	27	D	31.1	29	C	30.1	C	15	E	35.9	17	E	35.2	E	9	F	38.6	10	E	38.2	E
Orientation	45 Deg	15	E	36.2	19	E	34.4	E	8	F	39.1	11	E	37.8	E	6	F	39.8	8	F	39	F
Orientation	315 Deg	17	E	35.2	15	E	36.2	E	9	F	38.4	8	F	39.2	F	7	F	39.3	6	F	39.8	F
Window Size	0 Deg_Min Win	17	E	35.2	34	B	28.2	D	3	F	41.2	20	D	33.8	E	6	F	39.9	13	F	37	F
Window Size	0 Deg_Max Win	20	D	34	6	F	39.7	E	8	F	38.9	−1	F	42.7	F	−1	F	42.8	−7	F	45.4	F
Window Size	45 Deg_Min Win	9	F	38.3	22	D	33	E	−1	F	42.9	12	E	37.2	F							
Window Size	45 Deg_Max Win	−3	F	43.6	−8	F	45.8	F	−6	F	45	−9	F	46.2	F							
Window Size	315 Deg_Min Win	8	F	39	21	D	33.4	E														
Window Size	315 Deg_Max Win	5	F	40.4	−17	F	49.7	F														
Window Size	0 Deg_Min Win South	26	D	31.3	33	C	28.3	C														
Shading	0 Deg_Int Shd	27	D	31	34	B	28	C	16	E	35.5	21	E	33.6	E							
Shading	0 Deg_Ext Shd	26	D	31.2	31	C	29.2	D	16	E	35.4	23	D	32.5	E							
Shading	0 Deg_ImIns_LowE_FixExtShd	−3	F	42.3	19	E	27.6	E	−4	F	44.1	21	D	33.6	E							
Detailed Design Parameters	0 Deg_ImIns_LowE	31	C	29.2	38	B	26.3	B	19	E	35.4	24	E	32.5	E							
Detailed Design Parameters	0 Deg_LowE_IntShd_MaxWin	28	C	30.3	31	C	29.3	C	15	E	35.8	18	E	34.7	E							
Detailed Design Parameters	0 Deg_ImIns_LowE_IntShd	30	C	29.5	40	A	25.3	B	19	E	34.4	26	E	31.5	D							
Detailed Design Parameters	0 Deg_ImIns_LowE_ExtShd	29	C	30.2	39	B	25.8	C														
Detailed Design Parameters	0_Deg_ImIns_LowE_IntShd_ MaxWin	28	C	30.3	31	C	29.3	C														
Detailed Design Parameters	0 Deg_IntShd_Min Win South	27	D	31.1	35	B	27.4	C														
Detailed Design Parameters	0 Deg_SI_LowE_Min Win South	28	C	30.6	37	B	26.7	C														
Detailed Design Parameters	0 Deg_ImIns_Min Win South	29	C	30	35	B	27.5	C														
Detailed Design Parameters	0 Deg_ImIns_LowE_Min Win South	31	C	29.3	40	A	25.6	B														
Detailed Design Parameters	0 Deg_ImIns_LowE_IntShd_ Min Win South	30	C	29.5	40	A	25.6	B														
Detailed Design Parameters	0 Deg_ImIns_LowE_Min Win South_nNV	38	B	26.3	48	A+	22.1	A														
Detailed Design Parameters	0 Deg_ImIns_LowE_Min Win South_mNV	32	C	28.9	43	A	24.3	B														
Detailed Design Parameters	0 Deg_ImIns_LowE_Min Win South_mNV_ComfV	37	C	26.5	49	A+	21.8	A														
Detailed Design Parameters	0 Deg_ImIns_LowE_Min Win South_LC2S	34	B	28	41	A	24.9	B														
Detailed Design Parameters	0 Deg_ImIns_LowE_Min Win South_LCdim	39	B	25.8	46	A+	23	A														

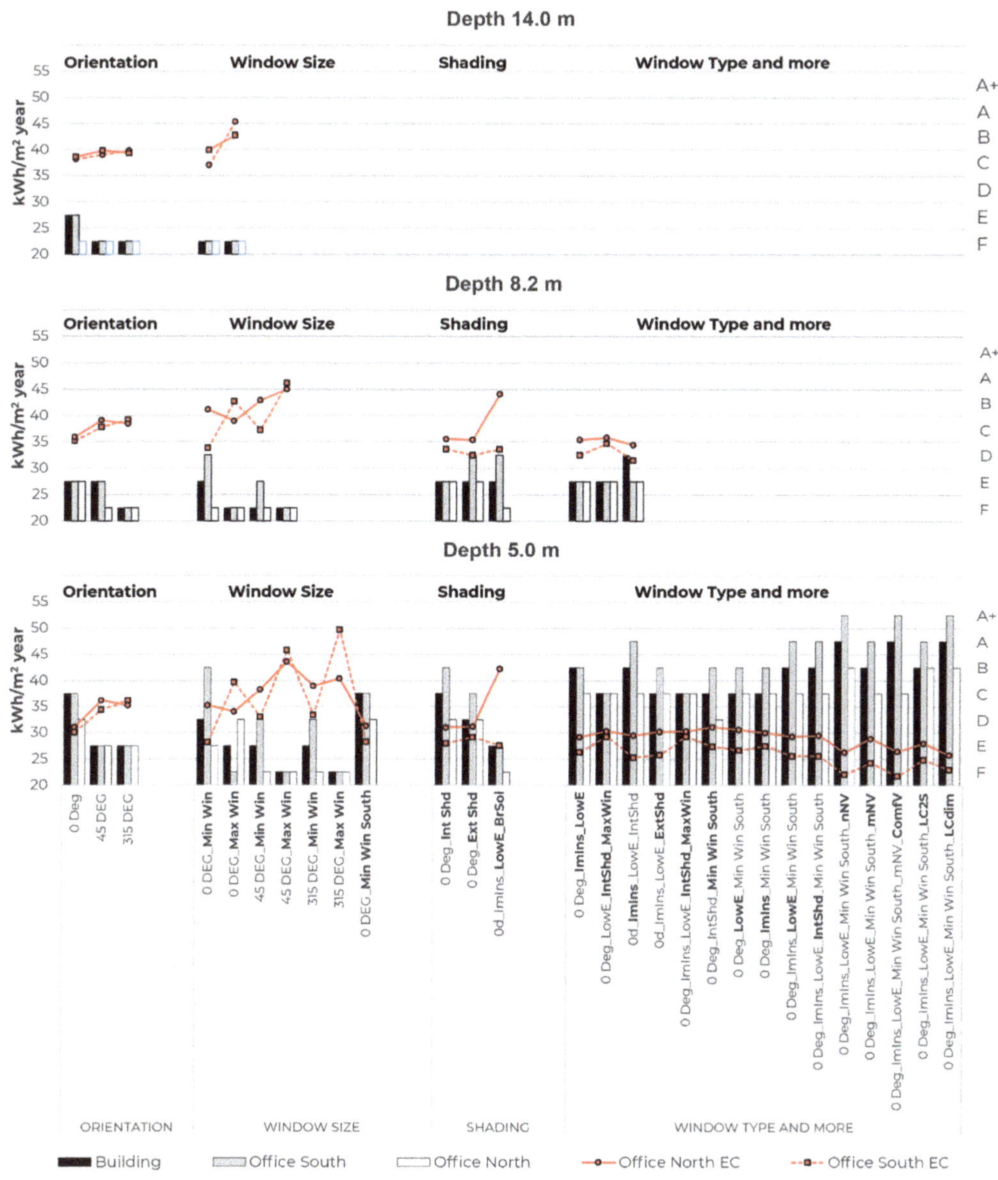

Figure 10. Case study building 2: Energy consumption and rating of design alternatives and decision making during the design process. **Top**: Deep offices (14 m); **Middle**: Standard depth (8.2 m); **Bottom**: Shallow depth (5 m). In bold: parameter changed in each alternative.

The second stage of the exploration tests the influence of window sizing on the performance of the building. This is a stage in the design process where basic decisions still need to be taken, as in the previous one. Three strategies for window size were evaluated, i.e., reducing the size to a minimum, enlarging the size to a maximum, and reducing the size only for the south-oriented offices (denoted as "min win", "max win", and "min win south", respectively). At this stage, the main improvements in the rating were obtained with the clean north-south orientation (0 deg), especially on the 5 m depth option. The 8.2 m depth shows a slightly better performance compared with the previous stage and the 14 m depth shows no improvement at all. This is probably due to the fact that the lighting consumption is heavily affected by the deep spaces, and therefore it was decided not to further pursue this design option. For the next stages, only the north-south options were kept for further exploration.

Once the more basic decisions were made, the third stage implements the use of shading strategies for the openings. They included dynamic shading devices, internal, and external (IntShd and ExtShd) activated when the amount of solar radiation reaching the office space overcomes a predetermined threshold. Fixed devices were also considered as a design option (FixExtShd). The internal shading performed better for the shallow building while the external was slightly better than the internal for the 8.2 m depth. The fixed shading type did not represent any improvement compared to the previous stage and was discarded. As a result, both dynamic shades were kept for the next design stage.

The fourth stage is intended for a more detailed level of decision-making in advanced stages of design. Improving the glazing type or using dynamic shade led to significant performance improvement in the 5 m depth office building. For the 8.2 m depth case, however, it showed no significant rating and performance improvement from the previous stage, maybe even worsening at the single office level. For this reason, it was decided to drop this building option and to pursue only the shallow-depth office. Significant improvement in the performance of the building and midfloor offices occurs when using low-emissivity (LowE) glazing, reducing the window size in the south-oriented offices, and improving the insulation of the opaque envelope. With one or more than those parameters, the building reached an overall B grade and the offices C (north) and A (south). Implementing some sort of ventilation (natural/mechanical night ventilation or comfort ventilation, provided by ceiling fans) or dimmer light control led to an A rating for the building and A+ for the south-oriented offices, which is the best one achieved for this case study (see Table 4 and Figure 10).

3.2. Residential Building

The second case study presents a residential tower building, 11-floor height (Sanhedrin Building), designed in the city of Ramat Gan, as well in the coastal plan climatic zone. Figure 11 shows two different design alternatives for this building.

In this building, due to design limitations and programmatic constraints, several fundamental design parameters were fixed and cannot be considered for change during the process, as building orientation. Therefore, the exploration path focused on design parameters related to the external building envelope. The case study assesses two design alternatives that allows for understanding the impact of the proposed changes on the performance of both the whole building and the apartments on a typical middle 2-apartment floor (dark grey in Figure 11). The first alternative shows a conservative approach, where the building uses the more common construction technologies available in the Israeli market, and the aesthetics are guided by modern traditional customs (Figure 11 top). The second alternative shows a more "fashionable" aesthetics of the building envelope, where constructors, entrepreneurs, and even users prefer to have larger windows with extensive use of glazing, mainly in the social areas of the residential units (Figure 11 bottom).

Table 5 presents a description of the different simulated stages for both design alternatives, a total of 4 for both alternatives. The table shows the summary of the results obtained in the simulations while changing different design variables for Apartments 1 and 2 and for the whole building.

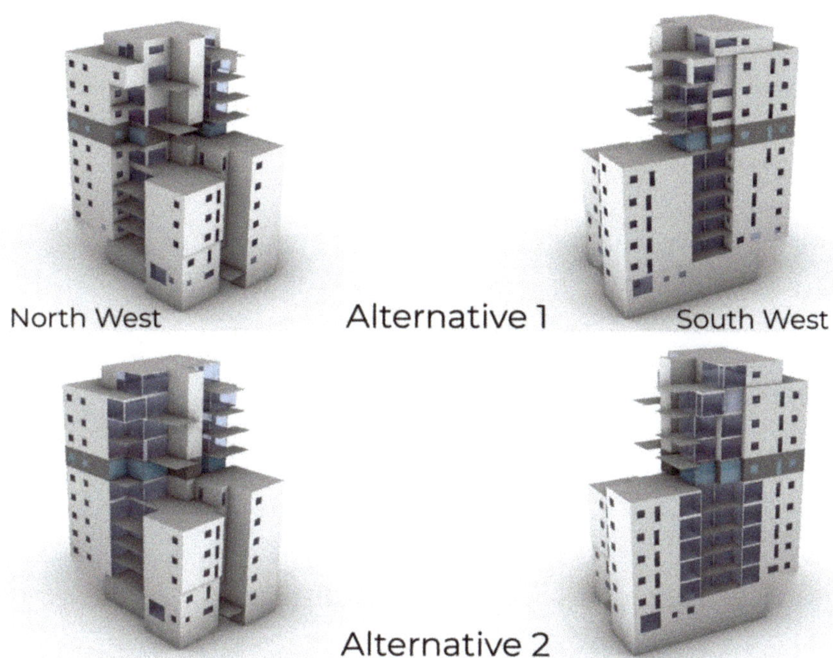

Figure 11. Case study building 2. View from northwest (**left**) and southwest (**right**). Top: Design Alternative 1; Bottom: Design Alternative 2 with larger windows.

Alternative 1: The building was first analyzed by implementing basic requirements for the insulation of the building's envelope (roof, walls), as currently required by local standards (Figure 11 top and Figure 12 left). In addition, windows with clear single glazing were designed at the beginning without any solar protection. This is despite the design tradition in Israel, which applies roller shutters to windows in residential buildings for privacy, allowing dynamic shading as needed as well as improving window insulation at night. According to IS5282, this building ranks the lowest rating possible F, as well as apartment 1, while Apartment 2 rates E (Alternative 1, Stage 1).

As a way to improve the rating, it was proposed to improve the insulation of the opaque envelope and to use double glazing clear in the windows instead. These changes led to the improvement of the rating of the whole building to the basic level of D, while Apartments 1 and 2 rate E and D, respectively (Alternative 1, Stage 2). Both apartments remain low-rated despite the improvements, probably due to the relatively large opening towards the south, west, and north directions, the lack of any shading protection for the windows, and a relatively large external surface area.

Adding dynamic external shutters that can be fully open and close as well as providing dynamic shading and improving night insulation proved to be a very efficient and important factor for energy savings. These changes changed the building rate to B, while the apartments rated C (Alternative 1, Stage 3).

Table 5. Design path showing improvements of two design alternatives.

| | | Description | Typical Middle Floor | | | | | | Whole Building |
| | | | Apartment 1 | | | Apartment 2 | | | |
		EC_{ref} Residential: 41.78 kWh/m² year	Improvement %	Rating	EC_{des} (kWh/m² year)	Improvement %	Rating	EC_{des} (kWh/m² year)	Rating
Alternative 1	1	Basic Insulation Single Glazing Clear No Shading	−6	F	44.3	5	E	39.9	F
	2	Improved Envelope Insulation IS5282-Level 2 Double Glazing Clear	6	E	39.3	16	D	35	D
	3	Full Opening External shutters	24	C	31.7	24	C	31.7	B
	4	Night Ventilation	32	A	28.2	31	A	28.6	A
Alternative 2	1	Increase Glazing—social areas	−9	F	45.7	−24	F	51.9	F
	2	LowE Glazing	5	E	39.7	−10	F	45.9	F
	3	Full Opening External shutters	23	C	32.2	14	D	36.1	C
	4	Night Ventilation	31	A	28.7	22	C	32.6	A

Allowing for night ventilation during summer nights (when the outside temperature is lower than the inside temperature) improved the final rating of the building and both apartments to A (Alternative 1, Stage 4). Further improvement could have been achieved on its energy performance by updating geometry, window size, or orientation; these changes were not tested in this alternative.

Alternative 2: As noted above, this alternative reflects design choices related to market preferences, namely in encouraging window enlargements mainly in the public areas of the apartments. For this alternative, the glazing ratio was significantly increased for the social areas of each apartment (Figure 11 bottom and Figure 12 right). This substantial change affected the performance of the apartments and the whole building for the worst. The apartments and the building rated F (Alternative 2, Stage 1). To deal with the enlargement of the windows, in the next stage we tried to improve the quality of the glazing, both in terms of insulation and solar heat gain coefficient (U value = 3.14, SHGC = 0.2). Changing the glazing type to LowE slightly improved the overall results (% improvement) but not the ratings. Only Apartment 1 is rated now at E (Alternative 2, Stage 2). In the next stage, as in Alternative 1, the addition of dynamic external shutters proved to be especially important to improve the performance of the building. This led to the building being rated at C and the apartments at C and D (Alternative 2, Stage 3). Lastly, implementing night ventilation, which is beneficial in this climate, contributed to bringing the final rating of the building and Apartment 1 to A while Apartment 2 received a C rating (Alternative 2, Stage 4).

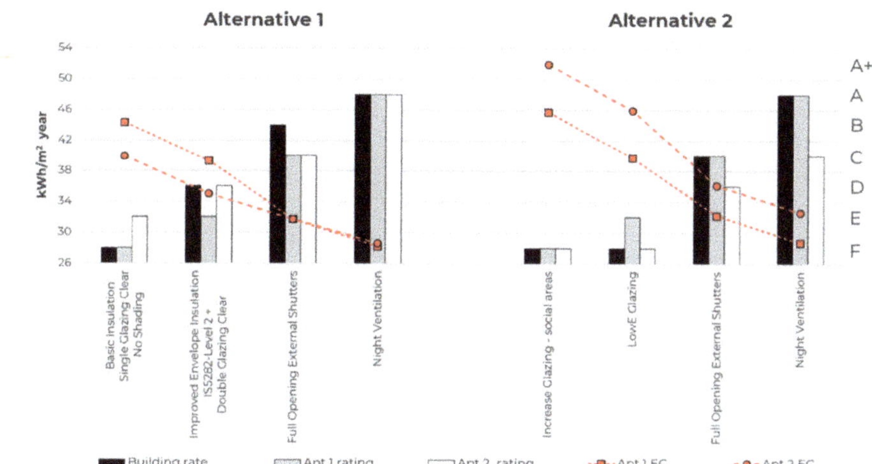

Figure 12. Case study building 2: Energy consumption and rating of design alternatives during the design process.

4. Conclusions

A simple model ENERGYui that allows for the use of the simulation program Energy Plus for improving design performance during the process of the rating of buildings was demonstrated. The model was applied at different stages through the design process of an office and a residential building. Significant improvements were obtained in the final ranking in both cases studied. With the help of the model, beyond obtaining the rating of the building, which is its main objective, it was possible to explore different alternatives at all design stages, including the early ones, and make decisions according to the qualities of each one of them. In this way, the model allows for increased awareness of energy implications regarding design decisions, and it diversifies the available design options. The proposed model focuses on information from designers related mainly to architectural parameters. The non-architectural ones are suggested by the tool itself based on smart default values. ENERGYui is expected to make a great impact not just among green building consultants, but also among practitioners in their architectural practice since it is the recommended tool to be used for checking compliance with IS5282.

Author Contributions: Conceptualization, A.Y. and I.G.C.; formal analysis, A.Y. and I.G.C.; funding acquisition, A.Y. and I.G.C.; investigation, A.Y. and I.G.C.; methodology, A.Y. and I.G.C.; project administration, A.Y.; resources, A.Y.; software, A.Y. and I.G.C.; validation, A.Y. and I.G.C.; visualization, A.Y. and I.G.C.; writing—original draft, A.Y. and I.G.C.; writing—review and editing, A.Y. and I.G.C. All authors have read and agreed to the published version of the manuscript.

Funding: The research was funded by the Israel Ministry of National Infrastructures, Grant 2004243.

Institutional Review Board Statement: Not applicable.

Informed Consent Statement: Not applicable.

Data Availability Statement: The data presented in this study are available on request from the corresponding author. The data are not publicly available due to privacy issues.

Acknowledgments: The authors express their gratitude to Gerstenfeld Company, Chen Architects, Arch Gady Heller, and Eng. Meira Bet El for their consent to use the case study building.

Conflicts of Interest: The authors declare no conflict of interest.

References

1. U.S. Green Building Council (USGBC). *LEED-NC for New Construction*; Reference Guide Version 4.0.; USGBC: Washington, DC, USA, 2013.
2. European Energy Performance of Buildings Directive (EPBD). Available online: https://ec.europa.eu/energy/topics/energy-efficiency/energy-efficient-buildings/energy-performance-buildings-directive_en (accessed on 19 October 2020).
3. Yezioro, A.; Shapir, O.; Capeluto, I.G. A Simple User Interface for Energy Rating of Buildings. In Proceedings of the Building Simulation Conference (IBPSA), Sidney, Australia, 14–16 November 2011; pp. 1293–1298.
4. Yezioro, A.; Shapir, O.; Capeluto, I.G. A Design Tool for the Determination of the Energy Rating of Buildings. In Proceedings of the 28th International PLEA Conference, Lima, Peru, 7–9 November 2012.
5. Marmolejo-Duarte, C.; Berrio, S.S.; Del Moral-Ávila, C.; Méndez, L.D. The Relevance of EPC Labels in the Spanish Residential Market: The Perspective of Real Estate Agents. *Buildings* **2020**, *10*, 27. [CrossRef]
6. Olofsson, T.; Meier, A.; Lamberts, R. Rating the Energy Performance of Buildings. In *The International Journal of Low Energy and Sustainable Buildings*; 2004, Volume 3. Available online: https://www.diva-portal.org/smash/get/diva2:208041/FULLTEXT01.pdf (accessed on 19 October 2020).
7. Crawley, D.B.; Hand, J.W.; Kummert, M.; Griffith, B.T. Contrasting the Capabilities of Building Energy Performance Simulation Programs. *Build. Environ.* **2008**, *43*, 661–673. [CrossRef]
8. Shady, A.; Hensen, J.L.M.; Beltrán, L.; De Herde, A. Selection Criteria for Building Performance Simulation Tools: Contrasting Architects' and Engineers' Needs. *J. Build. Perform. Simul.* **2012**, *5*, 155–169.
9. Capeluto, I.G. The meaning and Value of Information for Energy-Conscious Architectural Design. In Proceedings of the Building Simulation Conference (IBPSA), Sidney, Australia, 14–16 November 2011; pp. 1892–1898.
10. Bragança, L.; Vieira, S.M.; Andrade, J.B. Early Stage Design Decisions: The Way to Achieve Sustainable Buildings at Lower Costs. *Sci. World J.* **2014**, *2014*, 365364. [CrossRef] [PubMed]
11. IS5281. *Buildings with Reduced Environmental Impact ("Green Buildings")*; The Standards Institution of Israel (SII): Tel Aviv, Israel, 2005.
12. IS5281 Part 2. *Sustainable Buildings (Green Buildings): Requirements for Residential Buildings*; The Standards Institution of Israel (SII): Tel Aviv, Israel, 2016.
13. SII. IS1045 Part 1. *Thermal Insulation of Buildings: Residential Buildings*; The Standards Institution of Israel (SII): Tel Aviv, Israel, 2011.
14. SII. IS1045 Part 3. *Thermal Insulation of Buildings: Office Buildings*; The Standards Institution of Israel (SII): Tel Aviv, Israel, 2005.
15. SII. IS5282 Part 1. *Energy Rating of Buildings: Residential Buildings*; The Standards Institution of Israel (SII): Tel Aviv, Israel, 2017.
16. Shaviv, E.; Yezioro, A.; Capeluto, I.G. Energy Code for Office Buildings in Israel. *Renew. Energy* **2008**, *33*, 99–104. [CrossRef]
17. US Department of Energy. EnergyPlus. Available online: https://energyplus.net/downloads (accessed on 1 February 2021).
18. Yezioro, A.; Drori, D.; Shaviv, E. Passivhaus vs. Passive Solar—What is Better for the Israeli Mediterranean Climate? In Proceedings of the International Conference of Passive and low Energy Architecture (PLEA), Bologna, Italy, 9–11 September 2015.

Article

Factors Influencing Green Building Development in Kazakhstan

Daniyar Assylbekov [1], Abid Nadeem [1,*], Md Aslam Hossain [2,*], Gulzhanat Akhanova [1] and Malik Khalfan [3,4,*]

[1] Department of Civil and Environmental Engineering, Nazarbayev University, Nur-Sultan 010000, Kazakhstan; daniyar.assylbekov@nu.edu.kz (D.A.); gulzhanat.akhanova@nu.edu.kz (G.A.)
[2] School of Engineering, Monash University Malaysia, Jalan Lagoon Selatan, Bandar Sunway 47500, Selangor, Malaysia
[3] School of Property, Construction and Project Management, RMIT University, Melbourne, VIC 3000, Australia
[4] Plymouth Business School, University of Plymouth, Plymouth PL4 8AA, UK
* Correspondence: abid.nadeem@nu.edu.kz (A.N.); mdaslam.hossain@monash.edu (M.A.H.); malik.khalfan@rmit.edu.au (M.K.)

Citation: Assylbekov, D.; Nadeem, A.; Hossain, M.A.; Akhanova, G.; Khalfan, M. Factors Influencing Green Building Development in Kazakhstan. *Buildings* **2021**, *11*, 634. https://doi.org/10.3390/buildings11120634

Academic Editor: David Arditi

Received: 23 October 2021
Accepted: 6 December 2021
Published: 10 December 2021

Abstract: Green buildings have been actively spreading as a solution for sustainability issues of the construction industry in at least the last two decades. As green building practices unfold in developing countries, the need of identifying factors that both hinder and drive its spread rises. Multiple studies reveal a general inconsistency among results in different parts of the world, caused by each country's environmental, economic, and social conditions. Taking into account the experience of developing international green buildings and the current state of green building development in Kazakhstan, this study aims to spread the understanding of the factors that hinder and have the potential to drive the development of green buildings in Kazakhstan. A questionnaire survey was carried out among 38 industry experts in Kazakhstan to accomplish study objectives. Multiple data analysis methods were used to identify correlations among groups of experts and rank the factors. The results revealed a lack of skill/experience, a lack of government support, and the high cost of sustainable materials and products as the most crucial barriers. Water and energy efficiency, improved health of occupants, comfort, and satisfaction were identified as the most influential drivers. By expanding knowledge on factors affecting the implementation of green buildings, the study uncovered common trends in the responses of professionals, providing valuable information for field professionals and suggesting future research recommendations.

Keywords: green buildings; drivers; barriers; funneling technique; PESTLE

1. Introduction

1.1. Background

The global building industry has a significant impact on the environment, economy, and society. The construction phase, including material manufacturing, is responsible for 10% and the operation phase of the buildings responsible for another 28% of global CO_2 emissions [1]. Essential needs such as heating or cooking require the use of carbon-intensive sources of energy such as oil, gas, and coal, consuming around 60% of the global electricity used just for building operation purposes [1]. The energy consumption of residential and nonresidential buildings takes up to 30%, and the building construction industry represents 5% of global energy use as of 2019 [1]. Statistics represent a steady increase in energy consumption of 7%, with increased total floor area and population in the last nine years [2]. Emission rates related to the construction industry are increasing at a slow but steady pace [2]. Furthermore, the construction industry accounts for up to 40% of the world's materials consumption, almost 30% of the use of timber, and approximately 15% of the total water consumption [3]. On average, 40–60% of all landfill wastes are generated during construction processes [4]. Moreover, the construction industry is a significant contributor to global warming, resource depletion, and air and water pollution, and is the cause of various natural hazards [5–7].

Active measures are taken in the face of sustainable development strategies to diminish the harmful effects of the global building industry. Yudelson [5] defined a green building as a high performance property that reduces its impact on the environment and humans throughout its life cycle. It is intended to use less water and energy. It aims to improve the built environment radically. It considers preserving non-renewable energy sources and promotes renewable sources of energy, advancing the existing technologies and construction methods. Moreover, the green building gravitates toward a healthy environment for the occupants by enhancing indoor air quality and nontoxic materials.

Many countries are successfully implementing green practices, and some are in the process of embracing them. However, despite the rapid growth of the green building concept, numerous impediments prevent its adoption worldwide [8,9]. Moreover, the barriers that prevent the spread of green building vary from country to country. Factors that are more important in one place can be less critical in a different place due to country-specific characteristics such as demography, culture, economy, and location [10,11]. This discrepancy arises from reconsidering and readjusting existing green building practices to the needs and capabilities of the country. There are also risks and uncertainties related to implementing the green building concept that must be investigated [12]. Therefore, it is crucial to identify the drivers and barriers of green building to develop a proper approach for successfully promoting and implementing its practices.

1.2. Research Aim and Objectives

Based on the international experience of factors for green building development, the study considers the current state of Kazakhstan's green building development. This paper aims to spread understanding of the factors that hinder and the factors that can drive green building development in Kazakhstan. Consequently, the following objectives are established:

1. To examine the importance of green buildings in general and for Kazakhstan
2. To identify the factors (barriers and drivers) that influence the development of green buildings.
3. To evaluate the drivers and barriers to green building in Kazakhstan.

The research objectives are organized in a systematic approach called the "funneling technique", effectively utilized in many aspects of questioning, including research [13]. The "funneling technique" idea narrows the general information into practical and operable solutions [13].

The findings of this study will contribute to the existing body of knowledge regarding both green building drivers and barriers in the context of a developing country. In addition, the identified green building barriers and drivers were ranked using the PESTLE technique in order to provide a broader view of the factors. In this regard, the most and least significant factors were shown and ranked in terms of the Political, Economic, Sociocultural, Legal, and Environmental categories, and can be helpful for practitioners intending to take actions in mitigating the barriers of green building technology and using drivers to its adoption. The main novelty of the research is the implementation of the PESTLE method to understand the green building's barriers and drivers.

Additionally, this study can be valuable by expanding knowledge about factors that affect the implementation of green building, providing valuable information for field practitioners, and suggesting future research recommendations.

The paper is structured as follows: The next section presents a review of the literature on green buildings. It provides a summary of the concept of a green building, its drivers, and barriers. Furthermore, the current state of green buildings in Kazakhstan is described in this section. In the third section, an explanation of the four-step methodology of the study is given. The fourth section demonstrates the results, while the fifth section discusses the research results and findings. The last section concludes the research paper.

2. Literature Review

2.1. Green Buildings

The concept of a green building or sustainable construction is the step of our community toward sustainable development. It is the care for the future and future generations, as sustainability is often interpreted as utilizing resources to meet the needs of the present without compromising the future generations' ability to meet their own needs [14]. Ahn et al. [6] identified the benefits of sustainable construction, dividing them into three main pillars as presented in Table 1.

Table 1. Environmental, social, and economic benefits of sustainable construction.

Environmental Benefits	Social Benefits	Economic Benefits
<ul><li>Protecting air, water, and land ecosystems</li><li>Conserving Natural Resources</li><li>Preserving animal and genetic diversity</li><li>Protecting the Biosphere</li><li>Using renewable natural resources</li><li>Minimizing Waste Production or Disposal</li><li>Minimizing CO_2 Emissions</li><li>Pursuing active recycling</li><li>Maintaining the integrity of the environment</li><li>Preventing global warming</li></ul>	<ul><li>Improving the quality of life for individuals and society as a whole</li><li>Alleviating poverty</li><li>Satisfying human needs</li><li>Optimizing social benefits</li><li>Improving health, comfort, and well-being</li><li>Minimizing cultural disruption</li><li>Providing education services</li><li>Promoting harmony among human beings and between humanity and nature</li></ul>	<ul><li>Improving economic growth</li><li>Reducing energy consumption and costs</li><li>Raising Real Income</li><li>Improving productivity</li><li>Lowering infrastructure costs</li><li>Decreasing environmental damage costs</li><li>Reducing water consumption and costs</li><li>Decreasing health costs</li><li>Decreasing absenteeism in organizations</li><li>Improving Return on Investments (ROI)</li></ul>

It is important to note that sustainable development has limitations. According to Barbier [11], the three pillars of sustainability (environmental, economic, and social) cannot be utilized to their full potential concurrently. The meaning of development must overcome a series of continuous trade-offs, such as the trade-off between increased productivity and the degradation of the environment [11]. Further, the trade-offs are regularly changing due to the intense nature of development and the various ecological, economic, and social conditions [11]. Therefore, sustainable development demands have different levels of importance in other places; they are never constant and change over time. This difference directly applies to the concept of green building as part of sustainable development. Therefore, there is no guarantee that successful practices in one of the ecological, economic, and social dimensions will be similarly effective in other dimensions.

2.2. Barriers of Green Building

It is convenient to better use the PESTLE method to understand the factors affecting the development of green buildings, distributing various aspects according to political, economic, sociocultural, technological, legal, and environmental categories (PESTLE). Moreover, the PESTLE method provides a bird's eye view and an organized look at the factors [15].

There are no negative impacts on the environment caused by factors related to green buildings, as the concept of green building is based on minimizing the negative effects on the environment. Therefore, barriers affecting the spread of green buildings can be distributed only among political, economic, sociocultural, technological, and legal categories. Furthermore, the factors that affect the spread of green buildings are very interrelated. Some elements can correlate with several PESTLE categories, such as "lack of market demand", identified as one of the fundamental barriers by Chan et al. [16], which can be underlined in the economic category and partly in the sociocultural category. Market demand can arise from sociocultural circumstances, even mainly being an economic factor. However, factors were classified according to their primary attributes, not their origin, to avoid uncertainties in this study.

(1) Political barriers. Lack of governmental support and promotion can be classified as political factors. Chan et al. [16], a Ghanaian professional surveyor, identified the lack of government incentives as one of the top three most critical barriers to the development of green construction, highlighting the role of government as a crucial part.

The promotion of sustainable construction resulted in the advancement of low carbon technologies that reduce the impact on the environment in the construction phase; as pointed out in a study carried out on existing green buildings by Eichholtz et al. [17], a lack of promotion is the cause of the slow spread of green practices.

(2) Economic barriers. In many studies, the cost is the most critical barrier to green construction, as it requires more initial investment than traditional buildings [6]. Perception of higher costs causes the market to withdraw from green projects, as noted by Ahn & Pierce [18]. However, studies in the US and UAE show that cost is not the most crucial barrier [12,19].

An extended payback period is another substantial factor in the economic category, delaying the spread of green buildings, and is often ranked as the second most important barrier after cost. According to Lam et al. [20], the additional time required for a green project is a crucial factor affecting stakeholders' decisions on par with higher costs.

Darko et al. [12] also pointed out other barriers such as lack of market demand and risks and uncertainties involved in the implementation of new technologies as crucial factors in the study conducted in the USA.

(3) Sociocultural barriers. The literature represents lack of knowledge and awareness as a critical barrier to consider, as some studies suggest resolving it might solve multiple issues at once [9,19]. However, it might require much effort to raise awareness among stakeholders as it is directly tied to government incentives and educational programs [9].

Darko et al. [12] identified resistance to change as the most critical barrier in their study, followed by a lack of the benefits of knowledge and awareness of sustainable construction benefits. Further, the study stated that resistance to change could determine the success of green buildings in the US [12].

(4) Technological barriers. An extended construction period is another factor related to time, similar to more extended payback periods that affect the spread of green buildings. However, the underdevelopment of technologies in the area is the leading cause of longer construction periods [20], which puts it in this category. Langdon [20] emphasized that the extended construction period is due to soft costs (additional planning and design).

Furthermore, Darko et al. [12] highlighted other significant factors: a lack of experienced staff, educational programs, databases, and information.

(5) Legal barriers. Aktas & Ozorhon [21] emphasized the importance of green building regulation in their study carried out in Turkey. It was one of the factors that affected the decision-making of owners and the top management support. Additionally, Ulubeyli et al. [22] pointed out difficulties in adapting legislation and laws regarding green construction in Turkey.

Green labeling is another critical factor [12,21], as the lack of green building rating certifications can cause difficulties in adopting green projects [12].

2.3. Drivers of Green Buildings

Drivers of green buildings are classified similarly to barriers according to the PESTLE method.

(1) Political drivers. As lack of government support can be a critical factor affecting the spread of green buildings [9,16], contrary government incentives towards adopting green buildings can be a determining factor [9,23]. Darko et al. [23] suggested that government support could compensate stakeholders for the additional cost of building green, promoting

green construction. Similarly, Alsanad [9] had drawn the same conclusion examining factors in Kuwait.

Several studies have stated the importance of company image and reputation when choosing green projects [24,25].

(2) Economic drivers. The common perception that although green buildings have higher implementation costs, they also possess lower operational costs, reducing overall lifecycle expenses, has driven the market long [23]. Studies in Australia and New Zealand revealed the reduced lifecycle cost of green buildings as the most critical driver [26]. A similar study presented this factor in Ghana in the list of top five most influential factors [27].

Love et al. [28], examining an office building in Australia, pointed out several critical drivers, including the attraction of premium clients and high rental returns. High rental returns, reduced operational costs, and lower turnover variability lead to improved building value, which is itself a significant driver of green buildings [29].

(3) Sociocultural drivers. In addition to environmental benefits, green buildings improve the health, comfort, and satisfaction of occupants compared to traditional buildings [30]. It was also rated the second most important factor in Ghana [26]. Moreover, an improved environment for the occupants can attract quality employees [26]. By itself, the attraction of quality employees is an influential driver of green buildings [12].

Unlike lack of awareness being a critical barrier to the spread of green buildings, increased understanding can be a determining driver. Regulations, policies, and educational programs toward green buildings can improve the level of awareness [23].

(4) Technological drivers. Green building practices advance conventional technologies, improving the efficiency of construction processes and management practices. Although Darko et al. [10] revealed the low impact of improved construction efficiency as a driver, it is worth considering the improvements green practices provide. In addition, green projects require more technology, and participants are more likely to be in an integrated work environment [31], which brings construction management processes to another level.

(5) Legal drivers. Andelin et al. [25] noted that the number of governmental regulations and urban policies is constantly increasing and is expected to increase in the future. Such steps are essential in promoting green practice.

Another crucial factor that affects the spread of green buildings is the rating systems, such as LEED or BREEAM. The findings show that in addition to affecting the decision-making of stakeholders, the green design of the project undergoes changes depending on the requirements of the rating system [32], showing the importance and influence of certification systems.

(6) Environmental drivers. A green building is designed to minimize its harm to the environment, efficiently using water and energy resources, and considering human health and comfort [4]. Additionally, green practices encourage reducing construction and demolishing wastes.

Based on an international survey of green building experts carried out by Darko et al. [10], energy and water efficiency were the second and third most important factors driving the adoption of green buildings, respectively. Furthermore, Ulubeyli et al. [22] revealed the importance of Turkey's energy infrastructure and efficiency, ecological sustainability, and waste management. Gathering the environmental benefits of the green concept is tremendous and influential to its spread.

2.4. Green Building in Kazakhstan

Kazakhstan's annual CO_2 emissions have increased steadily since 1999, reaching 318 million tons in 2019 [33]. The country's energy consumption in 2019 was around 75 Mtoe (Mega tonnes of oil equivalent), increasing from 55 Mtoe in 2015 [34]. It is important to note that Kazakhstan's economy is profoundly dependent on coal, oil, and gas, having a massive potential in renewable energy in the face of small hydro, solar, wind, geothermal, biomass, and waste recycling [35]. However, coal-based electricity generation was 70%, natural gas 20%, and the other 10% were renewable energy sources (including hydroelectricity) in 2019 [33]. The electricity generated based on renewable energy sources did not grow since 2015 [33]. There was a decline compared to 2016 when renewable energy sources were 12%, and 11% in 2017 [33]. Furthermore, the actual utilization of renewable energy was only 1.4% in 2018 [34].

Energy consumption only in residential buildings was 27%, the 2nd highest after the industrial sector in Kazakhstan as of 2019 [33]. The annual rate of overall floor area increase of residential buildings was around 10% [33]. According to the Bureau of National Statistics of the Agency for Strategic Planning and Reforms of the Republic of Kazakhstan, 42,282 buildings were put into operation in 2019, 90% of which are residential [34]. The number of buildings in operation increases annually, with a growth rate ranging from 7 to 20% [34].

Kazakhstan has already taken the first steps toward sustainable development and has set ambitious goals. In 2015, Kazakhstan adopted the 2030 Agenda for Sustainable Development Goals at the UN Headquarters [35]. In 2013, in response to achieving green/sustainable construction goals, the Kazakhstan Green Building Council (KazGBC) was formed. In addition, KazGBC began devising the national certification system for residential buildings in 2017 [36,37]. Nonetheless, there are only 74 green-certified buildings in the country, mainly located in Nur-Sultan and Almaty [8]. They are rated according to BREEAM and LEED certification, and most buildings achieve the lowest acceptable score [8]. Consequently, there are still obstacles to overcome, which arise when identifying barriers and potential drivers to spread green building practices.

3. Research Methods

This research investigates the factors that affect the development of green buildings in Kazakhstan. Therefore, a quantitative research method was used to conduct the research, including a literature review, online questionnaire survey, data analysis, and the PESTLE method.

3.1. Review Method

The first step of the literature review was to identify articles based on their titles, abstracts, and keywords in Scopus. The keywords used for the search were "green building", "green construction", and "sustainable construction". The results showed more than 3237 articles related to green buildings with a constant growth rate of around 20% from 2010 to 2020. These numbers represent a significant interest in the area. The next layer of filtering was to add the keywords "barriers" and "drivers" to limit the number of papers according to their relevance to the topic. The final number consisted of 73 papers between 2010 and 2020. Skimming the abstracts allowed verification of around ten articles similar to studies carried out in other countries. Another 21 were identified as eligible for review, as they were closely related to the topic, had a good citation count, and were published in top-tier journals.

Several more articles were identified within the reference list of similar studies dating back to the early 2000s. However, the information from those papers is still relevant, as researchers constantly cite the work up to date. The identified articles were sufficient to cover the first and second objectives of the study. Nevertheless, very few research articles cover Kazakhstan's level of green building development in the Scopus database. Some part of the information was obtained through "The Green Building Information Gateway", "The Bureau of National Statistics of the Agency for Strategic Planning and Reforms of the Republic of Kazakhstan", and other open source databases.

3.2. Questionnaire Survey Preparation

An online questionnaire survey was conducted to collect professional perceptions of factors affecting the implementation of Kazakhstan's green buildings. A questionnaire survey is a standard method used in similar studies. Additionally, an online questionnaire survey is a time-efficient approach to collecting a massive amount of data without a researcher's presence. Additionally, with the existing variety of questionnaire tools available online, it has become more convenient to use this method. The questionnaire survey was conducted via the Qualtrics web survey tool. The survey was administered from February to March 2021. First, eligible respondents were identified. In this regard, the authors used the Green Building Information Gateway website to find the list of green-certified buildings [8]. Afterward, contractors of the green-certified buildings were found and contacted through phone calls, social networks such as Linkedin, Facebook, and personal visits to the companies' offices. Linkedin, in particular, allowed finding and contacting the experts easily.

Moreover, it was very convenient to have a conversation with the experts and collect their feedback. Multiple experts were also kind by sharing the contacts of their colleagues who contributed to the study. Combined, at least 70 invitations were sent from various sources.

In preparing the survey questionnaire, standard drivers and barriers were compiled through a literature review. A total of 36 barriers and 45 drivers of green buildings were identified. However, it was essential to simplify and refine the list of 81 factors since there were some similar factors (or derivatives from other factors) listed in different studies. Furthermore, the official Qualtrics recommendations on a successful survey state that surveys over 12 min can be tiresome and have low response rates [38]. The entry of the list of 81 factors alongside basic information questions was estimated to take 16 min on average, according to the built-in Qualtrics estimation application. Therefore, a mapping table was used to identify the literature factors' frequency of occurrence to identify their significance [35]. A factor in the list was removed if it had a low occurrence rate and was a derivative of a more significant factor. For example, "average income per capita" was mentioned only once and can be considered as part of the "economic state of the country" factor. However, some factors having a relatively low occurrence rate are still kept in the list, such as the "economic state of the country", due to its importance identified in previous studies [7].

Moreover, "Green building (GB) rating systems" and "difficulties adapting to the certification system" have relatively low occurrence rates. However, they are potentially influential factors due to the lack of a national certification system in Kazakhstan. Therefore, these factors remained on the final list. The final lists of barriers and drivers used for the questionnaire survey are provided in Tables 2 and 3, respectively.

Table 2. A compiled list of barriers based on the literature.

ID	Barriers	Frequency of Occurrence	References
B01	Lack of government support	5	[9,12,16,19,39]
B02	Higher costs of Green building technologies (GBTs)	9	[8,9,12,16,19,22,38–40]
B03	Lack of market demand	2	[12,16]
B04	Risks and uncertainties involved in implementing new technologies	6	[7,8,12,41,42]
B05	Economic state	2	[9,22]
B06	Long pay-back periods	6	[6,12,16,19,40,41]
B07	Lack of knowledge and awareness of GBTs and their benefits	8	[6,9,12,16,19,22,40,41]
B08	Conflicts of interests among various stakeholders in adopting GBTs	3	[12,16,41]
B09	Resistance to change	4	[9,12,16,21]
B10	Lack of GBTs databases and information	5	[12,16,19,21,41]
B11	Lack of reliable GBTs research and education	5	[12,16,19,21,41]
B12	Lack of skilled/experienced staff	8	[9,12,16,19,22,41,42]
B13	Longer construction period	4	[19,22,40,42]
B14	Lack of reliable and available and reliable GBTs suppliers	7	[6,12,19,21,22,41,42]
B15	High cost of sustainable materials and products	5	[6,21,22,41,42]
B16	Complexity and rigid requirements involved in adopting GBTs	5	[12,16,21,41,42]
B17	Fewer GB regulations available	6	[9,12,16,22,41,42]
B18	Insufficient GB rating systems and labeling programs available	3	[12,16,22]
B19	Difficulties adapting of the certification system	2	[22,42]

Besides asking respondents to rate the factors on a five-point Likert scale, basic information was collected to analyze the results further. The online questionnaire survey contained questions about the company, profession, years of experience of the respondents in the building industry, and whether they have experience in green building projects.

Table 3. Compiled list of drivers based on frequency of occurrence.

ID	Drivers	Frequency of Occurrence	References
D01	Government support	7	[6,9,12,23,31,40,43]
D02	Company image and reputation	6	[12,19,21,23,31,43]
D03	Reduced lifecycle costs	8	[6,10,12,19,23,31,42,43]
D04	Attract premium clients	6	[10,12,22,23,42,43]
D05	High rental returns	7	[10,12,19,23,31,42,43]
D06	Improvement in the national economy	3	[10,12,43]
D07	Increased building value	7	[10,12,19,23,31,42,43]
D08	Improved occupant health, comfort, and satisfaction	8	[10,12,19,22,23,31,42,43]
D09	Attract quality employees and reduce employee turnover	4	[10,12,23,41]
D10	Facilitation of practice sharing	4	[10,12,19,39]
D11	Educational programs	4	[6,9,23,39]
D12	Commitment to social responsibility	5	[6,10,12,23,42]
D13	Increase of awareness	4	[6,22,23,39]
D14	Efficiency in Construction Processes and management practices	5	[10,12,23,31,43]
D15	Construction standards/Urban planning policies	4	[6,9,12,39]
D16	GB rating systems	2	[6,23]
D17	Energy-efficiency	9	[6,10,12,19,22,23,31,42,43]
D18	Water-efficiency	8	[9,10,12,19,22,23,42,43]
D19	Low environmental impact	7	[10,12,19,23,31,42,43]
D20	Better indoor environmental quality	8	[9,10,12,21,23,31,42,43]
D21	Reduced construction and demolishing wastes	8	[9,10,12,19,22,23,42,43]
D22	Preservation of natural resources	7	[9,10,12,22,23,42,43]

3.3. Data Analysis

Data analysis was performed using descriptive statistics. These statistical methods include: (1) the reliability test using Cronbach's Alpha; (2) the concordance test using Kendall's W; (3) the correlation test using Spearman's rank correlation; and (4) the mean score (M).

Cronbach's Alpha was used to examine the reliability of the collected data, testing internal consistency. Cronbach's Alpha coefficient is based on calculating the average of all possible split-half reliability coefficients ranging from 0 (no internal reliability) to 1 (absolute internal reliability) [44]. Some studies consider a value of 0.6 reliable [45], while some suggest using the rule of thumb, meaning alpha values of 0.8 or higher are acceptable [46]. However, Alpha values above 0.7 are generally considered reliable [9,46]. Cronbach's Alpha coefficient is 0.815 for barriers and 0.895 for drivers, representing acceptable internal consistency and reliability in this study.

Kendall's coefficient of concordance (Kendall's W) represents the level of agreement among raters. The value ranges from 0 (no agreement) to 1 (perfect agreement) [47]. The null hypothesis (H0) for conducted tests is "the distributions of factors are the same". If Kendall's W is low in significance at $p < 0.05$, then the null hypothesis can be rejected, which means that there is no similarity within the distribution of drivers or barriers. Kendall's W is calculated to represent the agreement within different groups of respondents in this study.

The ranking of the mean score is widely used in studies related to green building to classify factors according to their significance [12]. In this study, the mean score ranking identifies the most significant barriers and drivers affecting the spread of green buildings. It is important to note that factors with identical mean scores were sorted according to standard deviation values. Less standard deviation represents higher consistency and, therefore, higher overall rank.

Spearman's rank correlation coefficient was also calculated for different respondents to display the level of association/correlation among their rankings of factors. The coefficient value ranges from -1 to $+1$, where $+1$ represents the perfect correlation of rank, 0 no correlation of ranks, and -1 perfect negative correlation. The null hypothesis for this test is that "there is no correlation between groups". Alpha (α) is set at 0.05, and if $p < 0.05$, then there is less than a 5% chance that the strength of the correlation occurred by chance; the null hypothesis was confirmed.

The responses were classified into several groups, according to the profiles of the respondents: consultant, contractor, government, and others, according to the type of company: engineer, project manager, consultant, and others, according to the profession, and two groups: experienced and not experienced in green building. Each group was tested on the level of agreement within its respondents using Kendall's concordance coefficient (Kendall's W). Additionally, group responses were ranked and analyzed with the Spearman rank correlation coefficient to represent the association level among various respondents.

4. Results

As mentioned earlier, over 70 survey invitations were distributed among practitioners, experts, and academics/researchers. Most of the respondents were local experts, except a couple of foreign professionals who consulted and assisted local green building projects. In total, 38 responses were collected, with a response rate of more than 50%. The results of the survey are described in the following subsections.

4.1. Respondents' Demographics

The majority of the respondents were contractors at 39%, followed by other companies at 21%, consultant companies at 16%, government companies at 13%, financial investment companies at 5%, and material supplier and architect companies at 3% each.

The survey revealed that most respondents were engineers at 40%, consultants and other disciplines at 18%, while 16% were project managers. Minor responses were from academics/researchers at 5% and architects at 3%. According to the survey results, 63% of the respondents had 1–5 years of experience in the construction industry, 26% had more than ten years of experience, and 11% had 6–10 years of experience in the industry.

There were a total of 21 (55%) respondents who claimed to have a green building experience, and 17 (45%) respondents did not have a green building experience (Table 4).

Table 4. Respondents' demographic information.

Company Type	Percentage	Profession	Percentage	Experience in Construction Industry	Percentage	Experience in Green Building	Percentage
Contractor	39	Engineer	40	1–5 years	63	Yes	55
Other	21	Consultant	18	6–10 years	11	No	45
Consultant	16	Other	18	>10 years	26		
Government	13	Project manager	16				
Financial investment	5	Academia	5				
Architect	3	Architect	3				
Material supplier	3						

A well-blended response from various participant groups ensures the reliability of the obtained results.

4.2. Agreement and Correlations among Respondent Groups

4.2.1. Company-Type Distribution

The questionnaire survey identified multiple groups of respondents. It is crucial to identify the differences and correlations between these groups.

There are a total of seven groups identified by the survey according to the type of respondents from the company. However, only four groups have a reasonable number of responses for the concordance test. Barrier factors' agreement within the particular group is represented in Table 5.

Table 5. Company type related groups and test of concordance for barriers.

	Consultant	Contractor	Government	Other
Kendall's W	0.277	0.074	0.304	0.172
Chi-Square	29.884	20.110	27.383	24.838
Degree of Freedom	18	18	18	18
Asymp. Sig.	0.039	0.327	0.072	0.129

Barriers. Kendall's W for the "consultant" group is 0.277, which is considered a low value, representing a low level of agreement. Additionally, asymptotic significance is identified as $p = 0.039$, meaning that the null hypothesis (the distribution of factors are the same) is rejected. Similarly, the "contractor", "government", and "other" groups have a low coefficient of concordance of 0.074, 0.304, and 0.172, respectively. However, the p-value is higher than 0.05, therefore retaining the null hypothesis. This result represents the similarity within responses and a nonsignificant level of agreement within these groups.

According to Spearman's rank correlation test of barriers rated by "consultant", "contractor", "government", and "other" groups, we can see a general trend of low correlation among these four groups. Furthermore, the significance level is higher than 0.05, which means that the null hypothesis is retained, and there is no significant correlation among group pairs. The correlation coefficient of the "consultant"–"other" pair is at -0.056, representing a slight negative correlation. However, the correlation coefficient of "contractor" and "government" is at 0.662, generally considered a reasonable association level, with a significance level of 0.002, showing an asymptotic significance.

Drivers. Table 6 represents Kendall's W of drivers rated by "consultant", "contractor", "government", and "other" groups. There is generally a low level of agreement within each group, with a significance level greater than 0.05. Therefore, the null hypothesis is retained. On the other hand, the "other" group has a significant level of 0.014, representing a significant difference in agreement.

Table 6. Groups related to company type and test of concordance for drivers.

	Consultant	Contractor	Government	Other
Kendall's W	0.239	0.084	0.217	0.224
Chi-Square	30.090	26.330	22.821	37.600
Degree Of Freedom	21	21	21	21
Asymp. Sig.	0.090	0.194	0.354	0.014

The groups' correlation coefficients are reasonable for most pairs, ranging from 0.493 to 0.778 with a significance level lower than 0.05. Only "contractor"–"other" and "consultant"–"contractor" pairs have a low association, with correlation coefficients of 0.177 and 0.261, respectively.

4.2.2. Profession Type Distribution

From the questionnaire survey, respondents can also be categorized according to their profession. Only four of the six categories are used for the agreement test (Kendall's W) and correlation test (Spearman's rank correlation), since other professions had too low responses to be considered for the test. The groups are: "engineer", "project manager", "consultant", and "other".

Barriers. Table 7 shows a low level of agreement in each group value ranging from 0.088 to 0.173 with a significance level higher than 0.05. Therefore, the null hypothesis is retained, showing nonsignificant similarities within the responses in each group.

Table 7. Groups related to profession types and test of concordance for barriers.

	Engineer	Project Manager	Consultant	Other
Kendall's W	0.088	0.173	0.128	0.156
Chi-Square	23.840	18.708	16.116	19.688
Degree Of Freedom	18	18	18	18
Asymp. Sig.	0.160	0.410	0.584	0.351

Each pair's significance levels in this category exceed the value of 0.05, which means that there are no significant similarities. The correlation coefficient value represents a similar trend showing a low association level, values ranging from 0.106 to 0.396.

Drivers. Profession-based groups do not show a significant level of concordance when rating drivers. Kendall's W stays within the range of 0.09 and 0.262, which is considered low. The p-value is higher than 0.05 for the "engineer", "project manager", and "other" groups, representing a nonsignificant level of similarity. The "consultant" group has asymptotic significance rejecting the null hypothesis (Table 8).

Table 8. Groups related to profession types and test of concordance for drivers.

	Engineer	Project Manager	Consultant	Other
Kendall's W	0.090	0.252	0.262	0.155
Chi-Square	28.352	31.770	38.474	22.798
Degree Of Freedom	21	21	21	21
Asymp. Sig.	0.130	0.062	0.011	0.355

The "consultant"–"other" pair shows a reasonable correlation at a value of 0.673 and a significance level of 0.001 when ranking the drivers. However, other group pairs share a low correlation, with p-values higher than 0.05.

4.2.3. Green Building Experience-Based Distribution

Responses from two respondents groups based on their experience in the green building were analyzed on the agreement within-group itself and correlation with the other group. Group 1 refers to respondents who have experience in green buildings. Group 2 refers to respondents who do not have experience in green buildings.

Barriers. Groups 1 and 2 have a Kendall's W of 0.133 and 0.098, respectively, which is considered low. However, the significance level is lower than 0.05 for both groups, representing asymptotic significance in the results (Table 9). According to Spearman's rank correlation test, Group 1 and Group 2 have a correlation coefficient of 0.274 with a significance of 0.257, which is considered a low association.

Table 9. Total mean barriers ranks and two response groups and test of concordance.

	Total	Group 1 (with Experience in GB)	Group 2 (No Experience in GB)
Kendall's W	0.081	0.133	0.098
Chi-Square	55.364	50.108	30.074
Degree Of Freedom	18	18	18
Asymp. Sig.	0.000	0.000	0.037

Drivers. Kendall's W for Group 1 is 0.142, with a high asymptotic significance level. The null hypothesis is rejected, and there is a significant difference in agreement within Group 1. Group 2 has a significance value of 0.533, meaning there are nonsignificant similarities, in addition to having a similarly low level of agreement with Group 1 (Table 10). It is important to note that the correlation coefficient between Groups 1 and 2 when ranking the driver factors is significantly higher when the same group rated the barriers. The correlation coefficient value is 0.633, with a significant level of 0.002.

Table 10. Total mean ranks of drivers and two respondent groups and test of concordance.

	Total	Group 1 (with Experience in GB)	Group 2 (No Experience in GB)
Kendall's W	0.085	0.142	0.055
Chi-Square	67.863	62.811	19.812
Degree Of Freedom	21	21	21
Asymp. Sig.	0.000	0.000	0.533

4.3. The Final Mean Rank of Factors Categorized According to PESTLE

The PESTLE method is used to provide a broad view of the factors. The total mean ranks of barriers and drivers are classified according to PESTLE in Tables 11 and 12, respectively. The critical thing to note is that there are some important trends in the factor distributions.

Table 11. Mean score with barriers ranks categorized according to PESTLE.

ID	Rank	Barriers	Mean	SD
B12	1	Lack of skilled/experienced staff	4.11	0.831
B01	2	Lack of government support	4.11	0.924
B15	3	High cost of sustainable materials and products	4.03	0.915
B05	4	Economic state	4.00	0.771
B02	5	Higher costs of GBTs	3.97	0.822
B03	6	Lack of market demand	3.87	1.044
B06	7	Long pay-back periods	3.84	1.151
B14	8	Lack of reliable and available and reliable GBTs suppliers	3.79	0.811
B07	9	Lack of knowledge and awareness of GBTs and their benefits	3.76	1.025
B17	10	Fewer GB regulations available	3.71	0.927
B04	11	Risks and uncertainties involved in implementing new technologies	3.68	0.904
B11	12	Lack of reliable GBTs research and education	3.66	0.966
B16	13	Complexity and rigid requirements involved in adopting GBTs	3.58	1.030
B09	14	Resistance to change	3.50	1.109
B19	15	Difficulties in adapting to the certification system	3.50	1.133
B08	16	Conflicts of interests among various stakeholders in adopting GBTs	3.47	1.133
B10	17	Lack of GBTs databases and information	3.39	1.104
B18	18	Sufficient GB rating systems and labeling programs are unavailable	3.37	1.025
B13	19	Longer construction period	3.24	1.025

Color Legend

Political	Economic	Socio-cultural	Technological	Legal

The results reveal that the three main barriers to green building are:
"Lack of skilled/experienced staff".
"Lack of government support".

"High cost of sustainable materials and products".

According to the respondents, the first most significant barrier is from the Technological category, while other Technological barriers are ranked low. The second most important barrier is the "lack of government support", the only Political factor. The third one is "high cost of sustainable materials and products", followed by other economic factors like "economic state", "higher costs of GBTs", "lack of market demand", and "long payback periods". Therefore, economic factors are found to be more prominent barriers. Sociocultural and legal barriers depict lower rankings than most technological, political, and economic factors.

Table 12. Mean score and driver ranks categorized according to PESTLE.

ID	Rank	Drivers	Mean	SD
D18	1	Water-efficiency	4.61	0.790
D17	2	Energy-efficiency	4.58	0.758
D08	3	Improved occupant health, comfort, and satisfaction	4.55	0.686
D19	4	Low environmental impact	4.47	0.797
D22	5	Preservation of natural resources	4.39	0.790
D15	6	Construction standards/Urban planning policies	4.37	0.751
D01	7	Government support	4.37	0.819
D20	8	Better indoor environmental quality	4.37	0.852
D02	9	Positive company image and reputation	4.29	0.835
D03	10	Reduced lifecycle costs	4.24	0.943
D21	11	Reduced construction and demolishing wastes	4.21	1.018
D14	12	Efficiency in Construction Processes and management practices	4.13	0.906
D16	13	GB rating systems	4.11	0.863
D12	14	Commitment to social responsibility	4.08	1.124
D13	15	Increase of awareness	4.05	1.064
D05	16	High rental returns	4.00	1.090
D07	17	Increased building value	4.00	1.090
D10	18	Facilitation of practice sharing	3.97	1.078
D11	19	Educational programs	3.95	1.114
D04	20	Attract premium clients	3.95	1.161
D06	21	Improvement in the national economy	3.92	1.075
D09	22	Attract quality employees and reduce employee turnover	3.89	1.134

Color Legend

Political	Economic	Socio-cultural	Technological	Legal	Environmental

According to the respondents, the top three most significant drivers of green building are:

"Water-efficiency".

"Energy-efficiency".

"Improved occupant health, comfort, and satisfaction of the occupants".

There are some noticeable trends in the drivers' mean score with the ranking list, as four out of the top five most significant drivers are from the Environmental category. The rest of the environmental drivers are also ranked relatively high at 8 and 11, respectively. "Improved health of occupants, comfort, and satisfaction" is the only Sociocultural driver located at the top of the list. The other five sociocultural drivers are contrarily ranked low. Political and Legal drivers are in the middle of the list after the Environmental drivers. The only technological driver, efficiency in construction processes and management practices, is also in the middle of the list and is of lesser importance than environmental, political, and legal drivers. According to experts' responses, it is noticeable that Economic and other Sociocultural drivers are less motivating factors than the rest.

5. Discussion

This research sought to identify the barriers and drivers that influence green building development in Kazakhstan. The results from the conducted questionnaire survey were analyzed and validated using descriptive statistics.

5.1. Validation of Survey Responses

Kendall's concordance coefficient revealed a generally low agreement among respondents of the groups, as it did not go over 0.3 for any of the groups. However, this does not necessarily mean that all results are nonsignificant. In multiple cases, the significance level was lower than the probability value of 5% ($p = 0.05$), representing statistical significance. In cases where the significance level was otherwise exceeding the p-value, there is still agreement, but it is considered nonsignificant. It is important to note that more significance is identified among the experienced/inexperienced groups than in company type or profession type groups. There was only one case, agreement within the group with no previous experience in green building rating drivers when the result had a nonsignificant outcome. However, this could be due to a smaller sample size for groups based on company type or profession than experienced and inexperienced groups.

Spearman's rank correlation coefficient tests showed a generally low to medium concordance level among group pairs. There is a noticeable trend within correlation tests when results favor driver factors representing a more consistent medium correlation coefficient with high significance values. The trend is shared for company-type groups, profession-type groups, and experienced and inexperienced groups. The mean values of the driver factors generally have a lower standard deviation. It is more prominent with experienced than inexperienced groups rating driver factors, where the experienced group had a lower standard deviation in the ratings. That is very reasonable as experienced people should have more consistent answers.

As factors were rated according to the Likert scale, statistically, the value of 3 (neither agree nor disagree) is considered neutral. If the mean values of the factors are statistically different from 3, then the result is considered significant. The mean value results showed that the barriers and drivers are different from 3, therefore, significant. However, driver factors have a higher minimum mean value of 3.89, where barrier factors are only 3.24. So, we can assume that the results of the driver factors are more significant than the results of the barrier factors.

Experienced vs. *inexperienced groups.* Comparing rankings of experienced and inexperienced groups on barriers revealed a low correlation with the nonsignificant result. We can notice this pattern in multiple rankings, such as for barriers, the experienced group ranked "lack of government support" as 1st (highest importance), and the inexperienced group ranked the same barrier as 15th out of 19. The "economic state" was ranked second by the experienced group and 10th by the inexperienced group. However, the ranks were not negatively correlated as the correlation value was low (0.274) but not zero or negative. The inexperienced group ranked "Lack of skilled/experienced staff", as an example, and the experienced one ranked it 1st, or the "high cost of sustainable materials and products" was ranked 2nd by the inexperienced group and 7th by the experienced one, showing correlation.

On the other hand, the same two groups ranking the driver had a medium level of correlation with a significance level of 0.002. The correlation coefficient between the groups was at 0.633. Notably, both groups ranked "energy efficiency" as the second most influential driver. Further, "improved occupants' health, comfort, and satisfaction" was ranked 1st by inexperienced and 3rd by experienced groups.

Such a difference between the results of barriers and drivers might have been caused by a relatively low sample size when completely different response patterns were used, resulting in a low consistency and correlation or the actual inconsistency in the respondents' knowledge and awareness, meaning respondents are generally less confident about barrier factors. The first theory, reasoning low correlation, and consistency of groups rating barriers being a low sample size seems more reasonable and favoring. However, it also contradicts relatively consistent and statistically significant results of driver factors. Nevertheless, the results for both barriers and drivers are reliable as the calculated Cronbach's Alpha was higher than 0.7.

5.2. Mean Scores and Rank

Barriers. "Lack of skilled/experienced staff" and "lack of government support" are rated as the first and second most important barriers, both sharing an identical mean value of 4.11. However, "lack of skilled/experienced staff" has a lower standard deviation than "lack of government support". The literature shows similar issues in the UAE [19] and Malaysia [6], where a lack of professionals was identified as similar in importance. This finding makes sense, since Kazakhstan is in its early stages of adopting the concept of green buildings. Therefore, the lack of experienced people in the industry is expected.

The "lack of government support" was rated as the second most important driver in this study. A study in Kuwait [9] identified the importance of government support, suggesting various incentives. Similar cases can be observed in Ghana [26] and Singapore [43], where the lack of government incentives was one of the top three most critical barriers.

The "high cost of sustainable materials and products" was rated as the third most crucial driver. It is closely related to the high cost of GBTs mentioned as critically important in other studies [6,18].

It is an interesting finding that the least influential barriers in this study are "longer construction period" alongside "insufficient GB rating systems and labeling programs available". However, the longer implementation time of green projects was considered one of Australia's most critical barriers [48], and Zhang et al. [40] point out that longer construction times often cause excessive and not attractive payback periods. The reasoning that "longer construction period" is rated low could be, that in general, all existing green-certified projects in Kazakhstan did not face the issue of more extended construction periods.

Although Kazakhstan does not have a national certification system, another low-rated barrier is "insufficient GB rating systems and labeling programs available". This barrier can be related to several successful implementation examples of LEED, BREEAM certifications in recent years. Before 2018, there were no LEED "gold-certified" projects in Kazakhstan, whereas today, there are 7 "gold-certified" buildings [8].

Drivers. Professionals rated "water efficiency" as the most important driver, with a mean value of 4.61. "Energy-efficiency" was rated the second most influential factor at a 4.58 mean value, and "improved occupants' health, comfort, and satisfaction" was the third most important factor at 4.55 mean value. All three factors have very similar mean values and low standard deviation values. Studies carried by Darko et al. [12] found these three factors on a similar significance level in the US. "Water-efficiency" alongside "energy-efficiency" were the top two ranked drivers revealed by Darko et al. [12], and the "improved occupants' health, comfort, and satisfaction" was the fourth most crucial driver. There are multiple benefits of green buildings, including reduced lifecycle costs. Around 40% of the reduced lifecycle costs of green buildings can be related to water and energy efficiency [4]. This observation justifies why local experts rated water and energy efficiency as the most significant drivers.

Energy efficiency was the highest-rated driver by Ahn et al. [6] and a similar study in Greece [48], as it is a high priority in many countries [12]. Energy efficiency is one of the most effective, cost-efficient approaches to mitigating climate change and improving air quality [12].

Other significant drivers rated by the respondents include "low environmental impact", "preservation of natural resources", "construction standards/urban planning policies", "government support", which are all commonly known benefits of green buildings.

Drivers like "attract quality employees and reduce employee turnover" and "improvement in the national economy" were ranked low compared to other drivers. However, they represent significant importance, as the mean values are close to 4.00, and their significance was mentioned in other studies [12].

It is hard to say that the highest rated factors in both barriers and drivers are drastically more important than the second or second than the third. However, we can observe trends in the rank distribution of the PESTLE categories. Using PESTLE distribution, we can now state the significance of the Political category. Also, Economic barriers are tightly clustered together, which shows the consistency in respondents' ratings. It shows the viability of using PESTLE analysis to categorize the factors, as it provides another perspective view on existing data and lets us draw interesting conclusions.

6. Conclusions

Green building was presented as a solution to multiple environmental, economic, and social problems. It is a progressively developing concept that is spreading throughout the world. However, green building is driven and obstructed by multiple factors that are constantly studied. The study results vary in different places due to the uniqueness of the area and the time the study is carried out, and the development of green buildings must compensate for continuous trade-offs. This study investigates barriers and drivers of green building implementation and their significance with Kazakhstan's perspective through a comprehensive questionnaire survey.

Ranking the barrier and driving factors provides valuable insight for green building implementation in Kazakhstan and suggests the importance of similar studies in other regions. As can be seen from the analysis obtained, the lack of skilled/experienced staff was considered by Kazakhstan professionals the most hindering barrier to green building. A possible reason for the decision of the respondents is that Kazakhstan is at an early stage of green building development, lacking qualified professionals in the area. The same barrier was similarly crucial in other countries such as the UAE and Malaysia.

Water efficiency and energy efficiency were rated as the most influential drivers of green building in Kazakhstan. Green buildings' water and energy efficiency tend to reduce lifecycle costs by around 40%, and they are generally known benefits of green buildings. Furthermore, the country's energy efficiency is considered low due to outdated technologies, which may have affected the respondents' opinions.

The application of PESTLE analysis on the existing data showed significant trends. It is recommended to enhance the analysis by increasing the sample size and normalizing the number of factors for each PESTLE category.

Government plays a vital role in the adoption of green practices, as discussed earlier. Furthermore, the results of the survey revealed that the lack of government support is the second most important barrier. Therefore, it is recommended that the government provide stronger incentives toward sustainable development. In addition, the government could resolve or stimulate the need for experienced employees by providing education programs or encouraging companies to do so. In addition, there is a lack of proper, user-friendly databases to observe the country's current state of sustainability. Information on existing green-certified buildings in the country was obtained through an international database. Obtaining the information in the area should not be difficult. In addition, contacting the experts or finding their contacts was relatively difficult. Although companies cannot publish their employees' personal information in open sources, it is recommended that local companies might provide viable alternatives to contact them as such change might promote more research in the area, advancing it.

Categorizing the responses according to company types and profession types of the respondents revealed valuable information. However, the results were generally nonsignificant, which might have been caused by the relatively low sample size for each category. It is recommended not to conduct a similar analysis due to low correlation among groups or increase the sample size and discover whether the correlation level was heavily affected by the sample size. Analyses of responses from experienced and inexperienced groups in green building revealed statistically significant results. It is recommended to enhance the analysis by increasing the sample size to increase confidence in the results. It is also recommended that the perspective of the owner or the client should also be recorded. In the present survey, this perspective is hidden in categories of company type such as financial investment, government, and others. In this regard, the present paper can be regarded as a precursor to a more comprehensive future study on the topic.

Comparing drivers to barriers results reveals a higher level of confidence among respondents rating the drivers, having higher mean values. Additionally, driver factors results were generally more correlated.

Further investigations are recommended as statistics reveal a relatively low progression level toward the sustainability of Kazakhstan. The results of the study are applicable to field professionals and further investigations in the area.

It is important to conduct a post-use evaluation of the existing green building in Kazakhstan to understand the efficacy of these buildings and the lesson learned. Future study is recommended to investigate the effectiveness and challenges of existing green buildings in Kazakhstan.

Author Contributions: Conceptualization, D.A. and A.N.; formal analysis, D.A.; funding acquisition, A.N.; investigation, D.A.; methodology, D.A.; project administration, M.A.H.; resources, A.N.; supervision, A.N. and M.A.H.; validation, D.A.; visualization, G.A.; writing—original draft, D.A.; writing—review & editing, A.N., M.A.H. and M.K. All authors have read and agreed to the published version of the manuscript.

Funding: The present research was supported by Nazarbayev University Competitive Grants Program (Funder Project Reference: 021220FD2251, Project Financial System Code: SEDS2021022).

Informed Consent Statement: Informed consent was obtained from all the participants before completing the survey.

Conflicts of Interest: The authors declare no conflict of interest.

References

1. Global Alliance for Buildings and Construction. Global Status Report for Buildings and Construction: Towards a Zero-Emission, Efficient and Resilient Buildings and Construction Sector. International Energy Agency and the United Nations Environment Programme. 2020. Available online: https://globalabc.org/sites/default/files/inline-files/2020%20Buildings%20GSR_FULL%20REPORT.pdf (accessed on 15 January 2021).
2. Global Alliance for Buildings and Construction. Global Status Report for Buildings and Construction: Towards a Zero-Emission, Efficient and Resilient Buildings and Construction Sector. International Energy Agency and the United Nations Environment Programme. 2019. Available online: https://www.unep.org/resources/publication/2019-global-status-report-buildings-and-construction-sector (accessed on 20 January 2021).
3. Pulselli, R.M.; Simoncini, E.; Bastianoni, S. Emergy analysis of building manufacturing, maintenance and use: Em-building indices to evaluate housing sustainability. *Energy Build.* **2007**, *39*, 620–628. [CrossRef]
4. Yudelson, J. *The Green Building Revolution*; Island Press: Washington, DC, USA, 2010.
5. WBCSD. Energy Efficiency in Buildings–Business Realities and Opportunities–Full Report. 2008. Available online: http://www.wbcsd.org (accessed on 10 January 2021).
6. Ahn, Y.H.; Pearce, A.R.; Wang, Y.; Wang, G. Drivers and barriers of sustainable design and construction: The perception of green building experience. *Int. J. Sustain. Build. Technol. Urban Dev.* **2013**, *4*, 35–45. [CrossRef]
7. Kibert, C.J. *Sustainable Construction: Green Building Design and Delivery*; John Wiley & Sons: Hoboken, NJ, USA, 2016.
8. The Green Building Information Gateway. 2020. Available online: http://www.gbig.org/places (accessed on 14 January 2021).
9. AlSanad, S. Awareness, Drivers, Actions, and Barriers of Sustainable Construction in Kuwait. *Procedia Eng.* **2015**, *118*, 969–983. [CrossRef]
10. Darko, A.; Chan, A.P.; Owusu-Manu, D.-G.; Ameyaw, E.E. Drivers for implementing green building technologies: An international survey of experts. *J. Clean. Prod.* **2017**, *145*, 386–394. [CrossRef]
11. Barbier, E.B. The Concept of Sustainable Economic Development. *Environ. Conserv.* **1987**, *14*, 101–110. [CrossRef]
12. Darko, A.; Chan, A.P.C.; Ameyaw, E.E.; He, B.-J.; Olanipekun, A.O. Examining issues influencing green building technologies adoption: The United States green building experts' perspectives. *Energy Build.* **2017**, *144*, 320–332. [CrossRef]
13. Boulton, E. The Research Funnel. Medium. 2018. Available online: https://medium.com/@emmaboulton/the-reseach-funnel-617b8333ad7f (accessed on 22 January 2021).
14. Brundtland, G.H.; Khalid, M.; Agnelli, S.; Al-Athel, S.; Chidzero, B. *Our Common Future*; Oxford University Press: New York, NY, USA, 1987; Volume 8.

15. Lam, P.T.I.; Chan, E.H.W.; Chau, C.K.; Poon, C.S.; Chun, K.P. Integrating Green Specifications in Construction and Overcoming Barriers in Their Use. *J. Prof. Issues Eng. Educ. Pract.* **2009**, *135*, 142–152. [CrossRef]
16. Chan, A.P.C.; Darko, A.; Olanipekun, A.O.; Ameyaw, E.E. Critical barriers to green building technologies adoption in developing countries: The case of Ghana. *J. Clean. Prod.* **2018**, *172*, 1067–1079. [CrossRef]
17. Eichholtz, P.; Kok, N.; Quigley, J.M. Doing Well by Doing Good? Green Office Buildings. *Am. Econ. Rev.* **2010**, *100*, 2492–2509. [CrossRef]
18. Ahn, Y.H.; Pearce, A.R. Green Construction: Contractor Experiences, Expectations, and Perceptions. *J. Green Build.* **2007**, *2*, 106–122. [CrossRef]
19. Yas, Z.; Jaafer, K. Factors influencing the spread of green building projects in the UAE. *J. Build. Eng.* **2020**, *27*, 100894. [CrossRef]
20. Langdon, D. Costing Green: A Comprehensive Cost Database and Budgeting Methodology. Lisa Fay Matthiesen and Peter Morris. 2004. Available online: https://vgbc.vn/wp-content/uploads/2018/12/Costing-Green_A-Comprehensive-Cost-Database-and-Budgeting-Methodology.pdf (accessed on 25 January 2021).
21. Aktas, B.; Ozorhon, B. Green Building Certification Process of Existing Buildings in Developing Countries: Cases from Turkey. *J. Manag. Eng.* **2015**, *31*, 05015002. [CrossRef]
22. Ulubeyli, S.; Kazancı, O.; Kazaz, A.; Arslan, V. Strategic Factors Affecting Green Building Industry: A Macro-Environmental Analysis Using PESTEL Framework. *Sak. Üniv. Fen Bilimleri Enstitüsü Derg.* **2019**, *23*, 1042–1055. [CrossRef]
23. Darko, A.; Zhang, C.; Chan, A.P. Drivers for green building: A review of empirical studies. *Habitat Int.* **2017**, *60*, 34–49. [CrossRef]
24. Serpell, A.; Kort, J.; Vera, S. Awareness, Actions, Drivers and Barriers of Sustainable Construction in Chile. *Technol. Econ. Dev. Econ.* **2013**, *19*, 272–288. [CrossRef]
25. Andelin, M.; Sarasoja, A.-L.; Ventovuori, T.; Junnila, S. Breaking the circle of blame for sustainable buildings—Evidence from Nordic countries. *J. Corp. Real Estate* **2015**, *17*, 26–45. [CrossRef]
26. Darko, A.; Chan, A.P.C.; Gyamfi, S.; Olanipekun, A.O.; He, B.-J.; Yu, Y. Driving forces for green building technologies adoption in the construction industry: Ghanaian perspective. *Build. Environ.* **2017**, *125*, 206–215. [CrossRef]
27. Worzala, E.; Bond, S. Barriers and drivers to green buildings in Australia and New Zealand. *J. Prop. Invest. Financ.* **2011**, *29*, 494–509.
28. Love, P.E.D.; Niedzweicki, M.; Bullen, P.A.; Edwards, D.J. Achieving the Green Building Council of Australia's World Leadership Rating in an Office Building in Perth. *J. Constr. Eng. Manag.* **2012**, *138*, 652–660. [CrossRef]
29. Devine, A.; Kok, N. Green certification and building performance: Implications for tangibles and intangibles. *J. Portf. Manag.* **2015**, *41*, 151–163. [CrossRef]
30. Paul, W.; Taylor, P.A. A comparison of occupant comfort and satisfaction between a green building and a conventional building. *Build. Environ.* **2008**, *43*, 1858–1870. [CrossRef]
31. Liu, J.Y.; Low, S.P.; He, X. Green practices in the Chinese building industry: Drivers and impediments. *J. Technol. Manag. China* **2012**, *7*, 50–63. [CrossRef]
32. He, Y.; Kvan, T.; Liu, M.; Li, B. How green building rating systems affect designing green. *Build. Environ.* **2018**, *133*, 19–31. [CrossRef]
33. Our World in Data. Kazakhstan CO_2 Country Profile. 2021. Available online: https://ourworldindata.org/co2/country/kazakhstan (accessed on 10 February 2021).
34. Agency for Strategic Planning and Reforms of the Republic of Kazakhstan Bureau of National Statistics. 2021. Available online: https://www.gov.kz/memleket/entities/aspr?lang=en (accessed on 10 February 2021).
35. Global Green Initiatives in Kazakhstan. 2019. Available online: https://kapital.kz/gosudarstvo/68657/globalnaya-zelenaya-iniciativa-doshla-do-kazahstana.html (accessed on 15 February 2021).
36. Akhanova, G.; Nadeem, A.; Kim, J.R.; Azhar, S. A multi-criteria decision-making framework for building sustainability assessment in Kazakhstan. *Sustain. Cities Soc.* **2020**, *52*, 101842. [CrossRef]
37. Electronic Government of the Republic of Kazakhstan. Sustainable Development Goals in the Republic of Kazakhstan. Public Services and Online Information. 2021. Available online: http://www.egov.kz (accessed on 17 February 2021).
38. Qualtrics. How to Increase Online Survey Response Rates. 2021. Available online: https://www.qualtrics.com/experience-management/research/tools-increase-response-rate/ (accessed on 2 February 2021).
39. Li, Y.; Song, H.; Sang, P.; Chen, P.-H.; Liu, X. Review of Critical Success Factors (CSFs) for green building projects. *Build. Environ.* **2019**, *158*, 182–191. [CrossRef]
40. Zhang, X.; Platten, A.; Shen, L. Green property development practice in China: Costs and barriers. *Build. Environ.* **2011**, *46*, 2153–2160. [CrossRef]
41. Hwang, B.G.; Zhu, L.; Ming, J.T.T. Factors Affecting Productivity in Green Building Construction Projects: The Case of Singapore. *J. Manag. Eng.* **2017**, *33*, 04016052. [CrossRef]
42. Windapo, A.O. Examination of Green Building Drivers in the South African Construction Industry: Economics versus Ecology. *Sustainability* **2014**, *6*, 6088–6106. [CrossRef]
43. Oguntona, O.A.; Akinradewo, O.I.; Ramorwalo, D.L.; Aigbavboa, C.O.; Thwala, W.D. Benefits and Drivers of Implementing Green Building Projects in South Africa. *J. Phys. Conf. Ser.* **2019**, *1378*, 032038. [CrossRef]
44. Bryman, A. *Social Research Methods*; Oxford University Press: New York, NY, USA, 2016.

45. Lau, R.; Zhao, X.; Xiao, M. Assessing quality management in China with MBNQA criteria. *Int. J. Qual. Reliab. Manag.* **2004**, *21*, 699–713. [CrossRef]
46. Pallant, J. *SPSS Survival Manual: A Step by Step Guide to Data Analysis Using IBM SPSS*; Routledge: Oxfordshire, UK, 2020.
47. Siegel, S. Nonparametric Statistics for the Behavioral Sciences. 1956. Available online: http://www.ru.ac.bd/stat/wp-content/uploads/sites/25/2019/03/303_11_Siegel-Nonparametric-statistics-for-the-behavioral-sciences.pdf (accessed on 1 March 2021).
48. Manoliadis, O.; Tsolas, I.; Nakou, A. Sustainable construction and drivers of change in Greece: A Delphi study. *Constr. Manag. Econ.* **2006**, *24*, 113–120. [CrossRef]

 *buildings*

Article

An Exhaustive Search Energy Optimization Method for Residential Building Envelope in Different Climatic Zones of Kazakhstan

Mirzhan Kaderzhanov, Shazim Ali Memon *, Assemgul Saurbayeva and Jong R. Kim

Department of Civil and Environmental Engineering, School of Engineering and Digital Sciences, Nazarbayev University, Nur-Sultan 010000, Kazakhstan; mirzhan.kaderzhanov@alumni.nu.edu.kz (M.K.); assemgul.saurbayeva@nu.edu.kz (A.S.); jong.kim@nu.edu.kz (J.R.K.)
* Correspondence: shazim.memon@nu.edu.kz; Tel.: +7-778-443-85-16

Abstract: Nowadays, the residential sector of Kazakhstan accounts for about 30% of the total energy consumption. Therefore, it is essential to analyze the energy estimation model for residential buildings in Kazakhstan so as to reduce energy consumption. This research is aimed to develop the Overall Thermal Transfer Value (OTTV) based Building Energy Simulation Model (BESM) for the reduction of energy consumption through the envelope of residential buildings in seven cities of Kazakhstan. A brute force optimization method was adopted to obtain the optimal envelope configuration varying window-to-wall ratio (WWR), the angle of a pitched roof, the depth of the overhang shading system, the thermal conductivity, and the thicknesses of wall composition materials. In addition, orientation-related analyses of the optimized cases were conducted. Finally, the economic evaluation of the base and optimized cases were presented. The results showed that an average energy reduction for heating was 6156.8 kWh, while for cooling it was almost 1912.17 kWh. The heating and cooling energy savings were 16.59% and 16.69%, respectively. The frontage of the building model directed towards the south in the cold season and north in the hot season demonstrated around 21% and 32% of energy reduction, respectively. The energy cost savings varied between 9657 to 119,221 ₸ for heating, 9622 to 36,088 ₸ for cooling.

Keywords: building envelope; optimization; OTTV; energy consumption; Brute Force Algorithm

Citation: Kaderzhanov, M.; Memon, S.A.; Saurbayeva, A.; Kim, J.R. An Exhaustive Search Energy Optimization Method for Residential Building Envelope in Different Climatic Zones of Kazakhstan. *Buildings* **2021**, *11*, 633. https://doi.org/10.3390/buildings11120633

Academic Editor: Roberto Alonso González Lezcano

Received: 29 October 2021
Accepted: 28 November 2021
Published: 10 December 2021

Publisher's Note: MDPI stays neutral with regard to jurisdictional claims in published maps and institutional affiliations.

1. Introduction

In Kazakhstan, the high demand for residential energy consumption is considered an important factor influencing economic development and domestic comfort. Energy-based economic relationships and domestic use constitute about 80% of the total energy distribution in Kazakhstan [1]. The electricity and heat generation is obtained from at least 40% of the total direct energy supply (TDES) and they are counted as one-third of the total final energy (TFEC) consumption. The ratio of TFEC to TDES, as an indicator of energy balance, shows the value as less than 50% for Kazakhstan and 69% for entire the world [2].

Kazakhstan's existing residential building stock comprises around 347.4 million square meters, almost the one third of which are outdated inefficient buildings that were constructed during the Soviet era [3]. In the 1960s, large-panel residential building projects for seismically active locations were planned and realized, while in the 1970s, the high-rise building development (from 9 to 12 stories) began. During this time, standard projects were developed that established a qualitatively new approach to standard design and expanded throughout Kazakhstan's cities [4]. The majority of the current housing stock is made up of multifamily structures that are connected to district heating via boiler houses or cogeneration stations, the rest is accounted for in unfamiliar and semi-detached houses [5,6].

Energy standards or regulations for buildings are considered an important factor in energy efficiency strategies. These codes contain details about building energy conservation methods, propose and evaluate the development of different types of energy-efficient building designs, create a specific regulation to assess building energy conditions, and propose new energy efficiency procedures for climate regions or countries [7]. Typical energy efficiency indices are perimeter annual load (PLA), envelope energy load (ENVLOAD), and overall thermal transfer value (OTTV) [6]. The PLA is the total yearly cooling and heating load in the perimeter of buildings per unit floor area; it comprises heat conduction through the envelope produced by temperature differences between outside and inside conditions, indoor heat gain, fresh air load, and solar radiation heat gain [8]. ENVLOAD is a multilinear regression equation and is commonly used for office buildings, with a focus on the thermal and optical properties of windows [9]. It is used to calculate the annual cooling load of the perimeter area and represents maximum allowable loads through the building envelope. The internal heat gains and efficiency requirements of the HVAC system contribute to controlling the cooling energy areas [10]. In the ENVLOAD method, the impact of indoor thermal comfort is not consolidated and the analysis of the energy performance of the building cannot be investigated for further changes and corrections. In addition, it is mostly applied to cooling predominant regions [11–14]. The overall thermal transfer value is a measurement of the average heat gain moving into a building through the building envelope that may be used to compare thermal performance between different building designs [15]. The OTTV is a measurement of the average heat gain moving into a building through the building envelope that may be used to compare the thermal performance between different building designs [15]. The main advantage of this method is the simplicity of calculation in terms of flexibility on the relationship between compositions of the building envelope. On the other hand, the negative aspects are that it does not consider the interaction between envelope, internal gains, and chiller efficiency [9]. However, OTTV can safely be used for evaluating the heating energy performance of the building model based on solar energy entering during the construction stage.

Zero-energy concept, low carbon future, and reduction of resources that affect construction, demolition, operation, and disposal of land-based factors are an important concept in residential buildings [7]. From an environmental point of view, energy efficiency has been mentioned as one of the most influential issues in building design. Discrete optimization methods and technological innovations can minimize the energy consumption of any energy-related system. One of such methods is called the brute force algorithm. Local search by brute-force (BF), also known as the exhaustive search method, is a global algorithmic approach for dealing with computationally multi-objective optimization problems [16]. The brute force approach has the advantage of searching all possible outcomes in the system [17], but can be computationally time-consuming when working with large and complex datasets. Despite this disadvantage, the BF approach is relevant for spatial analysis in the era of the massive data source in the context of demographic analysis or public health analysis [17]. This is exhaustive, as it becomes widespread when possibly scaling computer force to solve a problem on demand. It has been widely used for optimizing building envelope systems, as indicated in the literature [18].

Basically, construction standards determining the requirements regarding the energy efficiency of residential buildings are not commonly observed to construct a new building to decrease construction estimate costs. In addition, there are no mandatory energy standards in Kazakhstan regarding the maintenance and operation of the housing energy-related works, including those concerning the level of energy efficiency [6]. Thus, it was decided to use OTTV for the energy performance of the building model for the construction stage only by using the weather data of different cities in Kazakhstan. In this research, an OTTV-based building energy simulation model (BESM) integrated by using a brute-force optimizer was developed to significantly cut the energy consumption of residential buildings located in different cities of Kazakhstan. For this purpose, the heating and cooling energy demands were reduced by deeply investigating the performance of the following building envelope components: window-to-wall ratio, thermal and dimensional parameters of the wall, angle of the pitched roof, and depth of the overhang shading system. The validity of BESM was demonstrated by performing an energy analysis of a real townhouse building located in seven different cities of Kazakhstan (Nur-Sultan, Almaty, Karaganda, Aktobe, Atyrau, Kokshetau, and Semey). Finally, the comparative cost analysis of the base and the optimized cases obtained from the BESM was presented.

2. Methodology

This research focuses on the development of the OTTV-based building energy simulation model (BESM) for the reduction of the energy consumption through the envelope and by obtaining the optimal envelope configuration for residential buildings of Kazakhstan. This section presents the base and derived OTTV-based heat transfer analysis equations for heating and cooling; the characteristics of the building envelope; proposed equations, simulation, and optimization models; and case study. For the implementation of this approach, the modified mathematical model of OTTV was formulated, and the building energy simulation model was developed. After simulation, the brute force algorithm was applied to obtain the optimum solutions for heating and cooling energies. The reliability of the numerical model was verified against the simulation model, which was calibrated by comparison of the single-room building performance for heating energy consumption. Finally, the case study parameters, including building model and climate condition in selected cities, are presented. Figure 1 shows the framework of the proposed BESM and optimization method.

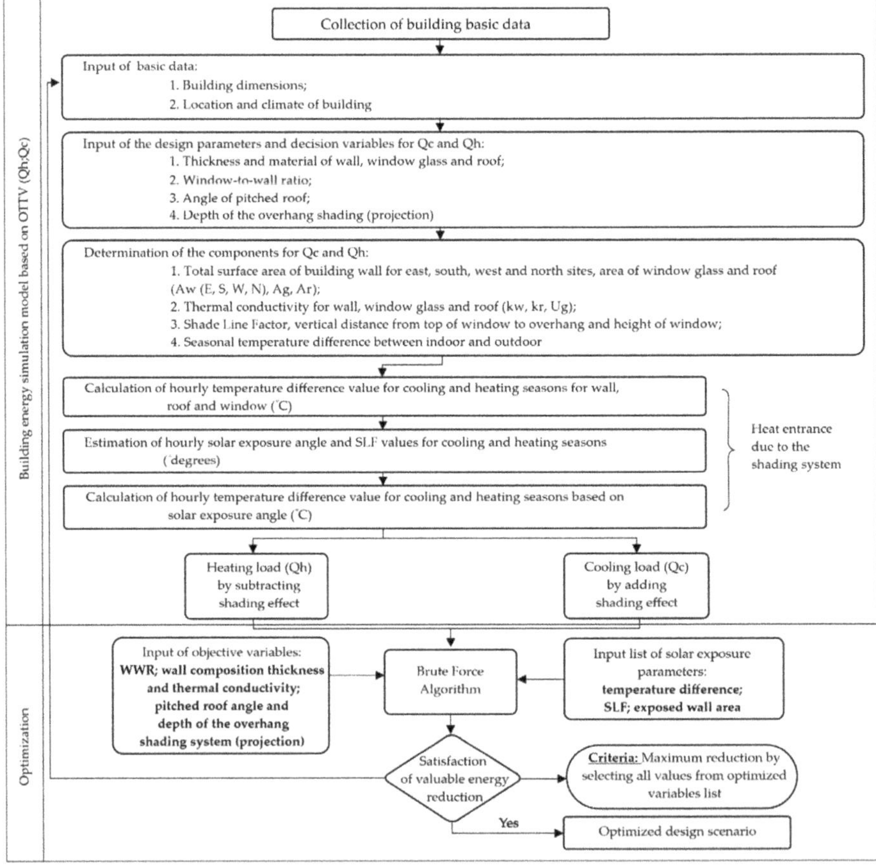

Figure 1. Flowchart of the BESM and optimization. Note: Q_c is a heat gain through the envelope for the cooling season; Q_h is a heat loss through the envelope for the heating season; A_w, A_g, and A_r are the surface area of the wall, window, and roof; k_w is the overall thermal conductivity of the wall; k_r is the thermal conductivity value of the roof; and U_g is the U-value of glazing.

2.1. The Overall Thermal Transfer Value (OTTV)

The OTTV standard is the maximum allowable heat gain value in cooling-dominant regions, calculated for a building or part of a building in W/m^2 [10]. Typically, two sets of OTTV are used, one for the external walls and the other for the roof [19]. For an exterior wall, the typical form of the OTTV equation isas follows:

$$OTTV_{wall} = \frac{Q_{wc} + Q_{gc} + Q_{gsol}}{A_{wall}} \tag{1}$$

where A_{wall} is the exterior wall's gross area, Q_{wc} is the conduction through the opaque wall, Q_{gc} is the conduction through the fenestration, and Q_{gsol} is the solar radiation through the fenestration. Heat transfer through an opaque wall is induced by a temperature difference between outside and indoor regions, as well as a solar radiation incident on the opaque wall Q_{wc} [19].

If the computed number from Equation (1) is equal to or less than an OTTV limit specified in an OTTV regulation, the OTTV requirement is deemed as being met. The basic assumption is that the higher the OTTV value, the greater the net heat gain through the building envelope and, as a result, the more energy required for cooling [20]. OTTV can provide building designers more flexibility for inventive design by considering the different components of building heat gain and allowing for trade-offs between them. In the energy-efficient building envelope design, OTTV is acknowledged as a simple and effective regulation [21]. The OTTV is used to measure average heat gain passing into the indoor area through the building envelope. Therefore, it can serve as an index of the impact of the envelope on the cooling energy utilized in air-conditioned buildings [20]. By changing the indoor setpoint temperature and reverse effect of the shading system compared with the cooling case, the heat loss equation was also developed for application during the heating season [21]. It consists of two measures, namely: envelope thermal transfer value (ETTV) and roof thermal transfer value (RTTV). The following equations demonstrate the calculation of ETTV and RTTV, respectively [22]:

$$ETTV = \frac{\sum Q}{\sum A} = \frac{Q_{wc} + Q_{gs}}{A_w + A_g} \tag{2}$$

$$Q_{wc} = A_w \times U_w \times \alpha_w \times TD_{eqw} \tag{3}$$

$$Q_{gs} = A_g \times SC \times ESM \times SF \tag{4}$$

$$RTTV = \frac{A_r \times U_r \times \alpha_r \times TD_{eqr}}{A_r} \tag{5}$$

where A_w, A_g, and A_r are the area of the wall, window, and roof, respectiy; Q is the a heat transfer through the envelope; U is the thermal transmittance of a specific material, α is the solar absorptivity constant related to the façade surface and color; TD is the equivalent temperature difference; SC is the shading coefficient of the glazing; ESM is the external shading multiplier; and SF is the solar factor (W/m^2).

As indicated in Equations (2)–(5), the areas (A_w, A_g, and A_r) and physical characteristics (U_w, U_r, α_w, α_r, and SC) of the building envelope, as well as the climate-dependent factors (TD_{eqw}, TD_{eqr}, and SF), are the key variables for determining the OTTV of a building [15]. OTTV enables the building designer to make trade-offs between various envelope characteristics such as U (value for the wall and roof (U_w and U_r, respectively)), shading coefficient (SC), etc.

2.2. Building Envelope Parameters

The building envelope is the component of the structure that physically divides the indoor and the outdoor environments [23]. It is reported that the majority of indoor cooling and heating loads are generated by heat gain and heat loss via building external envelopes such as walls, windows, and ceilings, which are caused by the difference in indoor and outdoor temperatures [24]. The temperature difference between indoor and outdoor conditions essentially affects the total energy consumption of the residential building in cold and moderate climate regions. The heating energy dominates rather than the cooling energy in the continental climate zone, such as Kazakhstan. Thus, for significant energy savings, while maintaining occupant thermal comfort, careful consideration of heat transfer through the envelope is critical [23]. The optimization of heat transfer based on envelope composition is investigated to reduce the building energy consumption. The details of the building model parameters including external sunshade, wall composition materials, and roof composition materials are provided and discussed in the next sections.

The heat gain and heat loss equations based on the OTTV method were developed by considering the boundary properties, such as indoor setpoint temperature, hourly outdoor dry-bulb temperature, hourly solar exposure, envelope material characteristics, and climate conditions [25].

Design Variables

In this research, the main variables used for OTTV determination are divided into two categories: endogenous and exogenous. Endogenous variables describe the particular effect on the model that is dependent on a variety of factors on the model [26]. These variables are the wall area, window glazing area, roof area, roof ceiling area, thermal resistance of ceiling, thermal resistance of the roof, vertical offset from shading to top of the window, height of the window, U-value of glazing, standard solar heat gain coefficient of glazing (SHGC), thermal conductivity, and window-to-wall ratio (WWR). Exogenous variables, in turn, are variables whose origin is outside of the model and whose purpose is to explain other variables or outcomes [27]. The temperature difference between the indoor set point and outdoor dry-bulb temperature, area of the wall exposed by solar radiation at specific times, shade line factor values based on the angle of solar radiation, heating and cooling loads, thickness of wall materials, angle of pitched roof, and depth of overhang shading (projection) are considered as exogenous variables in this research. The detailed chart regarding the connection between endogenous and exogenous variables and their consistency is shown in Figure 2.

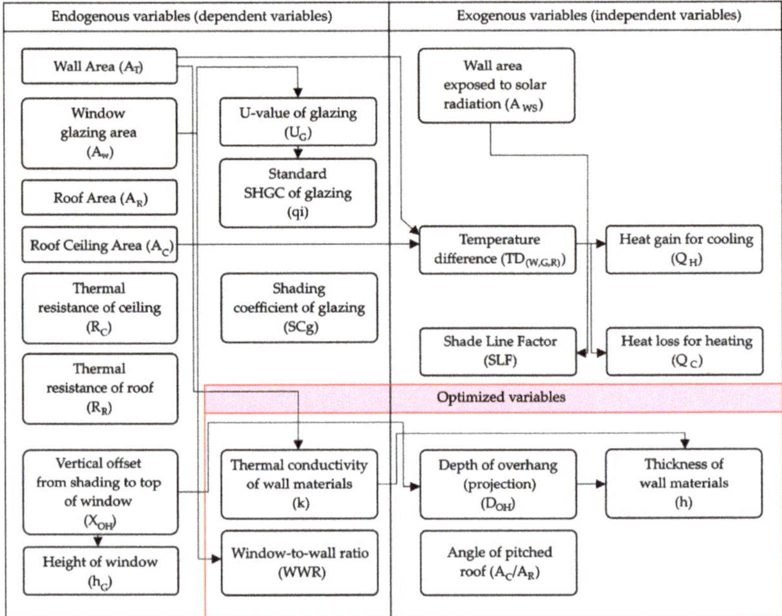

Figure 2. Components of the building energy simulation model and connection between the endogenous and exogenous variables.

2.3. Formulation of a Model

To provide the overall formulation of the analysis, the following assumptions were made in the model. (1) The wall materials were considered to be thermally isotropic and homogeneous. (2) The heat transfer mechanism was assumed to be two-dimensional, where the direction of the building's height was disregarded [28]. (3) The thermal conductivity value for all composition materials was constant. (4) Any sub-cooling or sub-heating activities were not included.

The heat gain and heat loss through the envelope were composed of four components: heat transfers through the walls, roof, and windows produced by differences in internal and external temperatures, and solar radiation heat transfer through the windows in terms of a shading system [19]. The following detailed mathematical models were developed from Equations (2)–(5) and present heat transfer through unit area of building envelope based on the temperature difference between indoor and outdoor for cooling and heating seasons:

$$Q_c = \sum_{i=1}^{n} k_{w_i} \times A_{w_i} \times \Delta TD_{w_c} + k_r \times A_r \times \Delta TD_{rc} + \sum_{i=1}^{m} U_g \times A_{g_i} \times \Delta TD_{gc}$$
$$+ \sum_{i=1}^{m} q_{ic} \times SC_{ic} \times A_{wsi} \times \Delta TD_{gc} \tag{6}$$

$$Q_h = \sum_{i=1}^{n} k_{w_i} \times A_{w_i} \times \Delta TD_{w_h} + k_r \times A_r \times \Delta TD_{rh} + \sum_{i=1}^{m} U_g \times A_{g_i} \times \Delta T_h$$
$$- \sum_{i=1}^{m} q_{ih} \times SC_{ih} \times A_{wsi} \times \Delta TD_{gc} \tag{7}$$

where Q_c is a heat gain through the envelope for the cooling season; Q_h is a heat loss through the envelope for the heating season; k_w is an overall thermal conductivity of the wall; TD_w, TD_r, and TD_g are equivalent temperature differences between indoor and outdoor surrounding space wall, roof, and glazing, reseptively; q is the standard solar heat gain factor; U_g is the U-value of glazing; A_w, A_g, and A_r are the surface area of the wall, window, and roof; A_{wsi} is wall area exposed to solar radiation; SC is the total shading coefficient; n and m are the numbers of external walls and windows; subscripts w, r, and g are external wall, roof, and external window, respectively; subscript i shows the various orientations; subscripts c and h represent the cooling and heating periods, respectively.

The overall thermal conductivity of the wall can be calculated as follows:

$$k_w = \frac{1}{\frac{h_1}{C_1} + \frac{h_2}{C_2} + \frac{h_3}{C_3} + \frac{h_4}{C_4}} \tag{8}$$

where h_1, h_2, h_3, and h_4 are the thicknesses of the wall composition layers, and C_1, C_2, C_3, and C_4 are the thermal conductivity values of wall composition layers.

$$A_T = A_{T_E} + A_{T_S} + A_{T_W} + A_{T_N} \tag{9}$$

The overall thermal conductivity of the roof can be calculated as follows:

$$k_r = \frac{1}{R_c + R_r \times \frac{A_c}{A_r}} \tag{10}$$

where k_r is the thermal conductivity values of the roof, R_c is the thermal resistance value for the roof ceiling, R_r is the thermal resistance value for the roof, A_c is the ceiling area, A_r is the roof area, and $A_T(E, S, W, N)$ is the total area building's wall.

$$A_w = (1 - WWR) \times A_T \tag{11}$$

$$A_g = WWR \times A_T \tag{12}$$

$$SC = SC_g \times \min\left(1, \max\left(0, \frac{SLF \times D_{oh} - X_{oh}}{h_g}\right)\right) \tag{13}$$

where SLF is the shade line factor, D_{oh} is the depth of overhang (projection), X_{oh} is the vertical offset from the top of the window to the overhang, h_g is the height of the window, WWR is the window-to-wall ratio, and SC_g is the shading coefficient of the glazing.

The constant dimensional values of the building envelope, such as the wall area for each building sector, roof area, wall material thickness, and thermal transmittance through the window were set to the model. Consequently, the building's total area (A_T) for the east, south, west, and north walls were 81.2, 62.4, 93.6, and 74.4 m², respectively, while the roof area was calculated as 154.2 m². The initial *WWR* was kept as 5% and the U-value of the window glass was 1.978 W/m²K. The area of windows was calculated by the multiplication of the total area and *WWR*. The solar heat gain coefficient (SHGC) is a ratio of solar radiation absorbed or transmitted through the window or door, to the heat that is released back (SHGC = 0.68). The total shading coefficient was calculated by multiplying the shading coefficient of the glazing and overhang shading system. The shading coefficient of the glazing (SC_g) was calculated by dividing the solar heat gain coefficient (SHGC) of the window by 0.86 to get 0.7988, while the standard solar heat gain factor (q) was a heat gain through 3 mm normal glass and was calculated as the ratio of the U-value and shading coefficient of the window to obtain 2.5 W/m²K. The sum of the hourly temperature difference for heating and cooling was calculated using the outside dry-bulb temperature and inside set-point temperatures that were obtained from DesignBuilder software by using the weather data purchased from the White Box Technologies, Inc, Moraga, CA, USA [29]. The weather data represent the typical year, the composition of which reflects the long-term average conditions for a location over a period of 12 up to 22 years. The heating season lasted from the middle of October to the middle of May, while the cooling season lasted from the middle of May to the middle of October. The following variables of the building envelope, such as *WWR*, wall composition layer thicknesses, thermal conductivity of wall composition materials, depth of the overhang normal to the building plane, and ratio of ceiling area to the roof area (angle of the pitched roof) were optimized by the proposed energy estimation model.

2.3.1. Heat Transfer through the Pitched Roof

The roof structure of the modeled building was assumed to be pitched with the unvented attic. Unvented attics are effective for the reduction of energy loss through ceiling facilities or leaky channels. The average value for energy savings by unvented attics was counted as 20% [30]. For the roof with the unvented attic, the heat transfer procedure occurred through three different mediums, namely the ceiling, roof, and attic itself. Thus, the overall thermal resistance value of the roofing system was directly related to the individual thermal resistance values of the roof and ceiling combined with the thermal resistance of the attic space. Attic space may be assumed as an air layer in the composition. In most cases, the practical influence of attic space accounted for the surface thermal resistance of the roof and ceiling adjacent to the space. The thermal resistance values for the roof and ceiling were determined separately using convection resistance case for still-air in the attic surface [31]. Thus, the overall R-value for the combination of ceiling−roof is expressed as follows:

$$R = R_{ceiling} + R_{roof} \times \frac{A_{ceiling}}{A_{roof}} \tag{14}$$

$$k_r = \frac{1}{R_{ceiling} + R_{roof} \times \dfrac{A_{ceiling}}{A_{roof}}} \tag{15}$$

where $A_{ceiling}$ and A_{roof} are the ceiling and roof areas, respectively, while $R_{ceiling}$ and R_{roof} are thermal resistance values for ceiling and roof and equal to 0.1328 m² K/W and 0.8733 m² K/W, respectively. If the area ratio is equal to one, the roof is flat, and it is less than one for pitched roofs. For the modeled building, the initial value of angle for the roof was 27°, and the thermal conductivity (k_r) of the entire roofing system was equal to 1.1247 W/mK. Additionally, the direction of heat flow was upward (heat loss) in the cold season and downward (heat gain) in the hot season. An adequate R-value for unvented roofing required the usage of appropriate materials with proper workmanship that met the standards. Otherwise, the usage of inappropriate materials and poor workmanship could lead to the R-value being different from the predicted one. The thermal resistance structure for a pitched roof−attic−ceiling case with an unvented attic is presented in literature [31].

2.3.2. Details of Overhang Shading Device Configuration

The location of the shading material plays an important role in the solar gain control process. The outdoor shading system can contribute to a substantial reduction in heat gains, but requires adequate regular maintenance and is difficult to install. On the contrary, an indoor shading system cannot be as effective as controlling solar heat gain quantity, but it is easier to install. Shading systems depend on glazing type, material, and shading properties [32,33]. The U-value of the fenestration system cannot be changed by the application of the shading system [34]. However, most of the shading systems can afford to insulate the heat and are used as an additional improvement for the thermal conductivity of the window, especially if they are tightly installed at a specific position with no air infiltration. Shading devices are intended to control daylighting and thermal control challenges of the building [35]. For fenestration systems of the modeled building, the simple overhang-shading device (Figure 3) was installed. The vertical offset from the top of the window to the shading device (X_{oh}) was kept as 0.2 m, while the depth of the overhang normal to the building plane (projection, D_{oh}) was 0.5 m and the fenestration height was 1.5 m. The shade line factor (*SLF*) based on the angle of solar exposure was calculated on an hourly basis for the entire season, by creating Table 1, representing the month, time, and subsequent solar exposure angle.

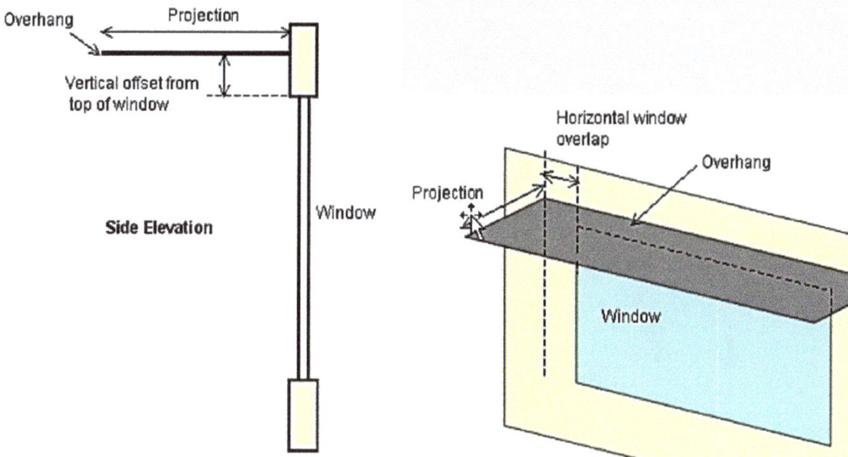

Figure 3. Details of the overhang shading device [36]. Reproduced with permission from DesignBuilder Software Ltd., (Gloucestershire, UK).

Table 1. The data collection example for the *SLF* estimation.

1st of January (Time)	Exposure Site	Angle of Sunlight (°)
9:00	southeast	4.73
10:00	southeast	10.82
11:00	southeast	15.15
12:00	southeast	17.45
13:00	southwest	17.39
14:00	southwest	14.99
15:00	southwest	10.4
16:00	southwest	3.08

The *SLF* is the ratio of the vertical distance of the shadow fall underneath the edge of the overhang to the depth of the overhang, although the shade line can be equal to the *SLF* times the depth of the overhang. Based on the angle of sunlight values evaluated in Table 1 and the shape line factors values of ASHRAE [37], the interpolation rule was used to calculate the *SLF* values for each time, month, and angle of solar exposure and to demonstrate the constant value for *SLF* at the hour with the highest solar intensity on exposures [25].

$$SCs = \min\left(\left(1, \max\left(0, \frac{SLF \times D_{oh} - X_{oh}}{h_g} \right)\right)\right) \tag{16}$$

where,

SC—total shading coefficient,
SLF—shade line factor from [37],
D_{oh}—depth of overhang (projection), m
X_{oh}—vertical offset from the top of the window to overhang, m
h_g—height of the window, m.

2.4. Building Design Optimization

2.4.1. Optimization Variables

In order to reach an optimal and practical design strategy, an optimization technique is required [18]. The first step in optimization is the identification of the input variables and their ranges. In this research, the optimization was performed by changing the ranges of the exogenous and endogenous design variables. The variables involved in envelope optimization include a *WWR* (4, 5, 10, 15, 20, 25, and 30%), the angle of a pitched roof (22, 27, 30, 34, 37, 40, 42, and 45°), the depth of the overhang shading system (0.5, 0.6, and 0.7 m), and the thermal conductivity and thicknesses of wall composition materials. The variable value ranges for *WWR*, the angle of a pitched roof and the depth of the overhang shading system were taken from the literature [38–40].

In this research, the wall composition consisted of four layers: exterior finish, material block (core layer), insulation layer, and interior finish (Figure 4), which were optimized. The materials utilized to construct the building envelope as well as their specifications were obtained from conventional architectural design schematics that fulfilled the requirements of Kazakhstan's building codes and standards [41–45]. Table 2 describes the detailed properties of the layers for the studied wall (thickness, thermal conductivity, specific heat, and density). The exterior finish consisted of different materials including ceramic brick, cement sand render, limestone mortar, burnt ceramic clay tile, dry ceramic clay tile, and ceramic glazed tile. The core layer consisted of masonry block, burnt brick veneer, aerated concrete block, brick veneer, reinforced concrete, and clay block. For the insulation layer: penoplex, glasswool, hydro isolation, mineral wool plate, extruded polystyrene, and cellulose were used, while for the interior finish, gypsum board, cement mortar, gypsum insulating plaster, plasterboard 1, plasterboard 2, and plasterboard 3 were used.

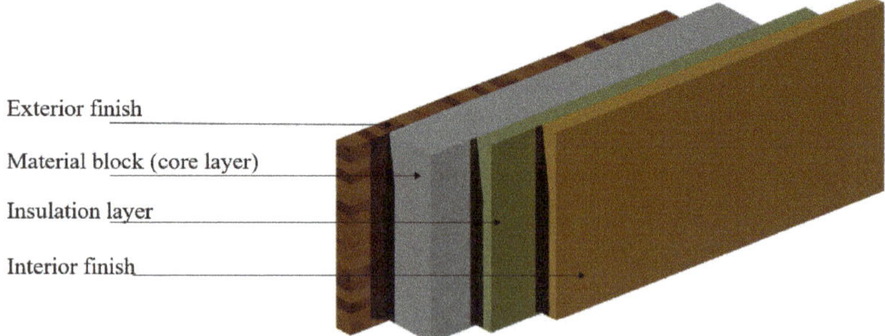

Figure 4. Cross-section view of the wall composition materials.

Table 2. Wall composition of the thermophysical properties analyzed.

Layers	Materials	Thickness, (h_1, h_2, h_3, h_4), (m)	Thermal Conductivity, (C_1, C_2, C_3, C_4), (W/mK)	Density (kg/m^3)	Specific Heat (J/kgK)
Exterior finish	Ceramic brick	0.5	0.59	1831	825
	Cement sand render	0.02	1	1800	1000
	Limestone mortar	0.02	0.7	1600	840
	Burnt ceramic clay tile	0.012	1.3	2000	840
	Dry ceramic clay tile	0.012	1.2	2000	850
	Clay block	0.0075	1.4	2500	840
Material Block (Core layer)	Masonry block	0.15	0.24	800	840
	Burnt brick veneer	0.15	0.74	1700	800
	Aerated concrete block	0.2	0.24	750	1000
	Brick veneer	0.15	0.547	1950	1000
	Reinforced concrete	0.15	0.5	1400	830
	Clay block	0.19	1.0	1800	920
Insulation layer	Penoplex	0.0795	0.030	30	1340
	Glasswool	0.012	0.039	20	840
	Hydro isolation	0.015	0.29	29	1210
	Mineral wool plate	0.1	0.036	70	810
	Extruded polystyrene	0.0795	0.03	43	1210
	Cellulose	0.2	0.04	48	1381
Interior finish	Gypsum Board	0.013	0.16	800	1090
	Cement mortar	0.012	0.72	1760	840
	Gypsum insulating plaster	0.013	0.18	600	1000
	Plaster board 1	0.012	0.72	840	1860
	Plaster board 2	0.012	0.25	600	1089
	Plaster board 3	0.012	0.35	817	1620

2.4.2. Optimization Algorithm

The second step is iteratively conducting the optimization. Before determining the optimal design solution, a brute force optimization method was utilized to analyze all design parameters. Brute force is an exhaustive search method and computational data analysis that is commonly counted as a general problem-solving method. The brute force method has an advantage in investigating all possible solutions from the list of the values but can take a long computational time with the complicated dataset [46]. Despite this, the brute force method is relevant for spatial coding analysis. As indicated in the literature [18], the method has been widely used for optimizing building envelope systems.

In order to facilitate the optimization, Python code was written to modify the design parameter. The analyzed variables in this numerical model were *WWR*, wall composition thicknesses, thermal conductivity of wall composition materials, and angle of the pitched roof. By investigating all of the existing possible values in the list of mentioned variables, the minimum value of heat gain (Q_c) and heat loss (Q_h) were proposed and analyzed for energy reduction compared with the value related to the initial building parameters. The detailed version of the analyzed mathematical model for the first part was computed and expressed below:

$$Q_1 = \left(A_{T_E} + A_{T_S} + A_{T_W} + A_{T_N}\right) \times (1 - WWR) \times \frac{1}{\dfrac{h_1}{C_1} + \dfrac{h_2}{C_2} + \dfrac{h_3}{C_3} + \dfrac{h_4}{C_4}} \times \sum_{i=1}^{n} \Delta TD_w$$

$$+ \frac{1}{R_c + R_r \times \dfrac{A_c}{A_r}} \times A_c \times \sum_{i=1}^{n} \Delta TD_r \tag{17}$$

$$+ U_g \times WWR \times \left(A_{T_E} + A_{T_S} + A_{T_W} + A_{T_N}\right) \times \sum_{i=1}^{n} \Delta TD_g$$

where TD_w, TD_r, and TD_g are the equivalent temperature differences between indoor and outdoor surrounding space wall, roof, and glazing, respectively; q is the standard solar heat gain factor; U_g is the U-value of glazing; h_1, h_2, h_3, and h_4 are the thicknesses of wall composition layers; C_1, C_2, C_3, and C_4 are the thermal conductivity values of the wall composition layers; R_c is thermal resistance value for roof ceiling; R_r is thermal resistance value for the roof; A_c is the ceiling area; A_r is roof area; $A_T(E, S, W, N)$ is the total area building's wall; and WWR is the window-to-wall ratio.

The analysis of the shading properties of the glazing is quite complicated, as the values for hourly temperature difference, the hourly orientation of solar exposure and hourly SLF values are considered in the second script. Primarily, to find the unique optimum value for WWR in both scripts, the second part of the equation is calculated without the WWR value. This partial equation is estimated by exporting the hourly values from the spreadsheet (Table 3) for the wall area of the entire building exposed by the solar radiation (A_{ws}), temperature difference, and SLF. Table 3 represents the hourly temperature difference, area of the wall exposed to sunlight, and corresponding shade line factor (SLF) value for the numerical model analysis.

Table 3. The data collection example of the hourly shading system performance.

1-st of January	Temperature Difference (°C)	Shade Line Factor	Wall Area Exposed to Solar Radiation (m^2)
9:00	37.58	2.43	144
10:00	36.98	2.29	144
11:00	36.2	2.18	144
12:00	35.4	2.13	144
13:00	34.6	2.13	156
14:00	33.88	2.19	156
15:00	33.4	2.30	156
16:00	33.53	2.47	156

The summation ofthe estimated values for each hour in each analyzed season are varied by the list of values of overhang shading depth, which means that for each value of overhang depth there are computed values of Q_2. The wall area exposed by solar radiation varies hour by hour, as the sunlight rises from the east and rests west. Each selected WWR value was multiplied by the resulting values of Q_2 to find the maximum and minimum values of Q_2 in the heating and cooling seasons, respectively. The maximum and minimum values of Q_2 were defined by the "$-$" and "$+$" signs before this expression, as an expression of Q_2 describes the reduction of heat entering into the indoor area by the shading system. The heat gain equation for cooling (Q_1) should add the minimum value of heat gain reduction value (Q_2), while the heat loss equation for heating (Q_1) should subtract the maximum value of heat gain reduction (Q_2) to get the overall value for heat transfer as minimum as possible to be optimized. The detailed version of the analyzed mathematical model for the heat gain reduction by the shading system was computed and is expressed below.

$$Q_2 = q_{ih} \times WWR \times (A_{ws}) \times SC_g \times \sum_{i=1}^{n} \Delta TD_{wi} \times \min\left(1, \max\left(0, \frac{SLF_i \times D_{oh} - X_{oh}}{h_g}\right)\right) \tag{18}$$

where q is the standard solar heat gain factor; WWR is the window-to-wall ratio; A_w is a surface area of the wall; SC_g is the shading coefficient of the glazing; TD_w is the equivalent temperature differences between indoor and outdoor surrounding space wall; SLF is shade line factor from; D_{oh} is the depth of overhang (projection); X_{oh} is vertical offset from the top of the window to overhang; and h_g is the height of the window.

Finally, the combination of the two parts appears in the following way: Q_1 for heating minus the maximum value of Q_2 for heating, and Q_1 for cooling plus the minimum value of Q_2 for cooling to obtain the minimum energy consumption in heating and cooling seasons. Based on the iterative summation of the second part due to the large data obtained from the spreadsheet, the running process of the coding takes an insignificant amount of time. The optimized results, by highlighting all optimum values of iterative variables and the final value of the entire energy equation heating and cooling, are shown in the Results and Discussion chapter.

2.5. Verification of Results between Building Energy Simulation Model and DesignBuilder Software

In this research, the reliability of the numerical model developed to calculate the energy transfer performance in Python software was validated. The heating load performance results between the simulation and numerical models were compared. For this purpose, a single-room building (Figure 5) with dimensions of 5.0 m × 5.0 m × 3.0 m, with an overhang shading system was modeled in Design Builder software. The material properties of the wall and roof are provided in Table 4. The *WWR* ratio was kept as 5% and the height of windows was 1.5 m. The overhang shading system was installed with a depth (projection) of 0.5 m. The weather data of Nur-Sultan was used for this verification. The numerical model-based analysis was also conducted for temperature conditions and sunshade performances developed for Nur-Sultan city.

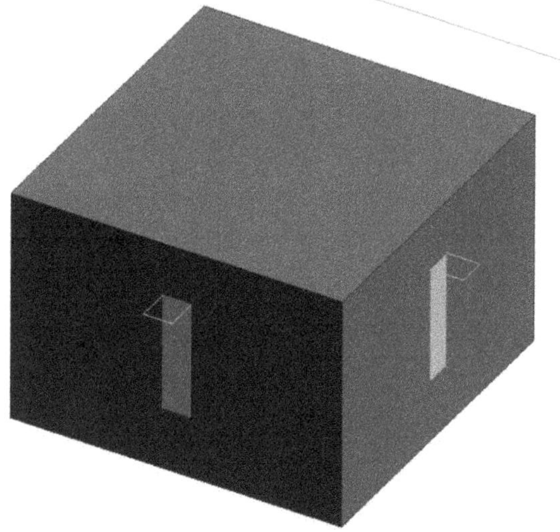

Figure 5. The model of the single-room building.

Table 4. Thermophysical properties of the building envelope materials.

Composition Materials	Layers	Material	Thickness, (h_1, h_2, h_3, h_4), (m)	Thermal Conductivity, (C_1, C_2, C_3, C_4), (W/mK)	Density (kg/m^3)	Specific Heat (J/kgK)
Wall	Exterior finish	Burnt ceramic clay tile	0.012	1.3	2000	840
	Core layer	Brick veneer	0.0165	0.542	1950	840
	Insulation	Glasswool	0.081	0.039	20	840
	Interior finish	Plaster board	0.0125	0.35	817	1620
Roof	Exterior finish	Roof tile	0.01	0.84	1900	800
	Core layer	Concrete slab	0.15	1.13	2000	1000
	Insulation	Polystyrene	0.2423	0.29	29	1210
	Interior finish	Roofing felt	0.005	0.19	960	837

The results of the numerical model and simulation are shown in Figure 6. According to the results, the maximum percentage difference between the numerical model and simulation-based results did not exceed 7.7%. Thus, it can be concluded that numerical model-based results with materials' thermal characteristics can be used and optimized for energy load performance.

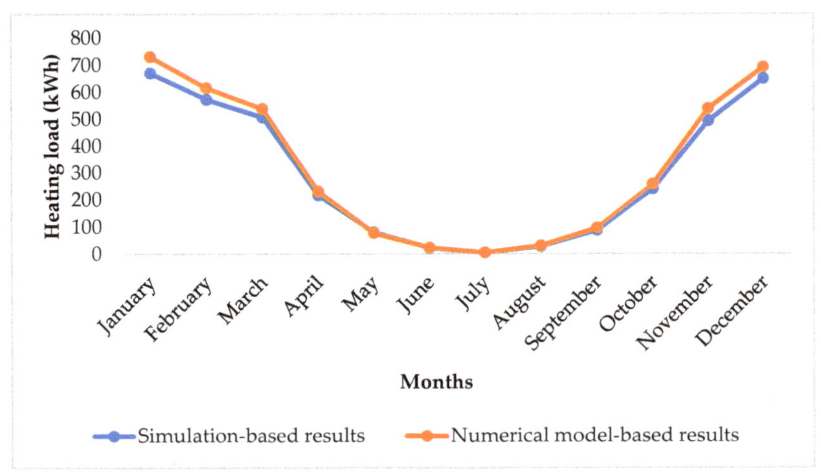

Figure 6. Simulation-based results vs. numerical model-based results.

2.6. Case Study

2.6.1. Climate Conditions

The Koppen climate classification is a widely used climate classification system that has been used since 1900. According to the background that was proposed over a century ago, the Koppen−Geiger world map was created based on monthly precipitation and temperature performance over a long period. The climates are defined by different letter symbols: A (tropical), B (arid), C (temperature), D (cold), and E (polar). Kazakhstan area has five different climate zones. There are hot-summer humid continental climate (Dfa), warm summer humid continental climate (Dfb), cold semi-arid climate (Bsk), cold desert climate (Bwk), and the Mediterranean influenced hot summer continental climate (Dsa) [45]. Based on these climate zones, seven cities were selected and shown in Figure 7. It is important to note that Kazakhstan has four seasons, namely winter (from December to February), spring (from March to May), summer (from June to August), and autumn (from September to November). January is counted as the coldest month based on the lowest average monthly temperature, while July is the hottest month. The details of the climate parameters, along with the average temperature in the hottest and coldest months are reflected in Table 5. The average temperature data were selected from the Climate Data website [47]. According to the cold season, Nur-Sultan, Karaganda, and Semey from the Dfb climate zone demonstrate average temperatures, with −18.3 °C, −14.2 °C, and −12.2 °C in the coldest month, respectively. Almaty and Aktobe are from the Dfa climate zone and have average temperatures of −8.4 °C and −16.5 °C, respectively. The average temperature in January for Atyrau from the Bwk climate zone and Kokshetau from the Bsk climate zone are −9.9 °C and −19.7 °C, respectively. During the hot season, the average temperatures of the hottest month in Nur-Sultan, Karaganda, and Semey from the Dfb climate zone are 20 °C, 18 °C, and 18 °C, respectively. Almaty and Aktobe cities from the Dfa climate zone have average temperatures in July measured as 24 °C and 27 °C, respectively. Atyrau city from the Bwk climate zone and Kokshetau city from the Bsk climate zone demonstrate the average outside temperatures in July of 27 °C and 20 °C, respectively.

Figure 7. Location of selected cities.

Table 5. Climate parameters of selected cities.

City	Latitude	Longitude	Climate Zone (According to Koppen Classification)	The Average Temperature in January (°C)	The Average Temperature in July (°C)
Nur-Sultan	51.18	71.45	Dfb	−18.3	20
Almaty	43.25	76.92	Dfa	−8.4	24
Karaganda	49.83	73.16	Dfb	−14.2	18
Aktobe	43.25	67.76	Dfa	−16.5	25
Atyrau	47.11	51.88	Bwk	−9.9	27
Semey	50.41	80.20	Dfb	−12.2	18
Kokshetau	53.28	69.39	Bsk	−19.7	20

2.6.2. Building Model

For this research, a townhouse was analyzed (Figure 8) in seven different cities (Nur-Sultan, Almaty, Karaganda, Aktobe, Atyrau, Semey, and Kokshetau) of Kazakhstan. This building was analyzed by facing the west direction with the front side. The dimensions of the building are 13.6 m × 12.4 m, with an elevation of 3 m for each story and 308.48 m^2 of living area. The overhang was installed as the main shading system with a depth (projection) of 0.5 m. The initial angle of the pitched roof was kept as 27°. The WWR was kept at 5% and the height of the windows was 1.5 m. Table 4 represents the thermophysical properties of the building envelope materials. This building was used as a base model for the energy analysis in all of the selected cities.

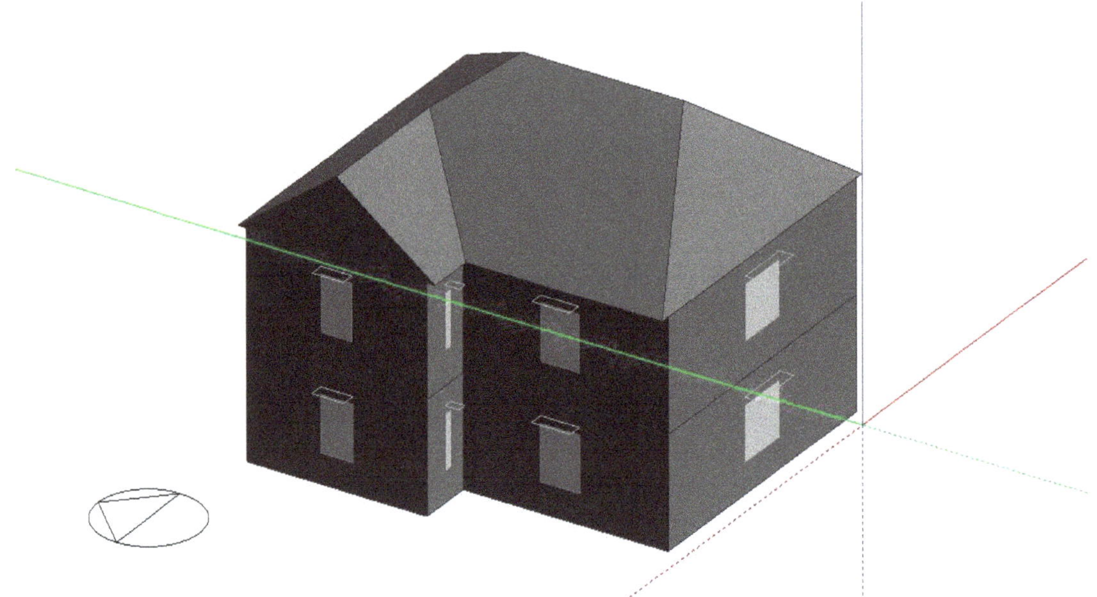

Figure 8. Model of a townhouse.

3. Results and Discussion

3.1. The Optimum Heating and Cooling Loads Reduction for Entire the Year

In this section, the numerical model for energy estimation reduced the heating and cooling loads by optimizing the building envelope parameters in seven different cities of Kazakhstan: Nur-Sultan, Almaty, Karaganda, Aktobe, Atyrau, Semey, Kokshetau. For analysis, various types of wall composition materials, angle of a pitched roof, depth of the overhang shading system and WWR were analyzed, and the HVAC system was turned off. Switching off the HVAC system allowed a detailed analysis of the building properties only in terms of heat transfer due to the climate conditions. The main purpose of this section is to demonstrate the effect of the thermal performance of building envelope components on the reduction of heating and cooling energy consumption by optimizing it based on the local climatic dataset. The results of seasonal heating (13 October–12 May) and cooling (13 May–12 October) energy for all cities are estimated and summarized in Table 6. From the obtained results, it is clearly seen that the amount of heating energy dominates almost in all cities and an average reduction of heating load for all cities was counted as 6157.8 kWh (or 5.29 Gcal) and an average reduction of building energy for cooling purposes was estimated as 1912.17 kWh (or 1.64 Gcal) due to the optimization of the numerical model. Since the optimized model is based on optimized variables, a cross-section of wall composition materials for the base and optimized cases are shown in Figure 9. The average percentage values for heating and cooling energy were reduced by 16.59% and 16.53% for heating and cooling seasons, respectively.

The maximum annual reductions of heating energy were witnessed in Koksetau city (6967.89 kWh), followed by Nur-Sultan city (6943.66 kWh). The lowest results in heating reduction were witnessed in Almaty (4758.35 kWh). On contrary, this city showed the maximum reductions in cooling energy (2107.94 kWh). The minimum reductions in the cooling energy were in Aktobe (1591.77 kWh). The overall maximum reductions achieved for optimal design to base design were fairly consistent among the selected cities and ranged between 16.59% and 16.53% for heating and cooling seasons, respectively.

Table 6. The heating and cooling energy results for the base case and optimized building model.

| List of Cities | Heating Load (kWh) | | Heating Load | | Cooling Load | | Cooling Load | |
	Base Design	Optimized Design	Reduction in (kWh)	Reduction (Gcal)	Base Design	Optimized Design	Reduction in (kWh)	Reduction (Gcal)
Nur-Sultan	41,848.71	34,905.05	6943.66	5.9701	11,538.66	9614.50	1924.16	1.6543
Almaty	28,673.86	23,915.51	4758.35	4.0912	13,453.56	11,345.62	2107.94	1.8124
Karaganda	40,564.44	33,833.53	6730.91	5.7872	12,280.95	10,233.19	2047.76	1.7606
Aktobe	38,536.64	32,142.22	6394.42	5.4979	9546.91	7955.13	1591.77	1.3686
Atyrau	30,500.17	25,439.52	5060.65	4.3511	12,213.90	10,179.55	2034.35	1.7491
Semey	37,651.78	31,402.88	6248.90	5.3728	10,304.82	8586.62	1718.19	1.4773
Kokshetau	41,996.78	35,028.89	6967.89	5.9909	11,759.43	9798.42	1961.01	1.6860
Average:	37,110.34	30,952.51	6157.82	5.2983	11,585.46	9673.29	1912.17	1.6453

Figure 9. Wall composition materials for (**a**) base case and (**b**) optimized building model.

The monthly results (Table 7) are summarized to identify the energy consumption (in kWh) of the building for each city. It can be observed from Table 7, that in January, Nur-Sultan and Kokshetau cities showed the largest amount of energy consumption in kWh. Although, in July, Atyrau and Almaty showed the highest performance in cooling energy demand. In May, as a month when the heating season ends and the cooling season starts, the average energy consumption of around 1000 kWh was obtained for the base and 900 kWh for the optimized building envelope case in each selected city. In the same way, the energy consumption for heating/cooling purposes in September shows the lowest energy performance in each city, approximately 500 kWh, representing the ending of the cooling season and starting of the heating season. In general, Table 7 demonstrated, that heating energy dominates in all analyzed cities of Kazakhstan, as energy consumption in January, February and December are the highest in each city. Thus, total energy consumption in all months for heating and cooling proves that the climatic zone of Kazakhstan is mostly heating-predominant.

The energy consumption reduction values in Kazakhstan cities can be considered as significant, especially for the construction stage of a building calculated per unit living area. The average energy consumption value for the base and optimized envelope models were 120 kWh/m^2 and 100 kWh/m^2, respectively, during the hot season. For the cool season, selected Kazakhstan's cities demonstrated an average value of 38 kWh/m^2 and 30 kWh/m^2 for the base and optimized building models, respectively. Several literature values are pertinent for this research concerning the existing Kazakhstani buildings and thermal demand. The average heating energy consumption for residential buildings in Switzerland was reported to be 101 kWh/m^2 in 2018 [48]. According to [49], the annual energy use performances across three Canadian cities, such as Quebec, Toronto, and Vancouver, are 126.08 kWh/m^2, 116.47 kWh/m^2, and 98.47 kWh/m^2 for the period from 1998 to 2014, respectively, and 126.03 kWh/m^2, 115.31 kWh/m^2, and 100.25 kWh/m^2, respectively, according to forecasting future climate conditions.

An orientation-related change can affect the optimized amount of the energy load of the building [50]. The building orientation affects the heat gains of the building, thus the diversity of solar radiation at different angles [50]. It is obviously observed, that on the current analyzed stage, the most valuable portion of the heat gain or loss mechanism of the building is carried out by solar exposure. The details of the HVAC system, namely as an effect of lightning, occupancy, hot water usage, and electronic devices, were not definitely influenced in this analysis, and the energy consumption results were based on climate impacted power usage to cool or heat the indoor area. The optimized energy consumption data for each orientation (west, north, east, and south) were obtained and the results are presented in Table 8. According to the table, when the models were directed to the south, the optimized buildings showed the highest reduction in energy consumption. On the other hand, in the cooling load-dominated season, the building directed towards the north side showed better performance for the optimized value of energy consumption. To be exact, directing the front side of the building towards the south in the cold season and north in the hot season demonstrated approximately 21% and 32% reductions, respectively, based on solar entrance into the building.

From the sun path diagram shown in Figure 10, it is seen that the north wall was less exposed to solar light within the daytime. Thus, for a better energy-saving performance in the cold season, the wall with the highest number of windows should be directed in the south direction. Alshboul and Alkurdi [51] discovered that orienting the largest glass area to the south allowed the building to gain the required heat in winter. In contrast, for the hot season, it was better to locate most of the living areas directed to the north side in a shaded direction. Finally, the suggested primary optimal design solution included suggested building orientation to the south, a combination of 0.7 m of the depth of the overhang shading system and a WWR of 4%, 12.30 mm ceramic brick as the exterior finish, 171.10 mm masonry brick as the core layer, 84.7 mm glasswool, and 13.10 mm plasterboard as the interior finish (Figure 9b).

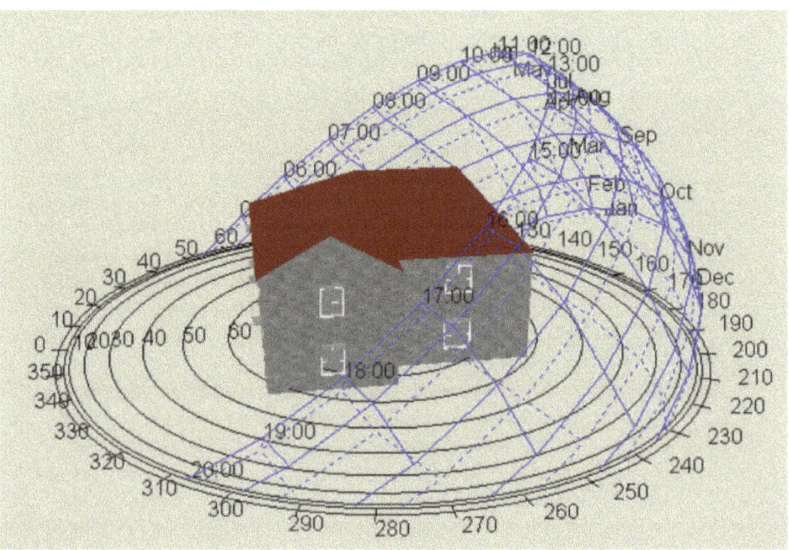

Figure 10. Sun path diagram.

Table 7. Monthly results of energy consumption for each city (kWh).

Months	Nur-Sultan		Almaty		Aktobe		Atyrau		Karaganda		Semey		Kokshetau	
	Base	Optimized	Base	Optimized	Base	Optimized	Base	Optimized	Base	Optimized	Base	Optimized	Base	Optimized
January	10,009	8349	6858	5720	9217	7688	7295	6085	9702	8092	9006	7511	10,045	8378
February	7436	6202	5095	4249	6847	5711	5419	4520	7207	6011	6690	5580	7462	6224
March	4671	3896	3200	2669	4301	3588	3404	2839	4528	3776	4203	3505	4688	3910
April	3050	2544	2090	1743	2809	2343	2223	1854	2957	2466	2745	2289	3061	2553
May	1239	1034	849	708	1141	952	903	753	1201	1002	1115	930	1244	1037
June	1846	1538	1119	932	1528	1273	1954	1629	1965	1637	1649	1374	1882	1568
July	4846	4038	2936	2447	4010	3341	5130	4275	5158	4298	4328	3606	4939	4115
August	4269	3557	2587	2156	3532	2943	4519	3766	4544	3786	3813	3177	4351	3625
September	577	481	350	291	477	398	611	509	614	512	515	429	588	490
October	2574	2147	1764	1471	2370	1977	1876	1565	2495	2081	2316	1931	2583	2154
November	4957	4135	3396	2833	4565	3807	3613	3013	4805	4008	4460	3720	4975	4149
December	7912	6599	5421	4522	7286	6077	5767	4810	7669	6397	7119	5937	7940	6623

Table 8. Optimized heating energy results of building oriented based on the front site direction.

List of Cities	Heating (kWh)					Cooling (kWh)				
	Base Design	Optimized Design				Base Design	Optimized Design			
		West	North	East	South		West	North	East	South
Nur-Sultan	41,848.71	34,905.05	33,405.05	36,405.05	33,205.05	11,538.66	9614.50	7914.50	11,114.50	8114.50
Almaty	28,673.86	23,915.51	22,415.51	25,415.51	22,215.51	13,453.56	5825.71	4125.71	7325.71	4325.71
Aktobe	38,536.64	32,142.22	30,642.22	33,642.22	30,442.22	9546.91	7955.13	6255.13	9455.13	6455.13
Karaganda	40,564.44	33,833.53	32,333.53	35,333.53	32,133.53	12,280.95	10,233.19	8533.19	11,733.19	8733.19
Atyrau	30,500.17	25,439.52	23,939.52	26,939.52	23,739.52	12,213.90	10,179.55	8479.55	11,679.55	8679.55
Kokshetau	41,996.78	35,028.89	33,528.89	36,528.89	33,328.89	11,759.43	9798.42	8098.42	11,298.42	8298.42
Semey	37,651.78	31,402.88	29,902.88	32,902.88	29,702.88	10,304.82	8586.62	6886.62	10,086.62	7086.62

The contribution of each building envelope component for the heat transfer was calculated in percentage rates, to simply point out the further investigation of the optimization goals for energy reduction. As Nur-Sultan and Kokshetau showed the highest results for the heating energy consumption, the energy consumption reduction by building envelope components are shown in Table 9, where 16.592% was the total heating energy reduction in Nur-Sultan city, while 6.37%, 1.61%, 8.64%, and 0.3% contribution rates of this reduction corresponded to the wall, window glazing, roofing, and shading systems, respectively. For Kokshetau, the contribution rate was almost identical to Nur-Sultan. For a cooling period, Almaty and Atyrau were selected for this analysis with the highest cooling energy reduction, showing 16.668% and 16.656%, respectively, corresponding to the base case performance. In Almaty, the contribution for cooling energy reduction was 5.88% from an opaque wall, 1.48% from window glazing, 8.14% from roofing, and around 1.2% from the shading system, while for Atyrau, the contribution for reduction was around 6.42% from the opaque wall, 1.62% from the window glazing, 8.36% from the roofing, and around 0.75% from the shading system (Table 9). This result indicates that the wall and the roof had a considerable impact on energy consumption in all of the analyzed cities. The energy reduction value for window glazing was similar in all cities. Shading systems could reduce energy consumption in cities located in the south and west parts of the country, while its impact was negligible in the cities located in the north part.

Overall, the results obtained in this research verified the effectiveness of the proposed BESM and optimization in enhancing the energy efficiency of the residential buildings. In addition, the results showed that the design variables and orientation were important and had a noteworthy impact on building energy efficiency, therefore energy consumption may be significantly reduced by selecting appropriate building orientation and wall composition materials.

Table 9. Energy reduction by building envelope components during the hot and cold seasons in Nur-Sultan, Kokshetau, Almaty, and Atyrau.

Envelope Components	Heating Energy Reduction (kWh) in Nur-Sultan			Heating Energy Reduction (kWh) in Koksetau		
	Base Case	Optimized Case	Percentage of Contribution for Energy Reduction	Base Case	Optimized Case	Percentage of Contribution for Energy Reduction
Wall	16,095.01	13,410.52	6.37	16,156.16	13,461.60	6.38
Window	4067.69	3385.79	1.61	4069.49	3397.80	1.61
Roof	22,309.55	18,590.43	8.84	22,384.28	18,652.88	8.83
Shading system	523.11	624.80	0.30	524.96	630.52	0.30
Total	41,848.71	34,905.05	16.592	41,996.78	35,028.89	16.591
	Cooling Energy Reduction (kWh) in Almaty			Cooling Energy Reduction (kWh) in Atyrau		
Wall	4802.92	4005.00	5.88	4702.35	3922.18	6.42
Window	1210.82	1008.63	1.48	1184.75	987.42	1.62
Roof	6713.33	5544.61	8.15	6143.59	5124.38	8.38
Shading system	980.76	816.88	1.20	549.63	458.08	0.75
Total	13,454	11,346	16.668	12,214	10,180	16.656

3.2. Economic Benefit of the Optimization

An economic assessment was conducted on the optimized buildings considering energy savings and economic benefits. According to the Ministry of Energy of the Republic of Kazakhstan, the average 3000 tenges (₸) per Gcal. Meanwhile, money spent on heating energy for residents of the Kokshetau was established as 1600 tenges per Gcal. In other regions, this tariff costs over 2000 tenges. The highest tariffs were observed in Almaty and

Atyrau, such as over 4000 tenges per Gcal. The average cost of electricity supply for the population reached 12.69 tenges per kWh. The lowest rates of electricity tariff are marked in the west and north regions of the country. At the same time, in the Atyrau region, the cost per kWh did not even reach 5 tenges. On contrary, this figure rate is 17.12 tenge per kWh in Almaty. The current electricity tariffs for the selected cities separately for central heating and electricity-based equipment for heating and cooling are shown in Table 10 [49].

Table 10. Economic analysis for the heating and cooling energy savings for the optimized building configuration.

			Heating Energy Reduction			
List of Cities	Energy Reduction (kWh)	Energy Reduction (Gcal)	Price Rate (₸ per Gcal)	Price Rate (₸ per kWh)	Economic Impact (₸ for the Period)	
					Centralized Heating	Equipment Heating
Nur-Sultan	6943.66	5.97	2176.76	11.93	12,996	82,838
Almaty	4758.35	4.09	4881.79	17.12	19,973	81,463
Karaganda	6394.42	5.49	2758.57	8.75	15,166	55,951
Aktobe	6730.91	5.79	3042.32	10.02	17,607	67,444
Atyrau	5060.65	4.35	4832.88	4.73	21,029	23,937
Semey	6967.89	5.99	1611.98	17.11	9657	119,221
Kokshetau	6248.90	5.37	3018.12	10.395	16,216	64,957
			Cooling Energy Reduction			
Nur-Sultan	1924.16	1.65	11.93		22,955	
Almaty	2107.94	1.81	17.12		36,088	
Karaganda	2047.77	1.76	10.02		20,519	
Aktobe	1591.77	1.36	8.75		13,928	
Atyrau	2034.35	1.74	4.73		9622	
Semey	1718.19	1.47	10.395		17,861	
Kokshetau	1961.00	1.68	17.11		33,553	

On the contrary, Atyrau and Kokshetau showed the lowest cost reduction for equipment-based heating and centralized heating, with 24,000 ₸ and 9700 ₸, respectively. Regarding the payment reduction for air-conditioning during the cooling season (from 13 May to 12 October), Almaty and Atyrau showed the highest and lowest reductions, namely, 36,000 ₸ and 9600 ₸, respectively. Compared with the base case, the optimal design in the envelope entails the energy efficiency for heating and cooling with a significant reduction in energy service charges. Thus, the energy demand can be affected by the heat gain and heat loss through building envelope during hot and cold seasons. In other words, the shading systems characteristics, wall and roof properties, and window-to-wall ratio were optimized using a single-objective optimization to reduce the fees charged for heating and cooling energy.

4. Conclusions

In this research, the energy performance of the townhouse for cold and hot seasons by optimizing the envelope characteristics in seven cities of Kazakhstan for energy consumption and cost reduction was investigated. To conduct the analysis, the selected cities were as follows: Nur-Sultan, Almaty, Karaganda, Aktobe, Atyrau, Kokshetau, and Semey. The OTTV-based numerical model was developed for the estimation of the energy transfer mechanism through the envelope components of the base case. The brute force algorithm was used to optimize the numerical model for reducing the value of energy performance in cold and hot seasons by searching the optimum variant of wall components, the thickness of the wall, WWR, depth of the shading system, and angle of the pitched roof. The energy reduction for hot (13 October–12 May) and cold (13 May–12 October) seasons were conducted by using brute force algorithm optimization with Python coding software. The OTTV-based numerical model was developed and implemented in exhaustive search optimization for energy-saving purposes by searching for the optimum envelope configurations. The main conclusions and recommendations are summarized below:

- This research has shown that proper selection of design variables can lead to notable energy savings in all cities. Due to the optimization of the numerical model analyzing the heat transfer through the envelope, the average annual heating reduction was 6156.8 kWh (or 5.29 Gcal), and the average cooling energy reduction was 1912.17 kWh (or 1.64 Gcal). In terms of percentage, heating and cooling energy were reduced by 16.59% and 16.69%, respectively. It is also concluded that the heating energy savings effect was more evident in the cities located in the northern part of Kazakstan (Nur-Sultan and Kokshetau), and the effect of the cooling energy savings was evident in the southern part (Almaty);
- Regarding monthly energy consumption, January, February, and December showed the highest energy consumption in each city. Overall energy consumption for heating and cooling throughout all months demonstrated that Kazakhstan's climatic zone is mostly heating-dominated;
- The results showed that proper selection of orientation is critical. The direction of the frontage of the building towards the south in the cold season and north in the hot season showed around 21% and 32% energy reduction, respectively, which were effective compared with an initial orientation of the building;
- The building orientation to the south, a combination of 0.7 m of the depth of the overhang shading system and the WWR of 4%, 12.30 mm ceramic brick as exterior finish, 171.10 mm masonry brick as core layer, 84.7 mm glasswool, and 13.10 mm plasterboard as the interior finish was found to be optimal design solution;
- From the economic analysis, it was found that equipment heating had a high-cost reduction compared to central heating. The highest cost reduction was observed in Atyrau with central heating and in Kokshetau with equipment-based heating, for cooling in Almaty. This highlights the fact that optimization of buildings brings significant economic benefits.

From the operational view point, it is suggested that for future work, the proposed analysis should take into account occupancy rate and the working schedule of HVAC considering the opening rate of external doors and windows. It is suggested that the numerical analysis should be done to increase the optimisation variables, e.g., roof composition materials and detailed properties of the windows in terms of the type and material configurations. The space total energy demand for heating and cooling for residential buildings in Kazakhstan cities is crucial to be investigated, as zero-energy buildings should be designed due to the forecast of shortage of electrical energy.

Author Contributions: Conceptualization, M.K. and S.A.M.; methodology, M.K. and S.A.M.; software, M.K.; validation, M.K. and S.A.M.; formal analysis, M.K.; investigation, M.K. and S.A.M.; resources, M.K.; data curation, M.K. and S.A.M.; writing—original draft preparation, M.K.; writing—review and editing, M.K., S.A.M. and A.S.; visualization, M.K. and S.A.M.; supervision, S.A.M.; project administration, S.A.M.; funding acquisition, S.A.M. and J.R.K. All authors have read and agreed to the published version of the manuscript.

Funding: This research was supported by Nazarbayev University Faculty development competitive research grants number 021220FD0651 and SOE2017004.

Institutional Review Board Statement: Not applicable.

Informed Consent Statement: Not applicable.

Data Availability Statement: The data presented in this research are available upon request from the corresponding author.

Conflicts of Interest: The authors declare no conflict of interest.

Abbreviations

TDES	Total direct energy supply
TFEC	Total final energy consumption
PLA	Perimeter Annual Load
ENVLOAD	Envelope Energy Load
OTTV	Overall Thermal Transfer Value
TRNSYS	Transient System Simulation Tool
GMDH	Grouped Method of Data Handling type neural network
NSGA-II	Non-Dominated Sorting Genetic Algorithm II
ASHRAE	The American Society of Heating, Refrigerating and Air-Conditioning Engineers
HVAC	Heating, ventilation, and air conditioning
PCM	Phase change material
WNN	Wavelet neural network
MLP	Multi-Layer Perceptron neural network
ACO	Ant Colony Optimization
ABC	Artificial Bee Colony
PSO	Particle Swarm Optimization
BF	Brute force
BESM	Building Energy Simulation model
ETTV	Envelope Thermal Transfer Value
RTTV	Roof Transfer Value
SHGC	Standard solar heat gain coefficient of glazing
WWR	Window-to-wall ratio
A_w, A_g, A_r	Surface area of the wall, fenestration, and roof
k_w	An overall thermal conductivity of the wall
k_r	Thermal conductivity values of the roof
U_g	U-value of glazing
A_{wall}	The exterior wall's gross area
A_{wsi}	Wall area exposed to solar radiation
Q_c	Heat gain through the envelope for the cooling season
Q_h	Heat loss through the envelope for the heating season
Q_{gsol}	Solar radiation through the fenestration
h_1, h_2, h_3, h_4	Thicknesses of wall composition layers
C_1, C_2, C_3, C_4	Thermal conductivity values of wall composition layers
R_c	Thermal resistance value for roof ceiling
R_r	Thermal resistance value for the roof
A_c	Ceiling area
A_r	Roof area
A_T	Total area building's wall
SLF	Shade line factor
D_{oh}	The depth of overhang (projection)
X_{oh}	Vertical offset from the top of the window to overhang
h_g	The height of the window
SC_g	Shading coefficient of the glazing
SC	Total shading coefficient
α	The solar absorptivity constant related to the façade surface and color
TD	Equivalent temperature difference
TD_w, TD_r, TD_g	Equivalent temperature differences between indoor and outdoor surrounding space wall, roof, and glazing
ESM	External shading multiplier
SF	Solar factor
q	The standard solar heat gain factor

References

1. Wang, X.; Zheng, H.; Wang, Z.; Shan, Y.; Meng, J.; Liang, X.; Feng, K.; Guan, D. Kazakhstan's CO_2 emissions in the post-Kyoto Protocol era: Production- and consumption-based analysis. *J. Environ. Manag.* **2019**, *249*, 109393. [CrossRef] [PubMed]
2. Sarbassov, Y.; Kerimray, A.; Tokmurzin, D.; Tosato, G.; De Miglio, R. Electricity and heating system in Kazakhstan: Exploring energy efficiency improvement paths. *Energy Policy* **2013**, *60*, 431–444. [CrossRef]

3. Assenova, B. What Is the Housing Stock in Kazakhstan—NewTimes.kz 2019. Available online: https://newtimes.kz/obshchestvo/96182-chto-iz-sebya-predstavlyaet-zhilishchnyj-fond-v-kazakhstane (accessed on 23 November 2021).
4. History of Housing Construction in Kazakhstan. 2020. Available online: https://google-info.org/7523418/1/istoriya-zhilishchnogo-stroitelstva-v-kazakhstane.html (accessed on 24 November 2021).
5. Agency for Strategic Planning and Reforms of the Republic of Kazakhstan Bureau of National Statistics n.d. Available online: https://stat.gov.kz/search (accessed on 24 November 2021).
6. Energy-Efficient Design and Construction of Residential Buildings. United Nations, Development Programme, PIMS No. 4133 Atlas Award 00059795, Project ID 0074950. Available online: https://www.kz.undp.org/content/kazakhstan/en/home/operations/projects/environment_and_energy/energy-efficient-design-and-construction-of-residential-building.html (accessed on 27 November 2021).
7. Imani, N.; Vale, B. A framework for finding inspiration in nature: Biomimetic energy efficient building design. *Energy Build.* **2020**, *225*, 110296. [CrossRef]
8. Yu, J.; Yang, C.; Tian, L.; Liao, D. Evaluation on energy and thermal performance for residential envelopes in hot summer and cold winter zone of China. *Appl. Energy* **2009**, *86*, 1970–1985. [CrossRef]
9. Huang, J.; Deringer, J. *Status of Energy Efficient Building Codes in Asia*; The Asia Business Council: Hong Kong, China, 2017; pp. 6–9.
10. Lin, Y.-H.; Tsai, K.-T.; Lin, M.-D.; Yang, M.-D. Design optimization of office building envelope configurations for energy conservation. *Appl. Energy* **2016**, *171*, 336–346. [CrossRef]
11. Yang, K.; Hwang, R.-L. Energy conservation of buildings in Taiwan. *Pattern Recognit.* **1995**, *28*, 1483–1491. [CrossRef]
12. Yang, K.H.; Hwang, R.L. The analysis of design strategies on building energy conservation in Taiwan. *Build. Environ.* **1993**, *28*, 429–438. [CrossRef]
13. Yang, K.-H.; Hwang, R.-L. An improved assessment model of variable frequency-driven direct expansion air-conditioning system in commercial buildings for Taiwan green building rating system. *Build. Environ.* **2007**, *42*, 3582–3588. [CrossRef]
14. Huang, K.T.; Lin, H.T. Design standard of energy conservation for building hvac system—A simplified method of chiller capacity estimation based on building envelope energy conservation index in taiwan. In Proceedings of the IBPSA 2005—International Building Performance Simulation Association 2005, Montréal, QC, Canada, 15–18 August 2005; pp. 435–442.
15. Chan, A.; Chow, T. Calculation of overall thermal transfer value (OTTV) for commercial buildings constructed with naturally ventilated double skin façade in subtropical Hong Kong. *Energy Build.* **2014**, *69*, 14–21. [CrossRef]
16. Guo, J.; Hermelin, D.; Komusiewicz, C. Local search for string problems: Brute-force is essentially optimal. *Theor. Comput. Sci.* **2014**, *525*, 30–41. [CrossRef]
17. Robinson, A.C.; Quinn, S.D. A brute force method for spatially-enhanced multivariate facet analysis. *Comput. Environ. Urban Syst.* **2018**, *69*, 28–38. [CrossRef]
18. Al-Saadi, S.N.J.; Al-Jabri, K.S. Energy-efficient envelope design for residential buildings: A case study in Oman. In Proceedings of the 2017 Smart Cities Symp Prague, SCSP 2017—IEEE Proceedings 2017, Prague, Czech Republic, 25–26 May 2017. [CrossRef]
19. Hui Sam, C.M. Overall thermal transfer value (OTTV): How to improve its control in Hong Kong. In Proceedings of the One-day Symposium on Building, Energy and Environment, Hong Kong, China, 16 October 1997; Volume 16, pp. 12-1–12-11.
20. Devgan, S.; Jain, A.; Bhattacharjee, B. Predetermined overall thermal transfer value coefficients for Composite, Hot-Dry and Warm-Humid climates. *Energy Build.* **2010**, *42*, 1841–1861. [CrossRef]
21. Chan, L.S. Investigating the thermal performance and Overall Thermal Transfer Value (OTTV) of air-conditioned buildings under the effect of adjacent shading against solar radiation. *J. Build. Eng.* **2021**, *44*, 103211. [CrossRef]
22. Chou, S.K.; Chang, W.L. Development of an energy-estimating equation for large commercial buildings. *Int. J. Energy Res.* **1993**, *17*, 759–773. [CrossRef]
23. Djamila, H.; Rajin, M.; Rizalman, A.N. Energy efficiency through building envelope in Malaysia and Singapore. *J. Adv. Res. Fluid Mech. Therm. Sci.* **2018**, *46*, 96–105.
24. Zhang, Y.; Long, E.; Li, Y.; Li, P. Solar radiation reflective coating material on building envelopes: Heat transfer analysis and cooling energy saving. *Energy Explor. Exploit.* **2017**, *35*, 748–766. [CrossRef]
25. Mustafaraj, G.; Marini, D.; Costa, A.; Keane, M. Model calibration for building energy efficiency simulation. *Appl. Energy* **2014**, *130*, 72–85. [CrossRef]
26. Kutlu, L.; Tran, K.C.; Tsionas, M.G. A spatial stochastic frontier model with endogenous frontier and environmental variables. *Eur. J. Oper. Res.* **2020**, *286*, 389–399. [CrossRef]
27. Lleras, C. Path Analysis. *Encycl. Soc. Meas.* **2005**, *3*, 25–30. [CrossRef]
28. Zhang, Y.; Sun, X.; Medina, M.A. Calculation of transient phase change heat transfer through building envelopes: An improved enthalpy model and error analysis. *Energy Build.* **2020**, *209*, 109673. [CrossRef]
29. White Box Technologies Weather Data, n.d. Available online: http://weather.whiteboxtechnologies.com/ (accessed on 24 November 2021).
30. Xing, F.; Mohotti, D.; Chauhan, K. Study on localised wind pressure development in gable roof buildings having different roof pitches with experiments, RANS and LES simulation models. *Build. Environ.* **2018**, *143*, 240–257. [CrossRef]
31. Çengel, Y.; Ghajar, R.A. Heat and Mass Transfer: Fundamentals & Applications. 2015. Available online: https://studylib.net/doc/8394668/heating-and-cooling-of-buildings (accessed on 8 December 2021).

32. Shen, H.; Tzempelikos, A. Daylighting and energy analysis of private offices with automated interior roller shades. *Sol. Energy* **2012**, *86*, 681–704. [CrossRef]

33. Barnaby, C.S.; Wright, J.L.; Collins, M.R. Improving load calculations for fenestration with shading devices. *ASHRAE Trans.* **2009**, *115*, 31–44.

34. Kotey, N.A.; Wright, J.L.; Barnaby, C.S.; Collins, M.R. Solar gain through windows with shading devices: Simulation versus measurement. *ASHRAE Trans.* **2009**, *115*, 18–30.

35. Bessoudo, M.; Tzempelikos, A.; Athienitis, A.; Zmeureanu, R. Indoor thermal environmental conditions near glazed facades with shading devices—Part I: Experiments and building thermal model. *Build. Environ.* **2010**, *45*, 2506–2516. [CrossRef]

36. DesignBuilder Software Ltd.—Previous Versions n.d. Available online: https://designbuilder.co.uk/download/previous-versions (accessed on 22 October 2021).

37. Residential Cooling and Heating Load Calculations. In *ASHRAE Handbook of Fundamentals*; SI Edition; ASHRAE: Atlanta, GA, USA, 2017; Chapter 17.

38. Ozmen, Y.; Baydar, E.; van Beeck, J. Wind flow over the low-rise building models with gabled roofs having different pitch angles. *Build. Environ.* **2016**, *95*, 63–74. [CrossRef]

39. Szcześniak, J.T.; Ang, Y.Q.; Letellier-Duchesne, S.; Reinhart, C.F. A method for using street view imagery to auto-extract window-to-wall ratios and its relevance for urban-level daylighting and energy simulations. *Build. Environ.* **2021**, *2021*, 108108. [CrossRef]

40. Xue, P.; Li, Q.; Xie, J.; Zhao, M.; Liu, J. Optimization of window-to-wall ratio with sunshades in China low latitude region considering daylighting and energy saving requirements. *Appl. Energy* **2019**, *233–234*, 62–70. [CrossRef]

41. GOST 530-2012. Interstate Standard. Ceramic Brick and Stone. General Specifications. 2012. Available online: https://docs.cntd.ru/document/1200100260 (accessed on 27 November 2021).

42. GOST 9573-2012. Interstate Standard. Thermal Insulating Plates of Mineral Wool on Syntetic Binder. Specifications. 2013, pp. 14–27. Available online: https://docs.cntd.ru/document/1200101613 (accessed on 27 November 2021).

43. SN RK 2.04-04-2011. State Standards in the Field of Architecture, Urban Planning and Construction. Thermal Protection of Buildings. 2011. Available online: http://www.hoffmann.kz/files/12_SN_RK_2-04-04-2011.pdf (accessed on 27 November 2021).

44. GOST 24992-81. State Standard of the Union of SSR. Masonary Structures. Method of Estimating Bonding Strength in Masonry. 1995. Available online: http://gostrf.com/normadata/1/4294853/4294853407.html (accessed on 27 November 2021).

45. GOST 31310-2015. Interstate Standard. Wall Three-Layer Reinforced Concrete Panels with Enerqy-Efficient Insulation. General Specifications. Interstate Council for Standardization, Metrology and Certification (isc). 2017. Available online: https://docs.cntd.ru/document/1200133283 (accessed on 27 November 2021).

46. Peel, M.C.; Finlayson, B.L.; McMahon, T.A. Updated world map of the Köppen-Geiger climate classification. *Hydrol. Earth Syst. Sci.* **2007**, *11*, 1633–1644. [CrossRef]

47. Climate Data for Cities Worldwide—Climate-Data.org n.d. Available online: https://en.climate-data.org/ (accessed on 22 October 2021).

48. Mutschler, R.; Rüdisüli, M.; Heer, P.; Eggimann, S. Benchmarking cooling and heating energy demands considering climate change, population growth and cooling device uptake. *Appl. Energy* **2021**, *288*, 116636. [CrossRef]

49. Tootkaboni, M.P.; Ballarini, I.; Corrado, V. Analysing the future energy performance of residential buildings in the most populated Italian climatic zone: A study of climate change impacts. *Energy Rep.* **2021**, *7*, 8548–8560. [CrossRef]

50. Albatayneh, A. Optimising the Parameters of a Building Envelope in the East Mediterranean Saharan, Cool Climate Zone. *Buildings* **2021**, *11*, 43. [CrossRef]

51. Alshboul, A.A.; Alkurdi, N.Y. Enhancing the Strategies of Climate Responsive Architecture, The Study of Solar Accessibility for Buildings Standing on Sloped Sites. *Mod. Appl. Sci.* **2018**, *13*, 69. [CrossRef]

 *buildings*

Article

Office Distractions and the Productivity of Building Users: The Effect of Workgroup Sizes and Demographic Characteristics

Maryam Khoshbakht [1,*], Eziaku O. Rasheed [2] and George Baird [3]

[1] Cities Research Institute, Griffith University, Gold Coast 4215, Australia
[2] School of Built Environment, Massey University, Auckland 0632, New Zealand; e.o.rasheed@massey.ac.nz
[3] School of Architecture, Victoria University of Wellington, Wellington 6011, New Zealand; george.baird@vuw.ac.nz
* Correspondence: m.kh@griffith.edu.au

Abstract: Knowledge workers are experiencing ever-increasing distractions or unwanted interruptions at workplaces. We explored the effect of unwanted interruptions on an individual's perceived productivity in various building types, user groups and workgroups. A case study of 68 buildings and their 5149 occupants using the Building Use Studies methodology was employed in this study. The database contains information on the occupants' perceptions of physical and environmental parameters, including unmined data on the frequency of unwanted interruptions. Pearson's correlation was used to test the correlation between the variables. In order to determine whether there are any statistically significant differences between the means of two or more independent (unrelated) groups, one-way ANOVA was employed to examine the significance of differences in mean scores between various user groups and workgroups. The evidence of clear correlations between the frequency of unwanted interruptions and perceived productivity is detailed in various user groups and in multiple building types. The Pearson correlation coefficients were -0.361 and -0.348 for sustainable and conventional buildings, respectively, demonstrating a lower sensitivity to unwanted interruptions in sustainable buildings. Females and older participants were more sensitive to unwanted interruptions and their productivity levels were reduced much more by unwanted interruptions. Comparing different sized workgroups, the highest sensitivity to unwanted interruptions for occupants in offices shared with more than 8 people was found. The findings of this study contribute to the understanding of different user needs and preferences in the design of workplaces.

Keywords: productivity; interruptions; workgroups; demographics; offices

Citation: Khoshbakht, M.; Rasheed, E.O.; Baird, G. Office Distractions and the Productivity of Building Users: The Effect of Workgroup Sizes and Demographic Characteristics. *Buildings* **2021**, *11*, 55. https://doi.org/10.3390/buildings11020055

Academic Editor: David Arditi
Received: 24 December 2020
Accepted: 2 February 2021
Published: 6 February 2021

Publisher's Note: MDPI stays neutral with regard to jurisdictional claims in published maps and institutional affiliations.

1. Introduction

Over recent decades, much consideration has been given to the effect of the physical environment of office buildings on the comfort, health and productivity of building users. In this regard, considerable effort has been expended to learn about the influence of the thermal, visual and acoustic aspects of buildings on occupant behaviours and expectations [1,2]. Building users and their productivity is influenced by various physical and behavioural components in an office environment [3]. Office distractions could be classified in the behavioural, environmental category as an integrated dimension of the office environment [4]. In a study of faculty research performance, an individual's research productivity was associated with a combination of individual and institutional characteristics [5]. Particularly, uninterrupted time to devote to scholarly activities was one of the major components of the institutional characteristics that led to higher perceived productivity [5].

Scholarly work has done much to highlight the detrimental effect of noise in sectors other than the knowledge sector across various demographics. For instance, Schneider et al. [6] documented extensive evidence that noise levels regularly exceed limit values in many sectors, such as agriculture, construction, engineering, the food and drink industry, woodworking, foundries and entertainment. The authors noted that in the Czech Republic, 75% of workers exposed to noise in textile production are observed to be female, followed by 5% in food production. Mohammadi et al. [7] observed a significant effect of occupational noise on the blood biochemical parameters of workers in an Iranian insulator manufacturing plant. A relationship between exposure to noise and significantly increased systolic blood pressure of 62 male workers in a sack manufacturing company was found in Nigeria [8].

However, the same effort has not been expanded on workers in the knowledge industry whose job involves the handling and use of information; and who spend most of their time in an office environment. Perhaps the reason for the lack of attention is that the noise levels experienced in office buildings are at a lower level compared to those observed on farms, factories and construction sites. Additionally, whereas the health effect of loud noise has been documented as significant [8,9], the impact of low noise has not. Thus, noise experienced in office buildings may not be regarded as detrimental.

As noted earlier, investigations of the effect of the physical aspects of buildings (in this case, the office environment) on productivity have hinged on importance of the worker to organisational success and productivity. While aspects such as temperature, lighting and facilities have been extensively studied, somewhat less attention has been given to the study of distractions or unwanted interruptions on the productivity of occupants in offices. That said, the potential disadvantages of noise, mostly in the form of distraction and unwanted interruptions on knowledge worker productivity, have been reflected in a few studies in the past [10,11]. Interestingly, some studies found contradictory results, showing that people can complete interrupted tasks in less time with no difference in quality, yet with more stress, higher frustration, and time pressure [12]. Another study demonstrated that different interruption moments have different impacts on user emotional states and positive social attributions [13]. The perception of too many unwanted interruptions along with the overall satisfaction with air quality were the most significant factors among all indoor environmental quality (IEQ) parameters in predicting perceived productivity according to Francis [14].

In his book *"Rewording the Brain"*, Astle [15] noted that you would need 23 min and 15 s to get back on track after a two-minute off-topic interruption. This struck a chord—not just because of the amount of time and the apparent precision of its duration, but because it echoed one's own experience. Knowledge workers need some time to recover from a distraction, which can potentially lead to errors [16]. It is estimated that interruptions in knowledge work cost 588 billion USD per annum in the United States alone [17]. It is also not by accident that so many authors seem to work on their novels in the garden shed or in a remote part of the house or the country; focus groups go into a retreat to hammer out policy, and academics pine for the mythical 'quality time' to conduct and write-up their research. All recognise the need to avoid off-topic interruptions to enable full focus on the task at hand.

While evidence is mounting that unwanted interruptions may be an important predictor of worker productivity in office environments, the part played by factors such as workgroup and demography is less studied. Offices can be categorised into several workgroups from cellular offices with a single occupant to open-plan offices with more than eight occupants [18]. Office design layout influences occupant satisfaction and perceived productivity in workplace environments [19]. Noise, distraction and privacy loss seem to have adverse effects on productivity in open-plan layouts [20]. Haynes et al. (2017) showed that the greatest impact on perceived productivity was the availability of a variety of physical layouts, control over interaction and the "downtime" offered by social interaction points. Depending on the workgroup sizes, office occupants may also experience various levels of visual and acoustic privacy, and consequently, different amounts of distraction and unwanted interruptions [21]. On the other hand, among all the environmental and behavioural components, in low-performance buildings, perceived productivity was strongly associated with building aesthetics and quality, and noise distraction and privacy; while in high-performance buildings, productivity was highly correlated with office layout, employees' working experience, and work hours [22]. Therefore, a hypothesis was developed that workgroup sizes, the type of use, and sustainability intention of office buildings might also be a significant indicator of the impact of noise on knowledge workers' productivity.

Our investigation commenced with (in retrospect rather belatedly) a pilot study of relevant data from the 30 or so buildings featured in Baird's survey of "Sustainable Buildings in Practice" [23]. This involved a simple comparison of the averages for each building of the "Unwanted Interruptions" scores versus the percentage increase or decrease in perceived productivity, this latter being adjudged to be the variable most likely to be influenced by unwanted interruptions (Table 1).

Table 1. List of buildings in "Unwanted Interruptions" order (on a 7-point Scale where 1 is best), together with scores for Perceived Productivity.

Building	Location	Type	Unwanted Interruptions Scores	Productivity % (Up or Down)
Torrent Research Centre (with AC)	Ahmedabad, India	commercial	2.87	20.88
Military Families Resource Centre	Toronto, Canada	institutional	2.90	20.00
Science Park in 2002	Gelsenkirchen, Germany	commercial	3.16	1.43
St Mary's Credit Union	Navan, Ireland	commercial	3.18	10.83
Min Energy Water & Communication	Putra Jaya, Malaysia	institutional	3.19	16.00
Nikken Sekkei Building	Tokyo, Japan	commercial	3.19	8.51
Torrent Research Centre (with PDEC)	Ahmedabad, India	commercial	3.20	13.66
Gifford Studios	Southampton, England	commercial	3.29	2.80
Science Park in 2006	Gelsenkirchen, Germany	commercial	3.46	−2.27
Tokyo Gas	Yokohama, Japan	commercial	3.47	5.62
NRG Systems Facility	Vermont, USA	commercial	3.50	19.51
Natural Resources Defense Council	California, USA	commercial	3.53	23.00
Institute of Languages	Sydney, Australia	institutional	3.65	0.48
40 Albert Road	Melbourne, Australia	commercial	3.77	10.00
Arup Campus	Solihull, England	commercial	3.82	4.47
Renewable Energy Systems	Kings Langley, England	commercial	3.89	5.77
Computer Science & Engineering	Toronto, Canada	institutional	4.01	2.54
National Engineering Yards	Vancouver, Canada	commercial	4.03	0.19
60 Leicester Street,	Melbourne, Australia	commercial	4.12	11.39
Landcare Research	Auckland, New Zealand	institutional	4.12	−2.18
ZICER Building	Norwich, England	institutional	4.23	−7.81
City Hall	London, England	institutional	4.41	−1.64
Found'n Building, Eden Project in 2006	St Austell, England	commercial	4.58	−7.00
AUT Akoranga	Auckland, New Zealand	institutional	4.71	3.64
Red Centre Building	Sydney, Australia	institutional	4.71	−5.00
Scottsdale Ecocentre	Tasmania, Australia	commercial	5.19	−4.29
Found'n Building, Eden Project in 2004	St Austell, England	commercial	5.30	−2.00
Erskine Building	Christchurch, New Zealand	institutional	5.39	9.80
Student Services Centre	Newcastle, Australia	institutional	na	−2.04
Institute of Technical Education	Bishan, Singapore	institutional	na	−10.61
General Purpose Building	Newcastle, Australia	institutional	na	−11.9
Menara UMNO	Penang, Malaysia	commercial	na	na

Note: in two cases, the same building was surveyed twice some years apart; while in another, different ventilation systems were used in separate parts of the building. Interruptions were not assessed in four of the buildings, but they are listed separately at the bottom for completeness. AC means air-conditioned; PDEC means passive down-draught evaporative cooling.

Twenty buildings reported increases in perceived productivity averaging +9.4% overall, corresponding to an unwanted interruptions score averaging 3.57 (on the 7-point scale where 1 signifies the least effect; and 7 the greatest). For the 9 buildings which reported average decreases of −5.0% overall, the unwanted interruptions score averaged 4.59. The highest frequency in unwanted interruptions were observed in commercial buildings (5.30; 5.19). The higher sensitivity of users in commercial buildings to unwanted interruptions was observed when the averages of unwanted interruptions for the buildings were plotted against perceived productivity for both commercial and institutional buildings. As shown in Figure 1, the R-square of the trendline of the linear equation for commercial buildings is closer to 1 than institutional buildings. The trend was clear, albeit on the basis of building averages rather than individual respondents. While building averages are useful to give an overview of performance, or when comparing buildings, the effect of interruptions requires study of the perceptions of individuals.

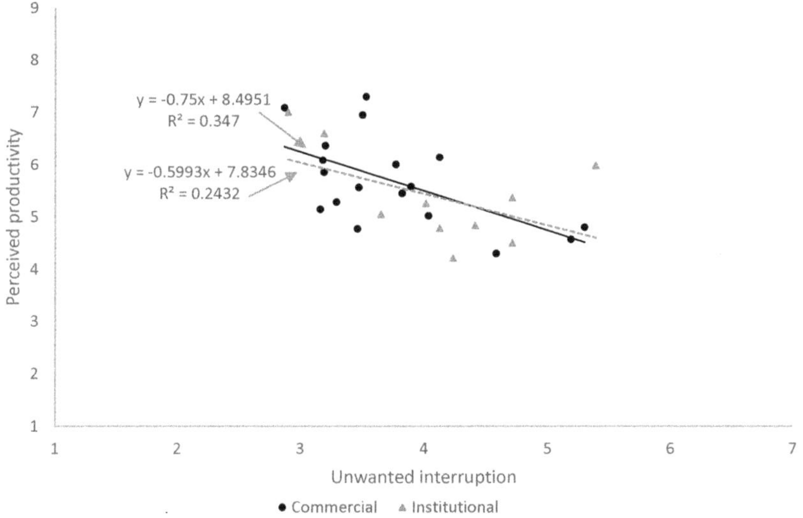

Figure 1. The scatter plot of the pilot study of sustainable building averages of unwanted interruptions against perceived productivity.

With the promising results of the pilot study, the decision was taken to expand the research to the full-scale with a database of 68 buildings and 5149 respondents. The aim of this research was to compare the tolerance of unwanted interruptions among several user groups and workgroup size in a systematic way. Thus, differences in sensitivity to unwanted interruptions between commercial and academic buildings, between sustainable and conventional buildings, between younger and older workers, between males and females, and between differently sized workplace clusters were explored in detail.

After the methodology section (Section 2), the main findings are presented as follows: The differences in sensitivity to unwanted interruptions found between different building use—commercial versus academic buildings in Section 3.1.; building design intent—sustainable versus conventional buildings in Section 3.2.; age groups—under 30 years versus over 30 years or over in Section 3.3.; genders—male and female in Section 3.4.; and workgroup sizes in Section 3.5.

2. Methodology

A database of post-occupancy evaluation (POE) studies and the BUS (Building Use Studies) methodology survey questionnaires were utilised for this research. The BUS methodology survey is a well-recognised survey tool, which has been effectively used in many research studies around the world. Full detail of the BUS methodology and the actual questionnaire are presented in earlier studies [24]. The survey consists of several questions regarding user background information and various satisfaction questions. From 68 buildings, 5149 survey responses were collected over a 12-year period. Table 2 provides a summary of the building description including countries, number of occupants in the buildings, sustainability credentials, and building use. The majority of data was collected from New Zealand (56.7%). The largest building had 342 occupants, and the smallest had only 11 occupants. There were a relatively equal number of green and conventional buildings. In terms of building use, 72% were commercial buildings and 28% were academic building.

Table 2. Studied buildings in the dataset and building specifications.

Variables	Dataset Distribution
Country	New Zealand (56.7%); Australia (17.3%); England (11.4%); USA (1.4%); India (3.3%); Ireland (0.3%); Japan (1.3%); Malaysia (3.1%); Canada (3.9%)
Number of occupants	Minimum = 11; Maximum = 342; Average = 75; Standard Deviation = 70
Sustainability credentials	Green buildings (55.7%); Conventional (44.3%)
Building use	Commercial (72%); Academic (28%)

A subset of the questionnaire responses was used to serve the purpose of this study (see Table 3). The utilised questions in this analysis included the age of participants, the gender of participants, workgroup sizes, perceived productivity, and unwanted interruptions. The age category was under 30 years for the younger groups and 30 years or over for the older groups. The productivity question asked survey participants to estimate how they think their productivity at work was decreased or increased by the environmental conditions in the building using a 9-point scale (where a 5 signifies conditions have no effect, less than 5 signifies a decrease and greater than 5 an increase). Respondents were asked to "Please estimate how you are affected by unwanted interruptions … " on a 7-point semantic differential scale ranging from "Not at all" (scoring 1) to "Very frequently" (scoring 7). The paper questionnaires were distributed in the buildings in person, and the responses were typically collected after five to seven days. A response rate greater than 75% was required to ensure results were representative. To verify whether a statistically significant linear relationship exists between two continuous variables, Pearson's correlation was used to test the correlation between the variables. In order to determine whether there are any statistically significant differences between the means of two or more independent (unrelated) groups, one-way ANOVA was also employed to examine the significance of differences in mean scores between various user groups and workgroups.

Table 3. The subset of the survey questionnaire used in this study.

Category	Topic	Questions
Background questions	Age	What is your age? Tick Under 30 or 30 or over
	Gender	What is your sex? Tick Male or Female
	Workgroups	Is your office or work area … ? Tick Normally occupied by you alone, Shared with one other, Shared with 2–4 others, Shared with 5–8 others, or Shared with more than 8 others
Satisfaction	Unwanted interruptions	Please estimate how you are affected by unwanted interruptions … ? Tick a number from 1 (for Not at all) to 7 (for Very frequently)
	Productivity	Please estimate how you think your productivity at work is decreased or increased by the environmental conditions in the building … ? Tick a number from 1 (for Decreased by 40%) to 9 (for Increased by 40%)

Statistical analysis included basic descriptive statistics, one-way ANOVA and Pearson correlations. Descriptive statistics consist of mean values, standard deviation and sample numbers in each group. One-way ANOVA is used to determine whether there are any statistically significant differences between the means of two groups of design intent, building use, age, gender and workgroup. One-way ANOVA tests the null hypothesis:

$$H_0:\ \mu_1 = \mu_2 = \mu_3 = \ldots = \mu_k \tag{1}$$

where μ is the group mean, and k is the number of groups. The null hypothesis for the one-way ANOVA is that there is no difference in the population means of the different groups. If one-way ANOVA reports a *p*-value less than 0.05, the null hypothesis is rejected, and it is confirmed that the two sample groups do not have the same mean.

Pearson correlation is the test statistics that measures the statistical relationship, or association, between two continuous variables. The Pearson correlation produces a sample correlation coefficient, *r*, which measures the strength and direction of linear relationships between pairs of continuous variables. The coefficient is a number between −1 and +1 that indicates to which extent two variables are linearly related and it is calculated from the following formula:

$$r_{xy} = \frac{\text{cov}(x, y)}{\sqrt{\text{var}(x)} \cdot \sqrt{\text{var}(y)}} \tag{2}$$

where cov(*x*, *y*) is the sample covariance of *x* and *y*; var(*x*) is the sample variance of *x*; and var(*y*) is the sample variance of *y*.

3. Findings

3.1. Building Use—Commercial Versus Institutional

The comparison analysis of descriptive statistics including mean scores and standard deviations showed that productivity was scored much better in commercial buildings with a mean score of 5.21 than in institutional buildings with a mean score of 4.96 (see Table 4); while unwanted interruptions were more frequent in institutional buildings than commercial buildings with mean scores of 4.28 and 4.06, respectively (Figure 2). One-way ANOVA tests showed a significant difference (*p*-value < 0.001) between the mean scores of productivity and unwanted interruptions in commercial and institutional buildings (Table 5).

Table 4. Descriptive statistics of productivity and unwanted interruptions in commercial and institutional buildings.

Parameters	Variables	Mean	Std. Deviation	*n*
Commercial buildings	Productivity	5.21	1.676	3331
	Unwanted interruptions	4.06	1.648	3377
Institutional buildings	Productivity	4.96	1.731	1407
	Unwanted interruptions	4.28	1.695	1374

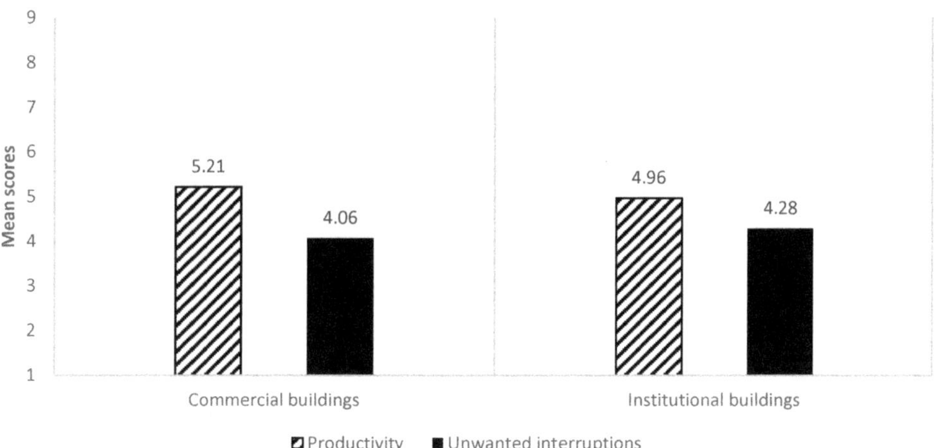

Figure 2. Comparative analysis of mean scores of productivity and unwanted interruptions between commercial and institutional buildings.

Table 5. One-way ANOVA analysis comparing productivity and unwanted interruptions in commercial and institutional building groups.

Parameters	Variables	Sum of Squares	df	Mean Square	F	*p*-Value
Productivity	Between Groups	59.829	1	59.829	20.897	<0.001 **
	Within Groups	13,559.334	4736	2.863	-	-
	Total	13,619.163	4737	-	-	-
Unwanted Interruptions	Between Groups	50.199	1	50.199	18.176	<0.001 **
	Within Groups	13,115.968	4749	2.762	-	-
	Total	13,166.167	4750	-	-	-

Note: ** Statistical significance *p*-value < 0.05.

The Pearson correlation analysis also showed negative coefficients of -0.384 and -0.324 with a statistical significance lower than 0.001 between productivity and unwanted interruptions in commercial and institutional buildings, respectively (see Table 6). Thus, in both commercial and institutional buildings, there seems to be a negative correlation between the frequency of unwanted interruptions and productivity. The negative effect of unwanted interruptions on productivity seemed slightly higher in commercial buildings than institutional buildings. Our result affirms that of the pilot study of sustainable buildings suggesting that irrespectively of the design intent, users in commercial buildings are more sensitive to unwanted interruptions than their counterparts in institutional buildings.

Table 6. Pearson correlation comparing productivity and unwanted interruptions in commercial and institutional building groups.

Parameters	Variables	Statistics	Productivity	Unwanted Interruptions
Commercial	Productivity	Pearson Correlation	1	−0.384 **
		Significance (2-tailed)	-	<0.001
		n	3331	3212
	Unwanted Interruptions	Pearson Correlation	−0.384 **	1
		Significance (2-tailed)	<0.001	-
		n	3212	3377
Institutional	Productivity	Pearson Correlation	1	−0.324 **
		Significance (2-tailed)	-	<0.001
		n	1407	1293
	Unwanted Interruptions	Pearson Correlation	−0.324 **	1
		Significance (2-tailed)	<0.001	-
		n	1293	1374

Note: ** Statistical significance *p*-value < 0.05.

3.2. Design Intent—Sustainable Versus Conventional

The comparison analysis of descriptive statistics, including mean scores and standard deviations showed that productivity was scored better in sustainable buildings than conventional buildings, with mean score values of 5.49 and 4.70, respectively (Table 7). However, unwanted interruptions were more frequent in conventional buildings than sustainable buildings, with the mean score values of 4.31 and 3.97, respectively (Figure 3). When comparing the two groups in design intent, the more frequent unwanted interruptions coincide with the worst productivity in conventional buildings. One-way ANOVA tests also showed a significant difference (*p*-value < 0.001) between the mean scores of productivity and unwanted interruptions in sustainable and conventional buildings (Table 8).

Table 7. Descriptive statistics of productivity and unwanted interruptions in sustainable and conventional buildings.

Parameters	Variables	Mean	Std. Deviation	*n*
Sustainable buildings	Productivity	5.49	1.706	2604
	Unwanted interruptions	3.97	1.701	2573
Conventional buildings	Productivity	4.70	1.581	2134
	Unwanted interruptions	4.31	1.602	2178

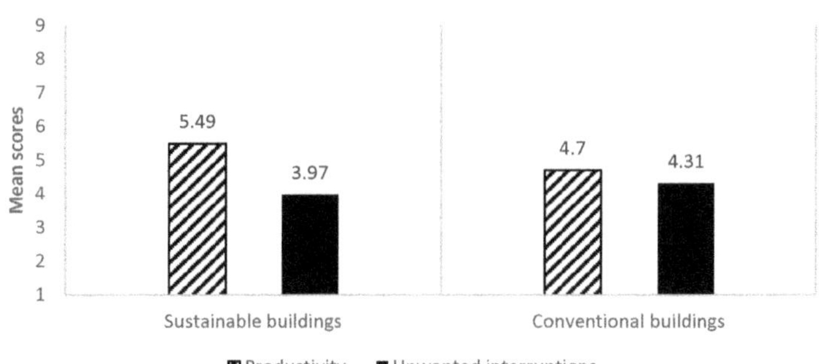

Figure 3. Comparative analysis of mean scores of productivity and unwanted interruptions between sustainable and conventional buildings.

Table 8. One-way ANOVA analysis comparing productivity and unwanted interruptions in sustainable and conventional building groups.

Parameters	Variables	Sum of Squares	df	Mean Square	F	*p*-Value
	Between Groups	714.707	1	714.707	262.301	<0.001 **
Productivity	Within Groups	12,904.457	4736	2.725	-	-
	Total	13,619.163	4737	-	-	-
Unwanted	Between Groups	136.968	1	136.968	49.923	<0.001
interruptions	Within Groups	13,029.199	4749	2.744	-	-
	Total	13,166.167	4750	-	-	-

Note: ** Statistical significance *p*-value < 0.05.

For both building groups, negative Pearson correlation coefficients were observed with statistical significance lower than 0.001 between productivity and unwanted interruptions in both building groups of sustainable and conventional buildings (Table 9). The Pearson correlation coefficients were -0.361 and -0.348 for sustainable and conventional buildings, respectively. This demonstrates a lower sensitivity of sustainable office occupants to unwanted interruptions.

Table 9. Pearson correlation comparing productivity and unwanted interruptions in sustainable and conventional building groups.

Parameters	Variables	Statistics	Productivity	Unwanted Interruptions
		Pearson Correlation	1	-0.361 **
	Productivity	Significance (2-tailed)	-	<0.001
Sustainable buildings		*n*	2604	2438
		Pearson Correlation	-0.361 **	1
	Unwanted Interruptions	Significance (2-tailed)	<0.001	-
		n	2438	2573
		Pearson Correlation	1	-0.348 **
	Productivity	Significance (2-tailed)	-	<0.001
Conventional buildings		*n*	2134	2067
		Pearson Correlation	-0.348 **	1
	Unwanted Interruptions	Significance (2-tailed)	<0.001	-
		n	2067	2178

Note: ** Statistical significance *p*-value < 0.05.

3.3. Age of Occupant—under 30 Years versus 30 Years or Over

Productivity was scored better among the younger participants with a mean of 5.39 compared to the older participants with a mean of 5.04 (see Table 10). For the younger participants, unwanted interruptions were scored better with the mean score of 3.85 in comparison with the older participants with a mean score of 4.22 (Figure 4). Therefore, lower unwanted interruptions and higher productivity were observed among the younger participants, while the older group projected higher unwanted interruptions and consequently lower perceived productivity. As shown in Table 11, the difference between the mean scores of the younger and older participants was statistically significant (p-value < 0.001) indicating that the null hypothesis was true and obtaining that difference in the mean values by chance is very small.

Table 10. Descriptive statistics of productivity and unwanted interruptions for under 30 years and 30 years or over.

Parameters	Variables	Mean	Std. Deviation	n
under 30 years	Productivity	5.39	1.729	1200
	Unwanted interruptions	3.85	1.668	1208
30 years or over	Productivity	5.04	1.672	3473
	Unwanted interruptions	4.22	1.649	3479

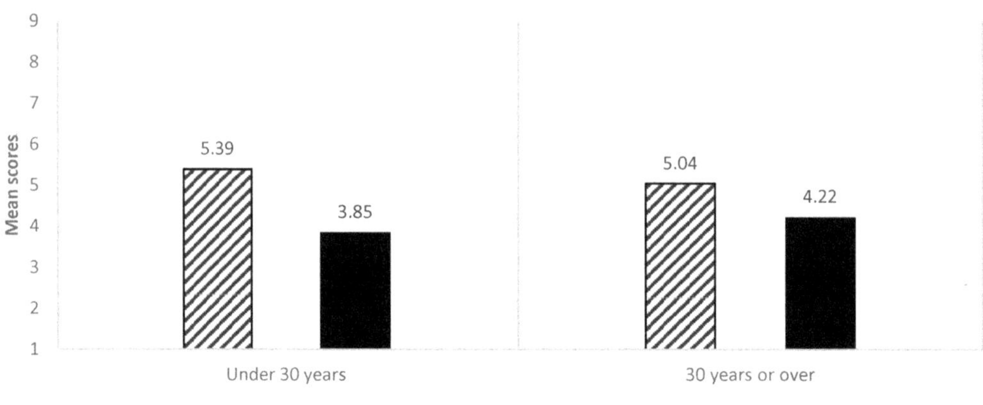

Figure 4. Comparative analysis of mean scores of perceived productivity and unwanted interruptions between under 30 years and 30 years or over.

Table 11. One-way ANOVA analysis comparing perceived productivity and unwanted interruptions for under 30 years and 30 years or over.

Parameters	Variables	Sum of Squares	df	Mean Square	F	p-Value
Productivity	Between Groups	104.669	1	104.669	36.788	<0.001 **
	Within Groups	13,289.805	4671	2.845	-	-
	Total	13,394.475	4672	-	-	-
Unwanted Interruptions	Between Groups	121.384	1	121.384	44.355	<0.001 **
	Within Groups	12,821.297	4685	2.737	-	-
	Total	12,942.681	4686	-	-	-

Note: ** Statistical significance p-value < 0.05.

As shown in Table 12, for under 30 years and 30 years or over, the Pearson correlation coefficients were negative with a statistical significance lower than 0.001, indicating that the null hypothesis was true and there was a measurable difference between the age groups. For the younger group, the Pearson correlation coefficients was −0.350 and for the older group the coefficient was −0.367. The Pearson correlation analysis showed that the older group was more sensitive to unwanted interruptions and the productivity scores were much lower than the younger group.

Table 12. Pearson correlation comparing productivity and unwanted interruptions for under 30 years and 30 years or over.

Parameters	Variables	Statistics	Productivity	Unwanted Interruptions
under 30 years	Productivity	Pearson Correlation	1	−0.350 **
		Significance (2-tailed)	-	<0.001
		n	1200	1148
	Unwanted Interruptions	Pearson Correlation	−0.350 **	1
		Significance (2-tailed)	<0.001	-
		n	1148	1208
30 years or over	Productivity	Pearson Correlation	1	−0.367 **
		Significance (2-tailed)	-	<0.001
		n	3473	3299
	Unwanted Interruptions	Pearson Correlation	−0.367 **	1
		Significance (2-tailed)	<0.001	-
		n	3299	3479

Note: ** Statistical significance *p*-value < 0.05.

3.4. Gender—Male versus Female

The mean score value of productivity for female participants was 5.00, which was worse than the male participants, who scored their productivity with a mean value of 5.25 (see Table 13). The unwanted interruptions were scored 4.05 for male participants and 4.21 by female participants (Figure 5). As shown in Table 14, a significant difference between the mean scores of productivity and unwanted interruptions was observed between male and female participants.

Table 13. Descriptive statistics of productivity and unwanted interruptions for different genders.

Parameters	Variables	Mean	Standard Deviation	*n*
Male	Productivity	5.25	1.650	2413
	Unwanted interruptions	4.05	1.605	2369
Female	Productivity	5.00	1.728	2247
	Unwanted interruptions	4.21	1.718	2301

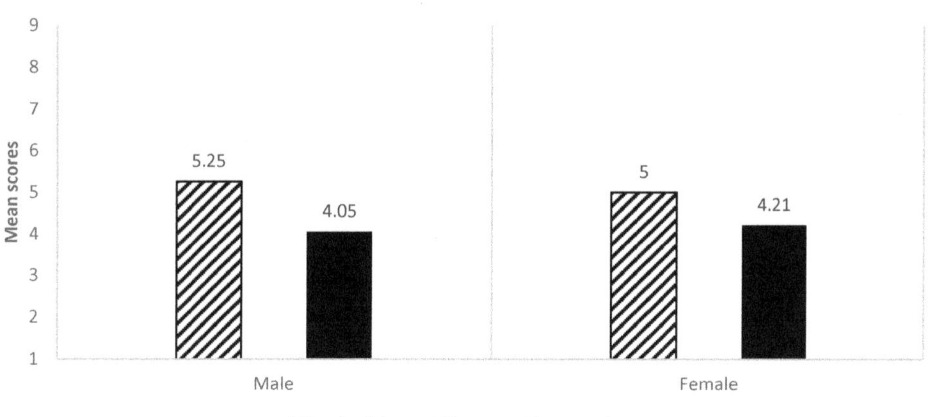

Figure 5. Comparative analysis of mean scores of productivity and unwanted interruptions between male and female gender groups.

Table 14. One-way ANOVA analysis comparing productivity and unwanted interruptions for male and female gender groups.

Parameters	Variables	Sum of Squares	df	Mean Square	F	*p*-Value
Productivity	Between Groups	68.197	1	68.197	23.935	<0.001 **
	Within Groups	13,271.737	4658	2.849	-	-
	Total	13,339.934	4659	-	-	-
Unwanted Interruptions	Between Groups	30.708	1	30.708	11.123	0.001 **
	Within Groups	12,887.738	4668	2.761	-	-
	Total	12,918.446	4669	-	-	-

Note: ** Statistical significance *p*-value < 0.05.

The Pearson correlation analysis showed negative coefficient values with statistical significance lower than 0.001 (see Table 15). For the male participants, the Pearson correlation coefficient was −0.359 and for the male group the coefficient was −0.371. This finding showed that female participants were more sensitive to unwanted interruptions and productivity levels were reduced much more by unwanted interruptions than with the male group.

Table 15. Pearson correlation comparing productivity and unwanted interruptions for male and female gender groups.

Parameters	Variables	Statistics	Productivity	Unwanted Interruptions
Male	Productivity	Pearson Correlation	1	−0.359 **
		Significance (2-tailed)	-	<0.001
		n	2413	2289
	Unwanted Interruptions	Pearson Correlation	−0.359 **	1
		Significance (2-tailed)	<0.001	-
		n	2289	2369
Female	Productivity	Pearson Correlation	1	−0.371 **
		Significance (2-tailed)	-	<0.001
		n	2247	2144
	Unwanted Interruptions	Pearson Correlation	−0.371 **	1
		Significance (2-tailed)	<0.001	-
		n	2144	2301

Note: ** Statistical significance *p*-value < 0.05.

3.5. Workgroups

The workgroup sizes included solo occupants, shared with one other, shared with 2 to 4, shared with 5 to 8, and shared with over 8 people. The solo group obtained the best mean score for productivity, with a mean value of 5.32. The worst productivity scores were achieved for the more than 8 group, with a mean score value of 4.99 (see Table 16). The best unwanted interruption scores were obtained in the shared with one other group, with the mean score value of 3.85. The worst unwanted interruption score was found in the shared with more than 8 group, with a mean value of 4.25 (Figure 6).

Table 16. Descriptive statistics of productivity and unwanted interruptions for different workgroups.

Parameters	Variables	Mean	Std. Deviation	*n*
Solo occupant	Productivity	5.32	1.607	1056
	Unwanted interruptions	4.09	1.675	1028
Shared with one other	Productivity	5.21	1.637	367
	Unwanted interruptions	3.85	1.670	357
Shared with 2–4	Productivity	5.13	1.740	781
	Unwanted interruptions	4.09	1.603	798
Shared with 5–8	Productivity	5.13	1.726	685
	Unwanted interruptions	4.08	1.696	702
Shared with more than 8	Productivity	4.99	1.724	1720
	Unwanted interruptions	4.25	1.668	1740

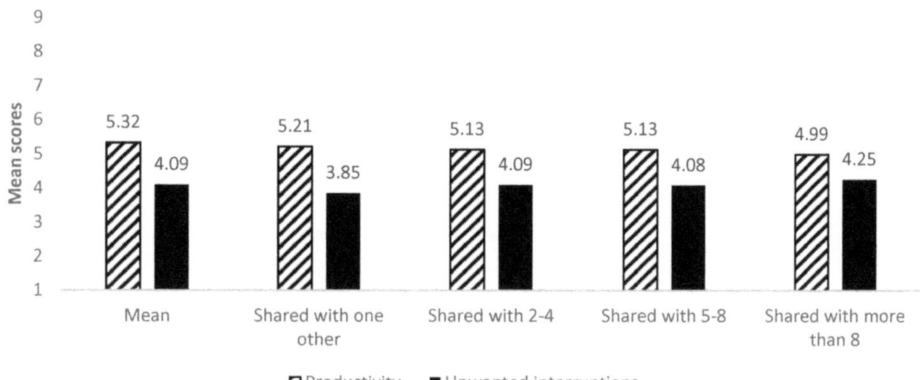

Figure 6. Comparative analysis of mean scores of productivity and unwanted interruptions among various workgroups.

Significant differences were found between the mean scores for both perceived productivity and unwanted interruptions when comparing various workgroups, as illustrated in Table 17. Except for solo offices, productivity and unwanted interruptions had negative correlations, meaning that the worse were the unwanted interruptions, the worse the productivity scores. Although unwanted interruptions were worse in solo offices in comparison with the shared with one other offices, the productivity scores were better in offices with a solo occupant in comparison to offices shared with one another. This indicated that there was an anomaly in the correlations between unwanted interruptions and productivity when small sized offices are compared.

Table 17. One-way ANOVA analysis comparing productivity and unwanted interruptions for different workgroups.

Parameters	Variables	Sum of Squares	df	Mean Square	F	*p*-Value
	Between Groups	77.330	5	15.466	5.390	<0.001 **
Productivity	Within Groups	13,230.293	4611	2.869	-	-
	Total	13,307.624	4616	-	-	-
	Between Groups	58.633	5	11.727	4.245	0.001 **
Unwanted Interruptions	Within Groups	12,782.466	4627	2.763	-	-
	Total	12,841.099	4632	-	-	-

Note: ** Statistical significance *p*-value < 0.05.

Multiple comparisons and Tukey honestly significant difference (HSD) analysis for productivity showed that only when comparing the smallest workgroup and the largest workgroup was there a significant difference in productivity values (Table 18). For unwanted interruptions only when comparing the second smallest workgroup and the largest workgroup was there a significant difference (Table 19).

Table 18. Multiple comparisons and Tukey honestly significant difference (HSD) analysis for productivity scores among different workgroups.

(I) Work-Group	(J) Work-Group	Mean Difference (I–J)	Std. Error	*p*-Value	95% Confidence Interval	
					Lower Bound	Upper Bound
Solo occupant	Shared with 1 other	0.115	0.103	0.874	−0.18	0.41
	Shared with 2–4	0.194	0.080	0.147	−0.03	0.42
	Shared with 5–8	0.192	0.083	0.190	−0.04	0.43
	Shared with more than 8	0.331 **	0.066	0.000 **	0.14	0.52
Shared with 1 other	Solo occupant	−0.115	0.103	0.874	−0.41	0.18
	Shared with 2–4	0.079	0.107	0.977	−0.23	0.38
	Shared with 5–8	0.077	0.110	0.981	−0.24	0.39
	Shared with more than 8	0.216	0.097	0.231	−0.06	0.49
Shared with 2–4	Solo occupant	−0.194	0.080	0.147	−0.42	0.03
	Shared with 1 other	−0.079	0.107	0.977	−0.38	0.23
	Shared with 5–8	−0.002	0.089	1.000	−0.25	0.25
	Shared with more than 8	0.137	0.073	0.420	−0.07	0.35
Shared with 5–8	Solo occupant	−0.192	0.083	0.190	−0.43	0.04
	Shared with 1 other	−0.077	0.110	0.981	−0.39	0.24
	Shared with 2–4	0.002	0.089	1.000	−0.25	0.25
	Shared with more than 8	0.139	0.077	0.458	−0.08	0.36
Shared with more than 8	Solo occupant	−0.331 **	0.066	0.000 **	−0.52	−0.14
	Shared with 1 other	−0.216	0.097	0.231	−0.49	0.06
	Shared with 2–4	−0.137	0.073	0.420	−0.35	0.07
	Shared with 5–8	−0.139	0.077	0.458	−0.36	0.08

Note: ** Statistical significance *p*-value < 0.05.

Table 19. Multiple comparisons and Tukey HSD analysis for unwanted interruptions among different workgroups.

(I) Work-Group	(J) Work-Group	Mean Difference (I–J)	Std. Error	*p*-Value	95% Confidence Interval	
					Lower Bound	Upper Bound
Solo occupant	Shared with 1 other	0.240	0.102	0.176	−0.05	0.53
	Shared with 2–4	−0.005	0.078	1.000	−0.23	0.22
	Shared with 5–8	0.009	0.081	1.000	−0.22	0.24
	Shared with more than 8	−0.163	0.065	0.125	−0.35	0.02
Shared with 1 other	Solo occupant	−0.240	0.102	0.176	−0.53	0.05
	Shared with 2–4	−0.244	0.106	0.191	−0.55	0.06
	Shared with 5–8	−0.231	0.108	0.268	−0.54	0.08
	Shared with more than 8	−0.403 **	0.097	<0.001 **	−0.68	−0.13
Shared with 2–4	Solo occupant	0.005	0.078	1.000	−0.22	0.23
	Shared with 1 other	0.244	0.106	0.191	−0.06	0.55
	Shared with 5–8	0.013	0.086	1.000	−0.23	0.26
	Shared with more than 8	−0.159	0.071	0.223	−0.36	0.04
Shared with 5–8	Solo occupant	−0.009	0.081	1.000	−0.24	0.22
	Shared with 1 other	0.231	0.108	0.268	−0.08	0.54
	Shared with 2–4	−0.013	0.086	1.000	−0.26	0.23
	Shared with more than 8	−0.172	0.074	0.189	−0.38	0.04
Shared with more than 8	Solo occupant	0.163	0.065	0.125	−0.02	0.35
	Shared with 1 other	0.403 **	0.097	<0.001 **	0.13	0.68
	Shared with 2–4	0.159	0.071	0.223	−0.04	0.36
	Shared with 5–8	0.172	0.074	0.189	−0.04	0.38

Note: ** Statistical significance *p*-value < 0.05.

For all workgroups, negative Pearson correlation coefficients were identified with statistical significance lower than 0.001. The Pearson correlation coefficients for solo occupants, shared with one other, shared with 2–4, shared with 5–8, and shared with more than 8 were −0.304, −0.374, −0.291, −0.397 and −0.421, respectively, which indicates the highest sensitivity to unwanted interruptions for occupants in offices shared with more than 8 people.

4. Discussion

As organisations adopt larger workgroup environments to aid collaboration, workers may become overloaded with distractions [25]. Distracting workplaces can cost companies millions of dollars in lost productivity. Thus, creating a distraction free workplace will not only have major financial implication for employers, but also help employees with wellness, work-life balance, frustration, and stress.

Our research showed that respondents who were less troubled by unwanted interruptions were also more likely to experience the workplace as supportive for their productivity. In all categories of the investigated groups, significant correlations between unwanted interruptions and productivity were found. In all subsets of our dataset the more frequent the unwanted interruptions were, the worse the perceived productivity. This showed that the ability to concentrate has a substantial influence on perceived productivity [26]. This further reinforced that the physical environment and acoustic performance of the office support employee productivity. The negative correlations between unwanted interruptions and knowledge worker productivity provided further evidence that the acoustic design of the office environment has a decisive effect on user experience in buildings.

The present study investigated the sensitivity of various user groups to office distractions in multiple building types. It was found that commercial office workers had higher tolerance to unwanted interruptions when compared to institutional office occupants. Because of the differences in the nature of the work in commercial and institutional buildings, the expectations and occupant experiences differ when the two building users are compared [18]. Haynes [10] reported that office occupants with the higher variety of responsibilities in the office may be the least affected by distraction. As academics have multiple responsibilities, the lower tolerance of unwanted interruptions among institutional office users may be explained by the point that Haynes [10] makes.

Higher tolerance of interruptions was also found among occupants in sustainable buildings. This finding is similar to previous studies that showed user of sustainable buildings tend to tolerate deficiencies rather more than users of more conventional buildings [27]. The higher tolerance of sustainable building users to interruptions in our study may be a result of the buildings' sustainable performance overriding other aspects. As noted by Onyeizu [28], a sustainable office environment should have good acoustics to enable easy communication and an appropriate soundscape while reducing possible unwanted noise and disturbance. Another plausible reason may be that sustainable buildings have a better acoustic performance than conventional buildings, which create less disturbance and distractions to building users.

Occupants over 30 years of age showed higher sensitivity to unwanted interruptions in office environments. Likewise, female participants in our study showed higher sensitivity to unwanted interruptions, which is in line with previous studies that demonstrated genders have different responses to the negative aspects of open-plan offices in terms of distractions [21]. Consistent with our findings, Kalgotra, Sharda, and McHaney [29] demonstrated that interruptions caused by technologies significantly increased the task completion time particularly among young adult females and middle-aged males. The higher tolerance of interruptions among young adult female and middle-aged males may be associated with multi-tasking abilities [30]. One of the most important implications of this finding is that office designers need to account for the nature of the work and multitasking responsibilities for individuals when designing office spaces. Background noise and speech intelligibility in office environments particularly need further exploration when designing for more sensitive building users such as female and over 30 years old users.

This highlights the importance of matching work patterns with user preferences and requirements. The matching of office user needs with space provision can only be accomplished through understanding the way people work in office environments and identifying their specific requirements. What is important is building user involvement in the evaluation of building performances and in the creation of the space.

Our study also demonstrated that unwanted interruptions occurred more frequently in larger workgroups. The only exception was the solo offices in comparison with the shared with one other office. Although unwanted interruptions were worse in the solo offices than shared with one other, productivity scores were better in the solo offices. This exception may be due to the other desirable attributes that solo occupant offices have, such as privacy and bigger storage spaces, so even with more frequent unwanted interruptions, the productivity was boosted in solo offices. When comparing the five workgroup sizes, occupants in offices shared with more than 8 people had the lowest tolerance of unwanted interruptions. Open-plan offices may seem aesthetically pleasing, stimulate relationship-building interactions, and increase collaboration, yet offices shared with fewer people obtained higher perceived productivity [19]. As demonstrated by previous studies, self-interruption has a higher rate in larger open plan offices [31]. The performance loss in large workgroups is associated with the lack of speech privacy and reduced concentration as a result of overhearing other conversations [32]. Therefore, creating flexible open-plan offices that simultaneously enable both effective collaboration and undisturbed concentration may remain one of the highest aspirations of office design.

Employees rate their productivity based on individual productivity, rather than team productivity or organisational productivity [26]. Employees also acknowledge the physical environment of their office as an important, inspiring factor for individual productivity. A healthy workplace environment improves productivity and reduces employee-related costs. While organisations prefer open-plan layouts to facilitate collaboration and productive interaction, designers must remain aware that opportunities for concentration are of huge importance. Although some of these are universally relevant, it is essential to consider the particular work processes within the organisation when determining which specific aspects to focus upon.

Thoughtful design and practice can reduce the impact of unwanted interruptions on building users' lives and improve the quality of working environment. Productivity was a factor that was studied in this paper, yet, sleep, fatigue, irritability, headaches and stress are others that interfere with human life. Including rooms that have higher levels of control over noise pollution is a one sensible solution to create productive buildings for more sensitive building users.

Productivity in buildings is affected by various personal and environmental parameters, for example, productivity seems to be higher in better ventilated buildings. However, it was shown in this paper that unwanted interruption is one of those parameters that influences productivity. Our statistical analysis with a large sample showed a correlation between the two variables, meaning that unwanted interruption greatly influences productivity and that is one of the important building design measures for productive buildings. Using a controlled data sample to keep all parameters the same for comparative studies is extremely difficult in this type of studies. However, to increase the validity of results, using a larger sample is recommend to future studies.

5. Conclusions

We have shown that distractions influence perceived productivity in different ways, depending on the physical and behavioural environments. Further evidence regarding knowledge-worker needs and preferences was provided in this research. The correlation between productivity, and unwanted interruptions in all groups tested proved to be significantly negative. Higher tolerance to unwanted interruptions was found among commercial office workers and in sustainable buildings. Our findings indicate the need for extensive investigation of the acoustic performance of commercial and institutional buildings at the architectural design stage. It was also shown that the acoustic design of office environments directly influences perceived productivity and special consideration is needed when designing workplaces for more sensitive building users such as females and those over 30 years old. These results highlight the importance of the need to account for user demographics when designing office spaces. Our study demonstrated higher levels of unwanted interruptions in bigger workgroups. This finding particularly contradicted conventional wisdom that bigger workgroups improve overall user performance and productivity by enhancing collaborations and interactions among building users. This research also provided a simple, effective framework for measuring the influence of physical features and demographics on user perceptions and productivity—the results of this work yield progressive improvement strategies for the relationship between office users and buildings. One limitation of the study is the sample size for the two age groups. In our database, the numbers of 30 years or over were much higher than those under 30 years. It is suggested that smaller age categories should be used in survey questions to balance the number of participants in different age groups.

Author Contributions: Conceptualization, M.K., E.O.R. and G.B.; methodology, M.K., E.O.R. and G.B.; software, M.K., E.O.R. and G.B.; validation, M.K., E.O.R. and G.B.; formal analysis, M.K., E.O.R. and G.B.; investigation, M.K., E.O.R. and G.B.; resources, M.K., E.O.R. and G.B.; data curation, M.K., E.O.R. and G.B.; writing—original draft preparation, M.K., E.O.R. and G.B.; writing—review and editing, M.K., E.O.R. and G.B.; visualization, M.K., E.O.R. and G.B.; All authors have read and agreed to the published version of the manuscript.

Funding: This research received no external funding.

Institutional Review Board Statement: Various ethical approval was obtained for the data collection that was conducted over the 12-year period.

Informed Consent Statement: Informed consent was obtained from all subjects involved in the study.

Data Availability Statement: The data presented in this study are available on request from the corresponding author.

Conflicts of Interest: The authors declare no conflict of interest.

References

1. Jami, S.; Zomorodian, Z.S.; Tahsildoost, M.; Khoshbakht, M. The effect of occupant behaviors on energy retrofit: A case study of student dormitories in Tehran. *J. Clean. Prod.* **2020**, *278*, 123556. [CrossRef]
2. Gou, Z.; Khoshbakht, M.; Mahdoudi, B. The impact of outdoor views on students' seat preference in learning environments. *Buildings* **2018**, *8*, 96. [CrossRef]
3. Haynes, B. The impact of the behavioural environment on office productivity. *J. Facil. Manag.* **2007**, *5*, 158–171. [CrossRef]
4. Haynes, B. Office productivity: A theoretical framework. *J. Corp. Real Estate* **2007**, *9*, 97–110. [CrossRef]
5. Bland, C.J.; Center, B.A.; Finstad, D.A.; Risbey, K.R.; Staples, J.G. A theoretical, practical, predictive model of faculty and department research productivity. *Acad. Med.* **2005**, *80*, 225–237. [CrossRef] [PubMed]
6. Schneider, E. *Noise in Figures*; Risk Observatory Thematic Report; Commissioned by the European Agency for Safety and Health at Work; European Agency for Safety and Health at Work: Bilbao, Spain, 2005; ISBN 92-9191-150-X. Available online: https://osha.europa.eu/en/publications/report-noise-figures (accessed on 19 December 2020).
7. Mohammadi, H.; Alimohammadi, I.; Roshani, S.; Pakzad, R.; Abdollahi, M.B.; Dehghan, S.F. The Effect of Occupational Noise Exposure on Blood and Biochemical Parameters: A Case Study of an Insulator Manufacturer in Iran. *Electron. Physician* **2016**, *8*, 1740–1746. [CrossRef] [PubMed]
8. Ismaila, S.; Odusote, A. Noise exposure as a factor in the increase of blood pressure of workers in a sack manufacturing industry, Beni-Suef University. *J. Basic Appl. Sci.* **2014**, *3*, 116–121. [CrossRef]
9. Mostaghaci, M.; Mirmohammadi, S.; Mehrparvar, A.; Bahaloo, M.; Mollasadeghi, A.; Davari, M. Effect of Workplace Noise on Hearing Ability in Tile and Ceramic Industry Workers in Iran: A 2-Year Follow-Up Study. *Sci. World J.* **2013**, *2013*, 923731. [CrossRef] [PubMed]
10. Haynes, B. Impact of workplace connectivity on office productivity. *J. Corp. Real Estate* **2008**, *10*, 286–302. [CrossRef]
11. Stokols, D.; Clitheroe, C.; Zmuidzinas, M. Qualities of work environments that promote perceived support for creativity. *Creat. Res. J.* **2002**, *14*, 137–147. [CrossRef]
12. Mark, G.; Gudith, D.; Klocke, U. The cost of interrupted work: More speed and stress. In Proceedings of the SIGCHI Conference on Human Factors in Computing Systems, Florence, Italy, 5–10 April 2008.
13. Adamczyk, P.D.; Bailey, B.P. If Not Now, When? The Effects of Interruption at Different Moments within Task Execution. In Proceedings of the SIGCHI Conference on Human Factors in Computing Systems, Vienna, Austria, 24–29 April 2004.

14. Francis, M. Transforming spaces in higher education: Academic office environments and their influence on staff, a case-study from Melbourne, Australia. In Proceedings of the 53rd International Conference of the Architectural Science Association, Roorkee, India, 28–30 November 2019.
15. Astle, D. *Rewording the Brain: How Cryptic Crosswords Can Improve Your Memory and Boost the Power and Agility of Your Brain*; Allen & Unwin: Crows Nest, Australia, 2018.
16. Brumby, D.P.; Janssen, C.P.; Mark, G. How do interruptions affect productivity? In *Rethinking Productivity in Software Engineering*; Springer: Berlin/Heidelberg, Germany, 2019; pp. 85–107.
17. Spira, J.B.; Feintuch, J.B. *The Cost of Not Paying Attention: How Interruptions Impact Knowledge Worker Productivity*; Report from Basex; Basex: New York, NY, USA, 2005.
18. Khoshbakht, M.; Baird, G.; Rasheed, E.O. The influence of work group size and space sharing on the perceived productivity, overall comfort and health of occupants in commercial and academic buildings. *Indoor Built Environ.* **2020**, 1420326X20912312. [CrossRef]
19. Rasheed, E.O.; Khoshbakht, M.; Baird, G. Does the Number of Occupants in an Office Influence Individual Perceptions of Comfort and Productivity?—New Evidence from 5000 Office Workers. *Buildings* **2019**, *9*, 73. [CrossRef]
20. Haynes, B.; Suckley, L.; Nunnington, N. Workplace productivity and office type: An evaluation of office occupier differences based on age and gender. *J. Corp. Real Estate* **2017**, *19*, 111–138. [CrossRef]
21. Yildirim, K.; Akalin-Baskaya, A.; Celebi, M. The effects of window proximity, partition height, and gender on perceptions of open-plan offices. *J. Environ. Psychol.* **2007**, *27*, 154–165. [CrossRef]
22. Göçer, Ö.; Candido, C.; Thomas, L.; Göçer, K. Differences in Occupants' Satisfaction and Perceived Productivity in High-and Low-Performance Offices. *Buildings* **2019**, *9*, 199. [CrossRef]
23. Baird, G. *Sustainable Buildings in Practice: What the Users Think*; Routledge: London, UK, 2010.
24. Baird, G.; Leaman, A.; Thompson, J. A comparison of the performance of sustainable buildings with conventional buildings from the point of view of the users. *Archit. Sci. Rev.* **2012**, *55*, 135–144. [CrossRef]
25. Haapakangas, A.; Hongisto, V.; Hyönä, J.; Kokko, J.; Keränen, J. Effects of unattended speech on performance and subjective distraction: The role of acoustic design in open-plan offices. *Appl. Acoust.* **2014**, *86*, 1–16. [CrossRef]
26. Maarleveld, M.; De Been, I. The influence of the workplace on perceived productivity. In Proceedings of the EFMC2011 10th EuroFM Research Symposium: Cracking the Productivity Nut, Vienna, Austria, 24–25 May 2011.
27. Leaman, A.; Bordass, B. Are users more tolerant of 'green' buildings? *Build. Res. Inf.* **2007**, *35*, 662–673. [CrossRef]
28. Onyeizu, E. The delusion of green certification: The case of New Zealand green office buildings. In Proceedings of the 4th New Zealand Built Environment Research Symposium (NZBERS), Auckland, New Zealand, 14 November 2014.
29. Kalgotra, P.; Sharda, R.; McHaney, R. Understanding the impact of interruptions on knowledge work: An exploratory neuroimaging study. In Proceedings of the 2016 49th Hawaii International Conference on System Sciences (HICSS), Koloa, HI, USA, 5–8 January 2016.
30. Haynes, B. The impact of office comfort on productivity. *J. Facil. Manag.* **2008**, *6*, 37–51. [CrossRef]
31. Dabbish, L.; Mark, G.; González, V.M. Why do I keep interrupting myself? Environment, habit and self-interruption. In Proceedings of the SIGCHI Conference on Human Factors in Computing Systems, Vancouver, BC, Canada, 7–12 May 2011.
32. Roelofsen, P. Performance loss in open-plan offices due to noise by speech. *J. Facil. Manag.* **2008**, *6*, 202–211. [CrossRef]

Article

Increasing Green Infrastructure in Cities: Impact on Ambient Temperature, Air Quality and Heat-Related Mortality and Morbidity

Matthaios Santamouris [1,*] and Paul Osmond [2]

[1] Anita Lawrence Chair High Performance Architecture, School Built Environment, University of New South Wales, Sydney, NSW 2052, Australia

[2] School Built Environment, University of New South Wales, Sydney, NSW 2052, Australia; p.osmond@unsw.edu.au

* Correspondence: m.santamouris@unsw.edu.au

Received: 9 November 2020; Accepted: 2 December 2020; Published: 7 December 2020

Abstract: Urban vegetation provides undeniable benefits to urban climate, health, thermal comfort and environmental quality of cities and represents one of the most considered urban heat mitigation measures. Despite the plethora of available scientific information, very little is known about the holistic and global impact of a potential increase of urban green infrastructure (GI) on urban climate, environmental quality and health, and their synergies and trade-offs. There is a need to evaluate globally the extent to which additional GI provides benefits and quantify the problems arising from the deployment of additional greenery in cities which are usually overlooked or neglected. The present paper has reviewed and analysed 55 fully evaluated scenarios and case studies investigating the impact of additional GI on urban temperature, air pollution and health for 39 cities. Statistically significant correlations between the percentage increase of the urban GI and the peak daily and night ambient temperatures are obtained. The average maximum peak daily and night-time temperature drop may not exceed 1.8 and 2.3 °C respectively, even for a maximum GI fraction. In parallel, a statistically significant correlation between the peak daily temperature decrease caused by higher GI fractions and heat-related mortality is found. When the peak daily temperature drops by 0.1 °C, then the percentage of heat-related mortality decreases on average by 3.0% The impact of additional urban GI on the concentration of urban pollutants is analysed, and the main parameters contributing to decrease or increase of the pollutants' concentration are presented.

Keywords: green infrastructure; urban trees; heat mitigation; heat-related mortality

1. Introduction

Rapid urbanisation in combination with climate change cause serious environmental hazards such as increase of the urban temperature, elevated concentration of air pollutants, storms, increased droughts or excessive precipitation, and poses serious health problems in cities. In 2017, the global urban population reached 4 billion, representing almost 55% of the world population, and it is expected to increase to close to 68% by 2050 [1].

Urban overheating is the most documented phenomenon of climate change. Increased urban temperatures compared to the surrounding rural or suburban built environment are experimentally documented for more than 450 large cities in the world [2]. The magnitude of the urban overheating may be as high as 10 °C, with an average value close to 5–6 °C [3]. Recent research has shown that the magnitude of urban overheating increases considerably during heatwaves because of the important local climatic synergies [4].

Urban overheating negatively affects the cooling energy demand of buildings, peak electricity demand, concentration of air pollutants, heat-related mortality and morbidity and urban vulnerability levels [5]. It is well-documented that urban overheating is associated with an additional cooling energy penalty close to 0.7 kWh/m^2 of city surface and degree of temperature increase, while the additional required peak electricity demand is estimated close to 21 ($\pm$10.4) W per person and degree of temperature increase [6,7].

The impact of urban overheating on health is reasonably well-documented. Higher levels of ambient temperature are associated with increased mortality and morbidity as the human thermoregulation system cannot offset very high ambient temperatures [8]. As a result, numerous recent researches have shown that the health risk is substantially higher in urban than in rural environments, while the risk of heat-related mortality in warmer urban neighbourhoods is 6% higher than in cooler precincts [9,10]. Heat-related health issues combined with epidemic and non-communicable diseases, such as cancer, diabetes, mental disorders, cardiovascular and respiratory diseases, increase the health burden in cities and are expected to increase considerably in the low- and middle-income societies in the near future [11,12].

Urban overheating has a serious impact on air quality. Elevated ambient temperatures enhance photochemical reactions with hydrocarbons and NO_x generating ozone [13]. In parallel, the increased operation of power plants during the warm season considerably increases the emission of harmful pollutants [14].

Outdoor air pollution is the most serious environmental hazard for human health and citizens' wellbeing. Exposure to air pollutants like the ground-level ozone, particulate matter, NO_x and SO_2, may result in serious respiratory and cardiovascular health problems and increased mortality [15]. According to the World Health Organisation, outdoor pollution is responsible for about 8 million deaths per year, while in about 56% of cities in the developed world and 98% in the developing countries, the concentration of harmful pollutants like ozone, particles and NO_x exceeds the thresholds of the World Health Organisation (WHO) [14,16]. Ozone is a highly toxic pollutant strongly affecting human health, contributing in 2010 to about 1.2 million premature respiratory deaths, equivalent to 20% of total respiratory deaths [17]. While systematic environmental policies have resulted in a considerable decrease of the concentration of most of the atmospheric pollutants, the concentration of ground-level ozone continues to rise as a result of intensive urbanisation and temperature increase. It is characteristic that in 75 Chinese cities, ozone concentration increased from 69 to 75 ppbv between 2013 and 2015, while the percentage of non-compliant cities increased from 23% to 39% [18]. Urban overheating has a serious impact on the concentration of ground-level ozone [19,20]. Several studies have documented that during extreme climatic conditions, the ozone concentration increases up to 20%, while studies in Japan have shown that the long-term variability of the peak ground-level ozone is related to the variation of the ambient temperature and wind speed [21–23]. Considerably higher ozone concentrations and frequencies of extreme ozone episodes are forecast for the coming years because of the expected temperature increase, and the corresponding modification of the chemical reaction rates [24]. It is predicted that the average background concentration of ground-level ozone may increase in the mid-latitudes of the Northern Hemisphere, by close to 85 ppb in 2100 compared to the current levels of 35–50 ppb [25]. Much higher frequencies of high ozone events are also predicted by other studies. In particular, it is foreseen that in four major Canadian cities, the frequency of future ozone episodes may increase by 50% by 2050 and 80% by 2080, while in Tucson Arizona, the predicted increase by the end of the century may reach 400% [26,27].

To prepare efficient responses to the above threats and challenges and to minimise vulnerability and associated risks, actions involving efficient mitigation and adaptation policies based on multisector approaches are necessary [28]. Several mitigation technologies to counterbalance the impact of urban overheating have been proposed and implemented in numerous large-scale projects [29]. Mitigation technologies involve the use of reflecting and super-cool materials, additional greenery, use of evaporative and transpiration cooling systems, solar control and shading devices, and the use of low-temperature natural sinks [30–33]. Monitoring of a high number of large-scale mitigation projects has demonstrated that the implemented technologies can reduce the peak ambient temperature up to 2.5 °C and may contribute significantly to decrease building cooling energy demand, peak electricity load and heat-related mortality and morbidity [34–36].

There are many definitions of the term Green Infrastructure (GI). According to the European Environmental Agency, "Green infrastructure as a term is used for a network of green features that are interconnected and therefore bring added benefits and are more resilient" [37], while Connop et al. [38] define GI as "a network of natural and semi-natural green spaces such as forests, parks, green roofs and walls that can provide nature-based and cost-effective solutions". It is widely accepted that an increase of the green urban infrastructure, and particularly tree cover, improves urban resilience [39,40]. Trees provide urban overheating reduction, pollutant removal, carbon sequestration, retention and detention of stormwater runoff, while improving residents' health [41–45].

Given the multiple benefits of increased GI, large-scale projects aiming to increase the fraction of tree cover occur in many cities. Large tree planting initiatives are undertaken in cities like New York, Chicago, Los Angeles, Sydney, Melbourne, etc. [46–48]. However, there is an uncertainty on the magnitude of the benefits provided by increased GI policies, and the associated disservices, like the emission of biogenic volatile organic compounds (VOCs), the increased use of water and the additional economic cost [49]. There is a clear need to evaluate the extent to which increased GI mitigates urban heat, reduces urban pollution and decreases health problems; in parallel, it is important to provide additional holistic information and knowledge to optimise future planting strategies in cities.

The present article aims to investigate in a global way the biophysical effects on regional urban climate, air quality and heat-related mortality and morbidity triggered by the increase of urban green infrastructure and tree cover. In parallel, it aims to quantify the mitigation potential of increased tree cover during the daytime and night-time, the potential decrease of heat-related mortality and to analyse the impact of additional tree cover on air pollution.

2. The Nature of the Study

Twenty-nine articles investigating the impact of increased green infrastructure on the ambient temperature, air pollution and heat-related mortality and morbidity were reviewed and analysed. The studies reported 55 fully evaluated scenarios of increased GI for 29 cities and precincts (see the Map Figure S1 in the Supplementary Materials). Nineteen of the cities are in North America, eight in Australia, seven in Europe, three in Asia and one in South America. Two of the studies reported the combined impact of GI on ambient temperature levels and heat-related mortality and morbidity, 11 studies on the ambient temperature and heat-related mortality, 5 on ambient temperature and heat-related morbidity, 4 on ambient temperature and air quality, 10 on air quality while 25 studies investigated only the impact on the ambient temperature. All the characteristics and results of the 55 scenarios of increased GI are reported in Table 1.

Existing literature provides information on the climatic, air quality and health impact of potential additional green infrastructure. We considered and analysed all studies evaluating the specific impact of GI in quantitative terms. Unfortunately, no quantitative studies are available on the other potential benefits of GI.

The overall article is divided into three parts. The first part analyses and discusses the impact of the potential increase of urban tree cover on the ambient temperature. Specific analysis on the temperature drop during the daily peak at 15:00 and at night is performed. Concerning the few studies reporting the afternoon (17:00) daily temperature drop, local temperature data are used to estimate the peak daily temperature at 15:00. The second part of the analysis investigates the impact of additional GI on the concentration of urban pollutants, focusing mainly on ground-level ozone and particulate matter. Finally, the third part of the study investigates the relation between the additional GI and health and with the levels of heat-related mortality and morbidity.

3. The Impact of Increased Green Infrastructure on Ambient Temperature-Mitigation Potential

3.1. Identity of the Existing Studies

Trees contribute to decrease the ambient temperature through evapotranspiration and shading. In parallel, when impervious surfaces like asphalt and concrete are replaced by vegetation, the stored heat during the day and the emitted sensible heat during the day and night are seriously reduced. The potential temperature decrease because of the increased GI fraction depends on the difference of the thermal balance between the non-vegetated control/reference scenario and the vegetated one. It is mainly affected by the specific climatic conditions, the availability of soil moisture, the type of vegetation and the way it is distributed in a city. As a result, increased vegetation fractions may contribute to reduce the ambient temperature during the night or the day or both, or even cause some warming effects under specific conditions.

Forty-six studies evaluating the impact of increased tree fraction in 30 cities and locations were analysed and evaluated. The study refers to seven North American cities and their suburbs (New York, Philadelphia, New Orleans, Detroit, Dallas, Toronto and the North Eastern part of the US), five Australian cities (Melbourne, Sydney, Brisbane, Darwin and Parramatta), three European cities (Vienna, Austria, Bochum and Stuttgart, Germany), three Asian cities (Hong Kong, Singapore and Tehran), and Sao Paolo in South America. The considered tree cover varies between 1.3% in Bronx NY to 100% in Hong Kong and Melbourne, Australia, with an average GI value close to 16%. The average increase of the GI for all case studies is close to 22% and the average final GI is close to 38%. Most of the assessments are performed using mesoscale modelling techniques like the Weather Research Forecast (WRF), or MM5, while four studies are based on microscale modelling like ENVI-met and i-Tree. These studies considered either a homogeneous increase of the tree fraction across the city or have considered potential additional parks in specific urban zones. Although the assumptions and the computational methods may differ substantially between the studies, there is a minimum homogeneity concerning the outputs and the results permitting comparison of the assessments. All the relative details of the considered studies are given in Table 1.

The impact of central and pocket parks as well as of trees distributed in streets on the ambient temperature of a district in Central Sao Paolo, Brazil, is discussed in Reference [50]. The tree cover is considered to increase from zero to 11%. Simulations were performed using the ENVI-met microscale climatic model and it was found that during the summer, trees in urban canyons can reduce the average peak daily ambient temperature in the area by 0.6 °C, and in the central and pocket parks by 0.4 °C.

The potential impact of the tree cover increases from 24.9% to 26.2% in Bronx, New York, NY, USA, is discussed in Reference [51]. Using the i-Tree software, it is found that the average peak daily temperature decrease may not exceed 0.09 °C. A second study aiming to investigate the cooling potential of various mitigation measures in numerous urban zones of NY City USA is presented in Reference [52]. The study used the MM5 mesoscale model to assess the potential decrease of the peak ambient temperature when hard urban surfaces are replaced by trees. A spatially variable increase of the tree fraction among the various urban precincts, from 6.2% to 14.4%, was considered. It was found that at 15:00, additional tree cover may drop the ambient temperature between 0.2 to 0.5 °C. A third study for NY, USA, investigated the impact of increased tree fraction from 10% to 20% and 30% using the MM5 mesoscale climatic tool [53]. It found that the average temperature drop during the peak daily period is close to 0.15 and 0.4 °C for the two considered scenarios. A quite similar study for Dallas, TX, USA, estimated the impact when GI increases by 7.5% from a base value of 27.5%, and found that the peak daily temperature drop is close to 0.38 °C [54]. Simulations for several US cities in the North Eastern part of the country considering a 20% increase of the tree cover to almost 40% are reported in References [55,56]. Using the WRF mesoscale climatic model, it was calculated that the maximum temperature decrease is close to 0.4 °C. A study performed for two suburban areas of Toronto, Canada, is presented in Reference [57]. Mesoscale, WRF and a microscale models were used to simulate the climatic impact when the tree fraction increases from 18% to 27%. The estimated average drop of the peak ambient temperature was 0.2 and 0.3 °C for the two precincts, while for the high greenery scenario, the corresponding temperature drop was 1.61 and 1.29 °C, respectively. Considerably higher temperature drop values were estimated for the night period. Simulations using WRF were performed during heatwave periods, to assess the impact of the GI increase by 10% in the cities of New Orleans, Philadelphia and Detroit [58]. The initial tree cover varied between 15 and 25% and the final between 25% to 35%. It was found that the average afternoon temperature may decrease by 0.21 °C in New Orleans, 0.32 °C in Philadelphia and 0.1 °C in Detroit.

In Australia, the potential decrease of the ambient temperature in the City of Melbourne was investigated in three studies [59–61]. Five specific scenarios considering a final tree cover of 5%, 36%, 38%, 49% and 100%, compared with an initial tree cover of 15%, were analysed in Reference [59]. Simulations were performed using the mesoscale model UCM-TAPM. It was calculated that the average peak daily temperature drop may vary between 0.25 and 1.8 °C, while the decrease of the tree cover from 15% to 5% may raise the average peak daily temperature by 0.2 °C. The study reported in Reference [60] considered an increase of the mixed forest fraction in the city from zero to 20% and 50%. Simulations were performed for the January 2009 heatwave using the mesoscale climatic model WRF [60]. It was reported that during the night-time, the amplitude of the Urban heat island (UHI) decreased between 0.5 °C and 5.0 °C for a 20% to 50% rise of the GI. The daytime decrease of the ambient temperature as well as the impact on the daily magnitude of the UHI was found to be almost negligible. However, a very significant decrease of the ambient temperature ranging between 1 °C and 3.5 °C was calculated for the night period. A third study for the City of Melbourne evaluated the impact of increasing the trees fraction from an initial value of 24% to 28%, 32% and 40% during the period of several recent heatwaves [61]. It reported that the maximum temperature reduction occurs during the night-time because of the reduced daytime heat storage and the decreased release of sensible heat by the greenery during the night. The maximum temperature drop during the night period was close to 0.28, 0.38, and 1.08 °C for the three greenery scenarios. The temperature drop during the daytime was considerably lower and did not exceed 0.25 °C [61]. The impact of an increased urban trees fraction on the ambient temperature of a subtropical Australian city, Brisbane, is discussed in Reference [62]. The study considered a decrease of the tree cover from 45% to zero, 'no vegetation', during a ten-year period using the CCAM CSIRO mesoscale climatic model. It calculated that the maximum temperature decrease happens during the night-time and is close to 1.83 °C. The daily average temperature decrease was estimated close to 0.99 °C, while the decrease during the peak hours was close to 0.44 °C. Similar results were obtained for the suburban areas of Archerfield and Logan. Two studies aiming to investigate the mitigation potential of additional greenery in Sydney, Australia are reported in References [63,64]. Using the WRF climatic tool, it was estimated that when two million additional trees are added to the city, the decrease of the peak daily temperature in Parramatta, Western Sydney, is close to 1 °C [64]. The second study estimated the temperature drop in the City of Sydney considering an increase of the trees fraction by 55%, from 20% to 75% [63]. Using ENVI-met, it was calculated that the potential drop of the peak daily temperature was close to 1.2 °C. ENVI-met was also used to calculate the mitigation potential of increased green infrastructure, from 19% to 39% in the tropical Australian city of Darwin. The average estimated daily maximum drop of the peak temperature was close to 0.5 °C [65].

In Europe, the impact on the urban climate of a central park covering 25% of the city of Bochum, Germany, was assessed in Reference [66]. Using WRF simulations, it was estimated that the new park may decrease the average peak daily ambient temperature by 0.45 °C. A similar study for the city of Stuttgart, Germany, also based on WRF, found that an increase of the GI from 18% to 30% may decrease the average peak daily temperature by 0.13 °C, while close to the new parks, the temperature drop may be as high as 2.0 °C [67]. In a similar study carried out for the city of Vienna, the MICLIMA 3 tool was used to assess the impact of increasing the size of urban parks by 20% [68]. A maximum night-time temperature drop of 1 °C was found.

Simulations using the WRF model were carried out for the city of Singapore, to estimate the cooling capacity of a potential increase of the average urban forestry from 15% to 35%, or 40% of the dense urban areas [69]. During the average peak daytime hours, the temperature decrease was found close to 0.3 °C, while it was considerably higher during the night and close to 1.0–2.0 °C. A significant increase of the humidity levels was also calculated as a result of the higher tree cover. The climatic impact of a potential increase of tree fraction by 20%, from almost 8% to 28%, in Tehran, Iran, was assessed in Reference [70]. Using WRF, it was estimated that the maximum daily drop of the ambient temperature was close to 0.6 °C, while phenomena of night-time warming, compared to the reference scenario, were reported. In a similar study for the city of Hong Kong, China, it was calculated that an increase of tree cover from zero to 100% could decrease the peak ambient temperature by 1.6 °C [71].

3.2. Association of the Temperature Drop and GI

The analysis of all the studies provided information on the potential drop of the average peak daily and the night-time temperature for 40 and 25 case studies, respectively. Also, 20 studies provided information on the drop of the ambient temperature during both the day and night-time. We have selected to use the temperature drop at 15:00 as a proxy, given that it corresponds to the time of the day that presents the maximum ambient temperature and potentially the maximum overheating load. Such a proxy helps to identify the potential temperature drop caused by GI during the peak time. In parallel, this proxy is the most frequently reported parameter by the relevant studies.

The potential temperature drop during the night-time caused by the GI is of specific interest as it helps to cool down the city and reduce the ambient temperature during the next day. It also highly affects comfort and energy conditions during the night-time. To characterise the cooling potential of GI during the night-time, we propose to use as a proxy the maximum achieved temperature drop between the sunset and sunrise time. Such a proxy helps to identify the maximum cooling potential of GI during the night-time and is usually reported by all relevant articles.

The relation between the average peak daily temperature drop, ΔT_{15}, and the initial, GI_{in}, and final considered tree cover, GI_{fin}, can be described by a multiparameter correlation as below:

$$\Delta T_{15} = a + b\, GI_{in} + c\, GI_{fin}, \tag{1}$$

where $a = -0.0145$, $b = -0.01068$ and $c = 0.0167$. The R^2 of the correlation is equal to 0.83, the p-value for both independent parameters is very low, and the correlation is statistically significant. Figure 1 plots the reported against the predicted values, from Equation (1), of the average daily peak temperature drop. Supplementary Table S1 provides in detail all statistical indicators of the correlation given in Equation (1).

The relation between the reported drop of the average peak daily temperature and the increase of the tree cover is shown in Figure 2. As shown, the average peak daily decrease of the ambient temperature, ΔT_{15}, may be expressed as a linear function of the increase of the tree cover ΔGI:

$$\Delta T_{15} = d \cdot \Delta GI \tag{2}$$

where $d = 0.0181$. The R^2 of the correlation is equal to 0.90, and the p-value is very close to zero. Supplementary Table S2 provides all the statistical characteristics of Equation (2). To minimize the impact of outliers in the correlation, the available data of ΔT_{15} are grouped into four clusters of ΔGI, as shown in Figure 3. When the average value of ΔT_{15} for each cluster is correlated against the corresponding mean ΔGI value, the resulting correlation can be better presented by a second-degree polynomial:

$$\Delta T_{15mean} = 0.00013\, \Delta GI^2 + 0.0079\, \Delta GI \tag{3}$$

with an R^2 equal to 0.98. The statistical characteristics of Equation (3) are given in Supplementary Table S3.

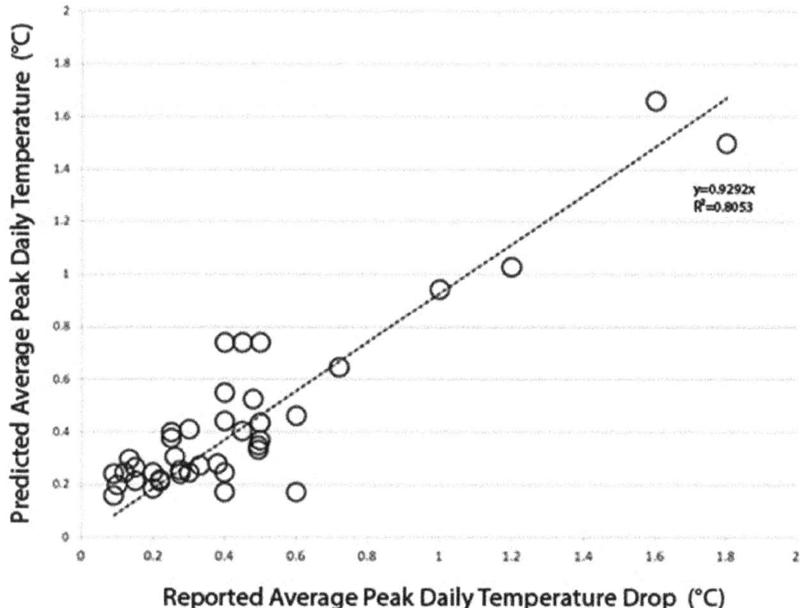

Figure 1. Predicted average peak daily ambient temperature predicted from Equation (1) against the reported values.

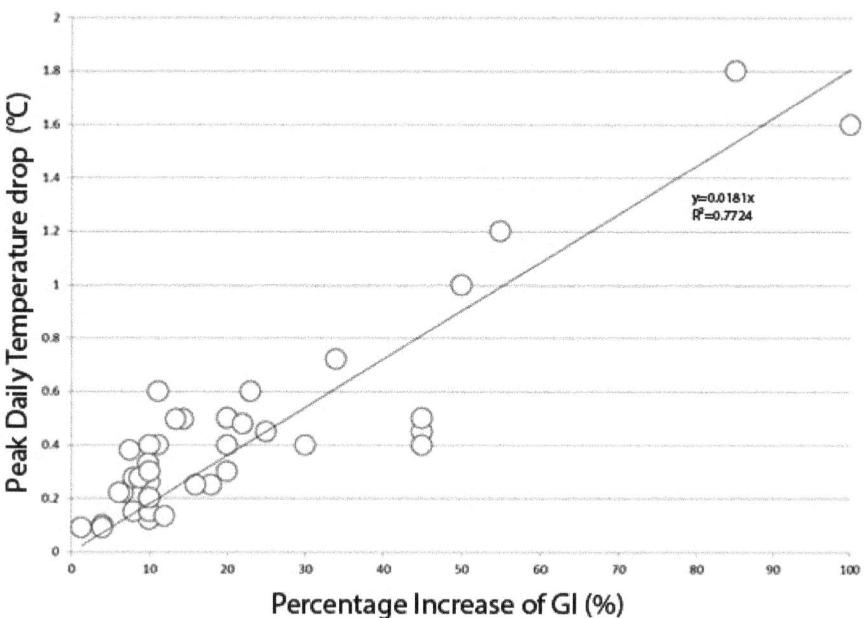

Figure 2. Correlation between the average peak daily temperature drop at 15:00 p.m. against the corresponding increase of the tree cover.

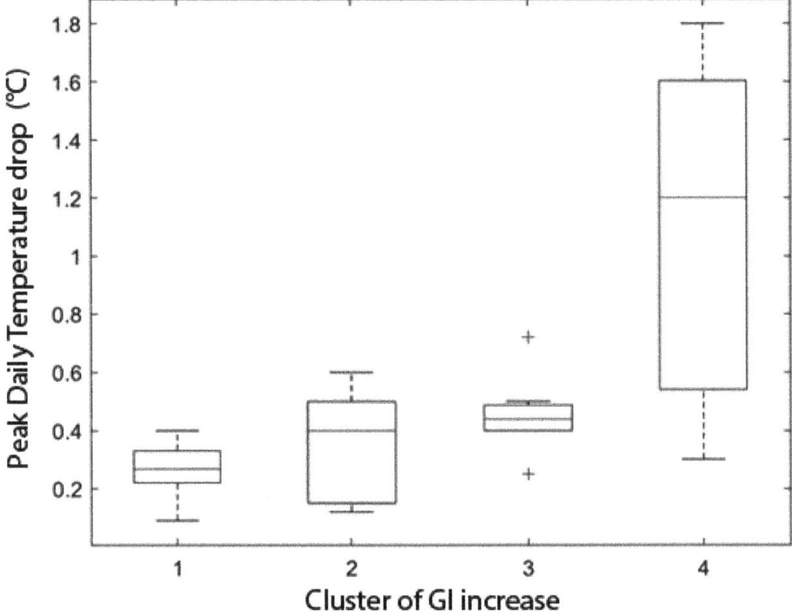

Figure 3. Box plots of the maximum temperature drop at 15:00 for four clusters of tree-covering increases. Cluster 1: 0 < GI < 10, Cluster 2: 10 < GI < 20, Cluster 3: 20 < GI < 30, Cluster 4: GI > 30.

The observed non-linearity between the temperature decrease and the tree cover is in agreement with the experimental results reported in Reference [72], where it was found that the temperature decrease caused by greenery was a nonlinear function of the increasing canopy cover, with the higher cooling when the tree cover exceeded 40%.

A similar expression as Equation (1) was also obtained between the average maximum temperature drop during the night-time period and the initial and final GI values. In particular,

$$\Delta T_{ngt} = a_1 + b_1\ GI_{in} + c_1\ GI_{fin} \tag{4}$$

where $a_1 = 0.4123$, $b_1 = -0.0355$ and $c_1 = 0.0262$. The R^2 of the correlation is equal to 0.86, the p-value for both independent parameters is very low, and the correlation is statistically significant. The statistical characteristics of Equation (4) are given in Supplementary Table S4.

Figure 4 presents the correlation between the average maximum drop of the ambient temperature at night, ΔT_{ngt}, and the corresponding increase of GI.

$$\Delta T_{ngt} = d_1 \cdot \Delta GI \tag{5}$$

where $d_1 = 0.0344$. The R^2 of Equation (3) is equal to 0.76, and the p-value is very close to zero. As observed, the slope of the ambient temperature drop is considerably higher during the night than the daytime. This is clearly shown in Figure 5 comparing the slope of the temperature drop during the day, left, and night-time, right. Figure 5 includes data only from case studies reporting both the day and night-time temperature drop.

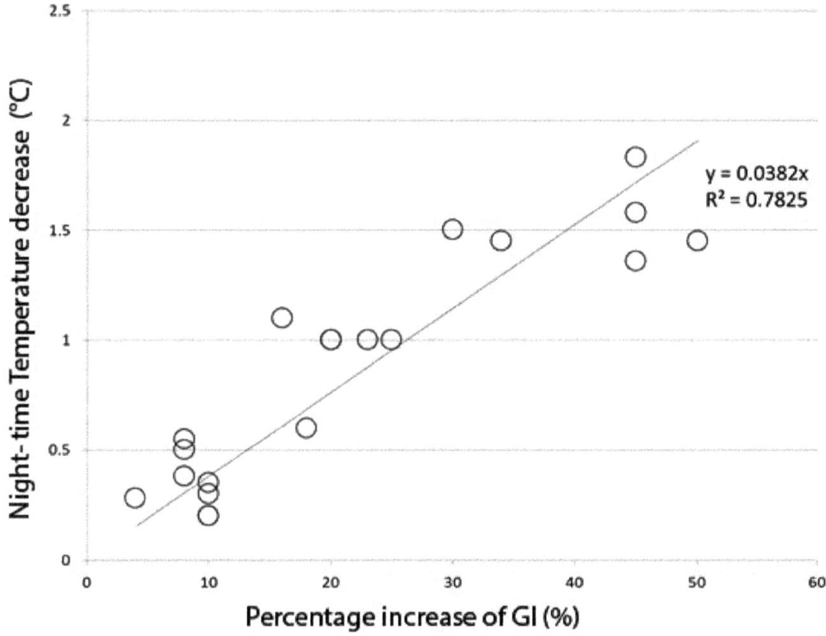

Figure 4. Correlation between the maximum drop of the ambient temperature at night against the corresponding increase of the tree covering.

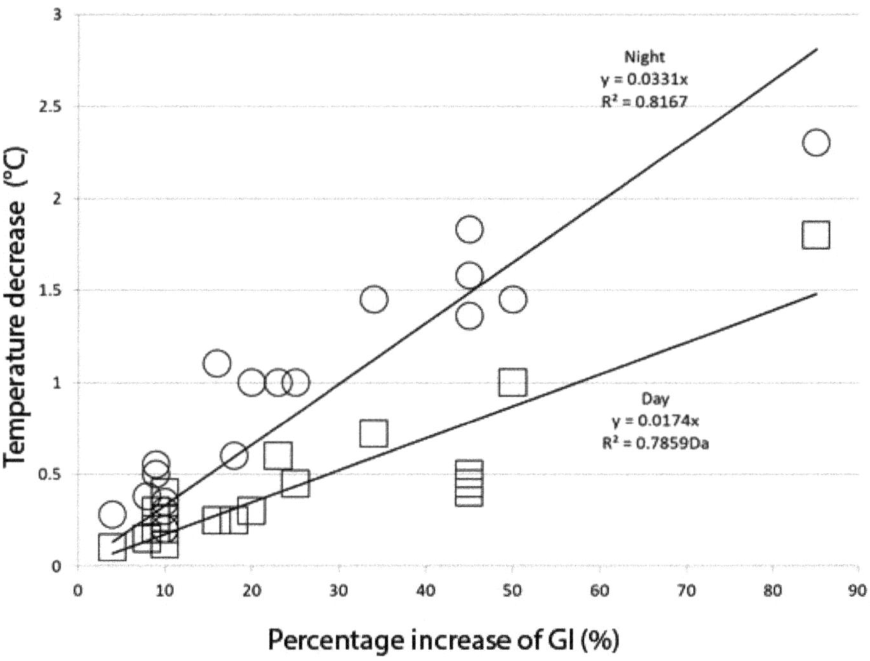

Figure 5. Daily peak and night-time temperature drop as a function of the considered increase of the GI. Circles correspond to the night-time temperature drop, and squares to the peak daytime decrease of the temperature.

3.3. Discussion and Conclusions on the Mitigation Potential of GI

Analysis of the case studies may lead to the following conclusions:

(a) The maximum potential drop of the average daily peak temperature caused by the increased tree cover in cities may not exceed 1.8 °C even if the green infrastructure increases up to 100%. For a reasonable increase of the GI by 20%, the average expected peak temperature drop is close to 0.3 °C.

(b) During the night, the maximum ambient temperature decrease corresponding to a GI rise of 80% may not exceed 2.3 °C, while the temperature drop for a GI increase by 20% is close to 0.5 °C.

(c) Of the 22 studies reporting both the daytime and night-time temperature drop, 19 studies reported a higher temperature drop during the night than the day. This is due to two main reasons: (a) increased tree cover considerably reduces the daytime stored heat in the ground and the corresponding release of heat by the ground during the night, and (b) the released sensible heat by the trees during the night is considerably lower than the corresponding heat released by impervious urban surfaces [59,61]. During the daytime, evapotranspiration is the main cooling mechanism, however its impact on the surface energy budget may not be enough to compensate and counterbalance heat fluxes due to advection and sensible heat released by the impervious surfaces. The calculated magnitude of evaporation losses depends highly on the humidity of the ground assigned by the models and the humidity content of the atmosphere [73,74].

(d) Several articles report warming effects during the night because of the increased tree cover [60,70]. Increase of the night-time temperature may be as high as 2 °C and is explained by the decreased sky view factor in the urban canyons where trees are located, reducing the escape of long-wave radiation [75].

(e) The calculated magnitude of the temperature drop highly depends on the considered urban landscape assumptions about the location and the sizing of the additional tree cover. Simulations considering patches of urban greenery fully covering several grid cells, when compared to simulations considering a mixture of impervious and green surfaces in the grid cells, present higher evapotranspiration and less heat storage during the day, and thus a considerably lower night-time temperature. In parallel, when urban trees are surrounded by other quite warm urban impervious surfaces and/or zones of high anthropogenic heat release, they present a considerably lower cooling potential [76]. However, it has to be recognised that it is difficult to evaluate sources of uncertainty regarding the simulation assumptions and accuracy around the coupling and evaluation of the specific land conditions in cities.

(f) Simulations carried out with the WRF mesoscale model are highly influenced by the considered parameterisation scheme. The tool offers four parameterisation schemes presenting different ways to characterise the interaction between the land surfaces and the lower atmosphere: The bulk urban parameterisation, the single layer and the multi-layer parameterisation schemes [77–79], and another multilayer scheme considering outdoor–indoor interactions [80]. Simulations carried out for the city of Stuttgart using different parameterisation schemes revealed important differences in the obtained results [67].

(g) Deployment of additional trees in cities may raise the levels of ambient relative humidity and deteriorate the levels of thermal comfort. Especially in tropical cities where relative humidity is considerably high, a further rise of the ambient humidity levels may be a serious problem. Simulations for Singapore have shown that deployment of additional greenery may increase the levels of relative humidity by up to 8% [69].

4. Impact of Increased Green Infrastructure on Urban Pollution Levels

4.1. Introduction and Identity of the Studies

Increase of urban tree cover is associated with multiple dynamic and chemical changes in the lower atmosphere. Vegetation accumulates particulate matter by sedimentation, interception and impaction, while it also absorbs gaseous pollutants such as NO_x and O_3 mainly through the stomata of the leaves [81]. Absorption of pollutants varies significantly between tree species, while removal of the pollutants depends on several parameters like the concentration of the pollutants, tree species traits such as leaf area, stickiness, porosity and roughness, length of growing season, the canopy structure, the size of pollutant particles, meteorological conditions and possible precipitation [82,83]. In parallel, trees emit biogenic VOCs (BVOCs), like isoprenoids, monoterpenes and sesquiterpenes, which are much more reactive than the anthropogenic VOC emissions, and about 2–3 times more reactive than the weighted average emissions from petrol combustion [84–86]. Isoprenoids interact with nitrogen oxides and seriously affect the concentration of ground-level ozone, while monoterpenes and sesquiterpenes increase the production of airborne particles during the hot summer period [70]. As already discussed, increase of tree cover results in lower ambient and surface temperatures, which may have either a positive or negative impact on pollutant concentration. Positive impacts include the reduction of BVOC emission rates [87,88] and the decrease of the rate of photochemical reactions [89]. Negative impacts are related to reduced buoyancy and turbulent mixing caused by trees, resulting in lower Planetary Boundary Layer Heights (PBLH), where pollutants are mixed, which may increase the concentration of the primary gaseous pollutants like NO_x, and primary airborne particles [90].

Apart from the reduced surface temperature, trees may decrease wind speed in the built environment, vertical mixing and atmospheric dispersion, increasing the concentration of pollutants in the urban canyons and canopies, while it may affect the circulation of sea breezes in coastal areas and the transport of precursors and pollutants in the city domain [91–93].

It is widely accepted that additional urban green infrastructure has a non-linear feedback on urban pollution levels, while the impact may be either positive or negative depending on the urban form and vegetation characteristics.

Eleven studies assessing the potential impact of increased urban green infrastructure on pollutant concentrations were analysed [51,55,56,93–98]. The main characteristics and results of the studies are depicted in Table 1. Seven of the studies refer to USA cities or regions (Bronx, NY, Kansas City, New York, Atlanta, California South Coast Air Basin and two studies for North Eastern cities), three studies are for UK cities and regions (Greater London area, West Midlands and Glasgow), and one study for Melbourne, Australia. Seven of the studies analyse the potential impact of increased urban green infrastructure on ground-level ozone, six on particulate matter (PM_{10}, $PM_{2.5}$) and one on NO_2, SO_2 and CO. The characteristics and the findings of the studies are analysed below.

4.2. Impact of Additional Greenery on Particulate Matter

Relevant studies can be classified in two groups: Those focusing on the estimation of the removal rate of particulate matter mainly because of the deposition of the particles [51,83,98], while the second group of studies focuses on the impact of the additional trees on the spatial concentration of particles in a city or region taking into account the global phenomena influencing pollutant concentration [84,97].

The potential impact of enhanced urban green infrastructure in Bronx, New York, NY, USA, was investigated in Reference [51]. A well-defined planting program for each census block of the city was presented and its impact on the actual and future (2030) $PM_{2.5}$ concentration was evaluated for three different tree mortality scenarios: no mortality, 4% and 8% annual mortality. While the current tree cover in the city is 2470 ha, or 22.7% of the total land, it is estimated that by 2030, it will increase up to 24.9%. The i-Tree assessment tool was used to estimate and characterise the benefits of the additional green infrastructure [99]. It was reported that during 2010, trees removed 5.1 tonnes/year of $PM_{2.5}$, while the increase of the green infrastructure by 2030 may raise the removal of pollutants to between

5.6 and 6.2 tonnes/year, corresponding to an increase of between 9.8% and 21.6% depending on the selected tree mortality scenario.

The impact of increased tree cover from 20 to 80 trees per ha on six atmospheric pollutants including $PM_{2.5}$ and PM_{10}, in the Brooklyn industrial precinct of Melbourne, Australia, was assessed in Reference [83]. Calculations were also performed using the i-Tree assessment tools. It was estimated that the existing trees remove up to 7 kg of $PM_{2.5}$ and 225 kg of PM_{10} per year. When new trees were considered, the removed quantities rise to 43 kg/year of $PM_{2.5}$ and 1474 kg/year of PM_{10}. This corresponds to an increase of the removed pollutants close to 514% for $PM_{2.5}$ and 555% for PM_{10}.

The impact of an increase of tree cover from 20% to 30% on the removal of PM_{10} in the Greater London Area, UK, was investigated in Reference [98]. Calculations were performed using the Urban Forest Effect Model, UFORE. It is calculated that by 2050, between 1109 to 2379 tonnes of particles can be removed by the urban greenery. This corresponds to a removal rate of between 1.1% and 2.6%.

The potential impact of enhanced green infrastructure on the concentration of $PM_{2.5}$ and ground-level ozone in Kansas City, USA, was investigated in Reference [84]. A detailed greening scenario was considered and evaluated using the coupled WRF-CMAQ mesoscale climatic simulation model. It was calculated that the increase in tree cover may raise the concentration of $PM_{2.5}$ in the city's downtown area by about 10% or 1.1 µg per cubic meter during the night-time. The calculated increased concentration of $PM_{2.5}$ is associated with the decrease of surface temperature caused by the additional trees that lowers the height of the Planetary Boundary Layer (PBL), limits the ventilation potential and increases the concentration of primary $PM_{2.5}$ close to the ground. During the daytime, the concentration of airborne particles may decrease in specific parts of the city because of the slower rate of chemical reactions caused by the lower ambient temperature.

The potential impact of additional greenery on the concentration of PM_{10} in West Midlands and Glasgow in the UK, was investigated in Reference [97]. Simulations of the pollutant concentration were performed using the atmospheric transport model FRAME. Two greening scenarios were investigated for West Midlands, involving increase of the tree cover from 3.75% to 16.5% and 54% For the low greening scenario. It was calculated that about 110 tonnes/year of airborne particles are removed from the atmosphere, while the average concentration of the primary pollutants decreases by 10% from 2.3 mg/m^3 or 18.0 mg m^{-3} of total PM_{10}, to 2.1 mg/m^3 of primary PM_{10}. When the tree cover increases to 54%, it was calculated that about 200 tonnes/year of PM_{10} are removed from the atmosphere and the total concentration may decrease by 26%.

Two scenarios of greenery increase from 3.75% to 8.0% and 14% were also investigated for the city of Glasgow. It was calculated that the moderate increase of greenery may contribute to remove 4 tonnes/year of PM_{10}, resulting in a reduction of the average PM_{10} concentration by 2%. Under the high greenery increase scenario, about 13 tonnes/year of PM_{10} are removed and average concentration decreases by 7%. Most of the pollution reduction is estimated to happen in the outskirts of the city, while a quite limited decrease of the PM_{10} concentration is calculated for the city centre because of the limited space for additional tree planting.

Analysis of the above studies leads to the following conclusions:

(a) Additional urban green infrastructure substantially increases the removal of particles from the atmosphere mainly through deposition processes. The magnitude of the removal depends on several parameters like the type and size of trees, the type of tree cover, the concentration of the particles in the atmosphere and the specific transport and climatic conditions. Particle size determines the magnitude and the process of deposition. While large particles, >10 mm in diameter, fall in the soil below the trees by sedimentation, particles with diameter between 1 to 10 mm are deposited as trees force the air flow to bend, while ultrafine particles below 1 mm are deposited by diffusion [100,101]. Numerical simulations have shown that increased greenery in canyons contributes to reduce the concentration of airborne particles in canyons up to 60%, which seems a quite high value [102].

(b) Dispersion because of the aerodynamic impact of trees may considerably decrease the concentration of airborne particles. Simulations carried out for the city of Leicester, UK, showed that aerodynamic dispersion resulted in a 9.0% reduction of $PM_{2.5}$ concentration, while deposition on trees resulted in a reduction close to 2.8% [103]. However, several studies have reported that in urban canyons, trees may change the roughness properties, resulting in a considerable increase of the concentration of particulate matter [104–106].

(c) Although additional tree cover is associated with a higher removal of particles, the concentration of particulate matter in the atmosphere may increase as additional greenery reduces the surface temperature, affecting the height of the PBL and trapping the particles in the lower atmosphere, thereby reducing ventilation and transport of pollutants. Increase of particle concentration was predicted for Kansas City but not for West Midlands and Glasgow. Analysis of relevant data from Montreal has also found that increase of the PBLH resulted in a negligible increase of the $PM_{2.5}$ daily average concentration [107]. It is evident that meteorological as well as landscape parameters determine the potential decrease of the PBL and the corresponding rise of pollutant concentration in a city. Given that most of the information is from simulation studies and not systematic in situ measurements, the accuracy of the used models to describe the complex phenomena is a serious issue to be considered.

4.3. Impact of Additional Greenery on Ground-Level Ozone

Of the seven studies evaluating the impact of additional urban green infrastructure on ground-level ozone, one study focuses on the potential removal of O_3 in Melbourne, Australia, while the rest of the studies analysed the global impact on the atmospheric concentration of ozone [83]. The Australian study found that an increase of tree cover at the Brooklyn industrial area of Melbourne from 20 to 80 trees per hectare may improve the removal of the ground-level ozone from 246 to 1885 kg per year [83].

Evaluation of the potential impact of additional greenery on ground-level ozone in Kansas City, USA, concluded that the ozone concentration may decrease in the downtown area up to 2.0 ppbv (PARTS PER BILLION BY VOLUME) during the daytime [84]. This happens as the higher vegetation fraction increases dry deposition and decreases the ambient temperature, resulting in slower rates of the relevant chemical reactions. Much higher reductions of the ozone concentration, up to 5.2 ppbv, are calculated during the night, mainly because of the titration phenomena associated with the increased NO_x concentration caused by the significant reduction of the PBLH. Increase of the ozone concentration was also calculated for some areas of the city and some parts of the overall calculation domain.

The potential impact of increased greenery cover, from 20% to 40%, on the concentration of ground-level ozone in North-Western cities of USA, and in particular from Washington DC to central Massachusetts, was investigated in References [55,56]. Simulations were performed using the Colorado State University Mesoscale Model (CSUMM) and Urban Airshed Model (UAM-IV). Calculations demonstrated a decrease of the average daytime ozone concentration close to 1 ppb, or 2.4%, and a peak decrease close to 2.4 ppb, or 4.1%. In parallel, some domains of the model calculated an increase of the average ozone concentration of up to 0.26 ppb. During the night, the ozone concentration was found to increase because of the higher deposition of NO_x and reduced wind speeds. During the day, it was calculated that both the concentration of NO_x and BVOCs increased, while the height of the PBL decreased slightly following the variation of the ambient temperature and wind speeds. Despite the increase of BVOCs caused by the higher tree cover, the increase of deposition as well as the slower chemical reactions and the higher NO_x concentrations contributed to lower the ozone concentration.

The OZIPM4 Computer Program was used to calculate the impact of tree losses in Atlanta, GA, USA, by 20% [85]. It found that decreased tree cover in the city may slightly raise the ozone concentration.

Two scenarios considering an increase of the tree cover by 10% and 30% were investigated for the city of New York, USA. Simulations were performed using the MMA mesoscale climatic model [86]. A slight decrease of the average ozone concentration, up to 4 ppb, was calculated for both the moderate and high greenery scenarios, while in several parts of the computational domain, the ozone concentration slightly increased. Given that the average benefits are similar for both the moderate and high scenarios, the authors concluded that an increase of tree cover above 10% may not offer additional benefits if the criteria are limited to the 1 h ozone maximum values.

The impact of increased tree cover on the ozone concentration in California's South Coast Air Basin, USA, was investigated in Reference [89]. Two scenarios considering an increase of the tree cover by 6% and 12% were studied. Simulations were performed using the Colorado State University Mesoscale Model (CSUMM) and the Urban Airshed Model (UAM). It was calculated that increase of tree cover contributes to a net reduction of ozone concentration up to 30 ppb, provided that the new trees are low emitters of biogenic hydrocarbons. On average, the domain-wide population-weighted exceedance exposure to ozone above the threshold of the California standard decreased by up to 14% during the afternoon hours. While most of the simulated areas presented a decrease of ozone concentration, several domains exhibited a rise of ozone concentration of up to 20 ppb. Simulations also showed that when moderate and high emission trees are considered, the ozone concentration increases considerably because of the higher BVOC emissions.

Analysis of the considered studies leads to the following conclusions:

(a) The chemistry and atmospheric dynamics determining the concentration of ground-level ozone is complex and the relationship between increased tree cover and ozone concentration is not simple. Ozone is generated as a result of photolysis of NO_2 when VOCs are present [85]. More urban green infrastructure increases the emission of BVOCs to the atmosphere. In parallel, it results in a higher dry deposition and absorption of ozone and NO_x and lower ambient temperatures that decrease the emission rate of BVOCs and slow the photochemical reactions. Lower surface temperature usually decreases the height of the PBL, blocking pollutants in the lower atmosphere like NO_x that may result in decreased ozone concentrations because of titration processes. However, an analysis carried out for Montreal, Canada, found that the decrease of the PBLH caused a negligible decrease of the ozone concentration [107]. More tree cover may result in a decrease of the wind speed in the lower atmosphere and a reduced removal of ozone.

(b) The magnitude of the dry deposition of ground-level ozone considerably affects its balance in the atmosphere [15]. It is estimated that during 2010, urban trees removed almost 523 kilotonnes/year of ozone in 55 US cities [108] and 12.87 kilotonnes/year in 87 cities in Canada [109]. The standardised removal rates of ozone can be as high as 0.4 $g/m^2/y/ppb$ [15], while the mean annual reduction of the ozone hourly concentration varies from 0.1% to 1.5% [15]. In parallel, trees absorb ozone and assimilate NO_x. Their assimilation capacity differs up to a factor of 122 between different species [110,111]. Although important models have been developed to assess the removal capacity of trees, there is considerable uncertainty regarding the reported values [15].

(c) The chemistry between NO_x and O_3 is quite complex and depends on their relative concentrations. High concentration of NO_x, mainly generated from motor traffic, react with O_3, converting it to O_2, decreasing its concentration through a titration effect. This is a potential source of uncertainties resulting in significantly contrasting results reported from experimental studies. Measurements carried out in parks and green zones found a low concentration of NO_x and a high or equal concentration of ozone compared to adjacent non-vegetated zones, while in some experiments, lower ozone concentrations were observed in green areas only after rainfall [81,112–115]. Contrary to the above studies, measurements performed in Baltimore, MD, USA, did not find substantial differences in NO_x concentration in green and adjacent open residential spaces, while the concentration of ozone was considerably lower in the green zones [116]. Similar conclusions were drawn in Reference [117], reporting measurements in Spain. As concluded in Reference [118], urban parks and forests do not significantly affect the concentration of NO_x, and a potential decrease of the ozone concentration may be the result of the increased absorption by the trees and also of the reduced surface temperature and solar radiation in parks that slows down the photochemical reactions. Differences of NO_x concentration between green and non-green zones may result because of the proximity to high-traffic areas.

(d) Chemical reactions associated with the titration of ozone are less temperature-sensitive than photochemical reactions generating the atmospheric ozone. According to Reference [55], titration is the dominant mechanism depleting ozone at the ground level, while at the upper levels of the atmosphere, ozone is created through photochemical reactions, and then is transferred to the ground level because of the vertical diffusion. It was observed that high vertical diffusion values correspond to days with the highest ozone concentration.

(e) Almost all studies concluded that increased urban tree cover results in a net decrease of the ozone concentration, however several urban zones may exhibit an increase. Most of the studies investigating the balance of ozone uptake and formation concluded that urban trees generally contribute to decrease the concentration of ground-level ozone [112,118]. However, numerous experimental studies have not found reduced ozone concentration in tree canopies compared to adjacent non-vegetated open zones [115,119].

(f) The rate of BVOC emissions from the additional trees seems to be the determinant factor regulating the atmospheric concentration of ozone. Low emitters, emitting under 2 µg/g/h of isoprene and 1 µg/g/h of monoterpenes, may result in a net decrease of the atmospheric ozone [89]. On the contrary, moderately or highly emitting trees may result in increased ozone concentrations.

5. The Impact of Increased Green Infrastructure on Health

5.1. Introduction and Identity of the Studies

Green infrastructure in the built environment supports human health mainly through urban temperature mitigation and modulation of heatwaves, decrease of pollutants and air quality improvements, aesthetic and psychological benefits, provision of spaces for social interaction and social activity [120].

There is an increasing number of studies aiming to evaluate the relationship between green infrastructure in the built environment and health [121–124]. Systematic reviews of numerous relevant articles concluded that natural environments have a positive impact on wellbeing, while there is a significant positive association between the characteristics of the green spaces around residences with perceived general and mental health [124,125].

In addition, several studies have investigated the impact of green infrastructure on mortality, morbidity and life expectancy [123,126–129]. According to Reference [122], there are four mechanisms explaining the beneficial effects of green infrastructure on mortality and life expectancy:

(a) The natural outdoor environment influences health and wellbeing through viewing and observing green outdoor spaces.

(b) Greener spaces are associated with lower pollution levels and reduced temperature in healthier environments that affect the human immune system.

(c) Natural and green outdoor spaces offer higher opportunities to perform physical activity.

(d) Natural and green outdoor spaces offer higher opportunities for social interactions.

Although the existing studies assessing the impact of green infrastructure on human mortality are based on different assumptions, assessment methodologies and the reported results are heterogenous. Systematic reviews and meta-analyses have shown that a higher proportion of residential green is associated with reduced risk of mortality from cardiovascular disease, while no impact of GI on lung cancer was found [122]. Conflicting conclusions are reported about the potential impact of green infrastructure on all-cause mortality. According to a meta-analysis of 10 relative studies, there is limited evidence of an association between reduction of all-cause mortality and residential green spaces. A second systematic review analysing six relevant articles concluded that there is limited or suggestive evidence of higher mortality risk in less green areas [130]. A third systematic review assessing five relevant papers concluded that there is strong evidence that residential greenness reduces all-cause mortality, while an analysis of the impact of neighbourhood-level factors on heat-related mortality in London, UK, concluded that the odds of mortality increase significantly in urban areas with less greenery cover [124,131]. Similar results are reported for Seoul, Korea [132].

The impact of greenery on air quality and its association with mortality and morbidity in the United States was assessed in References [133,134]. It was reported that the removal of particulate matter from trees was directly linked to health benefits and in particular, reduced mortality and morbidity, as well as reduced respiratory symptoms. It was estimated that the health benefits per hectare values are close to US$1600, while in Portland, Oregon, lower pollution levels caused by urban greenery considerably decreases respiratory problems, offering a total financial benefit of close to US$7 million/year. It was reported that one percent improvement of air quality caused by greenery may save 850 deaths per year and about 670,000 incidences of acute respiratory problems. Other similar studies have reported that urban forests in Rome decrease the concentration of ozone by 3% and contribute to save 4 deaths per year [135]. In Madrid, Spain, it was estimated that a peri-urban park contributes to decrease the concentration of ozone up to 10 μg/m^3, decreasing the relative risk of mortality by 0.9%, corresponding to about one premature death per year [136]. In London, UK, it was calculated that a greenspace of 1 $\times$ 10 km in Eastern London, contributes to remove 90.4 tonnes/y of PM$_{10}$, decreasing mortality and hospital admissions by 2 deaths per year [137]. A study focussing on the impact of pollutant removal by woodlands on heat-related morbidity and mortality in Britain has employed existing assessments on the capacity of trees to remove atmospheric pollutants like particulate matter, ozone and SO$_2$, and also identified correlations between morbidity levels and pollutant concentrations [138]. It was estimated that in Britain, woodland contributes to avoid between 4 to 6 hospital admissions and 5–7 deaths per year. Finally, a study in Melbourne, Australia, found that the odds of hospital admissions for heart disease or stroke were 37% lower among adults living in neighbourhoods with the highest tertiles of greenery compared to those living in the lowest tertiles [139].

5.2. Impact of Increased Green Infrastructure on Heat-Related Mortality

There are five studies assessing the impact of a potential increase or decrease of urban green infrastructure on heat-related mortality [58,59,64,65,140]. The impact of additional tree cover on health was assessed in terms of the potential temperature decrease, while the impact on air quality was not considered. The studies present 13 different scenarios for seven cities: New Orleans, Dallas, Philadelphia and Detroit, USA, and Melbourne, Parramatta and Darwin, Australia. The studies for New Orleans, Detroit and Philadelphia focus on the impact of increased tree cover during specific heatwaves, while the rest of the studies concentrate on the whole summer period. All studies assessed the potential decrease of the ambient temperature caused by the increased or decreased tree covering using microscale (Darwin) and mesoscale climatic modelling (all the rest). The accuracy of the mesoscale modelling highly depends on the exact description of the land use in the selected place, the size of the grid, the assumed boundary and initial conditions, the characteristics and accuracy of the sub-models used and the validation against existing climatic data in the area. Table 1 provides information on the type of mesoscale model used and the size of the considered grid. As shown, newer studies used a finer grid up to 500 $\times$ 500 m, while past studies were based on coarser grid sizes up to 2 $\times$ 2 km. All studies were validated against existing climatic data.

At a second phase, correlations between historical data of heat-related mortality and climatic parameters were used to assess the potential increase/decrease of mortality. Correlations excluded deaths not related to heat, while assessments were adjusted to consider demographic changes. The excess or anomalous daily mortality was then assessed as the difference against the expected mortality. Two different methods to calculate the excess mortality were used. For all the American cities, the mortality levels were evaluated during the oppressive hot and very hot synoptic conditions, and empirical correlations between the excess mortality and climatic parameters were obtained for each type of synoptic conditions [141]. The second method used in all Australian cities was based on global correlations between the excess heat-related mortality and the climatic conditions during the whole summer period. In both cases, the same empirical relations were used to assess mortality before and after the increase or decrease of the tree cover in each city.

All studies provided data on the potential decrease of the maximum or mean daily temperature as well as information on the mortality levels before and after the potential change of the tree cover in the cities. The results follow a relatively similar format and comparisons are possible.

Analysis of the 13 cases shows that increase of tree cover in cities may contribute to reduce the levels of heat-related mortality (HRM) between 1.5% and 49% (Supplementary Tables S1 and S6). The lowest decrease of HRM is observed in the city of Detroit, USA, and the maximum in Parramatta, Australia. A quite strong linear correlation is observed between the calculated decrease of the maximum daily temperature at 15:00 and the percentage of HRM decrease (Figure 6). On average, it is observed that decrease of the maximum ambient temperature by 1 °C reduces the HRM by about 30.5%, as in the following relation:

$$\text{Percentage Decrease of HRM} = 39.45 \times \text{DTmax} \tag{6}$$

where DTmax is the decrease of the maximum daily temperature. The correlation is statistically significant, $p = 2.9 \times 10^{-7}$ and $R^2 = 0.896$. Sensitivity analyses excluding one study at a time showed similar results, while the slope of the curve varied between 27.0 and 31.1. The statistical characteristics of Equation (6) are given in Supplementary Table S6.

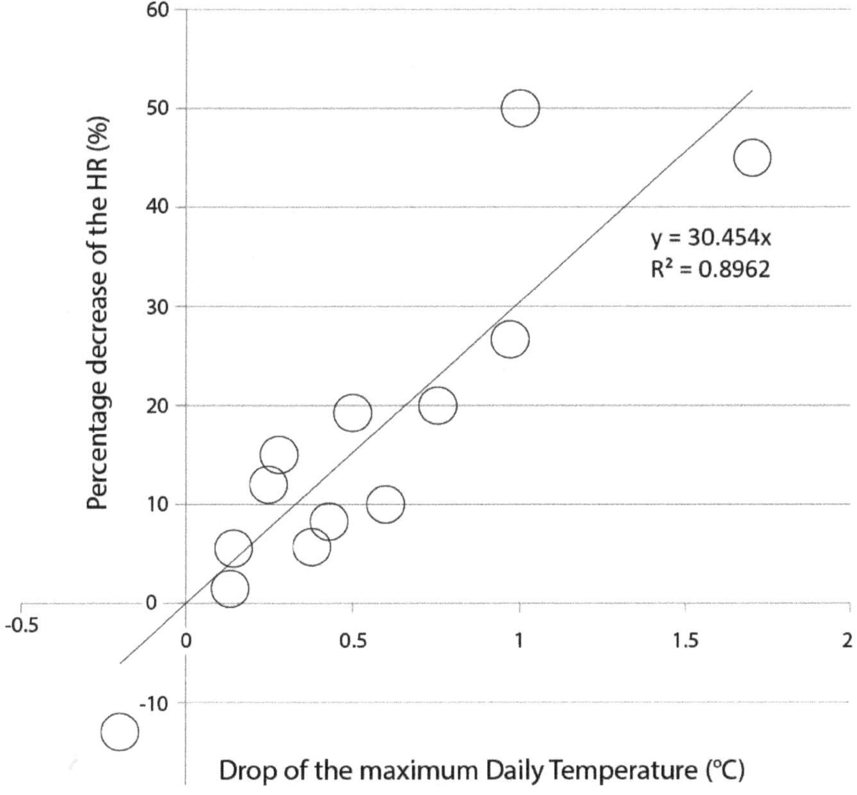

Figure 6. Correlation between the percentage decrease of the HRM against the corresponding decrease of the peak daily temperature at 15:00 p.m.

Previous research has found that socioeconomic and demographic parameters, like urban poverty and the size of the city, significantly affect the levels of heat-related mortality [142,143]. The possible association of the percentage decrease of the heat-related mortality caused by the increase of the GI was shown in Figure 6. A city with local poverty rates and population size was investigated, however, the association was not found to be strong or statistically significant.

While it is evident that there is a positive association between increased tree cover and mortality, several issues need to be considered to quantify in a more precise way the benefits in terms of mortality and morbidity.

(a) New green infrastructure in cities needs several years to grow and contribute substantially to lower urban temperatures and heat-related mortality and morbidity. According to the greening plans developed for several cities, the growing period may exceed 20 or even 30 years [140]. It is reasonable to consider that the future demographic and socioeconomic conditions will change considerably compared to the past and current times. Additionally, human technological and physiological adaptation may alter the existing relationship between the ambient temperature and heat-related mortality and morbidity [144,145]. Thus, the use of correlations between the ambient conditions and the heat-related mortality and morbidity based on past health data induces a very significant uncertainty regarding future health assessments. It is of considerable interest that new assessment studies should investigate and integrate the issues of altered socioeconomic, demographic and adaptation conditions in the global models.

(b) It is well-documented that the frequency of heatwaves and extreme summer weather events is increasing constantly [146]. In parallel, heatwaves have a synergetic effect with the urban heat island, further increasing urban temperatures and drought conditions [147,148]. Recent research has shown that during extreme or high summer ambient temperatures, the cooling potential of trees is seriously reduced or even minimised [41,149]. During excessive temperature and drought conditions, trees may close their stomata while the surface temperature of leaves increases, contributing more sensible heat to the atmosphere and increasing the ambient temperature [150,151]. Incorporation of the new stomata model of trees into global climate models has shown that the maximum ambient temperature during heatwaves may increase up to 5 °C, on top of the temperature increase caused by the increase of greenhouse gases [152]. The specific predictions highly exceed previous assessments on the potential contribution of greenery during extreme heat events [153,154]. Given the high growth period of new trees and the expected increase of the ambient temperatures, the existing relations between additional tree cover and ambient temperature may be seriously modified, altering the capacity of urban greenery to reduce heat-related mortality and morbidity. The development of new genetically modified tree species resistant to higher temperatures may offer additional cooling opportunities [155].

(c) Increased tree cover affects the concentration of harmful pollutants like O_3, NO_x, VOCs and particulate matter, while numerous studies have documented the impact of pollution levels on human health [108,109,156,157]. Existing studies evaluate the potential impact of additional green infrastructure on mortality and morbidity in terms of temperature or pollution decrease caused by the urban trees. However, it is evident that there is a synergetic impact that may enhance the magnitude of the potential benefits to health. It is characteristic that high ozone concentrations are usually observed during extreme heat events [157]. To our knowledge, no studies are available considering both the temperature and pollution variation in a synergetic way.

(d) Increase of urban greenery may affect the concentration of some pollutants in a negative way, like ground-level ozone, which could be detrimental to health [158]. It is characteristic that more than 21,000 premature deaths have been reported in the EU countries because of the increased ozone concentrations [159]. No studies are available on the potential negative impact of additional urban greenery on health, and although the impact may not be particularly high, it is important to be considered and analysed.

Table 1. Characteristics and results of all the analysed scenarios and cases of increased tree covering.

Part 1: Assessment of the Mitigation Potential, HR Mortality and Morbidity							
City	Scenario	Simulation Period	Simulation Tool	Impact on Temperature	Impact on Mortality	Impact on Morbidity	Reference
Darwin, Australia	Increase of tree cover tree cover from 19% to 39%.	Wet and dry seasons in 2016	ENVI-met 6 × 6 m	Decrease of the average peak daily temperature by 0.5 °C	Calculation of the HRM based on correlations between the local ambient temperature and mortality. Decrease of mortality by 19.3% against the base case.	Greenery reduces the annual excess hospital admissions of 40.14 to 27.51.	[65]
Paramatta, Australia	Increase of the tree cover from 20% to 70%.	Summer months 2002–2016	WRF Resolution: 500 × 500 m	Decrease of the average peak daily temperature by 1.0 °C	Calculation of the HRM based on correlations between the local ambient temperature and mortality. Decrease of mortality by 49% against the base case.	The daily excess HR morbidity decreases from 3.66 hospital admissions per day to about 2.6.	[64]

Part 2: Assessment of the Mitigation Potential and HR Mortality						
City	Scenario	Simulation Period	Simulation Tool	Impact on Temperature	Impact on Mortality	Reference
New Orleans, LA, USA	Increase of the tree cover from 25% to 35%	Heatwaves 31 May–2 June 1998 16–18 June 1998 14–8 July 2000	WRF Resolution 2 × 2 km	Reduction of the average afternoon ambient temperature by 0.21 °C	Calculation of the HRM for the offensive air mass types. Decrease of HRM by 0.82 deaths per day and 100,000 population, or 15% of the base case.	[58]
Philadelphia, PA, USA	Increase of the tree cover from 15% to 25%	Heatwaves 22-26 June 1997 3–8 July 1999 23–29 July 1999	WRF Resolution 2 × 2 km	Reduction of the average afternoon ambient temperature by 0.32 °C	Calculation of the HRM for the offensive air mass types. Decrease of HRM by 0.67 deaths per day and 100,000 population or 5.7% of the base case.	[58]
Detroit, MI, USA	Increase of the tree cover from 15% to 25%	Heatwaves 13–16 July 1995, 6–11 June 1999, 6–9 August 2001	WRF Resolution 1 × 1 km	Reduction of the average afternoon ambient temperature by 0.1 °C	Calculation of the HRM for the offensive air mass types. Decrease of HRM by 0.09 deaths per day and 100,000 population or 1.5% of the base case.	[58]
Philadelphia, PA, USA	Increase of the GI from 31% to 35%	Years 2020–2049	Based on past WRF simulations	Reduction of the average peak daily ambient temperature by 0.14 °C	Calculation of the HRM for the offensive air mass types. Decrease of the mortality by 5.6% compared to the base case.	[140]
Philadelphia, PA, USA	Increase of the tree cover from 31% to 52%	2020–2049	Based on past WRF simulations	Reduction of the average peak daily ambient temperature by 0.97 °C	Calculation of the HRM for the offensive air mass types. Decrease of the mortality by 26.6% compared to the base case. Reduction of deaths by 135–315 deaths over the period 2020 through 2049.	[140]
Melbourne CBD area Australia	Increase of the GI from 15% to 100%	2009–2050	(UCM-TAPM) Multiple one-way nesting procedure, steps of 30, 10, 3 and 1 km.	Decrease of the average peak daily temperature by 1.7 °C	Calculation of the HRM based on correlations between the local ambient temperature and mortality of elderly people. Decrease of mortality by 45% against the base case.	[59]

Table 1. *Cont.*

Part 2: Assessment of the Mitigation Potential and HR Mortality						
City	Scenario	Simulation Period	Simulation Tool	Impact on Temperature	Impact on Mortality	Reference
Melbourne CBD area Australia	Increase of the GI from 15% to 49%	2009–2050	(UCM-TAPM) Multiple one-way nesting procedure, steps of 30, 10, 3 and 1 km.	Decrease of the average peak daily temperature by 0.76 °C	Calculation of the HRM based on correlations between the local ambient temperature and mortality of elderly people. Decrease of mortality by 20% against the base case.	[59]
Melbourne CBD area Australia	Increase of the GI from 15% to 38%	2009–2050	(UCM-TAPM) Multiple one-way nesting procedure, steps of 30, 10, 3 and 1 km.	Decrease of the average peak daily temperature by 0.6 °C	Calculation of the HRM based on correlations between the local ambient temperature and mortality of elderly people. Decrease of mortality by 10% against the base case.	[59]
Melbourne CBD area Australia	Decrease of the GI from 15% to 5%	2009–2050	(UCM-TAPM) Multiple one-way nesting procedure, steps of 30, 10, and 1 km.	Increase of the average peak daily temperature by 0.2 °C	Calculation of the HRM based on correlations between the local ambient temperature and mortality of elderly people. Increase of mortality by 13% against the base case.	[59]
Melbourne CBD area Australia	Increase of the GI from 15% to 33%	2009–2050	(UCM-TAPM) Multiple one-way nesting procedure, steps of 30 m 10, 3 and 1 km.	Decrease of the average peak daily temperature by 0.25 °C	Calculation of the HRM based on correlations between the local ambient temperature and mortality of elderly people. Decrease of mortality by 12% against the base case.	[59]
Dallas, TX, USA	Increase of the GI from 27.5% to 35%	Summer months 2011	WRF Resolution: 500 × 500 m	Decrease of the average peak daily temperature by 0.38 °C	Exposure-response relationship between temperature and mortality. Decrease of mortality by 5.7% against the base case.	[54]

Part 3: Assessment of the Mitigation Potential and HR Morbidity						
City	Scenario	Simulation Period	Simulation Tool	Impact on Temperature	Impact on Morbidity	Reference
Phoenix, AZ, USA	Increase the GI by 5%	2002–2006	Zero-dimensional energy balance model	Decrease of the average daily temperature by 1.7%.	Heat-related emergency calls decreased by 17%.	[160]
Phoenix, AZ, USA	Increase the GI by 10%	2002–2006	Zero-dimensional energy balance model	Decrease of the average daily temperature by 3.6%.	Heat-related emergency calls decreased by 35%.	[160]
Phoenix, AZ, USA	Increase the GI by 15%	2002–2006	Zero-dimensional energy balance model	Decrease of the average daily temperature by 5.4%.	Heat-related emergency calls decreased by 53%.	[160]
Phoenix, AZ, USA	Increase the GI by 20%	2002–2006	Zero-dimensional energy balance model	Decrease of the average daily temperature by 7.2%.	Heat-related emergency calls decreased by 70%.	[160]

Table 1. *Cont.*

Part 3: Assessment of the Mitigation Potential and HR Morbidity

City	Scenario	Simulation Period	Simulation Tool	Impact on Temperature	Impact on Morbidity	Reference
Oslo, Norway	Zero GI in the city	Summer 2018	Satellite measured surface temperature data correlated against ambient air temperature	-	No relation between temperature and morbidity except for skin-related problems. Trees reduce the potential heat exposure for the elderly by 1.3 ± 0.1 heat risk person days.	[161]

Part 4: Assessment of the Air Quality and Mitigation Potential

City	Scenario	Simulation Period	Simulation Tool	Temperature Decrease	Impact on Air Quality	Reference
Bronx, NY, USA	Increase of GI from 24.9% to 26.2%	2010–2030	i-Tree assessment software	Decrease of the average peak daily temperature by 0.09 °C	Increase of the removal $PM_{2.5}$ by 9.8%, 13.7% and 21.6% for the high, average and low tree mortality rate.	[51]
North-eastern USA	Increase of the urban GI from 20% to 40%	Period of heatwaves	CSUMM tool	Decrease of the average peak daily temperature by 0.4 °C	Decrease in daytime hourly ozone concentrations of 1 ppb (2.4%) with a peak decrease of 2.4 ppb (4.1%). Increases in some parts of the computational domain.	[55,56]
New York, NY, USA	Increase of GI from 10% to 30%	Heatwaves	MMA mesoscale tool	Decrease of the average peak daily temperature by 0.4 °C	Domain-wide drop of about 4 ppb of ozone (132 to 128 ppb).	[86]
New York, NY, USA	Increase of GI from 10% to 20%	Heatwaves	MMA mesoscale tool	Decrease of the average peak daily temperature by 0.15 °C	Domain-wide drop of about 4 ppb of ozone (132 to 128 ppb).	[86]

Part 5: Assessment of the Air Quality

City	Scenario	Simulation Period	Simulation Tool	Impact on Air Quality	Reference
Brooklyn, Melbourne, AU	Increase of trees from 20 to 80 trees per hectare	Current	i-Tree assessment software	Increase of pollutant removal by 660%, from 577 to 4500 kg NO_2: from 68 to 964 kg; SO_2: from 22 to 125 kg; PM_{10}: from 225 to 1474 kg. $PM_{2.5}$: from 7 to 43 kg; O_3: from 246 to 1885 kg. CO: from 9 to 10 kg.	[83]
Atlanta, GA, USA	Reduction of Urban GI by 20%	Current	The OZIPM4 Computer Tool	Increase of ozone concentration (0% to 5%).	[85]
Kansas City, MO, USA	Non-quantified increase of GI	Current	WRF-CMAQ tool	Increase of $PM_{2.5}$ 10% or 1.1 $\mu g\ m^{-3}$ during the night period. Decrease of O_3 concentration by 2.0 ppbv during the daytime, and 5.2 ppbv during the night. Increase in some domains of the city.	[84]
West Midlands, UK	Increase of GI from 3.75% to 16.5%	-	Atmospheric FRAME model	Reduction of the average PM_{10} concentrations by 10% from 2.3 to 2.1 mg m^{-3}, removing 110 tonnes per year of primary PM_{10} from the atmosphere.	[97]
West Midlands, UK	Increase of GI from 3.75% to 54%	-	Atmospheric FRAME model	Reduction of the average PM_{10} concentration by 26%, removing 200 tonnes of primary PM_{10} per year.	[97]
Glasgow, UK	Increase of GI from 3.75% to 8.0%	-	Atmospheric FRAME model	Reduction of the primary PM_{10} concentrations by 2%, removing 4 tonnes of PM_{10} per year.	[97]
Glasgow, UK	Increase of GI from 3.75% to 21%	-	Atmospheric FRAME model	Reduction of the primary PM_{10} air concentrations by 7%, removing 13 tonnes of primary PM_{10} per year.	[97]

Table 1. *Cont.*

Part 5: Assessment of the Air Quality					
City	Scenario	Simulation Period	Simulation Tool	Impact on Air Quality	Reference
California's South Coast Air Basin, USA	Moderate Increase of greenery by 6%	-	CSUMM Tool	If low-emitting plants are used, the decrease of the population-weighted exceedance exposure to ozone above the Californian and National thresholds are up to 14% during peak afternoon hours, respectively.	[89]
California's South Coast Air Basin, CA, USA	High increase of greenery by 12%	-	CSUMM Tool	If low-emitting plants are used, the decrease of the population-weighted exceedance exposure to ozone above the Californian and National thresholds is up to 22% during peak afternoon hours, respectively.	[89]
Greater London Area, UK	Increase of GI from 20% to 30%.	-	The Urban Forest Effects Model (UFORE)	Deposition of 1109–2379 tonnes of PM_{10} (1.1–2.6% removal) by 2050.	[98]

Part 6: Assessment of the Mitigation Potential					
City	Scenario	Simulation Period	Simulation Tool	Temperature Decrease	Reference
Sao Paolo, Brazil	Increase of GI from zero to 11%. Street trees	Summer 2014	ENVI-met	Decrease of the peak daily temperature by 0.6 °C	[50]
Sao Paolo, Brazil	Increase of GI from zero to 11%. Pocket parks	Summer 2014	ENVI-met	Decrease of the peak daily temperature by 0.4 °C.	[50]
Brisbane, Australia	Increase of GI from zero to 45%	2000–2010	CCAM CSIRO	Decrease of the night temperature by 1.83 °C, average T by 0.99 °C, and peak temperature by 0.44 °C.	[62]
Archerfield, Australia	Increase of GI from zero to 45%	2000–2010	CCAM CSIRO	Decrease of the night temperature by 1.58 °C, average T by 0.94 °C, and peak temperature by 0.40 °C.	[62]
Logan, Australia	Increase of GI from zero to 45%	2000–2010	CCAM CSIRO	Decrease of the night temperature by 1.58 °C, average T by 0.94 °C, and peak temperature by 0.40 °C.	[62]
HK, China	Increase of the GI from zero to 100%	March 2000	MM5	Decrease of the peak daily temperature by 1.6 °C.	[71]
Melbourne, Australia	Increase of the mixed Forest from zero tree cover to by 20–50%	Heatwaves 27–30 January 2009	WRF	Decrease of the UHI intensity from 0.5 to 5 during the night-time. Non-significant differences during the daytime.	[60]
Melbourne, Australia	Increase of the trees cover by 5%, 10%, 40%. Average initial trees cover 24%, Final tree cover 28%, 32%, 40%.	12 heatwaves 1990–2014	WRF	Decrease of the night temperature by 0.28, 0.38, and 1.08 °C cooler than the control for the three scenarios.	[61]

Table 1. *Cont.*

Part 6: Assessment of the Mitigation Potential					
City	Scenario	Simulation Period	Simulation Tool	Temperature Decrease	Reference
Bochum, Germany	Increase of the GI by 25% in a specific urban zone. Initial tree cover: 9%, final tree cover: 25%	Heatwave of Summer 2010	WRF	Decrease of the peak daily temperature by 0.45 °C	[66]
Vienna, Austria	Increase of the size of urban parks by 20%	1981–2010	MUCLIMA 3	Maximum decrease of the night-time ambient temperature by 1 °C	[68]
New York City, NY, USA Mid-Manhattan West, Lower Manhattan East, Fordham Bronx, Maspeth Queens, Crown Heights, Ocean Parkway	Initial tree cover: around 20%. Additional tree cover from 6.2% to 14.4%	Heatwaves 2002	MM5	Average decrease at 3 p.m. close to 0.22 °C	[52]
		Heatwaves 2002	MM	Average decrease at 3 p.m. between 0.22 to 0.5 °C	[52]
Singapore	Increase of the trees cover on the dense and industrial zones from 5% to 60%	April and May period	WRF	Decrease of the average peak daily temperature close to 0.3 °C, and close to 1.5 °C during the night.	[69]
Stuttgart, Germany	Increase of the tree cover from 18% to 30%	Heatwaves August 2003	WRF	Decrease of the average peak daily temperature by 0.13 °C. Maximum decrease up to 2 °C.	[67]
Tehran, Iran	Increase of the tree cover from 8% to 28%	June 2016	WRF	Decrease of the average peak daily temperature up to 0.6 °C. Increase of the night-time temperature up to 1.5 °C.	[70]
Brampton, Toronto, ON, Canada	Increase of tree cover from 18% to 27%	Heatwaves 2018	WRF	Decrease of the average peak daily temperature by 0.2 °C.	[57]
Caledon, Toronto, ON, Canada	Increase of tree cover from 18% to 27%	Heatwaves 2018	WRF	Decrease of the average peak daily temperature by 0.3 °C.	[57]

5.3. Impact of Green Infrastructure on Heat-Related Morbidity

There is a substantial lack of information on the impact of greenery on heat-related morbidity. Very few studies have investigated the impact of increasing green infrastructure on heat-related morbidity [64,65,160,161]. These studies assessed the impact of the ambient temperature reduction caused by urban green infrastructure on heat-related morbidity. The characteristics and the main results of the five considered scenarios are summarised in Table 1. All studies concluded that existing greenery and potential additional tree cover contribute towards a substantial decrease of heat-related morbidity. However, the research question, as well as the methodology followed by the studies and the format of the reported results, are completely different, are not compatible and cannot be compared.

An analysis of the potential increase of the green infrastructure in Phoenix, Arizona, USA, by 5%, 10%, 15% and 20%, on ambient temperature and heat-related mortality, is presented in Reference [160]. A zero-dimensional energy balance model was used to simulate the distribution of the climatic parameters in the city for the period 2002–2006. It was calculated that when the 30.4% baseline fraction of vegetated area increases by 5%, 10%, 15% and 20%, the average 24 h temperature decreases by 1.7%, 3.6%, 5.4% and 7.2%. Unfortunately, the absolute decrease of the ambient temperature is not reported. Data on emergency calls were collected for the same period and correlations between the emergency calls and the ambient temperature and a heat index were developed. Based on these models, it was estimated that the number of heat-related emergency calls may decrease by 17%, 35%, 53% and 70%, when the fraction of the vegetation in the city increases by 5%, 10%, 15% and 20%, respectively.

The impact of urban greenery on the morbidity levels in Oslo, Norway, was analysed in Reference [161]. The ambient temperature distribution in the city during the summer of 2018 was derived using satellite-measured surface temperature data correlated against the corresponding ambient air temperature. The potential impact of high ambient temperatures on the morbidity levels of the population over 75 years old was investigated and no relation between temperature and morbidity was found except for skin-related problems. Analysis of the combined land use, ambient temperature and skin-related morbidity data, concluded that the threshold of health risk may increase from 23% to 29% in all neighbourhoods of Oslo exceeding 30 °C, in a situation where trees are replaced by the most common non-tree cover. In the case of a non-tree covered city, it was calculated that the potential heat risk exposure rises to about one day for one elderly person per removed tree.

The study reported in Reference [65] analysed and evaluated the impact of overheating as well as of increased tree cover in the tropical city of Darwin, Australia. Analysis of the existing morbidity data revealed that hospital admissions increased by 7% for every 1 °C increase of the daily maximum temperature above 27 °C. The study used microscale modelling methods and found that a possible increase of the greenery by 20% can decrease the peak ambient temperature by 0.5 °C. Using correlations between the ambient temperature and peak daily temperature, it is found that additional tree covering reduces the annual excess hospital admissions of 40.1 to 27.5 per 100,000 population.

The study reported in Reference [64] evaluated the impact of 2 million additional trees on heat-related morbidity for the city of Parramatta and the Greater Sydney area in Australia. Using epidemiological data, it was estimated that increase of the daily maximum temperature by 1 °C, increases the heat-related hospital admissions by 1.1%, while it reaches 4.6% for the days with maximum temperature above 27 °C. Using mesoscale simulations, it was found that the addition of 2 million new trees in the area may decrease the peak ambient temperature by 1 °C. Correlations between the ambient temperature and the excess heat-related morbidity showed that when 2 million trees are added, the daily excess HR morbidity decreases from 3.66 hospital admissions per day to about 2.6 hospital admissions per day and 100,000 population.

6. Conclusions

To counterbalance the continuous increase of the amplitude of urban overheating as well as of the frequency and magnitude of heatwaves, efficient urban mitigation and adaptation policies involving natural and man-made cooling technologies have to be developed and implemented.

Vegetation, and in particular, trees, offer undeniable benefits to the urban climate and urban wellbeing. Among many other benefits, they contribute to decrease ambient temperatures, reduce the concentration of harmful pollutants and limit heat-related mortality and morbidity. There is an enormous literature of experimental and theoretical studies on the potential impact of urban vegetation on urban temperature, thermal comfort, pollution control and health. Although most of the studies conclude that urban greenery offers numerous benefits, very little is known about the holistic and global contribution of urban greenery on urban climate, environmental quality and health, the possible synergies and trade-offs, while the potential problems arising from the deployment of additional greenery in cities is usually overlooked or neglected.

Existing literature provides a plethora of data and information. However, because of a serious heterogeneity in the research question asked, and the methodology followed by each study, the spectrum of the provided answers, results and conclusions is very wide, while in some cases, it is conflicting even for the same city. Thus, it is a serious source of misconceptions leading to an inappropriate sizing and establishment of tree cover in cities, potentially aggravating urban climatic, environmental and health problems.

Analysis of almost all known relevant studies performed by the present study has shown that there is an acceptable correlation between additional tree cover and the temperature drop during the peak day period and at night. The average maximum drop of the peak daily temperature may not exceed 1.8 °C, when the tree fraction reaches its maximum, while the average night-time maximum mitigation potential is much higher and close to 2.3 °C. Given the serious spatial limitations in cities, a potential increase of the tree cover by 20% may initiate a temperature decrease close to 0.3 and 0.5 °C during the daytime and night-time, respectively. The local landscape and climate conditions highly affect the potential temperature drop in a city; however, the order of magnitude of the temperature drop remains close to the previously mentioned levels. Given that most of the cooling demand is during the peak daytime period, the shift of the maximum mitigation potential of trees during the night reduces the expected cooling contribution. Potential problems of night warming and serious increase of humidity levels reported by numerous studies should not be neglected, especially when additional trees are deployed in urban canyons, reducing the sky view factor, and when additional GI is planned in humid climates.

Although the available computational tools, and in particular the mesoscale and microscale climatic prediction models, have evolved significantly, there is a high uncertainty regarding the simulation assumptions and accuracy around the coupling and evaluation of the specific land conditions in cities. Further developments are necessary to improve the description of the specific energy flows in cities, improve spatial resolutions and improve accuracy.

Additionally, for the generation of lower ambient temperatures, urban trees contribute to seriously decrease the concentration of several harmful pollutants but may increase the concentration of some others, like ground-level ozone, because of the released BVOC emissions. Although there are some available studies quantifying the impact of urban trees on heat-related mortality because of the pollutant decrease, the synergetic impact of the temperature drops and pollution control is not yet adequately evaluated and quantified. Future studies should concentrate on the holistic impact of urban greenery on heat-related mortality, also considering the potential negative impact caused by the considerable increase in the concentration of ground-level ozone.

Urban vegetation is the source of serious dynamic chemical changes in the lower atmosphere and affects the concentration of pollutants both in a positive and negative way. It decreases their concentration by accumulating and absorbing pollutants, while it emits very reactive biogenic VOCs that may raise the concentration of ground ozone and decrease the PBLH, resulting in a potential rise of NO_x and particulate matter. Existing studies evaluating the impact of additional urban greenery on the concentration of particles provide conflicting findings. Although the absorption and deposition of particles seems to considerably reduce their concentration in many cities, the potential increase of the PBLH caused by the temperature drop is found to raise the concentration in other cities. The potential positive or negative impact of urban greenery on urban particulate matter concentration depends on the specific climatic and landscape parameters in the considered area. However, as most of the reported data are from simulation studies, the considered assumptions as well as the accuracy of the used tools highly determine the final results and conclusions. It is evident that more precise experimental studies are necessary to improve knowledge in this field.

The impact of increased urban greenery on ground-level ozone is the result of complex chemistry and atmospheric dynamics. Higher vegetation fractions increase the emission of BVOCs and the corresponding potential risk for ozone creation. On the other hand, additional greenery increases the deposition and the absorption of ozone and NO_x, while because of the temperature decrease, the emission of BVOC is reduced, the speed of the photochemical processes is slowed down and titration processes are more effective, contributing to a decrease of the ozone concentration. While most of the existing theoretical studies conclude that in general, additional greenery in cities contributes to decrease the concentration of ground-level ozone (although with some parts of the cities exhibiting a substantial increase), the conclusions are not always confirmed and validated by existing field measurements. The rate of BVOC emission of the trees seems to determine the local concentration of the pollutants, with moderately or highly emitting vegetation resulting in a considerable increase of the ozone concentration. It is evident that additional experimental studies should be carried out to better understand the impact of urban greenery on ozone levels.

Additional trees in cities considerably decrease the levels of heat-related mortality and morbidity. Analysis of all the available studies has shown that there is a statistically significant correlation between the maximum daily temperature drop caused by greenery and the percentage decrease of the heat-related mortality. On average, it is found that when the peak daily temperature decreases by 0.1 °C, then the percentage of heat-related mortality drops on average by 2.94%. Very few studies are available on the impact of additional urban greenery on heat-related morbidity and solid conclusions cannot be drawn. There is an important need for such studies and statistical data.

It is important to acknowledge that the methodologies employed to estimate the impact of lower ambient temperatures on HRM caused by an increase of the urban GI are of a static nature. Given that the trees need 20 to 30 years to fully mature, and also the continuous engineering and physiological adaptation of humans to higher temperatures, dynamic prediction algorithms are necessary to estimate the potential impact of additional greenery on health. In parallel, estimations have to be performed considering the future climatic conditions and possibly the availability of more heat-tolerating tree species.

Increase of the vegetation fraction in cities offers serious benefits to the urban climate, pollution control and health. It is also associated with adverse phenomena that may affect air quality in cities. Successful deployment of additional greenery in cities requires a full knowledge of the phenomena and complete assessment of the potential benefits and drawbacks performed using advanced and detailed tools. In parallel, scientific knowledge should considerably improve, mainly through detailed and precise experimental studies.

Supplementary Materials: The following are available online at http://www.mdpi.com/2075-5309/10/12/233/s1, Figure S1: Map of the world with cities considered in the study, Table S1: Statistical parameters of Equation (1), Table S2: Statistical characteristics of Equation (2), Table S3: Statistical Characteristics of Equation (3), Table S4: Statistical Characteristics of Equation (4), Table S5: Statistical Characteristics of Equation (6), Table S6: Maximum Daytime Decrease of the Temperature and corresponding percenage of Heat Related Mortality.

Author Contributions: M.S. has collected the material and performed the analysis. P.O. proposed conclusions and review ctitically the whole analysis and the global conclusions. The authors have equally contributed to the preparation of the paper. All authors have read and agreed to the published version of the manuscript.

Funding: This research received no external funding.

Conflicts of Interest: The authors declare no conflict of interest.

References

1. World Bank—World Development Indicators. 2018. Available online: http://data.worldbank.org/datacatalog/world-development-indicators (accessed on 14 September 2020).
2. Santamouris, M. Analyzing the heat island magnitude and characteristics in one hundred Asian and Australian cities and regions. *Sci. Total Environ.* **2015**, *512–513*, 582–598. [CrossRef] [PubMed]
3. Santamouris, M. Regulating the damaged thermostat of the Cities—Status, impacts and mitigation strategies, energy and buildings. *Energy Build.* **2015**, *91*, 43–56. [CrossRef]
4. Founta, D.; Santamouris, M. Synergies between urban heat island and heat waves in Athens [Greece], during an extremely hot summer. *Sci. Rep. Nat.* **2017**, *7*, 10973. [CrossRef] [PubMed]
5. Santamouris, M. Recent progress on urban overheating and heat island research. Integrated assessment of the energy, environmental, vulnerability and health impact synergies with the global climate change. *Energy Build.* **2020**, *207*, 109482. [CrossRef]
6. Santamouris, M.; Cartalis, C.; Synnefa, A.; Kolokotsa, D. On the impact of urban heat island and global warming on the power demand and electricity consumption of buildings—A review. *Energy Build.* **2015**, *98*, 119–124. [CrossRef]
7. Santamouris, M. On the energy impact of urban heat island and global warming on buildings. *Energy Build.* **2014**, *82*, 100–113. [CrossRef]
8. Johnson, H.; Kovats, R.S.; McGregor, G.; Stedman, J.; Gibbs, M.; Walton, H.; Cook, L.; Black, E. The impact of the 2003 heat wave on mortality and hospital admissions in England. *Health Stat. Q.* **2005**, *25*, 6–11. [CrossRef]
9. Ho, H.C.; Knudby, A.; Walker, B.B.; Henderson, S.B. Delineation of spatial variability in the temperature-mortality relationship on extremely hot days in greater Vancouver, Canada. *Environ. Health Perspect.* **2017**, *125*, 66–75. [CrossRef]
10. Schinasi, L.H.; Benmarhnia, T.; De Roos, A.J. Modification of the association between high ambient temperature and health by urban microclimate indicators: A systematic review and meta-analysis. *Environ. Res.* **2018**, *161*, 168–180. [CrossRef]
11. Vos, T.; Barber, R.M.; Bell, B.; Bertozzi-Villa, A.; Biryukov, S.; Bolliger, I.; Charlson, F.; Davis, A.; Degenhardt, L.; Dicker, D.; et al. Global, regional, and national incidence, prevalence, and years lived with disability for 301 acute and chronic diseases and injuries in 188 countries, 1990–2013: A systematic analysis for the Global burden of disease study 2013. *Lancet* **2015**, *386*, 743–800. [CrossRef]
12. Van den Bosch, M.; Sang, Å.O. Urban natural environments as nature-based solutions for improved public health—A systematic review of reviews. *Environ. Res.* **2017**, *158*, 373–384. [CrossRef] [PubMed]
13. Li, W.L.; Wan, L.C. Air quality influences by urban heat island coupled with synoptic weather patterns. *Sci. Total Environ.* **2009**, *407*, 2724–2732. [CrossRef] [PubMed]
14. Meier, P.; Holloway, T.; Jonathan, P.; Harke, M.; Ahl, D.; Abel, D.; Schuetter, S.; Hackel, S. Impact of warmer weather on electricity sector emissions due to building energy use. *Environ. Res. Lett.* **2017**, *12*, 064014. [CrossRef]
15. Sicard, P.; Agathokleous, E.; Araminiene, V.; Carrari, E.; Hoshika, Y.; De Marco, A.; Paoletti, E. Should we see urban trees as effective solutions to reduce increasing ozone levels in cities?*. *Environ. Pollut.* **2018**, *243*, e163–e176. [CrossRef] [PubMed]
16. World Health Organization. *WHO's Urban Ambient Air Pollution Database—Update 2016*; Public Health, Social and Environmental Determinants of Health Department: Geneva, Switzerland, 2016.
17. Malley, C.S.; Henze, D.K.; Kuylenstierna, J.C.I.; Vallack, H.W.; Davila, Y.; Anenberg, S.C.; Turner, M.C.; Ashmore, M.R. Updated global estimates of respiratory mortality in adults 30 Years of age attributable to long-term ozone exposure. *Environ. Health Perspect.* **2017**. [CrossRef] [PubMed]
18. Wang, Y.; Shen, L.; Wu, S.; Mickley, L.; He, J.; Hao, J. Sensitivity of surface ozone over China to 2000–2050 global changes of climate and emissions. *Atmos. Environ.* **2013**, *75*, 374–382. [CrossRef]
19. Stathopoulou, E.; Mihalakakou, G.; Santamouris, M.; Bagiorgas, H.S. Impact of temperature on tropospheric ozone concentration levels in urban environments. *J. Earth Syst. Sci.* **2008**, *117*, 227–236. [CrossRef]
20. Pyrgou, A.; Hadjinicolaou, P.; Santamouris, M. Enhanced near-surface ozone under heatwave conditions in a Mediterranean island. *Sci. Rep.* **2018**, *8*, 9191. [CrossRef]

21. Diem, J.E.; Stauber, C.E.; Rothenberg, R. Heat in the southeastern United States: Characteristics, trends, and potential health impact. *PLoS ONE* **2017**, *12*, e0177937. [CrossRef]

22. Zhang, G.; Rao, S.T.; Daggupaty, S.M. Meteorological processes and ozone exceedances in the Northeastern United States during the 12–16 July 1995 episode. *J. Appl. Meteorol.* **1998**, *37*, 776–789. [CrossRef]

23. Ooka, R.; Khiem, M.; Hayami, H.; Yoshikado, H.; Huang, H.; Kawamoto, Y. Influence of meteorological conditions on summer ozone levels in the central Kanto area of Japan. *Procedia Environ. Sci.* **2011**, *4*, 138–150. [CrossRef]

24. Mika, J.; Forgo, P.; Lakatos, L.; Olah, A.B.; Rapi, S.; Utasi, Z. Impact of 1.5 K global warming on urban air pollution and heat island with outlook on human health effects. *Curr. Opin. Environ. Sustain.* **2018**, *30*, 151–159. [CrossRef]

25. IPCC. Intergovernmental panel on climate change: Summary for policy- makers. Climate Change 2014: Impacts, Adaptation and Vulnerability. In *Contribution of Working Group II to the Fifth Assessment Report of the Inter- governmental Panel on Climate Change*; Cambridge University Press: Cambridge, UK, 2014.

26. Jacob, D.J.; Darrell, A. Winner: Effect of climate change on air quality. *Atmos. Environ.* **2009**, *43*, 51–63. [CrossRef]

27. Wise, E.K. Climate-based sensitivity of air quality to climate change scenarios for the southwestern United States. *Int. J. Clim.* **2009**, *29*, 87–97. [CrossRef]

28. McMichael, A.J. Population health: A fundamental marker of sustainable development. In *Routledge International Handbook of Sustainable Development*; Redclift, M.S.D., Ed.; Taylor and Francis Inc.: Oxford, UK, 2015.

29. Akbari, H.; Cartalis, C.; Kolokotsa, D.; Muscio, A.; Pisello, A.L.; Rossi, F.; Santamouris, M.; Synnefa, A.; Wong, N.H.; Zinzi, M. Local climate change and urban heat island mitigation techniques—The state of the art. *J. Civil Eng. Manag.* **2016**, *22*, 1–16. [CrossRef]

30. Gao, K.; Santamouris, M. On the cooling potential of irrigation to mitigate urban heat island. *Sci. Total Environ.* **2020**, *740*, 139754. [CrossRef]

31. Santamouris, M.; Young Yun, G. Recent development and research priorities on cool and super cool materials to mitigate urban heat island. *Renew. Energy* **2020**, *161*, 792–807. [CrossRef]

32. Santamouris, M.; Synnefa, A.; Karlessi, T. Using advanced cool materials in the urban built environment to mitigate heat islands and improve thermal comfort conditions. *Sol. Energy* **2011**, *85*, 3085–3102. [CrossRef]

33. Santamouris, M. Cooling the Cities—A Review of reflective and green roof mitigation technologies to fight heat island and improve comfort in urban environments. *Sol. Energy* **2014**, *103*, 682–703. [CrossRef]

34. Santamouris, M.; Ding, L.; Fiorito, F.; Oldfield, P.; Osmond, P.; Paolini, R.; Prasad, D.; Synnefa, A. Passive and active cooling for the outdoor built environment—Analysis and assessment of the cooling potential of mitigation technologies using performance data from 220 large scale projects. *Solar Energy* **2017**, *154*, 14–33. [CrossRef]

35. Haddad, S.; Paolini, R.; Ulpiani, G.; Synnefa, A.; Hatvani-Kovacs, G.; Garshasbi, S.; Fox, J.; Vasilakopoulou, K.; Neild, L.; Santamouris, M. Holistic approach towards urban sustainability: Co-benefits of urban heat mitigation in a hot humid region of Australia. *Sci. Rep.* **2020**, *10*, 14216. [CrossRef] [PubMed]

36. Santamouris, M.; Paolini, R.; Haddad, S.; Synnefa, A.; Garshasbi, S.; Hatvani-Kovacs, G.; Gobakis, S.; Yenneti, K.; Vasilakopoulou, J.; Feng, K.G. Heat Mitigation Technologies can improve sustainability in Cities. An holistic experimental and numerical impact assessment of urban overheating and related heat mitigation strategies on energy consumption indoor comfort vulnerability and heat related mortality and morbidity in cities. *Energy Build.* **2020**, *217*, 110002.

37. European Environmental Agency [EEA]. *Green Infrastructure and Territorial Cohesion*; European Environmental Agency [EEA]: Copenhagen, Denmark, 2011.

38. Connop, S.; Vandergert, P.; Eisenberg, B.; Collier, M.J.; Nash, C.; Clough, J.; Newport, D. Renaturing cities using a regionally-focused biodiversity-led multifunctional benefits approach to urban green infrastructure. *Environ. Sci. Pol.* **2016**, *62*, 99–111. [CrossRef]

39. Davies, H.J.; Doick, K.J.; Hudson, M.D.; Schreckenberg, K. Challenges for tree officers to enhance the provision of regulating ecosystem services from urban forests. *Environ. Res.* **2017**, *156*, 97–107. [CrossRef]

40. Jayasooriya, V.M.; Ng, A.W.M.; Muthukumaran, S.; Perera, B.J.C. Green infra- structure practices for improvement of urban air quality. *Urban For. Urban Green.* **2017**, *21*, 34–47. [CrossRef]

41. Santamouris, M.; Ban-Weiss, G.; Cartalis, C.; Crank, C.; Kolokotsa, D.; Morakinyo, T.E.; Muscio, A.; Ng, E.; Osmond, P.; Paolini, R.; et al. Progress in urban greenery mitigation science—Assessment methodologies advanced technologies and impact on cities. *J. Civ. Eng. Manag.* **2018**, *24*, 638–671. [CrossRef]
42. Selmi, W.; Weber, C.; Rivie're, E.; Blond, N.; Mehdi, L.; Nowak, D.J. Air pollution removal by trees in public green spaces in Strasbourg city, France. *Urban For. Urban Green.* **2016**, *17*, 192e201. [CrossRef]
43. Anav, A.; De Marco, A.; Proietti, C.; Alessandri, A.; Dell'Aquila, A.; Cionni, I.; Friedlingstein, P.; Khvorostyanov, D.; Menut, L.; Paoletti, E.; et al. Comparing concentration-based [AOT40] and stomatal uptake [PODY] metrics for ozone risk assessment to European forests. *Glob. Chang. Biol.* **2016**, *22*, 1608–1627. [CrossRef]
44. Pataki, D.E.; Carreiro, M.M.; Cherrier, J.; Grulke, N.E.; Jennings, V.; Pincetl, S.; Pouyat, R.V.; Whitlow, T.H.; Zipperer, W.C. Coupling biogeochemical cycles in urban environments: Ecosystem services, green solutions, and mis- conceptions. *Front. Ecol. Environ.* **2011**, *9*, e27–e36. [CrossRef]
45. Gascon, M.; Triguero-Mas, M.; Martínez, D.; Dadvand, P.; Rojas-Rueda, D.; Plasència, A.; Nieuwenhuijsen, M.J. Residential green spaces and mortality: A systematic 'review. *Environ. Int.* **2016**, *86*, 60–67. [CrossRef]
46. MillionTrees NYC. About MillionTrees NYC. 2017. Available online: http://www.milliontreesnyc.org/ (accessed on 17 June 2017).
47. Chicago Region Trees Initiative. ChicagoRTI. 2018. Available online: http://chicagorti.org/ (accessed on 22 March 2018).
48. City Plants. About City Plants. 2018. Available online: http://www.cityplants.org (accessed on 22 March 2018).
49. Bottalico, F.; Travaglini, D.; Chirici, G.; Garfì, V.; Giannetti, F.; De Marco, A.; Fares, S.; Marchetti, M.; Nocentini, S.; Paoletti, E.; et al. A spatially- explicit method to assess the dry deposition of air pollution by urban forests in the city of Florence, Italy. *Urban Green.* **2017**, *27*, e221–e234. [CrossRef]
50. Denise, H.S.; Duarte, P.S.; Dos Santos Gusson, C.; Abrahão, C.A. The impact of vegetation on urban microclimate to counterbalance built density in a subtropical changing climate. *Urban Clim.* **2015**, *14*, 224–239.
51. Nyelele, C.; Kroll, C.N.; Nowak, D.J. Present and future ecosystem services of trees in the Bronx, NY. *Urban For. Urban Green.* **2019**, *42*, 10–20. [CrossRef]
52. Rosenzweig, C.; Solecki, W.D.; Slosberg, R.B.; Parshall, L.; Lynn, B.; Cox, J.; Goldberg, R.; Hodges, S.; Gaffin, S.; Savio, P. *Mitigating New York City's Heat Island with Urban Forestry, Living Roofs, and Light Surfaces*; New York State Energy Research and Development Authority: Albany, NY, USA, 2006; p. 6681.
53. Christopher, J.L.; Bond, J. *A Report to North East State Foresters Association. A Plan Integrate Management of Urban Trees into Air Quality Planning*; Davey Resource Group, New York State Department of Environmental Conservation: Albany, NY, USA; Division of Lands and Forests and Division of Air Resources USDA Forest Service, Northeastern Research Station: Syracuse, NY, USA, 2002.
54. *Urban Heat Island Management Study*; Texas Trees Foundation: Dallas, TX, USA, 2017.
55. Belle Hudischewskyj, A.; Douglas, S.G.; Lundgren, J.R. *Meteorological and Air Quality Modeling to Further Examine the Effects of Urban Heat Island Mitigation Measures on Several Cities in the Northeastern*; Global Programs Division USA Environmental Protection Agency Ariel Rios Building (6205J): Washington, DC, USA, 31 January 2001; SYSAPP-01-001.
56. Nowaka, D.J.; Civerolo, K.L.; Rao, S.T.; Sistla, G.; Luley, C.J.; Crane, D.E. A modeling study of the impact of urban trees on ozone. *Atmos. Environ.* **2000**, *34*, 1601–1613. [CrossRef]
57. Berardi, U.; Jandaghian, Z.; Graham, J. Effects of greenery enhancements for the resilience to heat waves: A comparison of analysis performed through mesoscale [WRF] and microscale [Envi-met] modelling. *Sci. Total Environ.* **2020**, *747*, 141300. [CrossRef]
58. Kalkstein, L.S.; Sheridan, S. *The Impact of Heat Island Reduction Strategies on Health-Debilitating Oppressive Air Massive in Urban Areas*; U.S. EPA Heat Island Reduction Initiative: Berkeley, CA, USA, 2003.
59. Chen, D.; Wang, X.; Thatcher, M.; Barnett, G.; Kachenko, A.; Prince, R. Urban vegetation for reducing heat related mortality. *Environ. Pollut.* **2014**, *192*, 275–284. [CrossRef]
60. Imran, H.M.; Kala, J.; Ng, A.W.M.; Muthukumaran, S. Effectiveness of vegetated patches as Green Infrastructure in mitigating Urban Heat Island effects during a heatwave event in the city of Melbourne. *Weather Clim. Extrem.* **2019**, *25*, 100217. [CrossRef]

61. Jacobs, S.J.; Gallant, A.J.E.; Tapper, N.J.; Li, D. Use of cool roofs and vegetation to mitigate urban heat and improve human thermal stress in Melbourne, Australia. *J. Appl. Meteorol. Climatol.* **2018**, *77*, 1748. [CrossRef]

62. Chapman, S.; Thatcher, M.; Salazar, A.; Watson, J.E.M.; McAlpine, C.A. The Effect of Urban Density and Vegetation Cover on the Heat Island of a Subtropical City. *J. Appl. Climatol. Meteorol.* **2017**, *57*, 2532. [CrossRef]

63. Santamouris, M.; Haddad, S.; Saliari, M.; Vasilakopoulou, K.; Synnefa, A.; Paolini, R.; Ulpiani, G.; Garshasbi, S.; Fiorito, F. On the energy impact of urban heat island in Sydney: Climate and energy potential of mitigation technologies. *Energy Build.* **2018**, *166*, 154–164. [CrossRef]

64. Haddad, S.; Barker, A.; Yang, J.; Ilamathy, D.; Kumar, M.; Garshasbi, S.; Paolini, R.; Santamouris, M. On the Potential of Building Adaptation Measures to Counterbalance the Impact of Climatic Change in the Tropics. *Energy Build.* **2020**, *229*, 110494. [CrossRef]

65. Yenneti, K.; Ding, L.; Prasad, D.; Ulpiani, G.; Paolini, R.; Haddad, S.; Santamouris, M. Urban overheating and cooling potential in Australia: An evidence-based review. *Climate* **2020**, *8*, 126. [CrossRef]

66. Middendorp, M. Mesoscale Modelling the Influence of Urban Vegetation on the Urban Temperatures and Human Thermal Comfort, Supervisors: Steeneveld, G.J., Theeuwes, N.E. Master's Thesis, Meteorology and Air Quality Wageningen University, Wageningen, The Netherlands, October 2013.

67. Fallmann, J.; Emeis, S.; Suppar, P. Mitigation of urban heat stress—A modelling case study for the area of Stuttgart, DIE ERDE. *J. Geogr. Soc. Berl.* **2013**, *144*, 3–4.

68. Žuvela-Aloise, M.; Koch, R.; Buchholz, S.; Früh, B. Modelling the potential of green and blue infrastructure to reduce urban heat load in the city of Vienna. *Clim. Chang.* **2016**. [CrossRef]

69. Li, X.-X.; Norford, L.K. Evaluation of cool roof and vegetations in mitigating urban heat island in a tropical city. *Singap. Urban Clim.* **2016**, *16*, 59–74. [CrossRef]

70. Arghavani, S.; Hossein Malakooti, H.; Bidokhti, A.A. Numerical evaluation of urban green space scenarios effects on gaseous air pollutants in Tehran Metropolis based on WRF-Chem model. *Atmos. Environ.* **2019**, *214*, 116832. [CrossRef]

71. Tong, H.; Walton, A.; Sang, J.; Chan, J.C.L. Numerical simulation of the urban boundary layer over the complex terrain of Hong Kong. *Atmos. Environ.* **2005**, *39*, 3549–3563. [CrossRef]

72. Ziter, C.D.; Pedersen, E.J.; Christopher, J.; Kucharik, C.J.; Turner, M.G. Scale-dependent interactions between tree canopy cover and impervious surfaces reduce daytime urban heat during summer. *Proc. Nastl. Acad. Sci. USA* **2019**, *116*, 7575–7580. [CrossRef]

73. Gao, K.; Santamouris, M.; Feng, J. On the efficiency of transpiration cooling to mitigate urban heat. *Climate* **2020**, *8*, 69. [CrossRef]

74. Santamouris, M.; Kolokotsa, D. *Urban Climate Mitigation Techniques*; Routedge: London, UK, 2016.

75. Qiao, Z.; Tian, G.; Xiao, L. Diurnal and seasonal impacts of urbanization on the urban thermal environment: A case study of Beijing using MODIS data. *ISPRS J. Photogramm. Remote Sens.* **2013**, *85*, 93–101. [CrossRef]

76. Zulia, I.; Santamouris, M.; Dimoudi, A. Monitoring the effect of urban green areas on the heat island in Athens. *Environ. Monit. Assess.* **2009**, *156*, 275–292. [CrossRef] [PubMed]

77. Liu, Y.; Chen, F.; Warner, T.; Basara, J.J. Verification of a mesoscale data-assimilation and forecasting system for the Oklahoma city area during the joint urban 2003 field project—American meteorological society. *J. Appl. Meteorol. Climatol.* **2006**, *45*, 912–929. [CrossRef]

78. Kusaka, H.; Kondo, H.; Kikegawa, Y.; Kimura, F. A sim- ple single-layer urban canopy model for atmospheric models: Comparison with multi-layer and slab models. *Bound. Layer Meteorol.* **2001**, *101*, 329–358. [CrossRef]

79. Martilli, A.; Clappier, A.; Rotach, M.W. Rotach: An urban sur- face exchange parameterisation for mesoscale models. *Bound. Layer Meteorol.* **2002**, *104*, 261–304. [CrossRef]

80. Salamanca, F.; Krpo, A.; Martilli, A.; Clappier, A. A new building energy model coupled with an urban canopy parameterization for urban climate simulations—Part 1: Formulation, verification, and sensitivity analysis of the model. *Theor. Appl. Climatol.* **2010**, *99*, 331–344. [CrossRef]

81. Fantozzi, F.; Monaci, F.; Blanusa, T.; Bargagli, R. Spatio-temporal variations of ozone and nitrogen dioxide concentrations under urban trees and in a nearby open area. *Urban Clim.* **2015**, *12*, 119–127. [CrossRef]

82. Manes, F.; Vitale, M.; Fabi, M.A.; De Santis, F.; Zona, D. Estimates of potential ozone stomatal uptake in mature trees of Quercus ilex in a Mediterranean climate. *Environ. Exp. Bot.* **2007**, *59*, 235–241. [CrossRef]

83. Beckett, P.K.; Freer-Smith, P.; Taylor, G. The capture of particulate pollution by trees at five contrasting urban sites. *Arboric. J.* **2000**, *24*, 209–230. [CrossRef]

84. Carter, W.P.L. Development of ozone reactivity scales for volatile organic compounds. *Air Waste* **1994**, *44*, 881–899. [CrossRef]

85. Atkinson, R. Atmospheric chemistry of VOCs and NO_x. *Atmos. Environ.* **2002**, *34*, 2063–2101. [CrossRef]

86. Calfapietra, C.; Fares, S.; Manes, F.; Morani, A.; Sgrigna, G.; Loreto, F. Role of biogenic volatile organic compounds [BVOC] emitted by urban trees on ozone concentration in cities: A review. *Environ. Pollut.* **2013**, *183*, 71–80. [CrossRef] [PubMed]

87. Ghirardo, A.; Xie, J.; Zheng, X.; Wang, Y.; Grote, R.; Block, K.; Wildt, J.; Mentel, T.; Kiendler- Scharr, A.; Hallquist, M.; et al. Urban stress- induced biogenic VOC emissions and SOA-forming potentials in Beijing. *Atmos. Chem. Phys.* **2016**, *16*, 2901–2920. [CrossRef]

88. Fu, Y.; Liao, H.; Change, C. Impacts of land use and land cover changes on biogenic emissions of volatile organic compounds in China from the late 1980s to the mid- 2000s: Implications for tropospheric ozone and secondary organic aerosol. *Tellus B* **2014**, *66*, 1–17. [CrossRef]

89. Taha, H.; Douglas, S.; Haney, J. Mesoscale meteorological and air quality impacts of increased urban albedo and vegetation. *Energy Build.* **1997**, *25*, 169–177. [CrossRef]

90. Long, X.; Wu, J.; Li, X.; Feng, T.; Xing, L.; Zhao, S.; Cao, J.; Tie, X.; An, Z.; Li, G.; et al. Does afforestation deteriorate haze pollution in Beijing-Tianjin-Hebei [BTH], China? *Atmos. Chem. Phys.* **2018**, *18*, 10869–10879. [CrossRef]

91. Vos, P.E.; Maiheu, B.; Vankerkom, J.; Janssen, S. Improving local air quality in cities: To tree or not to tree? *Environ. Pollut.* **2013**, *183*, 113–122. [CrossRef]

92. Jeanjean, A.P.R.; Buccolieri, R.; Eddy, J.; Monks, P.S.; Leigh, R.J. Air quality af-fected by trees in real street canyons: The case of marylebone neighborhood in central London. *Urban. Urban Green.* **2017**, *22*, 41–53. [CrossRef]

93. Taha, H.; Douglas, S.; Haney, J.; Winer, A.; Benjamin, M.; Hall, D.; Hall, J.; Liu, X.; Fishman, B. *Modeling the Ozone Air Quality Impacts of Increased Albedo and Urban Forest in the South Coast Air Basin*; Rep. No. LBL-37316; Lawrence Berkeley National Laboratory: Berkeley, CA, USA, 1995.

94. Environmnetal Services City of Portland. *Portland's Green Infrastructure. Quantifying the Health, Energy and Commiunity Liveability Benefits*; ENTRIX: Portland, OR, USA, 2010. Available online: https://www. portlandoregon.gov/bes/article/298042 (accessed on 14 September 2020).

95. Zhang, Y.; Bash, J.O.; Shawn, J.; Roselle, S.J.; Repinsky, A.; Mathur, R.; Hogrefe, C.; Piziali, J.; Jacobs, T.; Gilliland, A. Unexpected air quality impacts from implementation of green infrastructure in urban environments: A Kansas city case study. *Sci. Total Environ.* **2020**, *744*, 140960. [CrossRef]

96. Cardelino, C.A.; Chameides, W.L. Natural hydrocarbons, urbanisation, and urban ozone. *J. Geophys. Res.* **1990**, *95*, 13971–13979.

97. McDonald, A.G.; Bealey, W.J.; Fowler, D.; Dragosits, U.; Skiba, R.I.; Smith, R.G.; Donovan, R.G.; Brett, H.E.; Hewitt, C.N.; Nemitz, E. Quantifying the effect of urban tree planting on concentrations and depositions of PM_{10} in two UK conurbations. *Atmos. Environ.* **2007**, *41*, 8455–8467. [CrossRef]

98. Tallis, M.T.; Taylor, G.; Sinnett, D.; Freer-Smith, P. Estimating the removal of atmospheric particulate pollution by the urban tree canopy of London, under current and future environments. *Landsc. Urban Plan.* **2011**, *103*, 129–138. [CrossRef]

99. Nowak, D.J.; Crane, D.E.; Stevens, J.C.; Hoehn, R.E.; Walton, J.T.; Bond, J. A ground-based method of assessing urban forest structure and ecosystem services. *Arboricult. Urban* **2008**, *34*, 347–358.

100. Janhil, S. Review on urban vegetation and particle air pollution e Deposition and dispersion. *Atmos. Environ.* **2015**, *105*, 130–137. [CrossRef]

101. Hinds, W.C. *Aerosol Technology: Properties, Behavior, and Measurement of Airborne Particles*, 2nd ed.; John Wiley & Sons: Hoboken, NJ, USA, 1999; p. 504.

102. Thomas, A.M.; Pugh, T.A.; Robert MacKenzie, A.R.; Whyatt, J.D.; Hewitt, C.N. Effectiveness of Green Infrastructure for Improvement of air quality in urban street canyons. *Environ. Sci. Technol.* **2012**, *46*, 7692–7699.

103. Jeanjean, A.P.; Monks, P.S.; Leigh, R.J. Modelling the effectiveness of urban trees and grass on $PM_{2.5}$ reduction via dispersion and deposition at a city scale. *Atmos. Environ.* **2016**, *147*, 1–10. [CrossRef]

104. Buccolieri, R.; Salim, S.M.; Leo, L.S.; Di Sabatino, S.; Chan, A.; Ielpo, P.; de Gennaro, G.; Gromke, C. Analysis of local scale treeeatmosphere interaction on pollutant concentration in idealized street canyons and application to a real urban junction. *Atmos. Environ.* **2011**, *45*, e1702–e1713. [CrossRef]

105. Salmond, J.A.; Tadaki, M.; Vardoulakis, S.; Arbuthnott, K.; Coutts, A.; Demuzere, M.; Dirks, K.N.; Heaviside, C.; Lim, S.; Macintyre, H.; et al. Health and climate related ecosystem services provided by street trees in the urban environment. *Environ. Health* **2016**, *15*, 95. [CrossRef]

106. Wania, A.; Bruse, M.; Blond, N.; Weber, C. Analysing the influence of different street vegetation on traffic-induced particle dispersion using microscale sim- ulations. *J. Environ. Manag.* **2012**, *94*, e91–e101. [CrossRef]

107. Touchaei, A.G.; Akbari, H.; Tessum, C.W. Effect of increasing urban albedo on meteorology and air quality of Montreal [Canada]—episodic simulation of heat wave in 2005. *Atmos. Environ.* **2016**, *132*, 188–206. [CrossRef]

108. Nowak, D.J.; Hirabayashi, S.; Bodine, A.; Greenfield, E. Tree and forest effects on air quality and human health in the United States. *Environ. Pollut.* **2014**, *193*, 119–139. [CrossRef]

109. Nowak, D.J.; Hirabayashi, S.; Doyle, M.; McGovern, M.; Pasher, J. Air pollution removal by urban forests in Canada and its effect on air quality and human health. *Urban For. Urban Green.* **2018**, *29*, 40–48. [CrossRef]

110. Abdollahi, K.K.; Sun, J.J.; Ning, Z.H. Relative ability of urban trees and shrubs in removing ozone [O-3]. In Proceedings of the 1996 Society of American Foresters Convention—Diverse forests, Abundant Opportunities, and Evolving Realities, Washington, DC, USA, 9–13 November 1996.

111. Takahashi, M.; Higaki, A.; Nohno, M.; Kamada, M.; Okamura, Y.; Matsui, K.; Kitani, S.; Hiromichi, M. Differential assimilation of nitrogen dioxide by 70 taxa of roadside trees at an urban pollution level. *Chemosphere* **2005**, *61*, 633–639. [CrossRef] [PubMed]

112. Harris, T.B.; Manning, W.J. Nitrogen dioxide and ozone levels in urban tree canopies. *Environ. Pollut.* **2010**, *158*, 2384–2386. [CrossRef] [PubMed]

113. Kuttler, W.; Strassburger, A. Air quality measurements in urban green areas—A case study. *Atmos. Environ.* **1999**, *33*, 4101–4108. [CrossRef]

114. Ciacchini, G.; Vincentini, M.; Notoli, R.; Giaconi, V. Ozone measurements in the urban and extra-urban areas of Pisa during the summer of 1995. *J. Environ. Pathol. Toxicol. Oncol.* **1997**, *16*, 111–117.

115. Grundström, M.; Pleijel, H. Limited effect of urban tree vegetation on NO_2 and O_3 concentrations near a traffic route. *Environ. Pollut.* **2014**, *189*, 73–76. [CrossRef]

116. García-Go'mez, H.; Aguillaume, L.; Izquieta-Rojano, S.; Valino, F.; Avila, A.; Elustondo, D.; Santamaría, J.M.; Alastuey, A.; Calvete-Sogo, H.; González-Fernández, I.; et al. Atmospheric pollutants in peri-urban forests of *Quercus ilex*: Evidence of pollution abatement and threats for vegetation. *Environ. Sci. Pollut. Res.* **2016**, *23*, e6400–e6413. [CrossRef]

117. Yli-Pelkonen, V.; Scott, A.A.; Viippola, V.; Heikki Setaala, H. Trees in urban parks and forests reduce O_3, but not NO_2 concentrations in Baltimore, MD, USA. *Atmos. Environ.* **2017**, *167*, 73–80. [CrossRef]

118. Nowak, D.; Patrick, M.; Ibarra, M.; Crane, D.; Stevens, J.; Luley, C. Modeling the Effects of Urban Vegetation on Air Pollution. *Air Pollut. Modeling Appl. XII* **1998**, 399–407. [CrossRef]

119. Yli-Pelkonen, V.; Setälä, H.; Viljami Viippola, V. Urban forests near roads do not reduce gaseous air pollutant concentrations but have an impact on particles levels. *Landsc. Urban Plan.* **2017**, *158*, 39–47. [CrossRef]

120. Roy, S.; Byrne, J.; Pickering, C. A systematic quantitative review of urban tree benefits, costs, and assessment methods across cities in different climatic zones. *Urban For. Urban Green.* **2012**, *11*, 351–363. [CrossRef]

121. Di Nardo, F.; Saulle, R.; La Torre, G. Green areas and health outcomes: A systematic review of the scientific literature. *Ital. J. Public Health* **2012**, *7*, 402–413.

122. Simpson, R.; Williams, G.; Petroeschevsky, A.; Best, T.; Morgan, G.; Denison, L.; Hinwood, A.; Neville, G.; Neller, A. The short-term effects of air pollution on daily mortality in four Australian cities. *Aust. N. Z. J. Public Health* **2005**, *29*, 205–212. [CrossRef]

123. Lachowycz, K.; Jones, A.P. Does walking explain associations between access to greenspace and lower mortality? *Soc. Sci. Med.* **2014**, *107*, 9–17. [CrossRef] [PubMed]

124. Van den Berg, M.; Wendel-Vos, W.; Van Poppel, M.; Kemper, H.; Van Mechelen, W.; Maas, J. Health benefits of green spaces in the living environment: A systematic review of epidemiological studies. *Urban. For. Urban Green.* **2015**, *14*, 806–816. [CrossRef]

125. Bowler, D.E.; Buyung-Ali, L.M.; Knight, T.M.; Pullin, S.A. A systematic review of evidence for the added benefits to health of exposure to natural environments. *BMC Public Health* **2010**, *10*, 456. [CrossRef]

126. Harlan, S.L.; Declet-Barreto, J.H.; Stefanov, W.L.; Petitti, D.B. Neighborhood effects on heat deaths: Social and environmental predictors of vulnerability in Maricopa County, Arizona. *Environ. Health Perspect.* **2013**, *121*, 197–204. [CrossRef]

127. Wilker, E.H.; Wu, C.-D.; McNeely, E.; Mostofsky, E.; Spengler, J.; Wellenius, G.A.; Mittleman, M.A. Green space and mortality following ischemic stroke. *Environ. Res.* **2014**, *133*, 42–48. [CrossRef]

128. Tamosiunas, A.; Grazuleviciene, R.; Luksiene, D.; Dedele, A.; Reklaitiene, R.; Baceviciene, M.; Vencloviene, J.; Bernotiene, G.; Radisauskas, R.; Malinauskiene, V.; et al. Accessibility and use of urban green spaces, and cardiovascular health:findings from a Kaunas cohort study. *Environ. Health* **2014**, *13*, 20. [CrossRef]

129. Jonker, M.F.; Van Lenthe, F.J.; Donkers, B.; Mackenbach, J.P.; Burdorf, A. The effect of urban green on small-area [healthy] life expectancy. *J. Epidemiol. Community Health* **2014**, *68*, 999–1002. [CrossRef]

130. Son, J.-Y.; Liu, J.C.; Bell, M.J. Temperature-related mortality: A systematic review and investigation of effect modifiers. *Environ. Res. Lett.* **2019**, *14*, 073004. [CrossRef]

131. Murage, P.; Kovats, S.; Sarran, C.; Taylor, J.; McInnes, R.; Hajat, S. What individual and neighbourhood-level factors increase the risk of heat- related mortality? A case-crossover study of over 185,000 deaths in London using high-resolution climate datasets. *Environ. Int.* **2020**, *134*, 1052922. [CrossRef] [PubMed]

132. Son, J.K.; Lane, K.J.; Lee, J.-T.; Bell, M.L. Urban vegetation and heat-related mortality in Seoul. *Environ. Res.* **2016**, *151*, 728–733. [CrossRef] [PubMed]

133. Nowak, D.J.; Hirabayashi, S.; Bodine, A.; Hoehn, R. Modeled $PM_{2.5}$ removal by trees in ten U.S. cities and associated health effects. *Environ. Pollut.* **2013**, *178*, 395–402. [CrossRef]

134. Rao, M.; George, L.A.; Rosenstiel, T.N.; Shandas, V.; Dinno, A.; Rao, M.; George, L.A.; Rosenstiel, T.N.; Shandas, V.; Dinno, A. Assessing the relationship among urban trees, nitrogen dioxide, and respiratory health. *Environ. Pollut.* **2014**, *194*, 96–104. [CrossRef]

135. Manes, F.; Incerti, G.; Salvatori, E.; Vitale, M.; Ricotta, C.; Costanza, R. Urban ecosystem services: Tree diversity and stability of tropospheric ozone removal. *Ecol. Appl.* **2012**, *22*, 349–360. [CrossRef]

136. Alonso, R.; Vivanco, M.G.; González-Fernández, I.; Bermejo, V.; Palomino, I.; Garrido, J.L.; Artíñano, B. Modelling the influence of peri-urban trees in the air quality of Madrid region [Spain]. *Environ. Pollut.* **2011**, *159*, 2138–2147. [CrossRef]

137. Tiwary, A.; Sinnett, D.; Peachey, C.; Chalabi, Z.; Vardoulakis, S.; Fletcher, T.; Hutchings, T.R. An integrated tool to assess @benefits: A case study in London. *Environ. Pollut.* **2009**, *157*, 2645–2653. [CrossRef]

138. Powe, N.A.; Willis, K.G. Mortality and morbidity benefits of air pollution [SO_2 and PM_{10}] absorption attributable to woodland in Britain. *J. Environ. Manag.* **2004**, *70*, 119–128. [CrossRef]

139. Pereira, G.; Foster, S.; Martin, K.; Christian, H.; Boruff, B.J.; Knuiman, M.; Giles-Corti, B. The association between neighborhood greenness and cardiovascular disease: An observ ational study. *BMC Public Health* **2012**, *12*, 466. [CrossRef]

140. Mills, D.M.; Laurence Kalkstein, L. Estimating reduced heat-attributable mortality for an urban revegetation project. *J. Heat Island Inst. Int.* **2012**, *7*, 18–24.

141. Kalkstein, L.S.; Greene, S.; Mills, D.M.; Samenow, J. An evaluation of the progress in reducing heat-related human mortality in major U.S. cities. *Nat. Hazards* **2011**, *56*, 113–129. [CrossRef]

142. Smargiassi, A.; Goldberg, M.S.; Plante, C.; Fournier, M.; Baudouin, Y.; Kosatsky, T. Variation of daily warm season mortality as a function of micro-urban heat islands. *J. Epidemiol. Community Health* **2009**, *63*, 659–664. [CrossRef] [PubMed]

143. Taylor, J.; Wilkinson, P.; Davies, M.; Armstrong, B.; Chalabi, Z.; Mavrogianni, A.; Symonds, P.; Oikonomou, E.; Bohnenstengel, S.I. Mapping the effects of urban heat island, housing, and age on excess heat-related mortality in London. *Urban Clim.* **2015**, *14*, 517–528. [CrossRef]

144. Hondula, D.M.; Balling, R.C.; Vanos, J.K.; Georgescu, M. Rising temperatures, human health, and the role of adaptation. *Curr. Clim. Chang. Rep.* **2015**, *1*, 144–154. [CrossRef]

145. Boeckmann, M.; Rohn, I. Is planned adaptation to heat reducing heat-related mortality and illness? A systematic review. *BMC Public Health* **2014**, *14*, 1112. [CrossRef]

146. Yao, Y.; Luo, Y.; Huang, J.; Zhao, Z. Comparison of monthly temperature extremes simulated by CMIP3 and CMIP5 models. *J. Clim.* **2013**, *26*, 7692–7707. [CrossRef]

147. Pyrgou, A.; Hadjinicolaou, P.; Santamouris, M. Urban-rural moisture contrast: Regulator of the urban heat island and heatwaves' synergy over a Mediterranean city. *Environ. Res.* **2020**, *182*, 109102. [CrossRef]

148. Khan, H.S.; Santamouris, M.; Riccardo Paolini, R.; Caccetta, P.; Kassomenos, P. Analyzing the local and climatic conditions affecting the Urban Overheating Magnitude during the Heatwaves [HWs] in a coastal city, a case study of the greater Sydney region. *Sci. Total Environ.* **2020**, *755*, 142515. [CrossRef]

149. Lu, J.; Li, C.-D.; Yang, Y.-C.; Zhang, X.-H.; Jin, M. Quantitative evaluation of urban park cool island factors in mountain city. *J. Cent. South. Univ. Technol.* **2012**, *19*, 1657–1662. [CrossRef]

150. Teskey, R.; Wertin, T.; Bauweraerts, I.; Ameye, M.; McGuire, M.A.; Steppe, K. Responses of tree species to heat waves and extreme heat events. *Plant. Cell Environ.* **2015**, *38*, 1699–1712. [CrossRef] [PubMed]

151. Zweifel, R.; Zimmermann, L.; Zeugin, F.; Newbery, D.M. Intra-annual radial growth and water relations of trees: Implications towards a growth mechanism. *J. Exp. Bot.* **2006**, *57*, 1445–1459. [CrossRef] [PubMed]

152. Kala, J.; De Kauwe, M.G.; Pitman, A.J.; Medlyn, B.E.; Wang, Y.-P.; Lorenz, R.; Perkins-Kirkpatrick, S.E. Impact of the representation of stomatal conductance on model projections of heatwave intensity. *Sci. Rep. Nat.* **2016**, *6*, 23418. [CrossRef] [PubMed]

153. Lorenz, R.; Davin, E.L.; Lawrence, D.M.; Stöckli, R.; Seneviratne, S.I. How important is vegetation phenology for European climate and heat waves? *J. Clim.* **2013**, *26*, 10077–10100. [CrossRef]

154. Stéfanon, M.; Drobinski, P.; D'Andrea, F.; de Noblet-Ducoudré, N. Effects of interactive vegetation phenology on the 2003 summer heat waves. *J. Geophys. Res.* **2012**, *117*, D24103. [CrossRef]

155. Harfouche, A.; Meilan, R.; Altman, A. Tree genetic engineering and applications to sustainable forestry and biomass production. *Trends Biotechnol.* **2011**, *29*, 9–17. [CrossRef]

156. Nuvolone, D.; Petri, D.; Voller, F. The effects of ozone on human health. *Environ. Sci. Pollut. Control Ser.* **2017**, *25*, 8074–8088. [CrossRef]

157. Krmpotic, D.; Luzar-Stiffler, V.; Herzog, P. Effects of urban ozone pollution on hospitalizations for exacerbation of chronic obstructive pulmonary disease. *Eur. Respir. J.* **2015**, *46*, PA3411.

158. Amorim, J.H.; Rodrigues, V.; Tavares, R.; Valente, J.; Borrego, C. CFD modelling of the aerodynamic effect of trees on urban air pollution dispersion. *Sci. Total Environ.* **2013**, *461*, 541–551. [CrossRef]

159. EEA. *Air Pollution in Europe 1990–2004*; EEA Report No2/2007; European Environment Agency: Copenhagen, Denmark, 2007.

160. Silva, H.R.; Phelan, P.E.; Golden, J.S. Modeling effects of urban heat island mitigation strategies on heat-related morbidity: A case study for Phoenix, Arizona, USA. *Int. J. Biometeorol.* **2010**, *54*, 13–22. [CrossRef]

161. Venter, Z.S.; Krog, N.H.; David, N.; Barton, D.N. Linking green infrastructure to urban heat and human health risk mitigation in Oslo, Norway. *Sci. Total Environ.* **2020**, *709*, 136193. [CrossRef] [PubMed]

Publisher's Note: MDPI stays neutral with regard to jurisdictional claims in published maps and institutional affiliations.

MDPI

Article

Evaluating Modular Healthcare Facilities for COVID-19 Emergency Response—A Case of Hong Kong

Wei Pan and Zhiqian Zhang *

Department of Civil Engineering, The University of Hong Kong, Pokfulam, Hong Kong 999077, China
* Correspondence: zzq007@connect.hku.hk

Abstract: In response to the COVID-19 pandemic, modular construction has been adopted for rapidly delivering healthcare facilities, but few have systematically explored the impacts of the pandemic and the contributions of modular construction. This paper aims to evaluate modular construction for delivering healthcare facilities in response to COVID-19, through the exploration of the challenges, strategies, and performance of using modular construction for emergency healthcare building project delivery. The study was conducted using 12 real-life healthcare building projects in Hong Kong with both within- and cross-case analyses. The results of the within-case study reveal critical challenges such as tight program but limited resources available and the corresponding strategies such as implementation of smart technologies. The results of the cross-case analysis indicate 106% improved time efficiency and 203% enhanced cost efficiency of using modular construction compared with conventional practices. Based on the multi-case studies, the paper develops an innovative framework which illustrates the roles of stakeholders, goals, engineering challenges, and management principles of using modular construction. Practically, the paper should assist both policymakers and industry stakeholders in addressing the critical challenges of delivering healthcare facilities under COVID-19 in an efficient and collaborative manner. Theoretically, it should set an exemplar of linking the building construction industry with emergency management and healthcare service systems to facilitate efficient response to pandemics.

Keywords: COVID-19; emergency response; healthcare facility; modular integrated construction; modular building

Citation: Pan, W.; Zhang, Z. Evaluating Modular Healthcare Facilities for COVID-19 Emergency Response—A Case of Hong Kong. *Buildings* **2022**, *12*, 1430. https://doi.org/10.3390/buildings12091430

Academic Editor: Krishanu Roy

Received: 14 August 2022
Accepted: 8 September 2022
Published: 11 September 2022

Publisher's Note: MDPI stays neutral with regard to jurisdictional claims in published maps and institutional affiliations.

1. Introduction

The fast spread of the COVID-19 pandemic has disrupted healthcare systems globally and has imposed great challenges on the construction industry [1,2]. Nevertheless, the pandemic may also accelerate the process of innovation adoption to address urgent social needs under the COVID-19 pandemic. Various strategies and innovations have been proposed to ensure the capacities of healthcare facilities during and after disasters, e.g., optimization of public hospital resources under calamitous situations [3], application of preparedness control measures such as communication and information management and training [4], and design of safe spaces for residential housing [5].

With the adoption of prefabricated construction, various emergency healthcare facilities have been rapidly delivered worldwide. For example, the Leishenshan hospital in China was delivered in only two weeks using prefabricated steel structures [6]; and an isolation hospital in Korea was built in 23 days using steel-framed modules manufactured in China [7]. The adoption of prefabricated systems (e.g., precast and modular construction) can speed up the project delivery process to provide isolation and curing places in the shortest time possible, and it can also mitigate the risks of cross infection during construction due to the minimized on-site labor [8].

Although these facilities were built in an ever-fast manner, the adoption of the modular approach was significantly challenging to the construction industry under the COVID-19 pandemic. On the one hand, modular construction normally involves intensive and complex module prefabrication [9]. On the other hand, the pandemic raises some new challenges such as uncertain cross-border logistics [10]. Nevertheless, most previous studies focused simply on the impact of the pandemic such as Prasad and Bhat [11], but they ignored how the industry responds to the pandemic. In addition, the modular approach has been widely explored in residential buildings [12], but with few investigations of its applications in emergency healthcare projects.

To comprehensively understand the impact of COVID-19 on the construction industry and to appreciate the contributions of modular construction to addressing the pandemic, this paper aims to evaluate the performance of modular construction-enabled healthcare facility delivery in response to COVID-19. The evaluation was conducted by systematically identifying the challenges, exploring the strategies, and measuring the time and cost efficiency of using modular construction for healthcare building project delivery. Multiple case studies were conducted by engaging 12 real-life building projects in Hong Kong, including modular quarantine camps and hospitals.

Following this introduction, the paper reviews the features of modular construction and the principles of emergency building project delivery, and develops a conceptual framework of modular construction for addressing COVID-19. The paper then elaborates on the methods of data collection and analyses, followed by the presentation of the identified challenges and strategies and the measured performance. Based on the results, the paper develops and discusses a systematic framework of efficient response to COVID-19 through modular construction. Finally, the paper draws its conclusions.

2. Literature Review

2.1. Features of Modular Construction

Modular construction represents the highest level of prefabricated construction technologies and was defined by Pan et al. [13] as an innovative approach to transforming fragmented site-based construction into integrated value-driven production and assembly. Modular construction is an instance of the application of modularity theory in the construction industry, which emphasizes product modularization and standardization and aims for productivity enhancement [14]. Globally, the modular approach has been widely adopted in building projects, e.g., modular integrated construction (MiC) in Hong Kong [15]. Compared with conventional construction, modular construction changes the project delivery process mainly in two aspects: spatially, volumetric modules are prefabricated in the factory, and then installed on site; temporally, prefabrication is carried out concurrently with the on-site installation [16]. The tempo-spatial transformation with modularization improves the construction performance of building projects. Both concrete and steel modular systems were demonstrated with multi-faceted benefits, e.g., faster construction, better product quality, improved environmental friendliness, reduced health and safety risks, and an improved industry image [17–19]. Nevertheless, different modularization schemes may have different construction performance. For example, highly modularized buildings with more work fabricated in factory can reduce the on-site labor consumption and increase the speed of superstructure construction [15].

The multi-faceted benefits demonstrate high potential of modular construction in response to COVID-19 by fast delivering healthcare facilities. Nevertheless, various constraints exist in prefabricated construction supply chain especially following a large-scale disaster, for example, the shortage of skilled workers [20], challenging just-in-time delivery of modules [9,21], the unsecured construction material procurement and delivery [22], and the complicated prediction of supply and demand [23]. To address the challenges along the construction supply chain, an innovative approach was designed to facilitate the procurement planning of construction materials following a large-scale disaster [24], and a dynamic model of prefabricated construction supply was developed to address the statistic constraints considering the multiple factories [22].

As the supply chain of modular construction is not as mature as that of conventional prefabrication, it is even more challenging for delivering emergency modular healthcare projects due to the tight program and limited resources. Therefore, it is critical to explore the challenges to, and identify the strategies for adopting modular construction in addressing the COVID-19 pandemic.

2.2. Emergency Project Delivery and Management

To explore the challenges and strategies of using modular construction for COVID-19, it is necessary to first examine the concept, principles and process of emergency project delivery and management. Emergency management is to apply science, technology, planning, and management to deal with extreme events that can cause extensive property damage and disrupt community life [25]. It addresses how humans and institutions interact and cope with hazards through a cycle with four major activities, i.e., mitigation, preparedness, response, and recovery [26,27]: mitigation includes actions taken to prevent or reduce the impact and consequences of disasters; preparedness involves planning and training activities for events that cannot be mitigated; response includes activities designed to address the immediate and short-term effects of an emergency or disaster; and recovery refers to long-term activities designed to return all systems to normal status. "Build Back Better" principles are normally introduced as an ideal reconstruction/recovery process to improve community's resilience following a disaster event, e.g., improved building codes and land-use plans [28]. This study focused on the emergency response, i.e., how modular construction contributes to the efficient response to the outbreak of COVID-19 through fast delivery of healthcare facilities.

Many researchers have examined the principles of emergency response and management. For example, Waugh Jr and Streib [26] and Bae et al. [29] elaborated on the importance of leadership and collaboration. Chen et al. [30] presented a set of design principles, e.g., resource monitoring and group decision-making. Cowick and Cowick [31] argued the effectiveness of using new technologies such as online coordination tools.

For emergency project delivery, Capolongo et al. [32] proposed some strategies such as strategic site selection, flexibility and user-centeredness. To accelerate the process of emergency project delivery, Schexnayder and Anderson [33] and Wang and Shi [34] summarized various techniques, e.g., working overtime, providing additional labor and equipment, and adopting innovative construction methods. To address the shortage of resources during an emergency, Chen et al. [35] suggested to develop logistics management and resource-sharing networks across local, national, and international levels. In addition, the importance of establishing an emergency response team with close collaboration was highlighted by McWilliams [36] and Gransberg [37].

However, the existing emergency response frameworks only specify the generic organizational roles and actions which cannot directly apply to the delivery of modular emergency healthcare facilities, e.g., that the HKSAR Government is committed to providing responses to emergency situations that threaten life, property and public security [38] and to convert suitable holiday camps into quarantine camps for COVID-19 [39]. In addition, the emergency project delivery strategies in the literature did not consider the features of modular construction and the waves of COVID-19. Therefore, this research was designed to also develop an innovative emergency response framework in the context of fast delivery of modular healthcare facilities in response to the COVID-19 pandemic.

2.3. Conceptual Framework of Modular Construction-Enabled Response to COVID-19

SWOT analysis is a strategic planning and management technique used to identify the internal strengths and weaknesses and the external opportunities and threats for a specific situation [40]. To guide the exploration of the challenges and strategies of modular construction-enabled response to the COVID-19, a conceptual framework (Figure 1) was developed based on a critical SWOT analysis. Modular construction has significant advantages over conventional construction such as improved speed of construction (strength) [17], and thus can address the urgent social needs on healthcare facilities (opportunity). Nevertheless, the modular construction itself is facing challenges such as cross-border logistics (weakness) [9], and the construction industry encountered new issues during the COVID-19 pandemic such as shortage of material supply (threat).

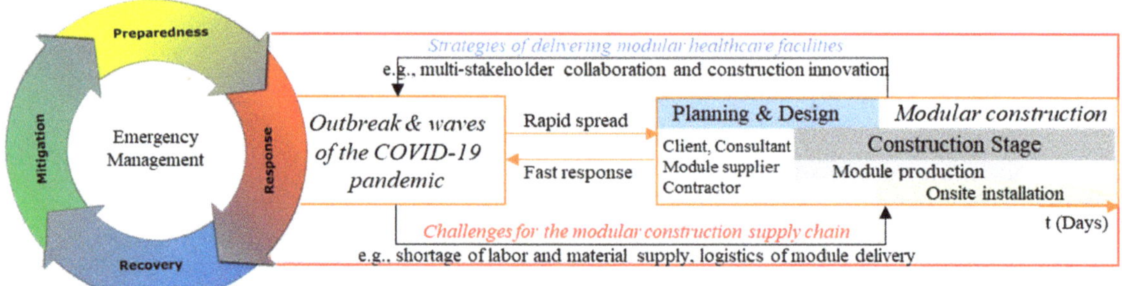

Figure 1. Conceptual framework of modular construction-enabled response to COVID-19.

Correspondingly, the framework integrates the potential challenges facing the construction industry during COVID-19 (e.g., shortage of material supply), the basic principles of emergency response (e.g., multi-stakeholder collaboration), and the process of modular project delivery (e.g., parallel module production and on-site installation). It illustrates the mutual impacts between modular construction and COVID-19: modular construction mitigates the impact of COVID-19 by rapidly delivering healthcare projects; COVID-19 greatly affects the modular construction supply chain. It also indicates the focus of this study, i.e., 'Response' of the four-stage cycle of emergency management [24]. Guided by the conceptual framework, the study has systematically identified both common and project-specific challenges, explored the corresponding strategies for better adoption of modular construction under the pandemic, measured how efficient the modular approach is in response to the pandemic, and finally proposed the framework of how modular construction addresses the COVID-19 pandemic.

3. Research Methodology

3.1. Overall Research Design

This research has adopted a multi-case study strategy using 12 case projects, and was carried out following the process shown in Figure 2. To start, a comprehensive literature review was conducted, and a conceptual framework was developed. Guided by the conceptual framework, a within-case study using 5 cases was conducted to identify and validate the challenges and strategies of using modular construction for addressing COVID-19; in parallel, a cross-case study using 12 cases was conducted to measure the performance of modular construction for healthcare project delivery. Based on the multi-case studies, the performance (e.g., time and cost efficiency) of modular healthcare facilities was evaluated, and the framework of modular construction-enabled efficient response to COVID-19 was developed.

The case studies were conducted in Hong Kong to demonstrate how an administrative region and its construction community have responded to COVID-19. The 12 case projects (referred to as Projects A-L) were selected by adopting the purposive sampling strategy, considering that (1) Hong Kong has established the supply chain of modular construction; (2) all projects were emergency healthcare facilities; (3) all major modular healthcare projects in Hong Kong were selected; (4) the projects covered conventional construction for benchmarking.

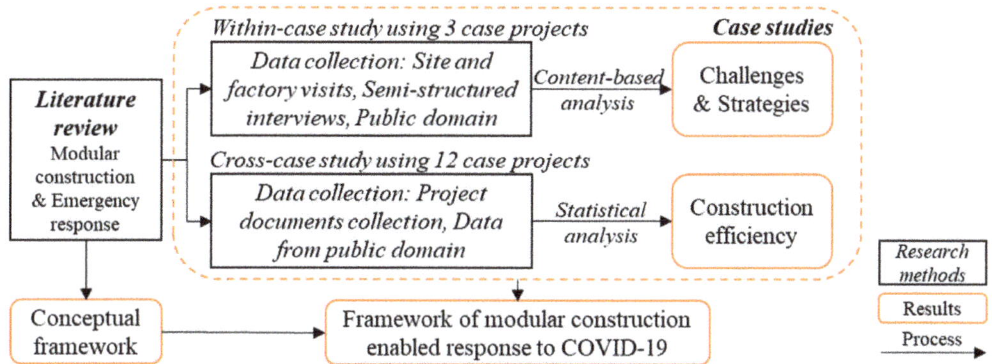

Figure 2. Overview of the research process with methods adopted and results derived.

The selected case projects included 10 quarantine camps (i.e., Projects A-I using modular construction, Project J using in-situ construction) and 2 hospitals (i.e., Project K using modular construction, Project L using in-situ construction). Project J was a scenario that was designed by the authors according to the expert interviews with the construction practitioners. To ensure the consistency of analysis, all projects adopted design-and-build contracts. All quarantine camps selected were completed in 2020. The basic project information is provided in Table 1. Projects A, B, C, H, and K were selected for the within-case study, while all 12 projects were used for the cross-case study. The timeline of project delivery and waves of COVID-19 are illustrated in Figure 3.

Table 1. Information of the selected case projects.

Cases	A	B	C	D	E	F	G	H	I	J	K	L
MR (%)	>95 (A–I)									*N/A*	>70	*N/A*
CFA (m2)	2052	5980	2000	3470	3000	13,158	13,125	15,938	15,938	*16,000*	44,000	*21,600*
No. of beds	118	234	120	198	110	700	700	850	850	*850*	816	*108*
Supplier	CN	CN	CN	SG	HK	CN	HK	CN	CN	*N/A*	CN	*N/A*
Experience	3–5	3–5	>10	3–5	1–3	3–5	3–5	>10	3–5		3–5	
Year of completion	2020 (A–I)									2020	2021	2007
Duration (days)	26	62	66	84	68	73	87	88	87	*300 (↑)*	~120	*~930 (↑)*
Cost (HK$M)	15	29.5	29.8	28	193.7	433	418	605.5	663	*663*	N/A	*964*

Notes: (1) 'Cost': contract sum; (2) MR (modularization rate) = modularized floor area/CFA; (3) Projects (J and L) in *italic*: in-situ construction method; (4) CN: China; HK: Hong Kong; SG: Singapore; (5) Experience: years in modular construction; (6) "↑": longer time consumed.

From Figure 3, it can be seen that all quarantine camps using modular construction were delivered within 3 months, and the design and construction of the modular hospital were completed within 4 months from the 4th quarter of 2020 to the 1st quarter of 2021. To accommodate the people who needed isolation (e.g., visitors from overseas), there was an urgency to deliver quarantine camps (e.g., Projects A to F) with sufficient beds as soon as possible since the outbreak of the pandemic. To address the future waves of the pandemic, it was also critical to build more isolation facilities for the locally confirmed cases, e.g., Projects G to I. In addition, to release the pressure of both the public and private hospital systems, a temporary hospital (Project K) was delivered between the peak of the 3rd and 4th waves, which provided both isolation and curing facilities. The rapid delivery of these facilities facilitated a timely response to the 4 waves of the pandemic in 2020.

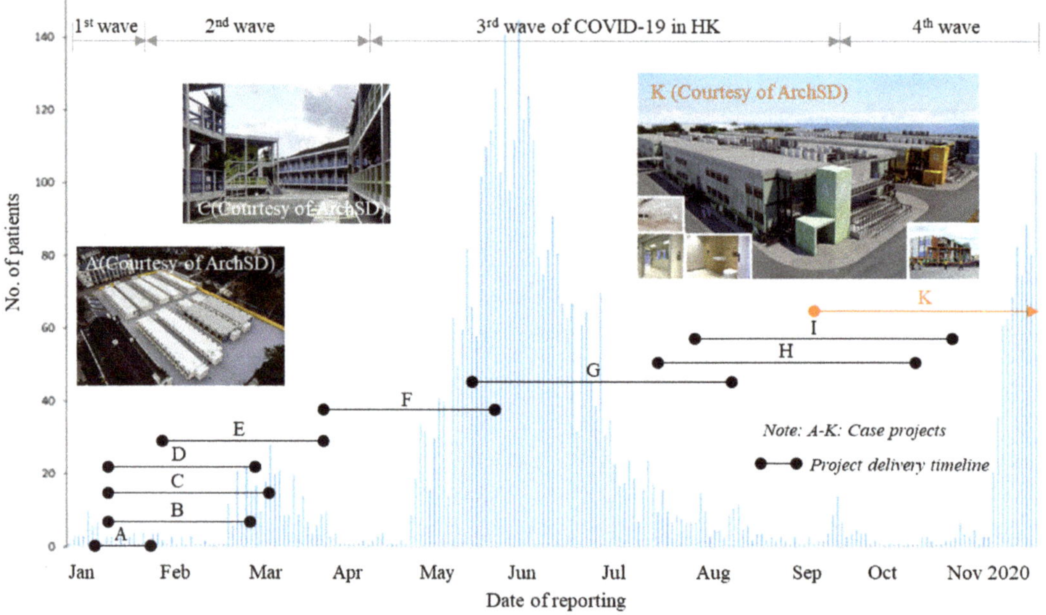

Figure 3. Timeline of project delivery and waves of COVID-19 in Hong Kong.

3.2. Methods of Within-Case Study

Considering the data availability, 5 case projects were used in the within-case study, i.e., Projects A, B and K for identification of the challenges and strategies, and Projects C and H for validation. Projects A, B, C and H were the typical quarantine camps in Hong Kong, and Project K was the only modular hospital that performed as an infection control center in response to COVID-19. To comprehensively identify the challenges and strategies and enhance the data validity, data were collected and verified through the triangulation of evidence sources [41]: site and factory visits, semi-structured interviews, and data from the public domain. Specifically, site and factory visits to Projects A, B and K were conducted to better understand the processes of site construction and factory production. Semi-structured interviews with project stakeholders were carried out to identify and validate the critical challenges and strategies. Information of the interviewees is summarized in Table 2.

Table 2. Information of the interviewees in the within-case study.

Interviews	Projects A, B and K	Projects C and H
Interviewees	Project Director (Client), Project Manager and Site Engineer (Main Contractor), Project Manager (Module Supplier)	Project Director (Client), General Manager and Project Manager (Main Contractor)

Informed by the conceptual framework (Figure 1), content-based analysis was adopted to summarize the challenges and the corresponding strategies. Explicitly, the identified challenges and strategies were categorized according to the major phases of project delivery (i.e., planning, design, and construction), and were classified as common ones that apply to all case projects and specific ones that only appeared in some of the projects.

3.3. Methods of Cross-Case Study

The cross-case study was conducted using 12 case projects with both quantitative and qualitative analyses (Figure 4). First, an *Excel* table was used to quantitatively measure the construction efficiency, followed by a comparative analysis using a scatter plot. Construction project efficiency is normally measured using time- and cost-efficiency [42]. The following equations were used: (1) Time efficiency (m^2/day) = CFA/Duration of project delivery; and (2) Cost efficiency (m^2/\$) = CFA/Cost of project development. The 'CFA' refers to the total construction floor area and was extracted from architectural drawings; 'Duration of project delivery' covers project design and construction and was extracted from the master program; and 'Cost of project development' is the contract sum approved by a client and was collected from public domain. Second, qualitative evaluation was conducted to comprehensively reflect the performance of modular construction in response to COVID-19. To enable like-to-like comparison, only projects of the same type were compared with each other, e.g., modular quarantine camp vs. conventional quarantine camp. The information was mainly collected from the project teams (e.g., design and construction documents) and public domain (e.g., government website and reports of public seminars), and analyzed under the three pillars of sustainability: economy, environment, and society.

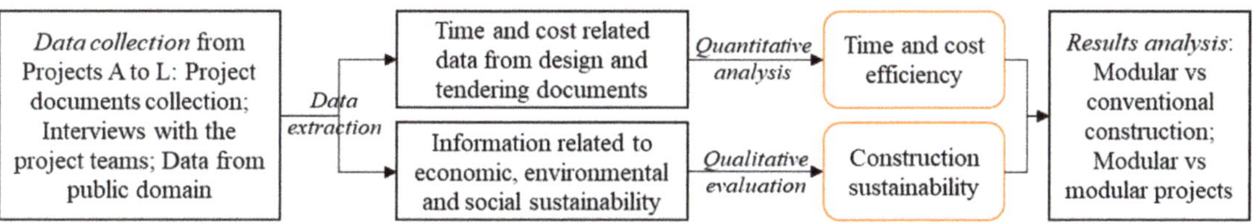

Figure 4. Research process of the cross-case study.

4. Results of Within-Case Study

Guided by the conceptual framework of modular construction-enabled response to the COVID-19 pandemic, the critical challenges and corresponding strategies were identified based on the within-case studies using Projects A, B and K, and supplemented and validated with Projects C and H. The results of the identified challenges and strategies are summarized in Table 3.

Table 3. Challenges and strategies of delivering modular healthcare facilities for COVID-19.

Process	Identified Challenges	Corresponding Strategies
Planning	Multi-faceted communication and coordination Tight program for planning	To enhance inter-government and cross-department collaboration To enable wide-industry partnership and early contractor involvement
Design	Multi-faceted communication and coordination Tight program for design Strict regulatory compliance Challenging modularization of hospital	To enable wide-industry partnership and early contractor involvement To follow the principle of Less is More To adopt professional and modularized design To design for production and transportation
Construction	Multi-faceted communication and coordination (between site and factory) Tight program but limited resources available (for both site and factory) High pressure on COVID-19 prevention (for both site and factory) Challenging logistics Project-specific site constraints	To enhance government-industry collaboration To count construction by hours and organize resources efficiently (for both site and factory) To take comprehensive infectious control measures To implement smart technologies To conduct systematic construction and production planning To take specific monitoring and control

Notes: Italic: project-specific challenges and strategies.

4.1. Identified Challenges

4.1.1. Multi-Faceted Communication and Coordination

Project clients and main contractors of all case projects were coordinating with multiple stakeholders to ensure efficient project delivery. For example, over 20 stakeholders were involved in Projects A and B, and over 26 in Project K. The multiple stakeholders included but were not limited to regulatory and works departments for design approval, sub-contractors for on-site activities, module suppliers for off-site logistics, and non-local governmental departments for factory production and cross-border transportation. It was also important but challenging for coordination between the site and factory teams, as fewer face-to-face meetings can be arranged due to the quarantine requirement. The efficiency of multi-faceted coordination determined the project success under COVID-19.

4.1.2. Strict Regulatory Compliance

It was challenging for the project teams to prepare statutory submissions in such a short time, e.g., approval-in-principal, detailed design approval, and shop drawings. These works determined the intensive coordination with relevant regulatory departments. In addition, the quarantine camps were designed not only for temporary purposes but also with life-cycle considerations, e.g., re-used for transitional housing. Although Project K was a temporary infectious control center, it was designed as a permanent hospital. The project teams had to prepare a feasible but efficient scheme to fulfill all regulatory requirements.

4.1.3. Tight Program but Limited Resources Available

The project teams were expected to complete their projects the soonest to address the urgent social needs. For Projects A and B, a one-month target was set to complete the design and construction. For Project K, design and construction were expected to be completed within 4 months to reach a permanent hospital standard, which was quite challenging for the industry even without COVID-19. However, there were limited resources available, e.g., challenging raw material procurement due to the restrictions at customs under the pandemic. The module suppliers were facing great challenges to deliver a large number of modules in such a short time, as most of the suppliers had insufficient experience on modular construction at that time.

4.1.4. High Pressure on COVID-19 Prevention

Facing the fast spread of the virus the construction teams were under high pressure on COVID-19 prevention and control. The increasing number of confirmed cases resulted in high risks of construction in a confined site. It was even more challenging to avoid such infections in a modular construction project with multiple transportation of modules from overseas. For example, in Project K over 300 management staff and workers were working and eating on site during the outbreak of the 4th wave of COVID-19 in Hong Kong.

4.1.5. Challenging Cross-Border Logistics

Logistics for module transportation from overseas was challenging in Projects A, B and K, as the border was under strict control and even closed in some countries under the COVID-19 pandemic. According to the interviews with Projects C and H, these challenges did exist in other healthcare projects. For example, Project C engaged a Malaysia module supplier, but the border was closed at the time of construction. They had to take additional efforts to negotiate with the Malaysian government for special arrangements.

4.1.6. Project-Specific Challenges

Apart from the common challenges above, the interviewees also mentioned some critical but project-specific challenges. For example, the modularized design for Project K (temporal hospital) is much more challenging than that for Projects A and B (quarantine camps), as it involved more types of modules (e.g., that for negative pressure ward) and many over-sized modules to satisfy hospital requirement (e.g., modules with the overhead ventilation system).

The modules had to be designed at 3.8 m in height, and thus the vehicle together with the modules approached the height limit of transportation (4.6 m) in Hong Kong.

In addition, various site constraints existed in some projects. For example, the site of Projects A and B was transformed from a football court on a mountain, and thus more time was needed for site formation, and it was challenging to ensure water and electricity supply. The site of Project K was close to the airport with strict height limits and also to an MTR line with requirements on noise and vibration control. Interviewees of Projects C and H also indicated the project-specific site constraints due to the urgent transformation of temporary lands for healthcare facilities.

4.2. Corresponding Strategies

4.2.1. Cross-Department Collaboration and Wide-Industry Partnership

To address the challenges of multi-faceted coordination, cross-department collaboration and wide-industry partnership were taken as a primary strategy. In Projects A and B, the strategic planning was mainly executed at two levels (Figure 5): (1) cross-department collaboration coordinated by a works department of the HKSAR Government with 16 supportive government agencies and regulatory departments; and (2) wide-industry partnership led by the main contractor with over 40 sub-contractors and 20 material suppliers involved. One more strategy in Project K was the inter-government collaboration between Shenzhen Municipal Government and HKSAR Government to facilitate efficient planning.

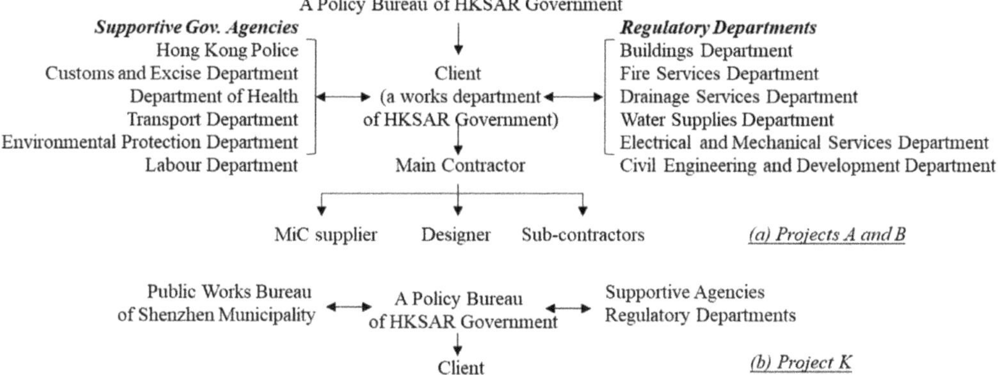

Figure 5. Inter-government and cross-department collaboration.

The cross-department collaboration assisted the project teams in fluent regulatory approvals thus facilitating fastest completion of design works. In addition, the clients of Projects A and B teamed up with the Transport Department and the Customs to make sure that the modules could be transported from the factory in Mainland China to Hong Kong within 4 h. The wide-industry partnership helped to address the problems of limited resources available, which made intensive construction realized within a shortest period possible.

4.2.2. Comprehensive Infectious Control

The design of modules in Projects A and B incorporated 12 infectious disease control criteria, mainly including clean and dirty zones for layout and unit arrangement, opposite orientation design of toilet units, use of anti-bacteria and easy-to-clean materials, natural ventilation for toilet, W-trap discharge pipe, double pipes system to eliminate the possible spread of virus and germs, and drainage system connections outside the units.

During construction, the project teams proposed infection control plans with various measures taken such as access monitoring, uniform arrangement of accommodations, and regular site training. Specific traffic control was taken for blocking the virus spread during cross-border transportation, e.g., transportation at night in Projects A and B.

4.2.3. Professional and Modularized Design

To deliver the project as fast as possible, 'less is more' was adopted as the design principle in Project A, which denoted to fulfill all functional requirements with minimized resources. After proposing 25 design schemes, an optimized design with repetitive units was selected. The optimized design was selected with the considerations as below: (1) it fulfilled the quarantine purposes, e.g., separation of clean and dirty zones; (2) it provided as many rooms and beds as possible; and (3) it involved the least materials that need to be procured from overseas. The project team streamlined all rooms into three types of modules. In Project K, extended modularized components were adopted to enable the fastest delivery of a high-quality hospital, e.g., modularized negative-pressure wards and building services modules.

4.2.4. Works Counting by Hours and Systematic Planning

To cater for the urgent social needs, all case projects adopted a 24-h working arrangement both on sites and in factories, with the principle of counting by hours. To ensure construction efficiency in such a tight program, systematic planning was conducted by the project teams both on site and in the factory. For example, the schedule and timeline of module delivery and installation were designed by minutes in Projects A and B. Modules were transported during nighttime and were installed successively from midnight to the early morning. The 'counting by hours' working arrangement made it possible for timely and fast project delivery (Figure 6), e.g., Project A was completed in 600 h and Project K in 120 days providing over 800 beds.

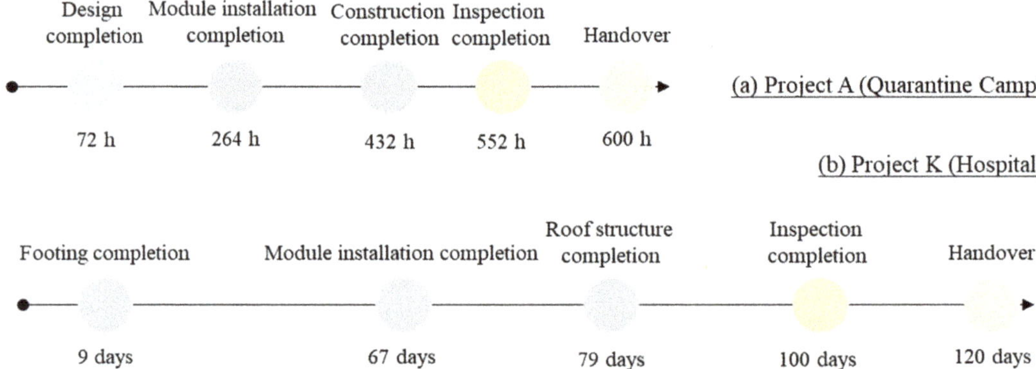

Figure 6. Timeline of project delivery.

4.2.5. Adoption of Smart Technologies

To address the challenging logistics and site constraints, the project teams adopted various smart technologies. For example, a cloud-based web portal was developed in collaboration with an academic research center for achieving real-time logistics monitoring in Projects A and B, which ensured the smoothness of the 24-h construction arrangement. In Project K, an AR-based building services checking system was developed by the main contractor, which allowed the construction team to easily do collision checking on site as hospitals normally involve complicated overhead ventilation pipes. Furthermore, an online quality checking platform was adopted in Project K to facilitate remote coordination between factory production and the project supervision team.

4.2.6. Project-Specific Strategies

To address the project-specific challenges, the project teams adopted corresponding strategies. For example, in Project K, vehicles with the super-low trailer were used, and transportation was arranged at night for the oversized modules. In Project H, the site was divided into several zones to facilitate efficient resource mobilization.

5. Results of Cross-Case Study

5.1. Measured Construction Efficiency

Cross-case studies on Projects A to L were conducted to evaluate the effectiveness and efficiency of emergency healthcare project delivery using modular construction. The time- and cost-efficiency were quantitatively measured and shown in Figure 7, and several interesting findings were identified from the measured results.

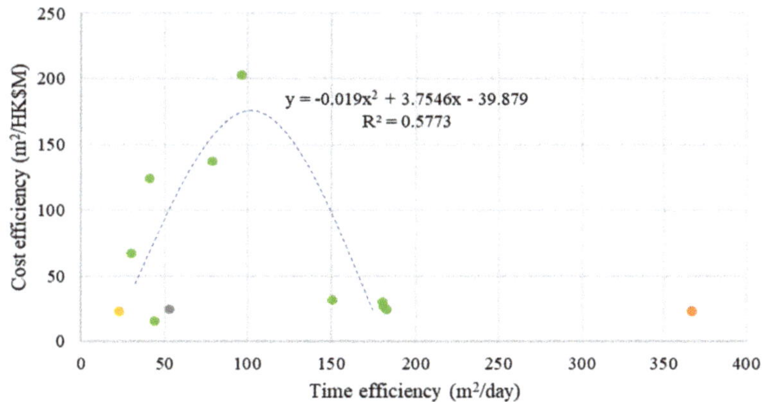

Figure 7. Measured time-efficiency and cost-efficiency. Notes: green dots—Projects A to I (modular QCs); yellow dot—Project J (conventional QC); orange dot—Project K (modular hospital); grey dot—Project L (conventional hospital).

The time and cost-efficiency of modular quarantine camps varied from each other, which was due to different project complexities (e.g., scales), teams, and site constraints. Nevertheless, compared with conventional construction (i.e., 53.3 m²/day and 24.1 m²/HK\$million of Project J), modular construction (i.e., 109.6 m²/day and 73.1 m²/HK\$million on average) increased the time and cost-efficiency by 106% and 203%, respectively.

In addition, the mathematical expression generated from the statistical analysis is a quadratic function, where there is an optimal point (i.e., optimized time and cost efficiency for modular quarantine camps). From Project A to E (CFA: 2000 m²–6000 m²), cost efficiency increased with the increase of time efficiency; while from Project F to I (CFA: 13,000 m²–16,000 m²), cost efficiency decreased with the increase of time efficiency. It was mainly due to that the complexity of the project (e.g., CFA to be built) affects the efficiency of project delivery. Therefore, each project should be designed with an appropriate scale (e.g., moderate CFA) to best mobilize the resources for achieving maximized efficiency. Nevertheless, the results were derived assuming that all projects were delivered in the same level of urgency.

Regarding the delivery of hospitals, the time efficiency was greatly enhanced by using modular approach (Project K: 366.7 m²/day) versus conventional construction (Project L: 23.2 m²/day). According to the interviews with the project teams, the cost efficiency of using modular construction should be around 20% higher than using conventional construction. The results suggest that with proper planning and design modular construction can greatly enhance the time- and cost-efficiency of various healthcare facilities under pandemics.

5.2. Qualitative Performance Analyses

Qualitative performance was analyzed in terms of not only economic performance but also environmental and social aspects. Economically, modular construction was proved efficient by speeding up the construction process but without cost increase, which echoes the quantitative measurement. In particular, modular construction greatly improved construction productivity and reduced delivery uncertainties, which addressed the challenges under the COVID-19 pandemic such as insufficient labor.

Environmentally, modular construction enhanced sustainability by reducing construction waste and pollution. For example, the waste generated from Project A was decreased significantly by using steel-framed modules; and noise generated in Project K was largely reduced with the mass adoption of prefabricated components. The reduced waste and pollution facilitated sustainable delivery of healthcare facilities under COVID-19.

Socially, modular construction addressed urgent social needs by delivering healthcare facilities in an ever-fast manner. In addition, the modular healthcare facilities could be easily disassembled after the pandemic and re-located for other purposes such as transitional housing, which could further address the severe housing shortage and unaffordability. In terms of lifespan, as interviewed with the case project teams, all these healthcare facilities comply with the design standard of permanent structures and should also have the same service life as that using conventional construction.

From the analysis above, modular construction can not only facilitate an efficient and sustainable response to COVID-19, but also help with the improvement of community recovery process through risk reduction of the built environment and effective integration of stakeholders along the construction supply chain to build back better.

6. Discussion

Derived from the results of the case studies and the evaluation, a systematic framework of modular construction-enabled response to COVID-19 was developed to facilitate efficient delivery of healthcare facilities. As is shown in Figure 8, the framework is process- and stakeholder-integrated, and involves the principles of stakeholder collaboration, professional and modularized design, early involvement of contractors, and the adoption of smart technologies.

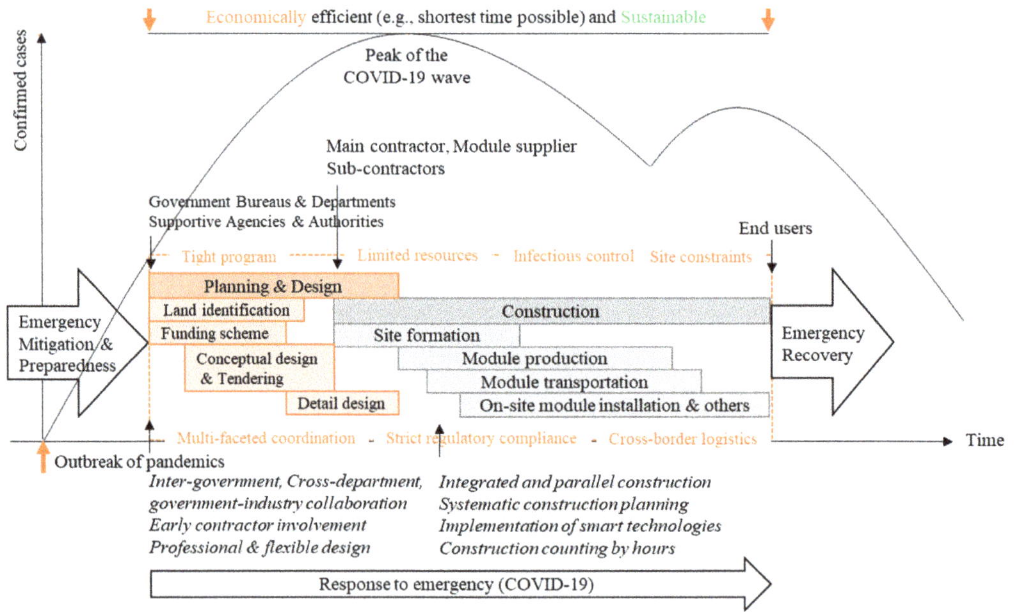

Figure 8. Framework of modular construction-enabled efficient response to COVID-19.

6.1. Major Principles of Modular Construction-Enabled Response to COVID-19

Collaboration mainly resides in the aspects of inter-government, cross-department, and government-industry collaboration. The three aspects echo the suggestions by Chen et al. [35] that joint effort should be made among public and private sectors across local, national, and international boundaries. The fast delivery of modular quarantine camps indicated the importance of cross-department and government-industry collaboration, e.g., to streamline the statutory submission and approval process. The importance of inter-government collaboration was reflected in Project K that the HKSAR Government teamed up with Shenzhen Government to rapidly set up the modular hospital delivery strategies. Multi-stakeholder collaboration and coordination is extremely important for modular construction compared with conventional practices [43]. For example, the guaranteed inter-government coordination in Project C ensured smooth cross-border logistics of module transportation. Government-industry collaboration is also necessary for emergency project delivery to facilitate efficient resource mobilization (e.g., water and electricity supply) considering the compressed time frame, which was fully reflected in Project B.

Professional and modularized design facilitates quality and fast delivery of emergency healthcare facilities. First, healthcare facilities for COVID-19 should be designed considering the infectious control criteria such as clean and dirty zones [8]. Second, the modularized design enables parallel factory and on-site construction, which accelerates the project delivery process and addressed the critical pandemic impacts such as labor scarcity identified by Rani et al. [44]. By incorporating the critical features of modular construction and healthcare facilities, these principles should outperform the existing design and construction strategies for conventional building projects, e.g., cost-effective design for commercial buildings [45], and should enhance the existing emergency design strategies proposed by Chen et al. [30] and Capolongo et al. [32].

The main contractor and module supplier should be involved in the early stage to design for manufacture and assembly, which was also suggested by Tan et al. [46]. The project team can then fix the design as early as possible and avoid late changes which are not allowed under COVID-19. The efficiency of early involvement of construction teams was proved in all case projects. For example, the design of Project A was completed within 72 h with contributions from the main contractor. In addition, the early contributions by the contractors and module suppliers in modular construction can reduce the late design changes which always occur in conventional building construction, and thus can minimize the time and cost uncertainties of project delivery.

As the activities of emergency construction are normally counted by hours, the use of smart technologies can help ensure construction efficiency, for example, a digital monitoring platform in Project A for coordinating off-site and on-site logistics [8] and a quality information management system for improving the efficiency of quality management process of module manufacturers [47]. Chen et al. [35] also suggested using smart technologies for emergency response, e.g., accurate time control with the assistance of sensor networks and GIS communication platforms. Apart from the enhancement of construction efficiency, the adoption of smart technologies in emergency healthcare project can also reduce the infection risks such as using tracking bracelets in Project B, and facilitate efficient design such as using a cloud-based synchronous collaboration platform in Project K.

6.2. Efficiency and Innovation of Modular Construction-Enabled Response to COVID-19

The results of the cross-case study indicated that the delivery of healthcare facilities using modular construction can enhance the cost-efficiency, which is inconsistent with Mao et al. [48] and Jang et al. [49] that prefabricated and modular construction was more expensive than conventional construction. Most importantly, the duration of building an emergency quarantine camp using conventional construction may take over a year, but only a few months by using modular approach. Nevertheless, to maximize both time- and cost-efficiency a large piece of land is suggested to be divided into a few for procurement, e.g., the development at Penny's Bay site was divided into Projects E to I. The framework was demonstrated efficient and effective in delivering community isolation facilities for addressing the 5th wave of the pandemic in Hong Kong, that 20,400 beds were delivered in 32 days to isolate the thousands of virus-infected cases [50].

Compared with the existing emergency response frameworks, e.g., that were developed by WHO [51] and FHB [39], the framework of modular construction-enabled response to COVID-19 is problem-driven (i.e., for pandemics), goal-oriented (i.e., economically efficient, socially and environmentally sustainable), stakeholder-integrated (i.e., governmental and industry stakeholders), and principle-explicit (i.e., principles concerning project planning, design and construction). By integrating modular construction into the emergency response process, the framework provides an exemplar for government-industry collaboration. Nevertheless, the effectiveness of the proposed framework and the time and cost efficiency of using modular construction can be further verified using more emergency healthcare building projects.

7. Conclusions

This paper has systematically evaluated the performance of modular construction for healthcare facility delivery in response to COVID-19. The evaluation was conducted based on the examination of the challenges to, strategies for, and efficiency of using modular construction for delivering emergency healthcare facilities. Multi-case studies were conducted using 12 real-life projects.

Within-case study revealed multi-faceted challenges to and corresponding strategies for the rapid delivery of modular healthcare facilities. The major ones are: (1) government-industry collaboration for addressing the limited resources available; (2) early contractor involvement and construction counting by hours for overcoming the tight program; (3) professional design for releasing the high pressure on COVID-19 prevention; and (4) inter-government collaboration and smart technologies for smooth cross-border logistics.

Cross-case analysis showed that modular construction can enable fast, cost-efficient and sustainable delivery of emergency healthcare facilities: (1) greatly improved economic efficiency, e.g., 106% improved time efficiency and 203% enhanced cost efficiency of the modular quarantine camps measured; and (2) enhanced environmental and social sustainability, e.g., reduced waste of materials.

Based on the multi-case analyses, a novel framework was developed to facilitate efficient delivery of modular healthcare facilities to address the issue of 'emergency response' in the circle of emergency/disaster management. Compared with the existing frameworks of emergency management (e.g., by WHO and Asian Disaster Reduction Center), it is innovative in three aspects. First, it integrates the multi-stakeholders along both the supply chain of modular construction (e.g., module supplier) and organizations for emergency response (e.g., Hospital Authority). Second, it involves a series of new principles such as inter-government collaboration to facilitate efficient logistics for module transportation. Third, it sets the goals of modular construction-driven emergency response, i.e., not only improved efficiency but also enhanced sustainability.

Practically, the identified challenges and strategies should assist both government and industry stakeholders in fighting COVID-19 by efficient delivery of modular healthcare facilities in a collaborative manner. Specifically, a joint working group could be formed with the involvement of building regulators, clients, contractors, and module suppliers to collaboratively deliver the healthcare projects as fast as possible. Theoretically, the developed framework should enhance the four-stage emergency management cycle by integrating modular construction into the stage of emergency response.

Although the study was conducted within the Hong Kong context, the paper should enlighten the emergency responses in other regions with established supply chains of modular construction. By exploring the contributions of the modular approach to addressing COVID-19, the paper should set an exemplar for linking the building construction industry with urban emergency management systems.

Author Contributions: Conceptualization, W.P.; data collection and analysis, W.P. and Z.Z.; writing—original draft preparation, Z.Z.; writing—review and editing, W.P.; supervision, W.P.; funding acquisition, W.P. All authors have read and agreed to the published version of the manuscript.

Funding: This research was funded by the Development Bureau of the HKSAR Government (Project No. 200009500) and the Strategic Public Policy Research Funding Scheme from the Policy Innovation and Co-ordination Office of the Government of the Hong Kong Special Administrative Region (HKSAR) (Project No. S2019.A8.013.19S).

Institutional Review Board Statement: The study was conducted in accordance with the Declaration of Helsinki, and approved by the Ethics Committee of The University of Hong Kong (Reference Number: EA1904016 and EA1909001; Date of approval: 26 April 2019 and 10 September 2019).

Informed Consent Statement: Informed consent was obtained from all subjects involved in the study.

Data Availability Statement: All data, models, or code generated or used during the study are available from the corresponding author by request.

Acknowledgments: Acknowledged are the project information provided by the Architectural Services Department of the HKSAR Government and China State Construction (Hong Kong) Limited and the individuals who participated in the case studies.

Conflicts of Interest: The authors declare no conflict of interest.

References

1. Rhodes, O.; Rostami, A.; Khodadadyan, A.; Dunne, S. Response Strategies of UK Construction Contractors to COVID-19 in the Consideration of New Projects. *Buildings* **2022**, *12*, 946. [CrossRef]
2. Pan, Y.; Zhang, L.; Yan, Z.; Lwin, M.O.; Skibniewski, M.J. Discovering optimal strategies for mitigating COVID-19 spread using machine learning: Experience from Asia. *Sustain. Cities Soc.* **2021**, *75*, 103254. [CrossRef] [PubMed]
3. Aghapour, A.H.; Yazdani, M.; Jolai, F.; Mojtahedi, M. Capacity planning and reconfiguration for disaster-resilient health infrastructure. *J. Build. Eng.* **2019**, *26*, 100853. [CrossRef]
4. Lestari, F.; Paramitasari, D.; Fatmah; Yani Hamid, A.; Suparni; EL-Matury, H.J.; Wijaya, O.; Rahmadani, M.; Ismiyati, A.; Firdausi, R.A.; et al. Analysis of Hospital's Emergency and Disaster Preparedness Using Hospital Safety Index in Indonesia. *Sustainability.* **2022**, *14*, 5879. [CrossRef]
5. Keenan, J.M. COVID, resilience, and the built environment. *Environ. Syst. Decis.* **2020**, *40*, 216–221. [CrossRef] [PubMed]
6. Chen, L.-K.; Yuan, R.-P.; Ji, X.-J.; Lu, X.-Y.; Xiao, J.; Tao, J.-B.; Kang, X.; Li, X.; He, Z.-H.; Quan, S.; et al. Modular composite building in urgent emergency engineering projects: A case study of accelerated design and construction of Wuhan Thunder God Mountain/Leishenshan hospital to COVID-19 pandemic. *Autom. Constr.* **2021**, *124*, 103555. [CrossRef] [PubMed]
7. MBI. *Innovating in Modular Construction: Broad's Holon Building*; Modular Advantage—September/October 2021 Edition; Modular Building Institute (MBI): Charlottesville, VA, USA, 2021.
8. Zhang, Z.; Pan, W.; Zheng, Z. Fighting Covid-19 through fast delivery of a modular quarantine camp with smart construction. *Proc. Inst. Civ. Eng. Civ. Eng.* **2020**, *2*, 89–96. [CrossRef]
9. Yazdani, M.; Kabirifar, K.; Fathollahi-Fard, A.M.; Mojtahedi, M. Production scheduling of off-site prefabricated construction components considering sequence dependent due dates. *Environ. Sci. Pollut. Res.* **2021**, 1–17. [CrossRef]
10. Yang, Y.; Pan, M.; Pan, W.; Zhang, Z. Sources of uncertainties in offsite logistics of modular construction for high-rise building projects. *J. Manag. Eng.* **2021**, *37*, 04021011. [CrossRef]
11. Prasad, K.V.; Bhat, N. Impact of the Covid-19 pandemic on construction organisations in India: A case study. *Proc. Inst. Civ. Eng. Civ. Eng.* **2022**, *175*, 17–21. [CrossRef]
12. Wang, Z.; Pan, W.; Zhang, Z. High-rise modular buildings with innovative precast concrete shear walls as a lateral force resisting system. *Structures* **2020**, *26*, 39–53. [CrossRef]
13. Pan, W.; Yang, Y.; Zhang, Z.; Chan, S. *Modularisation for Modernisation: A Strategy Paper Rethinking Hong Kong Construction*; CICID, The University of Hong Kong: Hong Kong, China, 2019; Available online: http://hdl.handle.net/10722/275575 (accessed on 20 August 2022).
14. Pero, M.; Stößlein, M.; Cigolini, R. Linking product modularity to supply chain integration in the construction and shipbuilding industries. *Int. J. Prod. Econ.* **2015**, *170*, 602–615. [CrossRef]
15. Pan, W.; Zhang, Z.; Xie, M.; Ping, T. *Modular Integrated Construction for High-Rises: Measured Success*; The University of Hong Kong: Hong Kong, China, 2020; ISBN 978-962-8014-29-3. Available online: https://www.miclab.hk/success (accessed on 20 August 2022).
16. Lawson, M.; Ogden, R.; Goodier, C. *Design in Modular Construction*; CRC Press: Boca Raton, FL, USA, 2014; ISBN 978-036-7865-35-1.
17. Kamali, M.; Hewage, K. Life cycle performance of modular buildings: A critical review. *Renew. Sustain. Energy Rev.* **2016**, *62*, 1171–1183. [CrossRef]

18. Loizou, L.; Barati, K.; Shen, X.; Li, B. Quantifying Advantages of Modular Construction: Waste Generation. *Buildings* **2021**, *11*, 622. [CrossRef]

19. Khan, A.; Yu, R.; Liu, T.; Guan, H.; Oh, E. Drivers towards Adopting Modular Integrated Construction for Affordable Sustainable Housing: A Total Interpretive Structural Modelling (TISM) Method. *Buildings* **2022**, *12*, 637. [CrossRef]

20. Masood, R.; Lim, J.B.P.; Gonzalez, V.A. Performance of the Supply Chains for New Zealand Prefabricated house-building. *Sustain. Cities Soc.* **2021**, *64*, 102537. [CrossRef]

21. Pan, W.; Pan, M.; Yang, Y. Implementing Modular Construction in High-rise High-density Cities: Perspectives in Hong Kong. *Build. Res. Inform.* 2022; *in Press*.

22. Gharib, Z.; Tavakkoli-Moghaddam, R.; Bozorgi-Amiri, A.; Yazdani, M. Post-Disaster Temporary Shelters Distribution after a Large-Scale Disaster: An Integrated Model. *Buildings* **2022**, *12*, 414. [CrossRef]

23. Salari, S.A.-S.; Mahmoudi, H.; Aghsami, A.; Jolai, F.; Jolai, S.; Yazdani, M. Off-Site construction Three-Echelon supply chain management with stochastic constraints: A modelling approach. *Buildings* **2022**, *12*, 119. [CrossRef]

24. Gharib, Z.; Yazdani, M.; Bozorgi-Amiri, A.; Tavakkoli-Moghaddam, R.; Taghipourian, M.J. Developing an integrated model for planning the delivery of construction materials to post-disaster reconstruction projects. *J. Comput. Des. Eng.* **2022**, *9*, 1135–1156. [CrossRef]

25. Drabek, T.E. Managing the emergency response. *Public Adm. Rev.* **1985**, *45*, 85–92. [CrossRef]

26. Waugh, W.L., Jr.; Streib, G. Collaboration and leadership for effective emergency management. *Public Adm. Rev.* **2006**, *66*, 131–140. [CrossRef]

27. WHO. *Environmental Health in Emergencies and Disasters: A Practical Guide*; World Health Organization: Geneva, Switzerland, 2002; ISBN 9-241-54541-0.

28. Mannakkara, S.; Wilkinson, S.; Francis, T.R. "Build Back Better" principles for reconstruction. *Encycl. Earthq. Eng.* **2015**, 328–338. [CrossRef]

29. Bae, Y.; Joo, Y.M.; Won, S.Y. Decentralization and collaborative disaster governance: Evidence from South Korea. *Habitat Int.* **2016**, *52*, 50–56. [CrossRef] [PubMed]

30. Chen, R.; Sharman, R.; Rao, H.R.; Upadhyaya, S. Design principles for emergency response management systems. *J. Inf. Syst. e-Bus. Manag.* **2007**, *5*, 81–98. [CrossRef]

31. Cowick, C.; Cowick, J. Planning for a disaster: Effective emergency management in the 21st century. Emergency and Disaster Management: Concepts, Methodologies, Tools, and Applications. *IGI Glob.* **2019**, 142–163. [CrossRef]

32. Capolongo, S.; Gola, M.; Brambilla, A.; Morganti, A.; Mosca, E.I.; Barach, P. COVID-19 and Healthcare facilities: A decalogue of design strategies for resilient hospitals. *Acta Bio. Med.* **2020**, *91*, 50–60.

33. Schexnayder, C.; Anderson, S. Emergency accelerated construction. In Proceedings of the Construction Research Congress 2010: Innovation for Reshaping Construction Practice, Banff, AB, Canada, 8–10 May 2010; pp. 837–848. [CrossRef]

34. Wang, Z.Z.; Shi, B. Decision making and economic analysis for accelerated bridge construction. *Appl. Mech. Mater.* **2013**, *423–426*, 2196–2201. [CrossRef]

35. Chen, R.; Sharman, R.; Rao, H.R.; Upadhyaya, S.J. Coordination in emergency response management. In *Communications of the ACM*; ACM: New York, NY, USA, 2008; Volume 51, pp. 66–73. [CrossRef]

36. McWilliams, M. Creating an effective emergency response team. *Prof. Saf.* **2020**, *65*, 66–70. Available online: https://www.proquest.com/docview/2409675256?pq-origsite=gscholar&fromopenview=true (accessed on 31 August 2022).

37. Gransberg, D.D. Early Contractor Design Involvement to Expedite Delivery of Emergency Highway Projects: Case Studies from Six States. In *Transportation Research Record*; Sage CA: Los Angeles, CA, USA, 2013; Volume 2347, pp. 19–26. [CrossRef]

38. SB. *Emergency Response System*; Security Bureau (SB), HKSAR Government: Hong Kong, China, 2020.

39. FHB. *Preparedness and Response Plan for Novel Infectious Disease of Public Health Significance*; Food and Health Bureau (FHB), HKSAR Government: Hong Kong, China, 2020.

40. GURL, E. SWOT analysis: A theoretical review. *J. Int. Soc. Res.* **2017**, *10*, 994–1006. [CrossRef]

41. Yin, R.K. *Case Study Research Design and Methods*; SAGE Publications, Inc.: New York, NY, USA, 2014; ISBN 978-145-2242-56-9.

42. Zhang, Z.; Pan, W. Multi-criteria decision analysis for tower crane layout planning in high-rise modular integrated construction. *Autom. Constr.* **2021**, *127*, 103709. [CrossRef]

43. Wuni, I.Y.; Shen, G.Q.; Osei-Kyei, R. Quantitative evaluation and ranking of the critical success factors for modular integrated construction projects. *Int. J. Constr. Manag.* **2020**, *22*, 2108–2120. [CrossRef]

44. Rani, H.A.; Farouk, A.M.; Anandh, K.S.; Almutairi, S.; Rahman, R.A. Impact of COVID-19 on Construction Projects: The Case of India. *Buildings* **2022**, *12*, 762. [CrossRef]

45. Tam, V.W.; Le, K.N.; Wang, J.Y. Cost implication of implementing external facade systems for commercial buildings. *Sustainability* **2018**, *10*, 1917. [CrossRef]

46. Tan, T.; Mills, G.; Hu, J.; Papadonikolaki, E. Integrated approaches to design for manufacture and assembly: A case study of huoshenshan hospital to combat COVID-19 in Wuhan, China. *J. Manag. Eng.* **2021**, *37*, 05021007. [CrossRef]

47. Shin, J.; Choi, B. Design and Implementation of Quality Information Management System for Modular Construction Factory. *Buildings* **2022**, *12*, 654. [CrossRef]

48. Mao, C.; Xie, F.; Hou, L.; Wu, P.; Wang, J.; Wang, X. Cost analysis for sustainable off-site construction based on a multiple-case study in China. *Habitat Int.* **2016**, *57*, 215–222. [CrossRef]

49. Jang, H.; Ahn, Y.; Roh, S. Comparison of the Embodied Carbon Emissions and Direct Construction Costs for Modular and Conventional Residential Buildings in South Korea. *Buildings* **2022**, *12*, 51. [CrossRef]
50. ArchSD. *Turning Impossible to Possible: Construction of Community Isolation Facilities*; CIC Power Talk: Hong Kong, China, 2022. Available online: https://citac.cic.hk/en-hk/news-and-events/events/past-events/past-events-details/363 (accessed on 30 June 2022).
51. WHO. *Emergency Response Framework (ERF)*; World Health Organization: Geneva, Switzerland, 2017; ISBN 978-924-1512-29-9.

Article

Green and Gold Buildings? Detecting Real Estate Market Premium for Green Buildings through Evolutionary Polynomial Regression

Domenico Enrico Massimo [1], Pierfrancesco De Paola [2,*], Mariangela Musolino [1], Alessandro Malerba [1] and Francesco Paolo Del Giudice [3]

[1] Department of Patrimony Architecture and Urbanism, Mediterranea University of Reggio Calabria, Melissari St., 89124 Reggio Calabria, Italy; demassimo@gmail.com (D.E.M.); mariangela.musolino@unirc.it (M.M.); malerbale@gmail.com (A.M.)
[2] Department of Industrial Engineering, University of Naples "Federico II", Vincenzo Tecchio Sq. 80, 80125 Naples, Italy
[3] Department of Architecture and Design, University of Rome "Sapienza", Borghese Sq. 9, 00186 Rome, Italy; francescopaolo.delgiudice@uniroma1.it
* Correspondence: pierfrancesco.depaola@unina.it

Abstract: This study concerns the determination of empirical evidence of a real estate market premium for Green Buildings and of an aware role of the private real estate market as driver to foster-up urban and architectural sustainability and energy efficiency. In real estate markets, there is growing relevance of Green Buildings, especially in cities where the greater part of residential buildings is built before the first regulations on energy performance. Through policies oriented towards sustainable practices, a twofold goal can be achieved: energy consumption mitigation respecting the historical value for existing buildings, direct economic impacts on real estate values. In some metropolitan or urban contexts, the "green premium" for buildings can be understood as a real "gold premium". This result has been highlighted and quantified with a real estate market analysis developed for a central area of an Italian mid-size city, pursued through the innovative tool of Evolutionary Polynomial Regression (EPR). The study highlighted a higher sale price for properties characterized by the best ecological characteristics and energy efficiency (+41.52%).

Keywords: green buildings; green premium; real estate market; hedonic price model; evolutionary polynomial regression

Citation: Massimo, D.E.; De Paola, P.; Musolino, M.; Malerba, A.; Del Giudice, F.P. Green and Gold Buildings? Detecting Real Estate Market Premium for Green Buildings through Evolutionary Polynomial Regression. *Buildings* **2022**, *12*, 621. https://doi.org/10.3390/buildings12050621

Academic Editor: Audrius Banaitis

Received: 30 March 2022
Accepted: 5 May 2022
Published: 7 May 2022

Publisher's Note: MDPI stays neutral with regard to jurisdictional claims in published maps and institutional affiliations.

1. Introduction

In real estate markets, there is a growing relevance and appreciation for Green Buildings, especially in EU countries, where the greater part of residential buildings is built before the first regulations on energy performance.

The European directive 31/2010/UE was imposed on member states to lower the energy consumption of buildings and provided the first definition of nZEB building ("nearly Zero Energy Building"); in Italy, this directive was received with Legislative Decree 63/2013, then converted into Law 90 on 3 August 2013, according to which, from 1 January 2019, new buildings occupied by public administrations and owned by the latter must be nZEB, including school buildings; from 1 January 2021, the above provision is extended to all new buildings and buildings undergoing major renovations, therefore, both public and private buildings. Based on the most recent surveys, Italy registers a 15.86% nZEB share for new construction [1].

Green Buildings have amply demonstrated to have not only lower energy bills but can improve indoor life and productivity, with their sustainable characteristics and design. All this determines, consequently, an increase in market value for buildings [2].

Over time, the building energy performance has become an effective assets valuation tool, as demonstrated by regulatory innovations in many cities or countries.

In Europe, the Energy Performance Certification (EPC) for buildings was provided for by the European directives 2002/91/CE and 2006/32/CE. However, this application has different implications in each of the European Member States, as reported by Buildings Performance Institute Europe (BPIE) [3,4]. The energy label is mandatory in order to purchase/sell a property in some countries (e.g., Spain or Italy), while, in other countries, there is not this legislative obligation (e.g., among these: Czech Republic or Netherland, for example) [5]. EPC may be based on a seven A–G ranks related to the energy classification system, with or without more subclasses (i.e., A and A+ as in Austria, Portugal and Ireland, for example [6]), using non-renewable primary energy or not (kWh/m^2 year) or CO_2 emissions as indicators to build reference levels for energy ranking. Some countries—Denmark and Hungary—do not use any indicator to describe building energy performance.

Based on the latest references provided by ENEA [7], in Italy, the buildings with high energy performance went from about 7% to 10% of the total in the period 2016–2019, thanks to the contribution of major renovations and new constructions; over 60% of the Italian housing stock is included in the least efficient energy classes (F–G) because it was mainly built between 1945 and 1972; new buildings represent only 3.4% of the Energy Performance Certificates (APE), and of these, more than 90% are of high energy performance (A–B). The non-residential sector, which accounts for 15% of the total APE, accounts for more than 50% of the certificates in the intermediate energy classes (C–D–E) and for more than 10% in the most efficient ones (A–B).

Several international experiences are focused on the topic of green premium for buildings, in terms of higher sale price for properties having better ecological characteristics and energy efficiency.

The growing interest in this theme is clear evidence that energy performance certification is an element having an increasingly greater weight in real estate investments.

The literature particularly recommends expanding the studies on real estate markets of mid-size cities and in marginal regions with low per capita output and income [8].

The main aim of this work is to highlight how, through policies oriented towards sustainable practices, direct economic impacts can be achieved on real estate values and how, in some urban contexts, the "green premium" for buildings can be understood as a real "gold premium".

Generally, sustainable interventions on buildings have economic impacts about two different premia: a technical-thermal premium in terms of better quality of indoor life and lower energy management costs and a market premium in terms of higher real estate value. Identifying and quantifying this latter aspect is particularly the main objective of the paper, to be pursued through the use of Evolutionary Polynomial Regression, a very innovative tool for analyzing real estate markets. In this sense, we also want to test the reliability and flexibility of using this innovative tool.

Existing buildings, built in a traditional way, use energy resources in an inefficient manner, and in their construction phases and management, produce waste, greenhouse gases and pollutants in large quantities. Green Buildings, unlike conventional buildings or "Brown Buildings", as well as trying to use the energy resources efficiently and avoid an excessive land consumption, aim to preserve environmental resources, together improving the properties' indoor life.

The definition of a "Green Building" has evolved over time. The U.S. Office of the Federal Environmental Executive (OFEE) defines it as *"the practice of (1) increasing the efficiency with which buildings and their sites use energy, water, and materials, and (2) reducing building impacts on human health and the environment, through better siting, design, construction, operation, maintenance, and removal—the complete building life cycle"* [9].

A Green Building is defined in a similar way by the Environmental Protection Agency (EPA): *"the practice of creating structures and using processes that are environmentally responsible and resource-efficient throughout a building's life-cycle from siting to design, construction, operation, maintenance, renovation and deconstruction. This practice expands and complements the classical building design concerns of economy, utility, durability, and comfort. Green building is also known as a sustainable or 'high performance' building"* [10].

The two above definitions refer to the concept of Life Cycle Assessment (LCA), which considers and values the impacts of a service or product under all the aspects (social, economic and environmental). LCA verifies whether a service or product used in a Green Building is truly green, taking into account solid waste generated in its phases of industrial production, exercise and final disposal; in other terms, it considers a wide range of environmental impacts, without even neglecting the issues related to global warming or water and air pollution associated with the service/product considered.

This article first highlights the main scientific literature about the impact of Energy Performance Certificate on real estate values, with exclusive reference to those papers which implemented market analysis using Hedonic Price Models (HPM). Afterwards, Section 3 delivers the methodological structure of Evolutionary Polynomial Regression. In Section 4, this tool was applied for the city of Reggio Calabria (Italy) in order to detect the market premium for Green Buildings. Section 5 contains some final reflections about the topic and a summary of the results.

2. Literature Review

As already mentioned, Green Buildings generate a whole-family benefits in terms of possible government incentives or tax deductions, better quality life and productivity, lower management costs, as many international researches suggest [11–15].

Partial equilibrium models have been adopted in these studies about real estate markets to analyze the short-term effects on the properties' rental rate considering architectural and technical features characteristics, among which the energy and ecological values added.

The increase in the demand curve for Green Buildings is demonstrated to the fact that if higher initial building costs for better ecological and family benefits are supported, this causes a consequent decline in demand for "brown" buildings (alternatives to formers).

These works assume that a "rent market premium" in the long-term exist related to innovations in order to favor the effectiveness of green measures and to reduce the ecological costs of building projects and preferences and behaviors tending to maintain adequate standards for the technological characteristics involved. A further "market premium on the sale price" is derived for Green Buildings by the combination of two factors: higher rent obtainable due to the better appeal of these buildings and the higher effect of intrinsic characteristics on real estate values (mainly due to lower risk for "negative premium" from the real estate market if this latter discloses a growing tend toward green buildings at the expense of existing non-green buildings; lower risk of future and possible carbon taxes; adaptation, in advance, to future and possible new environmental laws and regulations; lower management costs as those related to energy).

The EPC rating is one the most common information currently reported in every real estate advertisement, providing a description of building's energy performance mandatory by law. Because of this, many studies started to investigate whether and/or how much this information influences the choices in real estate markets.

Hedonic Price Models are the most suitable tools to analyze the EPC's impact on real estate values, in this direction a review of the literature was performed for European countries [16]. The main study cases refer to residential and commercial markets.

Looking far back in time, interest for energy efficiency and environmental awareness developed with the increase over time in hydrocarbons' prices: Johnson and Kaserman [17], Laquatra [18], Longstreth [19], Dinan and Miranowski [20] and Longstreth et al. [21] all tried to implement hedonic price models including among the real estate attributes as an energy variable. After these pioneering works, only in 2011 did studies start investigating the effect of "green" labels on commercial buildings such as ENERGY STAR and LEED [15]. For example, for downtown Chicago, Dermisi and McDonald [22] highlighted that only the

LEED-certified properties sold for a 23% price premium, while an Energy Star certification had no influence on real estate sale prices.

About more recent studies, in 2011, Brounen and Kok [23] highlighted for the Dutch residential market the appreciation of EPC in terms of faster time sales and increasing price. In 2012, Kok and Jennen [24] investigate on the EPC's effect on sale prices for Dutch cities, they highlighted that green office building rents were 6.5% higher with respect to those with lower EPCs by analyzing about 1100 rental transactions. Cajias and Piazolo [25] applied in 2013 a hedonic model to verify in Germany the relationship among sales and rentals in consequence of EPC: buildings with best energy classes (B, C and D) were rented more than the worst G-ranked buildings; similar results were confirmed for real estate sales (+32.8% in real estate values for buildings with low energy consumption). Also in 2013, for single-family homes in Sweden, Högberg [26] verified through 1073 real estate data a growing real estate prices when the energy class increases.

A report published in 2013 by the Directorate-General for Energy of European Community collected various studies referred to several European cities or countries (Austria, Belgium, France, England, Ireland) to elevate the EPC's effect on properties values [27]. This report highlighted that the EPC's impact on properties prices is often influenced from the time this certification was already mandatory in the countries analyzed: a significant effect on real estate values did exist where the EPC was a consolidated practice, while the influence of EPC was ineffective in those countries where the rule application was recent. Concluding, for all countries considered, the real estate values increase in correspondence to higher energy rankings.

In Ireland, Hyland et al. [28] detected in 2013 that A-ranked energy efficiency properties registered a rental price premium of about 2% and a sale price premium of 9%, if compared to D-ranked energy class. In addition, they highlighted that the scarcity of monetary resources to perform renovation interventions, determines for buildings that not require further investment on energy retrofitting a preference in the real estate market.

In Portugal, Ramos et al. [29] showed in 2014 that real estate units with better EPCranking (A, B, C), if compared with D-rank ones, were characterized by a 5.9% higher unitary price. A reduction of 4% in real estate prices was recorded for properties with low energy rankings (E, F, G). Subsequently, Evangelista et al. [30] confirmed in 2019 the results of Ramos et al. [29] for Portugal, with higher values of EPC appreciation (properties with A and B energy classes recorded a green premium of 12.5% for existing buildings and 13.1% for new buildings).

Studies in 2015 and 2016 developed by Fuerst et al. in U.K. [31,32] found a relationship between energy performance rankings and sale prices. Compared to D-rank properties, buildings with A- and B-rank recorded a 5% market premium, C-rank buildings recorded a 1.8% market premium, while buildings with lower energy classes (F, E, G) recorded a 1–7% reduction in sale prices [32].

For cultural aspects and, perhaps, for certainly milder climatic conditions, in Southern European countries, the EPC's impact on real estate prices is a topic on which little has been investigated.

Investigating on the premium price of dwellings with high energy rankings (A, B, C and D classes) in 2016, De Ayala et al. [33] detected in Spain a premium price for these real estate goods ranging between 5.4% and 9.8% compared to the less efficient ones. Also in Spain (Barcelona) in 2016, Marmolejo [34] highlighted a low effect on real estate prices, due to the fact that energy retrofitting upgrading costs are not sufficiently recuperable by sellers. Italian experiences are mainly attributable to Fregonara et al. [35] in 2017 and Bottero et al. [36] in 2018, evaluating the EPC's impact in the residential real estate market of Turin (Italy) through hedonic models. Their results suggested that, also in Italy, a significant appreciation for green buildings exists. Marmolejo and Chen [37,38] recurred to a spatial hedonic model detecting a significant increase in real estate prices related to the EPC ranking and how, in 2019, this economic impact varies in different housing segments in Barcelona: for more recent buildings the energy class does not have a significant impact on their prices, but they are instead relevant for all other properties. Again, Taltavull et al. [39] assessed in 2019 the green premium for buildings in the province of Alicante (Spain) in correspondence to various climate zones. In line with this latter study, and with reference to the metropolitan areas of Barcelona, Valencia and Alicante, Marmolejo and Chen [40] found in 2019 that the impact of energy performance was higher a scarcity of efficient homes in local real estate markets was evident.

In any case, it should be noted that some studies have shown no positive relationship between real estate prices and energy class. This because EPC is often used as "proxy" variable to include in it the impacts of more and omitted real estate characteristics. Among these studies, in 2017, Olaussen et al. [41] highlighted in Norway the possibility that energy efficiency may incorporate the effect of construction quality variable. In addition, Cerin et al. [42], analyzing 67,559 real estate transactions in Sweden in the time period 2009–2010, recorded a negative relationship among energy label and real estate prices, likely for the lack of an EPC classification reference value. However, the latter study is conflicting with the study of Högberg [26] that, for the same time period, verified in Stockholm a positive impact of better energy class in the property market.

The literature review confirms and recognizes that a market premium for buildings' green characteristics does exist. Hedonic price models have a higher degree of reliability and completeness. Certainly, the consumers' choices in building energy performance vary depending on location, economic factors, real estate stock, time and variation in climate zones.

3. Materials and Method: Evolutionary Polynomial Regression

Evolutionary Polynomial Regression (EPR) is a data-driven hybrid technique, where genetic programming is combined with numerical regression to develop flexible mathematical models, suitable for multiple application purposes.

The EPR approach overcomes the main limits of so-called "black-box" data-driven models. Often, the latter can be difficult to build or understood, or in other cases, they need many data that are difficult to be quantified or find. For example, artificial neural networks or genetic algorithms are effective to reproduce databases related to some observed phenomenon but have obvious limits in the model structure identification and overfitting. Instead, the stepwise regression is generalized in EPR by considering non-linear model components, although, with respect to regression parameters, these components are linear.

From this point of view, EPR is similar to non-linear global stepwise regression, since mathematical expressions of optimal models are searched taking into account a full set of available formulas by leveraging flexible modifications of the original mathematical structures. Some general expressions of EPR models can be represented as follows: in particular, they are "pseudo-polynomial" expressions because the parameters can be calculated as for a linear problem or for polynomial forms [43,44]:

$$\hat{Y} = a_0 + \sum_{j=1}^{m} a_j (X_1)^{ES(j,1)} \cdot \ldots \cdot (X_k)^{ES(j,k)} f\left((X_1)^{ES(j,k+1)}\right) \cdot \ldots \cdot f\left((X_k)^{ES(j,2k)}\right) \quad (1)$$

$$\hat{Y} = a_0 + \sum_{j=1}^{m} a_j f\left((X_1)^{ES(j,1)} \cdot \ldots \cdot (X_k)^{ES(j,k)}\right) \quad (2)$$

$$\hat{Y} = a_0 + \sum_{j=1}^{m} a_j (X_1)^{ES(j,1)} \cdot \ldots \cdot (X_k)^{ES(j,k)} f\left((X_1)^{ES(j,k+1)} \cdot \ldots \cdot (X_k)^{ES(j,2k)}\right) \quad (3)$$

$$\hat{Y} = g\left(a_0 + \sum_{j=1}^{m} a_j (X_1)^{ES(j,1)} \cdot \ldots \cdot (X_k)^{ES(j,k)}\right) \quad (4)$$

In the above equations, $\hat{Y}$ represents the vector of model predictions, m is the number of additive terms, the parameter a_j to be assessed is determined by a least squares (LS) method, X_k is candidate explanatory variables, the exponent (j, l) with $l = (1, \ldots, 2k)$ is the exponent of the *l*-th input within the *k*-th term (the exponents are chosen among candidate values, real numbers, which should include the value 0) and, at last, the function *f* is selected among a set of possible alternatives and may be exponential, tangent hyperbolic, natural logarithmic or others. Note that structure of Equation (4) requires the assumption that function *g* is invertible, due to the subsequent step of parameter estimation.

The LS method has the advantage of relating the "pseudo-polynomial" structure of the model with its coefficients; furthermore, it is possible to impose the LS to find for mathematical structures that only contain positive coefficients aj in particular modeling systems, where negative coefficient values are often used to balance the realization of specific errors related to the finite training dataset [45,46].

The structure of the model is searched by exploring the combinatorial space of exponents to be assigned to each candidate input. Any real number could be chosen as exponent values; however, they are coded as integers during the search procedure. Genetic algorithms and iterative implementation of LS method allow searching for EPR the statistically better function expressions that link the possible combinations of vectors of the explanatory variables (i.e., real estate characters) to the dependent variable (i.e., property sale price). Note that for EPR method implementation, an exogenous definition of the mathematical expression and a minimum number of parameters to fit the dataset in the best way are not required, since the optimal solution is directly provided by an iterative process related to genetic algorithm.

Two main phases characterize EPR: identification of the model structure by generating a set of polynomial expressions and traditional regression method to estimate the polynomial coefficients. At the basis of the algorithm used, there is the idea of generating a population of functional expressions considering their capacity to adapt to the available data. For this reason, the algorithm of EPR finds both the functional forms of the model and the values of the polynomial coefficients. All this without the identification a priori of a specific functional expression or with several inputs of the model, namely, the parameters and the exponents, preliminarily defined at the first stage of the method implementation.

The Coefficient of Determination (COD) allows us to check the statistical accuracy of each model provided by the EPR implementation, ranging between 0 and 1:

$$\text{COD} = 1 - \frac{N-1}{N} \cdot \frac{\sum_N (y_{estimated} - y_{detected})^2}{\sum_N [y_{detected} - mean(y_{detected})]^2} \tag{5}$$

where $y_{estimated}$ is the dependent variable value assessed by EPR, $y_{detected}$ is the collected values of the dependent variable and N is the dataset size. The model statistical accuracy is greater when the COD is close to the value 1.

A more recent version of EPR exploits Multi-Objective Genetic Algorithms (MOGA) to identify those models which maximize accuracy of data and parsimony of mathematical expressions simultaneously [47]. Then, EPR-MOGA provides an expression set with several accuracy to experimental data and different complexity degree of mathematical structure of models. The trade-off between accuracy and complexity allows an optimization strategy leading to a range of model solutions, among which the user could select the most appropriate one according to the specific requirements of interest and typology of experimental data considered.

The genetic algorithm underlying EPR-MOGA carries out a multi-objective optimization strategy based on the Pareto dominance criterion. These objectives aim to maximize the model accuracy with appropriate statistical criteria for verification of the model equation, maximization of the model's parsimony considering the minimization of the number of terms of the model equation and reduction in the model complexity by minimization of the number of explanatory variables in the model equation.

4. Discussion and Results

The aim of the case study is to highlight the capabilities of EPR-MOGA as an analysis tool and particularly to detect the market real estate premium for green buildings characterized by the higher degree of energy efficiency. Firstly, this application should demonstrate that EPR-MOGA is significantly helpful as a tool for data modeling and analysis; secondly, EPR-MOGA is tested on a wide real estate dataset aimed at determining the relationship between the selling price and a set of real estate characteristics influencing it (among them, the energy class of the building/residential unit).

To highlight the impact of Green Buildings in marginal economic regions, the case study examines a mid-size real estate market related to a non-metropolitan city (Reggio Calabria, about 180,000 inhabitants). In particular, for some its semi-central neighborhoods, EPR implementation involved a real estate market segment of newly built residential buildings. All features are unexplored so far in the real estate market analysis of Green Buildings.

The analysis concerns a sample of 515 residential properties located in the urban central area of Reggio Calabria (Southern Italy) and detected over 25 years. Only 24 are Green Buildings (energy class equal to A or B), but the data are not insignificant; on the contrary, it makes the research even more significant taking into account that the market for green buildings is practically in its infancy for the city of Reggio Calabria. Most of the observation of green buildings are located in zone 6, i.e., in a semi-peripheral area of a suburban district (see Table 1). This avoids any effects related to the context. We also excluded any possible interference due to the characteristic of its panoramic character.

Table 1. Variable description.

Variable	Description
Real estate price (PRC)	expressed in thousands of Euros
Property's age (AGE)	expressed retrospectively in no. of years
Sale date (DAT)	expressed retrospectively in no. of months
Internal area (AREA)	expressed in sqm
Number of services (BATH)	no. of services in residential unit
Positional Variable (ZONE)	expressed with a score scale (from 1 to 6, passing from more central areas to more peripheral areas)
Maintenance (MAIN)	expressed with a score scale (1 for bad conditions, 2 for mediocre conditions, 3 for good conditions, 4 for optimal maintenance state)
Floor level (FLOOR)	no. of floor levels of residential unit
Energy efficiency class (EN)	expressed with a score scale (1 for "A" or "B" energy efficiency class, 0 for "G" energy efficiency class)

Given the lack of transparency in the Italian real estate market, the methodology used in the research has provided data collection through the difficult and complex process of "elicitation", that is, the confidential confession of information to the researchers from: the direct actors (buyers, especially, and sellers), operators (promoters), intermediaries (realtors and agencies) and by notaries.

The quality and build type are the same for all sampled real estate units (apartments located in used multi-story buildings), and the central area of interest is homogeneous under the points of view regarding the qualification and distribution of main urban services. About the sample, georeferencing procedures have been used to verify the "Zone" variable to facilitate and support data building. For this purpose, with WebGIS tools, the data collection is facilitated by the fact that the geodatabase makes available, in a coordinated way and continuously, every real estate document specifically relating to each property, with further possibilities to integrate the information systems [48,49].

In Table 1, the variables considered are described, their acronyms given, their typologies provided (cardinal, ordinal or dummy) and the description and measurement units specified. Table 2 reports the statistical description of real estate variables.

Table 2. Statistical description of variables.

Parameter	PRC	AREA	FLOOR	MAIN	AGE	BATH	DAT	ZONE	EN
Mean	124,360.35	106.51	3.13	3.00	27.33	1.62	10.43	4.10	0.05
Median	115,000.00	105.00	3.00	3.00	24.00	2.00	10.00	5.00	0.00
Std. Dev.	150,000.00	120.00	3.00	3.00	5.00	2.00	8.00	5.00	0.00
Kurtosis	60,844.53	29.87	1.47	0.89	19.01	0.54	6.92	1.56	0.22
Asymmetry	4.71	5.16	−0.92	−0.66	1.07	2.04	−0.20	−1.23	15.82
Min	1.62	1.26	0.21	−0.47	1.04	0.42	0.45	−0.35	4.21
Max	13,000.00	39.00	1.00	1.00	0.00	1.00	0.00	1.00	0.00

EPR-MOGA methodology is iteratively implemented for the real estate sample by considering the structure of the generic expression as reported in the mentioned equation (1) without the function (f). Each additive monomial term of the mathematical expression is assumed to be a combination of the input variables raised to the proper numerical exponents. The candidate exponents selected from the research belong to the set (0; 0.5; 1; 2), and the maximum number l of additive terms in the final expressions is assumed to be 5. The implementation of the econometric technique for the real estate sample considered has generated several models. The optimal model to be analyzed for highlighting the relationship between the real estate characteristics and the selling price has been selected according to the statistical performance level, the complexity of the algebraic expressions and the coherence of the coefficients' signs under an empiric profile. The first two aspects are resolved through the COD associated by EPR-MOGA with each model, and the mathematical form is visible in the quantity of the terms of the equation and in the combination of the variables in each term. The empirical coherence of the functional relationships between the explanatory variables in each model and the selling price, is a less immediate operation with some complex aspects related to the presence of more variables combined in the terms of the equation and/or they occur repeated more times.

The application of EPR-MOGA has generated five equations (Table 3) classified—from the first to the fifth—according to the increasing statistical accuracy of the outputs in terms of COD and to the complexity of the models in relation to the number of terms, the number of selected explanatory variables and the combination of the explanatory variables that constitute each term.

Table 3. Models generated by EPR-MOGA.

Model	Mathematical Expression
I	$\mathrm{PRC} = +5.8514\mathrm{e} - 7\,\dfrac{\mathrm{AREA}^{0.5}\cdot\mathrm{MAIN}^{0.5}\cdot\mathrm{BATH}^{0.5}}{\mathrm{ZONE}^{0.5}\cdot\mathrm{EN}^{0.5}} + 58,251.253$
II	$\mathrm{PRC} = +693.3236\,\dfrac{\mathrm{AREA}\cdot\mathrm{MAIN}^{0.5}\cdot\mathrm{BATH}^{0.5}}{\mathrm{ZONE}^{0.5}} + 37,692.884$
III	$\mathrm{PRC} = -197,671.948\,\dfrac{\mathrm{DAT}^{0.5}}{\mathrm{AREA}^{0.5}} + 629.405\,\dfrac{\mathrm{AREA}\cdot\mathrm{MAIN}^{0.5}\cdot\mathrm{BATH}^{0.5}}{\mathrm{ZONE}^{0.5}} + 103,351.203$
IV	$\mathrm{PRC} = -28,398.709\,\dfrac{\mathrm{DAT}}{\mathrm{AREA}^{0.5}} + 638.164\,\dfrac{\mathrm{AREA}\cdot\mathrm{MAIN}^{0.5}\cdot\mathrm{BATH}^{0.5}}{\mathrm{ZONE}^{0.5}} + 203.749\,\dfrac{\mathrm{AREA}\cdot\mathrm{FLOOR}\cdot\mathrm{MAIN}\cdot\mathrm{EN}^{0.5}}{\mathrm{ZONE}} + 70,234.119$
V	$\mathrm{PRC} = -500,117.237\,\dfrac{\mathrm{ZONE}^{0.5}}{\mathrm{AREA}^{0.5}} - 32,825.256\,\dfrac{\mathrm{DAT}^{0.5}}{\mathrm{AREA}^{0.5}} + 523.570\,\dfrac{\mathrm{AREA}\cdot\mathrm{MAIN}^{0.5}\cdot\mathrm{;BATH}^{0.5}}{\mathrm{ZONE}^{0.5}}$ $+197.789\,\dfrac{\mathrm{AREA}\cdot\mathrm{FLOOR}\cdot\mathrm{MAIN}\cdot\mathrm{EN}^{0.5}}{\mathrm{ZONE}} + 206,216.537$

The models selected by EPR-MOGA are characterized by a different algebraic form complexity, with COD ranging from about 29% to about 91%. Then, under the point of view of statistical performance indicator associated by EPR-MOGA, for some models, there is a high statistical reliability in terms of the coherence of the detected data (models IV and V).

The reliability of the model chosen is checked by another calculated statistical index, i.e., the absolute percentage mean error which takes into account all the percentage errors measured for each detected prices respect to corresponding values estimated through the model.

Interpreting the results, the only use of the statistical criterion would lead to choose equation V (see Table 3) as the model that better replicate the analyzed phenomenon, as it is characterized by a COD next to unity and, therefore, by a very high degree of statistical reliability. This model consists of all the explanatory variables considered, with exception of "AGE" variable (Table 4). This is because the property's age for the various real estate sampled is sufficiently overlapping.

Table 4. Variables selected (green) or excluded (red) by EPR-MOGA for each model.

Model/Variables	PRC	AREA	FLOOR	MAIN	AGE	BATH	DAT	ZONE	EN
Model I	green	green	red	green	red	green	red	green	green
Model II	green	green	red	green	red	green	red	green	red
Model III	green	green	red	green	red	green	green	green	red
Model IV	green	green	green	green	red	green	green	green	green
Model V	green	green	green	green	red	green	green	green	green

Table 4 shows, for each model, the variables selected by EPR-MOGA reputed as the most relevant on the real estate sale prices. Regarding this aspect, it should be pointed out that the internal area, maintenance status, positional variable and the number of bathrooms are included in all models, whereas the property age variable is not included in any model. The energy class variable is relevant in three of the five models.

Note that the complexity of the terms of the mathematical expression V does not allow an immediate interpretation of the functional relationships among the explanatory variables. For this reason, the functional links of the i-th independent explanatory variable with the variation in the selling prices has been explained through an exogenous simplified approach that, instead of determining the partial derivative of the dependent variable with respect to the i-th variable, considers the values of the other variables in the model equal to their average values of the starting database and provides the analysis of the variations in value of the assessed changes of selling prices in correspondence to each i-th variable in the admissible range of its corresponding sample values.

Among the main objectives is that of highlighting the impact of the variable energy class, whose presence, in the model with greater statistical reliability, determines a variation of 41.52%. Confirming the significant weight that this variable assumes in valorization of residential properties.

The outputs of the elaborations carried out for all models have been represented in Figures 1–5, where X_{axis} represents the number of observations, and on the Y_{axis} of each model, the selling price determined by EPR-MOGA (PRC_{EPR}) is compared to the selling price expected in correspondence of each model (PRC_{exp}).

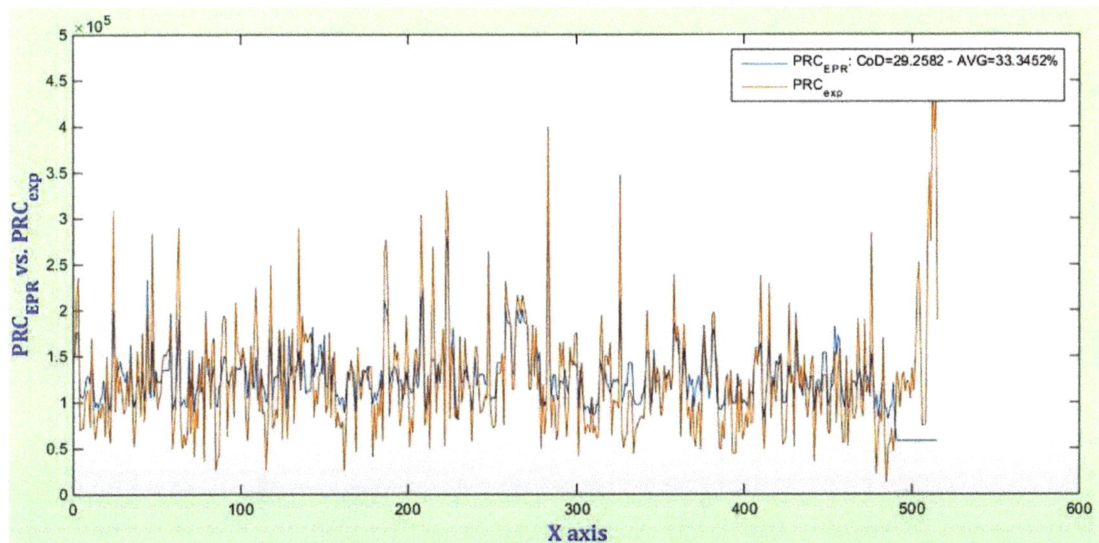

Figure 1. Selling price determined by EPR-MOGA (PRC$_{EPR}$) vs. selling price expected for model I (PRC$_{exp}$).

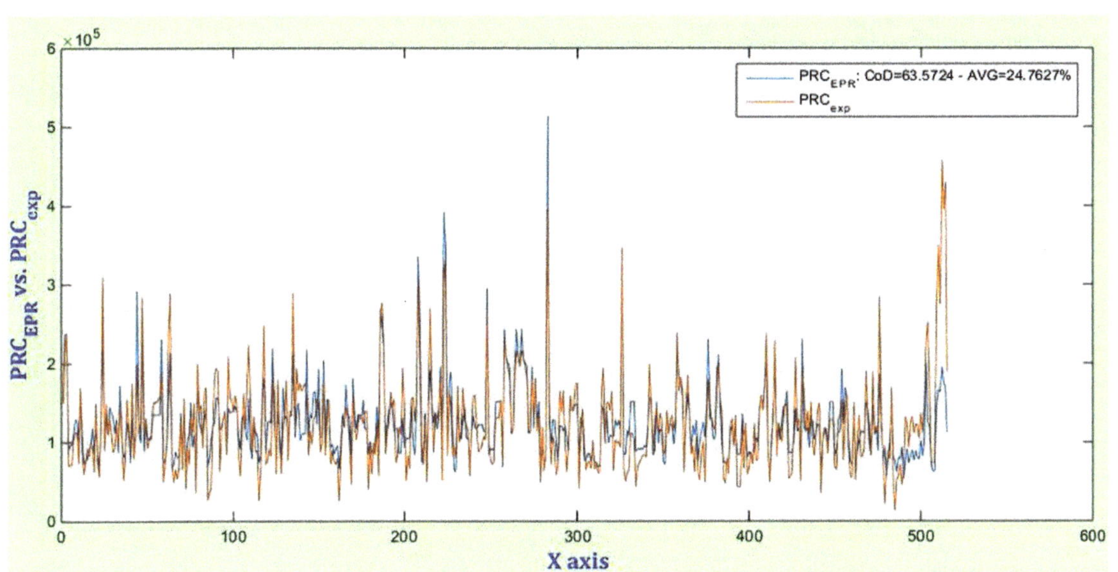

Figure 2. Selling price determined by EPR-MOGA (PRC$_{EPR}$) vs. selling price expected for model II (PRC$_{exp}$).

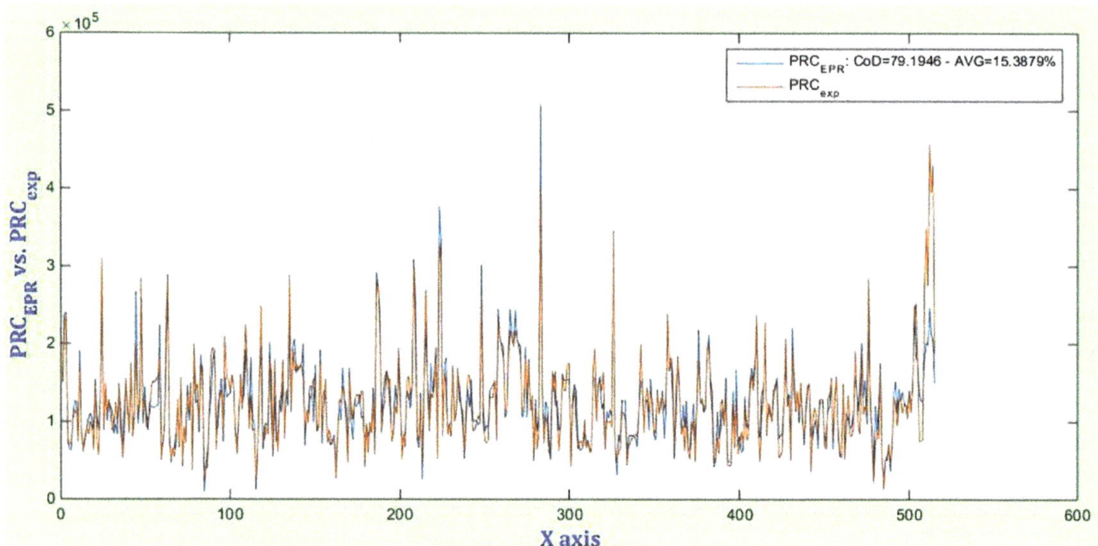

Figure 3. Selling price determined by EPR-MOGA (PRC_{EPR}) vs. selling price expected for model III (PRC_{exp}).

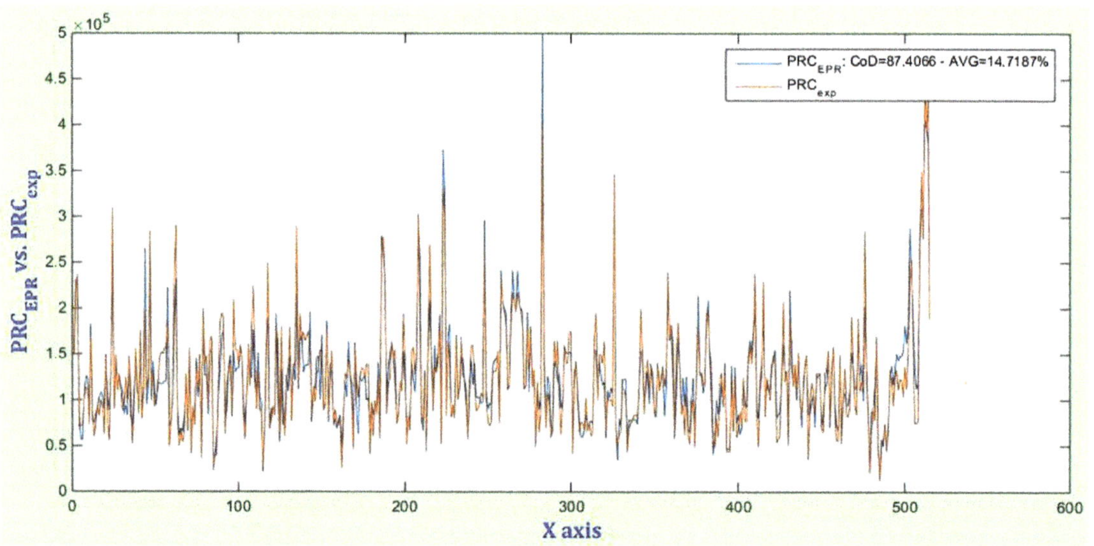

Figure 4. Selling price determined by EPR-MOGA (PRC_{EPR}) vs. selling price expected for model IV (PRC_{exp}).

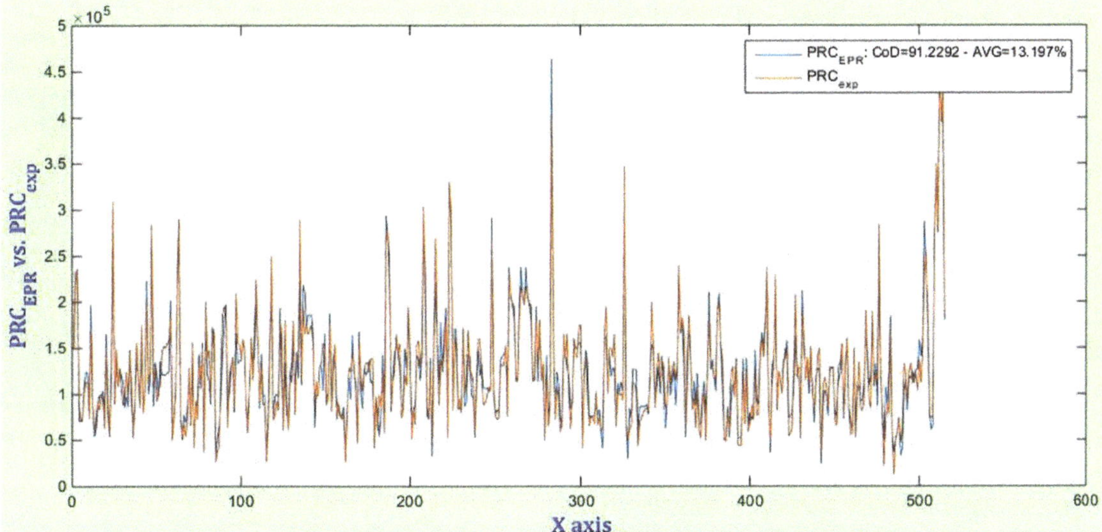

Figure 5. Selling price determined by EPR-MOGA (PRC$_{EPR}$) vs. selling price expected for model V (PRC$_{exp}$).

5. Conclusions

The research reached the objective to obtain empirical evidence of a real estate market premium for Green Buildings and of an aware role of the private real estate market as a driver to foster urban and architectural sustainability and energy efficiency.

The negative environmental impact of the construction and buildings sector has worsened over time due to the over-use of resources in the last decades. Pollution, mainly produced by energy over-consumption in buildings, has increased considerably due to wrong architectural design and urban management. A set of issues makes the adoption of general mitigation measures no longer able to be extended, creating to significant incentive toward building sustainability.

Urban and architectural policies, in fact, are increasingly oriented towards sustainability, energy efficiency, conservation, reuse, architectural retrofit in an ecological manner, re-vitalization of the existing city and a more effective management of historical and architectural heritage, including energy aspects.

Housing market analysis is the basis that defines links between housing characteristics and their market price.

In the study area, it has been detected that the first sale of apartments in some buildings with proven ecological characteristics carrying energy certification belonging to energy class A or B show a higher selling price than usual housing, which is unexpected in a small market of a poor city in a marginal region. This differential is due to the marginal price (i.e., market premium) paid for the ecological feature (i.e., energy efficiency).

Through a real estate market analysis carried out for the urban area of Reggio Calabria (Italy), the market premium for green buildings has been detected, i.e., the positive differential in terms of higher selling price for buildings having better energy efficiency and ecological characteristics. This research's goal has been pursued with an innovative tool: the Evolutionary Polynomial Regression.

The results obtained by the application of the proposed method suggest there is a percentage impact equal to 41.52% in relation to the incidence of the price of the properties on the presence of a good/excellent energy class (A or B).

Author Contributions: Conceptualization, D.E.M.; Data curation, M.M., A.M. and F.P.D.G.; Formal analysis, A.M. and F.P.D.G.; Investigation, P.D.P., M.M. and F.P.D.G.; Methodology, P.D.P.; Supervision, P.D.P.; Validation, P.D.P. and M.M.; Visualization, D.E.M.; Writing—review & editing, A.M. The paper is to be attributed in equal parts to all authors. All authors have read and agreed to the published version of the manuscript.

Funding: This research received no external funding.

Institutional Review Board Statement: Not applicable.

Informed Consent Statement: Not applicable.

Data Availability Statement: Not applicable.

Conflicts of Interest: The authors declare no conflict of interest.

References

1. Schimschar, S.; Hermelink, A.; Ashok, J. Zebra 2020—Nearly Zero-Energy Building Strategy 2020. Available online: http://bpie.eu/wp-content/uploads/2016/12/ZEBRA2020_Strategies-for-nZEB_07_LQdouble-pages.pdf (accessed on 19 March 2022).
2. Massimo, D.E.; Del Giudice, V.; Malerba, A.; Bernardo, C.; Musolino, M.; De Paola, P. Valuation of Ecological Retrofitting Technology in Existing Buildings: A Real-World Case Study. *Sustainability* **2021**, *13*, 7001. [CrossRef]
3. Buildings Performance Institute Europe (BPIE) Europe's Buildings under the Microscope. A Country-by-Country Review of the Energy Performance of Buildings. Available online: http://bpie.eu/wp-content/uploads/2015/10/HR_EU_B_under_microscope_study.pdf (accessed on 19 March 2022).
4. Buildings Performance Institute Europe (BPIE). *Energy Performance Certificates Across Europe: From Design to Implementation*; BPIE: Brussels, Belgium, 2010.
5. Andaloro, A.P.F.; Salomone, R.; Ioppolo, G.; Andaloro, L. Energy certification of buildings: A comparative analysis of progress towards implementation in European countries. *Energy Policy* **2010**, *38*, 5840–5866. [CrossRef]
6. Buildings Performance Institute Europe. Energy Performance Certificates across Europe. Available online: https://www.buildingrating.org/file/948/download (accessed on 19 March 2022).
7. ENEA—Agenzia Nazionale per le Nuove Tecnologie, L'energia e lo Sviluppo Economico Sostenibile. Available online: https://www.efficienzaenergetica.enea.it/vi-segnaliamo/online-il-rapporto-annuale-sulla-certificazione-energetica-degli-edifici-di-enea-cti.html (accessed on 19 March 2022).
8. Newell, G.; MacFarlane, J.; Walker, R. Assessing energy rating premiums in the performance of green office buildings in Australia. *J. Prop. Invest. Financ.* **2014**, *32*, 352–370. [CrossRef]
9. U.S. Office of the Federal Environmental Executive. Available online: https://www.naturalstoneinstitute.org/default/assets/File/consumers/historystoneingreenbuilding.pdf (accessed on 19 March 2022).
10. Environmental Protection Agency. Available online: https://archive.epa.gov/greenbuilding/ (accessed on 19 March 2022).
11. Quigley, J.M. Market Induced and Government Mandated Energy Conservation in the Housing Market: Econometric Evidence from U.S. *Rev. Urban Reg. Dev. Stud.* **1991**, *3*, 28–38. [CrossRef]
12. Quigley, J.M. *Why Do Companies Rent Green? Real Property and Corporate Social Responsibility*; University of California: Berkeley, CA, USA, 2009.
13. Fuerst, F.; McAllister, P. *Green Noise or Green Value? Measuring the Effects of Environmental Certification on Office Property Values*; SSRN eLibrary: Rochester, NY, USA, 2008.
14. Fuerst, F.; McAllister, P. *An Investigation of the Effect of Eco-Labeling on Office Occupancy Rates*; University of Reading, School of Planning, Reading: Berkshire, UK, 2009; Available online: https://papers.ssrn.com/sol3/papers.cfm?abstract_id=1140409 (accessed on 19 March 2022).
15. Fuerst, F.; McAllister, P. The impact of Energy Performance Certificates on the rental and capital values of commercial property assets. *Energy Policy* **2011**, *39*, 6608–6614. [CrossRef]
16. Marmolejo-Duarte, C.; Chen, A.; Bravi, M. Spatial Implications of EPC Rankings Over Residential Prices. In *Values and Functions for Future Cities*; Mondini, G., Oppio, A., Stanghellini, S., Bottero, M.C., Abastante, F., Eds.; Springer: Cham, Switzerland, 2020; pp. 51–71.
17. Johnson, R.; Kaserman, D. Housing market capitalization of energy saving durable good investments. *Econ. Inq.* **1983**, *21*, 374–386. [CrossRef]
18. Laquatra, J. Housing market capitalization of thermal integrity. *Energy Econ.* **1986**, *8*, 134–138. [CrossRef]
19. Longstreth, M. Impact of consumers' personal characteristics on hedonic prices of energy-conserving durables. *Energy* **1986**, *11*, 893–905. [CrossRef]
20. Dinan, T.M.; Miranowski, J.A. Estimating the implicit price of energy efficiency improvements in the residential housing market: A hedonic approach. *J. Urban Econ.* **1989**, *25*, 52–67. [CrossRef]
21. Longstreth, M.; Coveney, A.R.; Bowers, J.S. The effects of changes in implicit en- ergy costs on housing prices. *J. Consum. Aff.* **1985**, *19*, 57–73. [CrossRef]

22. Dermisi, S.; McDonald, J. Effect of "Green" (LEED and ENERGY STAR) designation on prices/sf and transaction frequency: The Chicago Office Market. *J. Real Estate Portf. Manag.* **2011**, *17*, 39–52. [CrossRef]

23. Brounen, D.; Kok, N. On the economics of energy labels in the housing market. *J. Environ. Econ. Manag.* **2011**, *62*, 166–179. [CrossRef]

24. Kok, N.; Jennen, M. The impact of energy labels and accessibility on o_ce rents. *Energy Policy* **2012**, *46*, 489–497. [CrossRef]

25. Cajias, M.; Piazolo, D. Green performs better: Energy e_ciency and financial return on buildings. *J. Corp. Real Estate* **2013**, *15*, 53–72. [CrossRef]

26. Högberg, L. The impact of energy performance on single-family home selling prices in Sweden. *J. Eur. Real Estate Res.* **2013**, *6*, 242–261. [CrossRef]

27. Lyons, R.; Cohen, F.; Lyons, R.; Fedrigo-Fazio, D. Energy Performance Certificates in Buildings and Their Impact on Transaction Prices and Rents in Selected EU Countries. Final Report Prepared for European Commission, DG Energy. Available online: https://ec.europa.eu/energy/sites/ener/files/documents/20130619-energy_performance_certificates_in_buildings.pdf (accessed on 19 March 2022).

28. Hyland, M.; Lyons, R.C.; Lyons, S. The value of domestic building energy efficiency—Evidence from Ireland. *Energy Econ.* **2013**, *40*, 943–952. [CrossRef]

29. Ramos, A.; Gago, A.; Labandeira, X.; Linares, P. The role of information for energy efficiency in the residential sector. *Energy Econ.* **2015**, *52*, S17–S29. [CrossRef]

30. Evangelista, R.; Ramalho, E.; Andrade e Silva, J. *On the Use of Hedonic Regression Models to Measure the Effect of Energy Efficiency on Residential Property Transaction Prices: Evidence for Portugal and Selected Data Issues*; REM 2019/64; REM—Research in Economics and Mathematics: Lisbon, Portugal, 2019.

31. Fuerst, F.; Oikarinen, E.; Harjunen, O. Green signalling effects in the market for energy-efficient residential buildings. *Appl. Energy* **2016**, *180*, 560–571. [CrossRef]

32. Fuerst, F.; McAllister, P.; Nanda, A.; Wyatt, P. Does energy efficiency matter to home-buyers? An investigation of EPC ratings and transaction prices in England. *Energy Econ.* **2015**, *48*, 145–156. [CrossRef]

33. de Ayala, A.; Galarraga, I.; Spadaro, J.V. The price of energy e_ciency in the Spanish housing market. *Energy Policy* **2016**, *94*, 16–24. [CrossRef]

34. Marmolejo Duarte, C. La incidencia de la calificación energética sobre los valores residenciales: Un análisis para el mercado plurifamiliar en Barcelona. *Inf. Constr.* **2016**, *68*, e156. [CrossRef]

35. Fregonara, E.; Rolando, D.; Semeraro, P. Energy performance certificates in the Turin real estate market. *J. Eur. Real Estate Res.* **2017**, *10*, 149–169. [CrossRef]

36. Bottero, M.C.; Bravi, M.; Dell'Anna, F.; Mondini, G. Valuing building energy efficient through Hedonic Prices Method: Are spatial effects relevant? *Valori Valut.* **2018**, *21*, 27–40.

37. Marmolejo-Duarte, C.; Chen, A. The evolution of energy e_ciency impact on housing prices. An analysis for Metropolitan Barcelona. *Rev. Constr.* **2019**, *18*, 156–166.

38. Marmolejo-Duarte, C.; Chen, A. The Uneven Price Impact of Energy E_ciency Ratings on Housing Segments and Implications for Public Policy and Private Markets. *Sustainability* **2019**, *11*, 372. [CrossRef]

39. Taltavull, P.; Anghel, I.; Ciora, C. Impact of energy performance on transaction prices. *J. Eur. Real Estate Res.* **2017**, *10*, 57–72. [CrossRef]

40. Marmolejo-Duarte, C.; Chen, A. La incidencia de las etiquetas energéticas EPC en el mercado plurifamiliar español: Un análisis para Barcelona, Valencia y Alicante. *Ciudad y Territorio* **2019**, *199*, 101–118.

41. Olaussen, J.O.; Oust, A.; Solstad, J.T. Energy performance certificates—Informing the informed or the indifferent? *Energy Policy* **2017**, *111*, 246–254. [CrossRef]

42. Cerin, P.; Hassel, L.G.; Semenova, N. Energy Performance and Housing Prices. *Sustain. Dev.* **2014**, *22*, 404–419. [CrossRef]

43. Giustolisi, O. Using multi-objective genetic algorithm for SVM construction. *J. Hydroinform.* **2006**, *2*, 125–139. [CrossRef]

44. Giustolisi, O.; Savic, D.A. A Symbolic Data-driven Technique Based on Evolutionary Polynomial Regression. *J. Hydroinform.* **2006**, *3*, 207–222. [CrossRef]

45. Giustolisi, O.; Doglioni, A.; Savic, D.A.; Webb, B.W. A multi-model approach to analysis of environmental phenomena. *Environ. Model. Softw.* **2007**, *5*, 674–682. [CrossRef]

46. Giustolisi, O.; Doglioni, A.; Savic, D.A.; di Pierro, F. An Evolutionary Multi-Objective Strategy for the Effective Management of Groundwater Resources. *Water Resour. Res.* **2008**, *44*, W01403. [CrossRef]

47. Giustolisi, O.; Savic, D.A. Advances in data-driven analyses and modelling using EPR-MOGA. *J. Hydroinform.* **2009**, *11*, 225–236. [CrossRef]

48. De Paola, P.; Del Giudice, V.; Massimo, D.E.; Forte, F.; Musolino, M.; Malerba, A. Isovalore Maps for the Spatial Analysis of Real Estate Market: A Case Study for a Central Urban Area of Reggio Calabria, Italy. In *Smart Innovation, Systems and Technologies*; Calabrò, F., Della Spina, L., Bevilacqua, C., Eds.; Springer: Cham, Switzerland, 2019; Volume 100, pp. 402–410.

49. Massimo, D.E.; Del Giudice, V.; De Paola, P.; Forte, F.; Musolino, M.; Malerba, A. GeographicallyWeighted Regression for the Post Carbon City and Real Estate Market Analysis: A Case Study. In *Smart Innovation, Systems and Technologies*; Calabrò, F., Della Spina, L., Bevilacqua, C., Eds.; Springer: Cham, Switzerland, 2019; Volume 100, pp. 142–149.

Article

BIM in the Center of Digital Transformation of the Construction Sector—The Status of BIM Adoption in North Macedonia [†]

Lihnida Stojanovska-Georgievska [1,*], Ivana Sandeva [1], Aleksandar Krleski [1], Hristina Spasevska [1], Margarita Ginovska [1], Igor Panchevski [1,2], Risto Ivanov [2], Ignasi Perez Arnal [3], Tomo Cerovsek [4] and Tomas Funtik [5]

[1] Faculty of Electrical Engineering and Information Technologies, Ss Cyril and Methodius University, Rugjer Boskovic 18, 1000 Skopje, North Macedonia; ivana@feit.ukim.edu.mk (I.S.); krleski@feit.ukim.edu.mk (A.K.); hristina@feit.ukim.edu.mk (H.S.); gmarga@feit.ukim.edu.mk (M.G.); igor@kreacija.org (I.P.)

[2] Kreacija Association of Business and Consultants Skopje, Nikola Parapunov 41, 1000 Skopje, North Macedonia; risto@kreacija.org

[3] BIM Academy, WITS Institute SL, Calle de Mallorca 346, 08013 Barcelona, Spain; ignasiperezarnal@bimacademy.es

[4] Faculty of Civil and Geodetic Engineering, University of Ljubljana, Kongresni trg 12, 1000 Ljubljana, Slovenia; tomo.cerovsek@fgg.uni-lj.si

[5] Faculty of Civil Engineering, Slovak University of Technology in Bratislava, Radlinského 11, 810 05 Bratislava, Slovakia; tomas.funtik@stuba.sk

* Correspondence: lihnida@feit.ukim.edu.mk; Tel.: +389-71-39-77-80

[†] This paper is an extended version of our paper published in the 9th Annual Edition of Sustainable Places (SP 2021) Conference, Rome, Italy, 29 September 2021; p. 8.

Citation: Stojanovska-Georgievska, L.; Sandeva, I.; Krleski, A.; Spasevska, H.; Ginovska, M.; Panchevski, I.; Ivanov, R.; Perez Arnal, I.; Cerovsek, T.; Funtik, T. BIM in the Center of Digital Transformation of the Construction Sector—The Status of BIM Adoption in North Macedonia. *Buildings* **2022**, *12*, 218. https://doi.org/10.3390/buildings12020218

Academic Editor: Bo Xia

Received: 24 December 2021
Accepted: 9 February 2022
Published: 15 February 2022

Abstract: The building sector nowadays has come to the stage where it needs a "digital" renovation. This is to be accomplished by an introduction of change into the methodology of construction and using new tools and technologies, such as BIM technology. This paper gives an insight into the status of BIM adoption in North Macedonia. It presents the threefold actions toward introduction of BIM in the national construction industry. These actions refer to scanning the current situation regarding digitalization of the sector, then taking promotional actions to express the benefits of BIM, and finally identifying and proposing the most suitable measures, summarized in the proposal of the National Roadmap for BIM adoption. The methods used consist of a brief literature review of the global status of BIM development. Then, the results of a survey conducted on more than 300 respondents representing a sample of building professionals in the country are discussed, and the barriers for successful BIM adoption are accordingly identified. The next step is to showcase the potential benefits of BIM for assessment of energy performance of buildings. As a final point, the conclusions drafted toward identification of the most important challenges are addressed in the proposed National Roadmap for BIM adoption.

Keywords: digitalization; BIM; construction; productivity; energy performance; BIM roadmap

1. Introduction

Construction serves as an integral part of the economy worldwide. About 6% of global GDP is generated by engineering and construction (E&C) activities, or approximately $10 trillion in the annual revenue. However, if we consider the level of productivity, the industry is failing to keep up with the gains made by other industries [1]. This disables wider growth of national economies, especially ones that rely more on construction, which is the case in North Macedonia. This industry has a very low level of adoption of new technologies, thus resulting in lower project efficiencies and a lower labor productivity. Digitalization has the potential to change this situation. The construction industry has continued to operate as in the past 50 years and remains the least digitized sector (except for agriculture) [2,3]. In this sector, manual labor is still mostly "on-board", together with commonly having workers with the type of knowledge based on "all-in-one" experience, without an insight into new technologies and established operating and business models, as is commonly the case in the Macedonian construction sector [4].

Embracing digitalization is necessary in order to increase the productivity of the sector and to bring innovations into the service of infrastructure and urban development (IU). Building information modeling (BIM) is a key first step toward these goals [5,6]. Concepts such as an increased productivity and efficiency, sustainability, an increased security, or saving financial resources are becoming synonymous with digitalization, but, over the last 20 years, productivity gains have only been about 1%. The implementation of BIM in construction is evidently a clear act toward digitalization, and it is expected that it will bring 13–21% savings in the design and implementation phase and 10–17% in the operation and maintenance within the global infrastructure market by 2025 [7–9].

BIM is the central piece of the industry's digital transformation based on a collaboration between different professional groups. It powers new technologies, such as mobile applications, automatization of processes, remotely driven equipment, and prefabrication, that are some of the "must-have" technologies when we speak about BIM. BIM is often perceived as a software for simulation of physical but also functional assets of an object, thus enabling new services and use of digital technologies [10–13].

1.1. Global Status of BIM Adoption

BIM as a new technology started its introduction in the construction sector over the last two decades, and it offers tools for transformation and enhancement of project performance and productivity by decreasing inefficiencies and lowering non-productivity, as well as by increasing the collaboration among different groups of project stakeholders. However, a number of barriers and obstacles have inhibited its more extensive implementation, despite the potential benefits [14]. BIM adoption means "the successful implementation whereby an organization, following a readiness phase, crosses the 'Point of Adoption' into one of the BIM capability stages, namely, modeling, collaboration, and integration". Over the past years, the adoption of BIM has significantly increased worldwide and particularly in the highly developed countries [15].

The United States are one of the pioneers in BIM development, whose adoption in the construction industry started from 2003 by launching the "National 3D–4D program" for gradual implementation of 3D, 4D, and BIM for all major public projects.

The first European legislative incentives toward BIM were released in 2014 with the European Commission directive 2014/24/EU, which recommended to member states "to use specific electronic tools such as BIM for public work contracts and design contests" (European Parliament, 2014). The UK is widely perceived to be a global leader in the adoption of BIM technology. After requiring BIM in all public works by 2016, the UK has moved toward BIM level 3, resulting in almost 70% of all professionals in the British construction industry using BIM by 2019 [16,17]. The Scandinavian countries are at the forefront of BIM adoption. In 2011, in the Netherlands, the Government Buildings Agency mandated the use of BIM for all public projects. Wide adoption of BIM is also shown in other EU countries, such as France, Germany, and Austria.

The situation with the introduction of BIM has changed in a more positive direction with the putting into place of BIM standards as European standards. In parallel, on the local level, there are national standards that are sometimes different from the standards at the European level in order to better reflect the national conditions [18–20].

The Republic of North Macedonia, as a part of the global community, is following the directions of the global movement toward digitalization in all sectors. However, the country is still in the very first segment of the road to more significant implementation of BIM. Therefore, the need to investigate the current status of BIM on the national level, to probe the attitude and preparedness of the sector toward its introduction, as well as to identify the main obstacles for wider BIM adoption on the national level arise as a necessary topic for profound exploration and study. A supporting consideration for justification of this need is the fact that also the literature review has identified a gap regarding sources related to the status or implementation of BIM at the national level. In fact, there are almost no publications and relevant sources in this topic, except for the references to the actions undertaken in the frame of the H2020 project TRAINEE, whose results will be a subject of elaboration in this work.

1.2. Benefits and Barriers for BIM Adoption

The literature review identifies many reported advantages of BIM adoption for executing construction projects over the traditional construction practices [21–24]. BIM benefits are identified in different building life cycle phases, as briefly summarized in Table 1.

Table 1. Benefits of BIM adoption.

Phases	Pre-Construction	Construction	Post-Construction
Benefits of BIM use	• Introduction of GIS (geographic information systems) data in digital infrastructure models • Improved accuracy of the data by using point cloud data of the site • Improving applicability of energy efficiency measures • Reducing design clashes in the early stage by visual representation of the model • Enhancing the accuracy of cost estimation • Checking of project constructability and sustainability	• Subsequent evaluation of the construction phases that enable accurate planning of resources • Exact planning of the storage and procurement of project resources • Prefabrication of building components • BIM allows better site utilization • Follow and allocate health and safety issues on the construction site • Real-time planning and financing dashboard completion • Connection to lean construction principles	• BIM as-built model after the construction effectively supports the operations, maintenance, repair, and replacement of appliances • Supports accurate and in-time management of the assets • Enables exact scheduling of maintenance operations and easy access to information • Preview of disassembling process for the end of the use of the facility

The literature review reports show that the barriers to BIM adoption are common for both developed and developing countries [25–27]. Table 2 summarizes some of the most frequently identified barriers for BIM adoption in the construction industry.

Table 2. Barriers for BIM adoption.

Origin	Personal	Corporate
Barriers for BIM introduction	• Lack of awareness regarding the potential of BIM • Low and inadequate offer of BIM trainings • No support from company's management structure • Longer time for model developing • Problem with software programs' interoperability • Nonexistent standardized tools and procedures	• High initial cost • Resistance to change current construction industry culture • Insufficient governmental support • Legal issues • Lack of interest from clients • Shortage of experienced BIM users • The issues related to intellectual property rights of the whole or parts of the model • Lack of demand from the contractors and sub-contractors • No legal requirements for BIM implementation

The brief scope into the global state of the art given above aims to give a general scope of the level of development reached by developed countries in order to identify the degree of lagging behind of the Macedonian construction sector and, even more importantly, to identify the good practice examples to follow in a proposed national approach. This also applies to the summarized information in Tables 1 and 2, which result from the summary of the most commonly identified benefits and barriers for BIM adoption worldwide. This extraction of the most frequently cited literature data will serve as a reference point to compare the identified issues from the corresponding categories after investigating the situation on the national level further in the study.

1.3. Key Insights and Recommended Actions for BIM Introduction

The World Economic Forum (WEF) report from 2016 notifies that the construction sector is evidently lagging behind most other industries in terms of embracing digital technologies. Other sectors have undergone tremendous changes during the last few decades and have experienced the benefits of the use of new processes and product innovations [5].

The increased efficiencies tackled with this change are becoming more and more recognized by governments across the globe [28,29]. Governments are learning from the good practice examples that strategic support of lean construction, BIM, and other innovations in the sector can help to address the rising problem with productivity in construction.

Increasing BIM adoption requires greater collaboration between different professional groups, higher motivation of the stakeholders, and expressing the right professional capabilities (Figure 1). In this process, the stakeholders from the construction industry should have the crucial role, but the first step is to understand how BIM benefits them and adds value to their projects. This should further develop new collaboration forms and should result in increased teamwork, integration of contracting processes, and implementation standards for sharing the project data among all of the relevant parties involved [30–32].

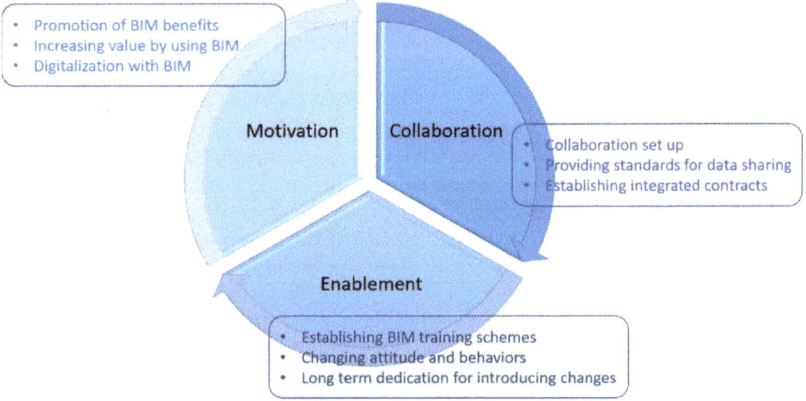

Figure 1. Fields of action toward BIM adoption (adopted by WEF), adapted from Ref. [6].

The approach proposed by WEF toward initiating adoption of digital transition is reflected in most of the countries that can evaluate their stage of BIM development as a successful one. It consists of a three-step organization of the target actions:

- Motivation
- Collaboration
- Enablement

2. Objectives and Aim of the Paper

The brief scope into the global status of BIM implementation given above does not aim to give a precise overview of the worldwide situation but rather to shed light on the level of development reached by developed countries in order to identify the degree of lagging behind of the Macedonian construction sector and to follow these achievements as targets in creating proposal and selecting measures to be undertaken on the national level further in the study. The previous paragraph reports on different levels of penetration of BIM in different countries, depending on the peculiarities of the national conditions, mainly in terms of the legal enablement and the procurement requirements, but they all have in common the unwavering determination to introduce digitalization into the national building sectors.

The Republic of North Macedonia, as a candidate country for EU membership, follows and transposes European legislation, including that regarding the use of digital tools in construction. However, the country is still at the start of enabling more significant implementation of BIM.

The current status of BIM, on a national level, regarding professionals' perception of it has not been studied elsewhere yet. This also applies to identification of the main obstacles and barriers for wider BIM adoption on a national level. These targets are the initial objectives of the research work presented in this paper. This paper, nonetheless, goes much further than just giving an insight into the attitude and awareness about BIM. Especially for a country with no particular advances in this topic such as North Macedonia, we find it necessary as well to consider a broader scope of actions to tackle the need for digital changes, to present some actions toward their enablement, and to propose measures and steps for a successful change in this situation.

In the following subsections, we will refer to the most commonly identified benefits and barriers for BIM adoption on a global level, including the successful actions taken to enable BIM implementation by other countries, identified through a literature review, in order to better explain the whole scope and background of the proposed actions. This will also serve for comparing these actions with the ones determined as appropriate on a national level. In addition, by referring to them, the selection of the most adequate measures will be governed so as to define an organized and structured approach to bring the national construction industry in track with contemporary achievements.

Specific Aim and Approach of the Study

The main objective of the paper has a broad scope—to give an extensive insight into the status of BIM adoption in North Macedonia and to propose a set of threefold actions toward introduction of BIM in the national construction industry. These threefold actions apply to an array of steps and organized efforts toward an introduction of BIM on the national level. Starting with scanning the current situation regarding digitalization of the sector, then taking promotional actions to express the benefits of BIM, and finally by identifying and proposing the most suitable measures, summarized in the proposal of the National Roadmap for BIM adoption, these actions are organized according to the three-step approach proposed by WEF, related to motivation, collaboration, and enablement.

In order to explain the starting points and the desired outcome of the whole process of initiating BIM, we propose this threefold approach, which can be expressed in terms of the process loop consisting of the three specific aims of this paper (Figure 2).

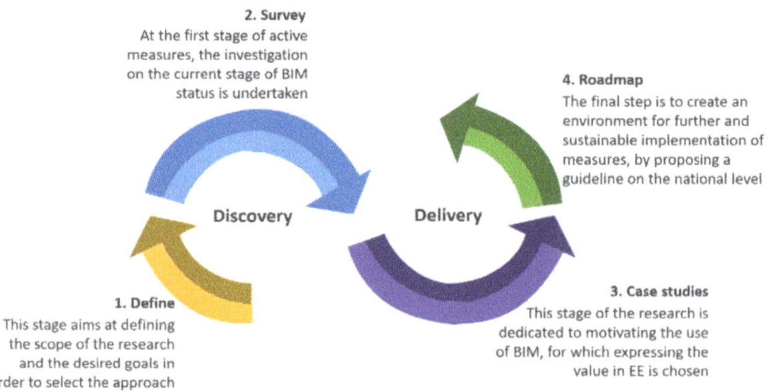

Figure 2. Specific aims of the paper presented as successive steps in a process loop.

After defining the scope of the paper and its impact, the three main consecutive actions are presented. These actions are consequently aligned, each dependent on the previous one. These three specific aims of the paper are aligned as successive steps, to lead to the desired goal and the enablement of more intense adoption of BIM. These specific aims are defined as:

1. Investigating the level of BIM awareness among building professionals in the Republic of North Macedonia—the method used is conducting a survey, as described in Section 3.1, while the survey results are analyzed in Section 4.1.
2. Demonstrating the feasibility of BIM adoption for energy consumption purposes—the method used is conducting case studies on the benefit of BIM for improving the energy performance of buildings, which is explained in Section 3.2, and the results of the case studies are then discussed in Section 4.2 of the paper.
3. Tracing the path for further BIM introduction in the construction sector—by proposing a document roadmap for national BIM adoption, whose approach is explained in Section 3.3, while the proposal itself is elaborated in Section 4.3.

Further in the paper, the following sections will explain the implemented methods in more detail, followed by the discussion and analysis of the obtained results.

3. Materials and Methods

Introducing digitalization in the construction sector in the Republic of North Macedonia is still at the beginner's level. Moreover, the experience of rare professionals and the sporadic implementation by a very few small design studios are a matter of ambitious devotion of the rare exceptions among construction engineers and architects who had embraced self-learning, aiming for further professional development. The remaining majority of the sector is still strongly traditional, both on an operational and management level. Thus, the initiatives for settling BIM implementation are easy to notice. Most of these initiatives are project-driven. Among them, one of the widest actions for the introduction of BIM was implemented by the H2020 project TRAINEE, https://trainee-mk.eu/en/ (accessed on 8 February 2022). The project TRAINEE (**T**oward market-based skills for sustainable energy-efficient construction) is built on the results of the previous national actions in the Build Up Skills Initiative in Pillars 1 and 2, concerning two priorities: creating a skilled building workforce and overcoming the barriers to implementation of EE measures in construction, operation, and maintenance. The overall objective of TRAINEE was to increase the number of skilled building professionals by addressing three topics:

(a) upgrading and developing qualifications and training schemes and setting up large-scale qualification for on-site workers, installers, and building professionals for energy efficiency measures in construction and renewable energy;
(b) improving the multidisciplinary approach to sustainable construction by initiating BIM at the national level as a tool to demonstrate reduction of the energy performance gap and for measurable energy saving resulting from improved construction skills; and
(c) market acceptance of developed construction skills by construction companies and professional trade associations through focusing on the higher value of the buildings, improved appreciation of the end user's satisfaction and quality of living, as well as increased market value of the companies that employ skilled professionals.

Having initialization of BIM as one of its main objectives, this project has taken the leading role to bring BIM technology closer to professionals in a wide range, thus acting as a promoter of digitalization in construction on the national level. Threefold actions were implemented: wide promotional activities and introductory campaign; development of training programs and training content and piloting the training schemes; and eventual drafting of the proposal for a national document to act as a guidebook for national BIM implementation. Thus, TRAINEE has been marked as a narrator of the new coming digital era in national construction.

The threefold actions discussed in this paper are undertaken in the frame of the TRAINEE project. They apply to the specific aims of the paper: to conduct a survey in order to study the level of BIM awareness; then, an exercise to demonstrate the value of BIM through expressing BIM benefits in achievement of energy efficiency; and, as an ultimate goal, to propose a detailed plan for the necessary actions to enable national BIM introduction.

3.1. BIM Survey for Level of Awareness

In order to weigh the penetration and acceptance of BIM practice in North Macedonia, a survey on the level of awareness, knowledge, and experience in implementing BIM practice was carried out within the project TRAINEE involving professionals from the Macedonian construction sector [33,34].

3.1.1. Selection of the Method and Survey Sample

According to the implemented methodology, as a technique for gathering data, a survey method was chosen consisting of questions distributed to different professional groups. The questionnaires were distributed via electronic communication by using the Google survey tool.

The respondents were selected by inviting them to participate in the survey, using the mailing lists of the registered professionals in the official Chamber of Certified Architects and Certified Engineers of Republic of North Macedonia and then the mailing lists of the members of other professional associations in the sector, such as the Engineering Institution of Macedonia; then, the companies working in the construction and energy sectors were invited to participate in the survey through the communication channels of the Economic Chamber of North Macedonia, and, finally, the representatives from the authorities and the public sector were reached using the communication network of the steering committee of the project.

An e-mail reminder was sent twice during the one-month period during which the survey was conducted.

Since building information modeling (BIM) aims to induce changes in both public and private sectors to work together in design, solving problems, and building better projects, faster and at less cost, the respondents involved in the survey were from three main sectors: AEC, public authorities, and academia. There were 1500 questionnaires distributed by e-mail among state officials, construction companies, and building professionals with job experience in AEC (architecture, engineering, and construction). From the total of 57,000 employees in the national construction sector, the sample size of 1500 questionnaires makes a 2.6% sample size percentage relative to the total population of interest. There were 312 valid responses obtained, which represents a 20,8% respondent answer rate. Responses were analyzed by using simple qualitative analysis of the obtained results because of the rather low answering rate relative to the total number of the sample size. Most of the total 312 respondents belonged to the AEC (architecture, engineering, and construction) sector, with over 78% share, or 244 respondents, while the public sector was represented with 55 respondents, or 18% share in the total number of valid responses, and academia was represented with 13 respondents, or 4%. The detailed structure of the companies to which the respondents belong is given in Table 3.

Table 3. Statistics about the profiles of the companies to which the respondents belong.

Number of Companies	Company Profile					
Sector	Building sector 230			Infrastructure 82		
Size of the company	1–4 91	5–10 48	11–20 33	21–49 41	50–99 18	over 100 81
Field	Construction 60	Investment 6	Public sector 18	Control 31	Design 177	University 13	Management 7

3.1.2. Preparation of Survey Questions and Structuring the Questionnaire

The exact formulation of the questions is presented in the graphical representation describing the survey's logic jumps, given in Figure 3 below.

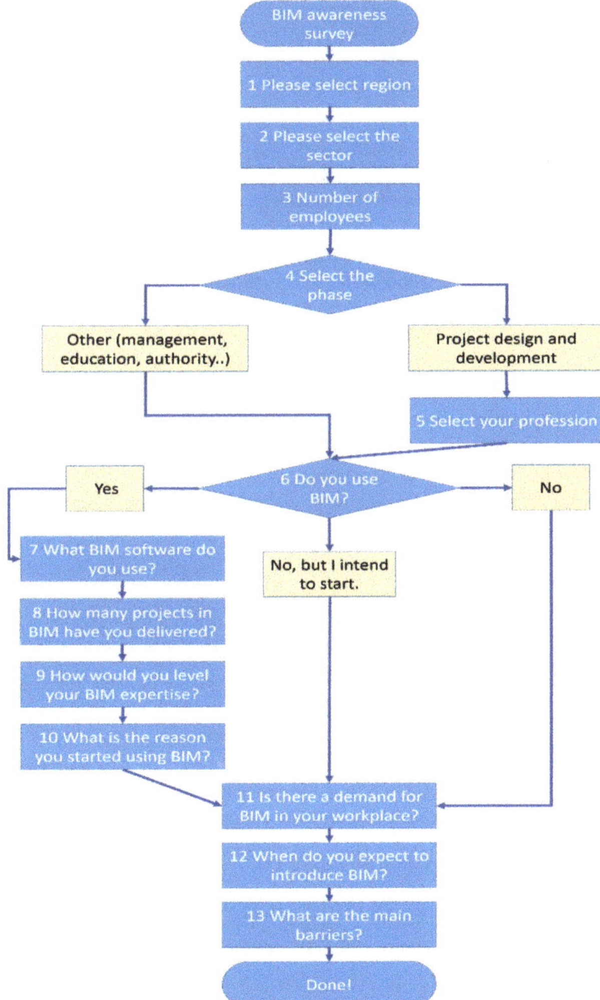

Figure 3. BIM awareness survey logic jumps.

The questions were close-end, and the questionnaire contained two parts.

- Part 1 included questions about respondents' geographical location, field of work in the construction sector, number of employees, current career stage (architect or engineer);
- Part 2 included questions about the skills of respondents in drawing, CAD tools, BIM, and use of software. This part also included the identified barriers for BIM implementation.

Questions from part 2 were precisely addressed with survey logic jumps, depending on the answers to the selection-making questions, that were used to determine the next questions in the survey (Figure 3).

3.2. Expresing the Value of BIM in Achieving Energy Efficiency in Construction

The second action undertaken in this work was realized in order to provoke stronger BIM penetration in practice, through a wide and comprehensive promotional action. Furthermore, to trigger more intense BIM implementation on the national level or introduction of BIM topics in academia and professional associations, the demonstration of the best practices and successful case studies appears to be the most indisputable approach for expressing the benefits. As explained above, in order to overcome the identified barriers that hamper BIM implementation on a national level, a set of measures were implemented by the H2020 project TRAINEE:

- wide public promotion of BIM involving six promotional workshops that gathered more than 300 interested building professionals and other stakeholders involved in the whole life cycle of construction [33];
- development of four qualification schemes for BIM for which series of pilot trainings were organized in order to make BIM benefits more understandable to the industry and the technicians involved;
- drafting a document that pathways the step-by-step introduction of BIM in a national construction industry. This document is the first of its kind in the country submitted for approval by the relevant authorities and aiming to support the implementation of the new national legislation for public procurement that should introduce the requirements for use of electronic tools in the procurement procedures, as encouraged by the Directive 2014/24/EU.

However, the strongest promotional impact comes from the demonstration of benefits of using BIM tools, as the certified professionals for BIM, through the pilot trainings, were engaged in piloting the use of a BIM software in order to express the potential of BIM for reduction of two gaps:

1. the first one between the projected and the actual energy consumption in the phase after construction, which can be reduced with better design of new buildings regarding their orientation and materials and minimizing energy needs; and
2. the second one between designed and actual energy performance of the building [33–36].

The first gap was demonstrated with a comparison of architectural design and the as-built object through analysis of construction materials used, the chosen products, systems installed, and energy efficiency solutions executed, which have made the difference between building qualities visible for both construction companies and consumers.

The second gap was demonstrated with a comparison between performance indicators during architectural design and the actual energy consumption after several years of usage.

Building Energy Modeling—BEM

Executing energy performance simulation of a building is a challenging task, especially as the building design usually experiences changes during the construction phase and dynamic changes in real time while it is being used. A building design process includes a series of activities and decisions in each design phase, but real occupancy, natural behaviors of its occupants, final thickness of used insulation materials, the accessibility to control devices, etc. make uncertain the complete accountability of the final energy state, even though it is made by professionals. In general, there are three phases of a new building project: concept or pre-design, final concept design, and development of executive detailed design and construction model. Thus, the modeling of building energy through BIM (in this case, known as building energy modeling—BEM) can be outdated if it is not fully connected with a building data model. This means that BIM and BEM procedures are to be executed simultaneously or, even better, in an integrated manner. Embedding the joint BIM+BEM process is a solution offering multiple iterations toward an optimized design in the environment of shared building data [35]. The difficulty in finding proper examples, investigation, and also specialized software and hardware demonstrates the value of the TRAINEE program.

There are three different methods for integrating BIM and BEM procedures [37,38]:

- Combined method: achieved by combining the BIM model and BEM sub-model, where both are completed using one or more tools. This requires a multi-specialized professional proficient in engineering and also in architectural tasks to complete the modeling and simulating in real time. The available software market does not provide a unique software that helps with that process, and it is necessary to use a combination of different softwares inside a suite or different software proceedings from different vendors.

- Central method: using shared environment for both procedures, where "readability" of specific data is ensured by an interoperability gateway, such as the IFC format. Again, it is necessary to use a combination of different softwares inside a suite or different software proceedings from different vendors, and so interoperability is a crucial requirement of the digital model.

- Distributed method: performed by transferring data between the design tool and simulation tools via a middleware.

Most preferable among them, according to the literature review but also confirmed with the records of executed projects, is the central method, and it corresponds to the chosen approach in the TRAINEE project in order to express the benefits of BIM toward energy efficiency in buildings.

The training, which has delivered more than 110 h of lectures to approximately 60 professionals and companies, has introduced the trainees to the advantages of the use of BIM software tools Edificius and TerMus PLUS, as well as SolariusPV for photovoltaic systems design, provided by the Italian software developing company ACCA Software [39,40]. It was a wining decision to select ACCA Software as the software provider, as it is a buildingSMART winner of a professional software award for 2019 [41,42] and CDE platform award for 2020. The importance of this selected software consists in the possibility of editing IFC data and that it is an affordable software compared to other software options, something that is of vital importance for massive adoption in a sector such as architecture and engineering.

3.3. Drafting National Roadmap for Impelementation of BIM

The third action in the threefold approach is the preparation of a document that will act as a guideline for enabling national implementation of BIM. Some of the main characteristics that this document should have are defined according to the best practice experiences for other countries, identified through the literature review [19–21]. They are given in continuation.

3.3.1. Strategic Goals of the National BIM Roadmap

The specific objectives that support this vision, to trigger the challenges identified by the BIM survey, are set out below:

- engaging the governmental bodies as an investor in public construction works, thus acting as a leader in supporting BIM adoption, is expected to act directly on the increasing of the demand;

- creation of a BIM certification body to support and enhance the training offer with defined training structure and content in terms of learning outcomes in order to ensure quality of the BIM skills offer;
- aligning proposed actions for digital transition of the national construction sector with EU and international regulation and guidelines will serve to facilitate adoption of international BIM standards, necessary to introduce a framework for a new digital approach in construction;
- involving all key enablers form different professional profiles to ensure a successful digital transition, thus supporting the building of a culture of collaborative projects.

These strategic goals will be evaluated upon achievement of key performance indicators (KPIs), defined in the official document of the National BIM Roadmap as quantifiable criteria for progress (success or failure). This strategic paper provides guidelines for the advancement of digital design and aims at optimization of the successful introduction of BIM.

3.3.2. Main Stakeholder Groups

Target groups of interested parties that should play a certain role in the process of implementation of BIM or take an advantage of digitization are grouped in three groups as follows:

- companies (construction companies, design studios, engineering firms, owners, developers ...);
- industry or professional bodies (manufacturing of construction materials and products, professional associations, engineering societies ...);
- government (public bodies, legal authorities, public education and training institutions ...).

It is to be mentioned that the achievement of the above-listed objectives shall be orchestrated in collaboration with relevant governmental and non-governmental stakeholders' engagement to ensure successful implementation of the proposed measures.

3.3.3. Types of Action and Fields of Action toward BIM Adoption

Based on the findings of the national BIM survey and knowing the current situation of the construction sector in the country regarding the level of digitization (the level of awareness of existing digital tools and the benefits of their use), as well as taking into account the available professional workforce and their interest and potential for adopting digital tools in their work (as expressed in the BIM survey), the identified necessary types of action toward introduction and wider BIM adoption are summarized as presented in Figure 4.

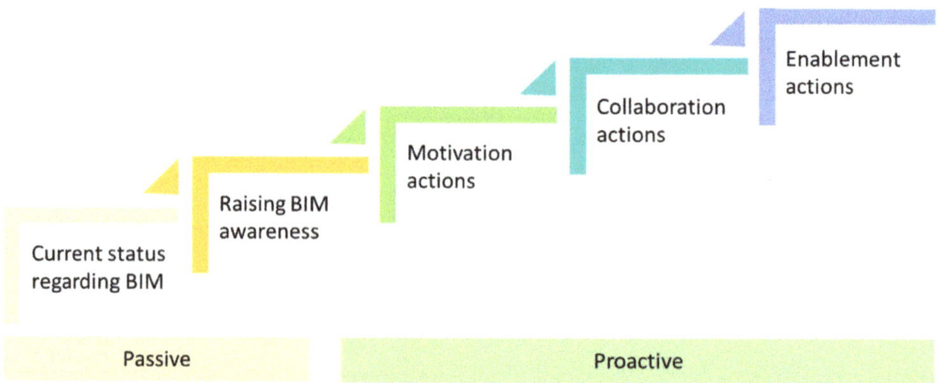

Figure 4. Proposed types of action toward BIM adoption in the National BIM Roadmap.

The actions proposed to achieve step-by-step introduction of BIM on national level are elaborated by types of action that correspond to WEF's proposed actions for BIM adoption [6]. These types of action can be distinguished into two groups: passive measures—not leading toward initialization of implementation of BIM but rather prerequisites for starting with implementation of the next group of proactive measures that occasionally will result in a systematic approach ensuring positive output. Here, awareness-raising activities are on the midway of the whole process, acting as a milestone and transition point between passive and proactive measures.

The above-mentioned types of action are to be taken in different "sectors of influence"—fields of action that will ensure coordinated driving force toward successful introduction of BIM on the national level. The necessary fields of action when pathwaying national BIM adoption, as defined by the WEF, are:

- Leadership
- Standards
- Education and training
- Procurement

4. Results and Discussion

4.1. The Analysis of Survey Data and Discussion

The data obtained through the survey were analyzed statistically and described by percentages, proportions, graphs, and charts for comparing the level of knowledge of AEC industry practitioners. The questionnaires aimed at gathering information in the following topics:

1. types of building professionals;
2. tools for BIM design; and
3. barriers/obstacles for increased BIM use.

The survey about penetration and acceptance of BIM practice shows that the most concerned by BIM software tools are architects and engineers, for their design purpose. Construction and renovation of buildings in residential and public sectors are shown to be more relevant, because over 70% of respondents belong to this sector, in comparison to the infrastructure sector. Half of the respondents have heard of BIM software tools, but only 2.5% are fully focused on delivering all of their designs in BIM. Most of these architects and engineers generally use two commercial software tools, Graphisoft ArchiCAD and Autodesk Revit. No openBIM tools or other alternatives were known to most of the respondents.

In the next section, descriptive statistical values for presenting the survey findings are given. The data gathered from the respondents presented in Table 4 demonstrate the descriptive statistics for the level of usage of BIM measured by a semi-Likert scale; illustrated are the mean and standard deviation of the findings based on the answers to questions related to the subject of usage of BIM. We introduce this modified application of the Likert scale to express the statements in terms of measurable values, marking with the highest value a strong agreement (or approval and acceptance) and with the lowest value the most negative attitude. Although this is not a standard use of the Likert scale, this analysis can give a quantitative impression of the attitude of the respondents and express trends of distribution of opposite statements.

Table 4 summarizes the conclusions that, in general, the respondents are interested in using BIM technology; however, its use is still not in place. From the respondents that are already familiar with the technology, most of them have declared themselves as advanced users, while the strongest deviation of the answers is shown for the question related to the origin of the initiative for starting with BIM implementation, where the majority have answered that it was on their own initiative.

Table 4. Statistics for the level of usage of BIM measured by a semi-Likert scale.

Semi-Likert Scale Value	3	2	1		
Do you use BIM?	Yes	Still not, but I am interested	No, not interested	Mean	SD
	24	201	87	1.79	1.32
Level of expertise	Expert	Advanced	Beginner		
	5	10	9	1.83	1.44
Behavioral attitude	My own initiative	Investor	Employer		
	19	3	2	2.71	2.24

The findings related to the behavioral attitude of the respondents regarding intentions for future use of BIM are presented in Figure 5a. The respondents show a strong belief that the introduction of BIM will come in the very near future (according to the answers, in the next 1–2 years), but the findings that can easily be listed as barriers to the introduction of BIM state very explicitly that currently the major problem for wider BIM adoption is the lack of demand by clients (Figure 5b).

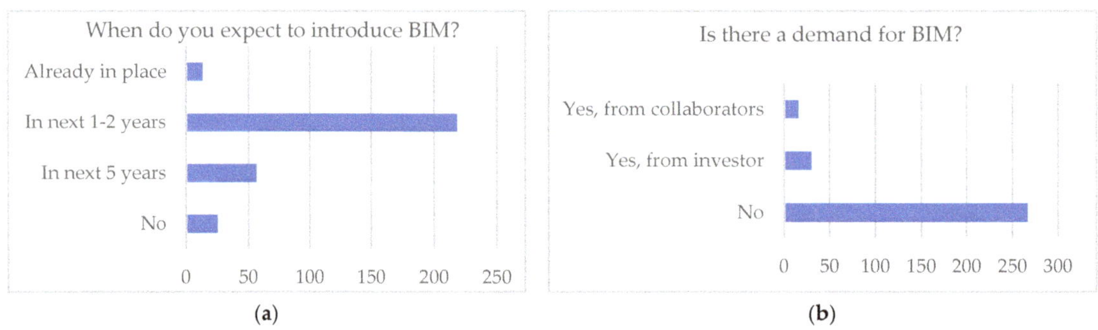

Figure 5. Attitude toward using BIM. (**a**) The intention to use BIM in the near future; (**b**) the existence of request for BIM.

Furthermore, the identified barriers for more intense introduction of BIM are identified and discussed. Figure 6 presents the radar chart of six sources that hamper BIM adoption, identified by the respondents of the survey, and their weight in terms of percentage of respondents that have chosen a certain option as the most important one. Here, it is to be emphasized that only one option identified as the most important could be selected.

The survey has identified challenges for further BIM adoption that correspond to the common conclusions on this topic, as is also the case also in other international studies. Maybe the most evident is the challenge to make a big switch from traditional to digitally driven construction. This requires changes in attitude, resulting from much harder introduction of new things in more traditional environments. The identified specific issues hampering the implementation of BIM are the lack of skilled professionals, the lack of BIM training, very low demand from market actors, and high initial cost for implementation of BIM standards and specifications. There are very limited courses teaching BIM-related software tools in North Macedonia. However, these training options are mostly commercial ones, available from one authorized training provider of Autodesk that offers training in only one specific BIM software, such as REVIT. Because the scope of this study did not cover exploring the presence of BIM in training programs in formal education (including vocational education for construction and architecture and academic studies in these fields), more profound study on the existing BIM programs in formal education was not conducted. However, the fact that there are no courses at the university level entirely devoted to BIM (or titled as so) is indisputable. There are only some parts of programs related to the issue of digital construction, covered with the syllabuses of other academic courses, mainly dealing with management of construction.

Figure 6. Barriers for BIM adoption identified during the BIM survey in North Macedonia, adapted from Ref. [34].

Main Conclusions and Recommendations upon Survey Results

The survey results gave a clear picture of where the focus should be placed in order to achieve successful introduction and acceptance of BIM practice.

Three major barriers that were detected through the survey are [35]:

1. lack of BIM training (33%, or 103 respondents);
2. insufficient requests from clients (no demand for BIM design) (21.2%, or 66 respondents);
3. missing standards and guidelines (15.4%, or 48 respondents)

The survey about penetration and acceptance of BIM practice shows that the most concerned by BIM software tools are architects and engineers, for their design purpose. Buildings are most appropriate for the introduction of BIM tools, because over 70% of respondents belong to this sector. Half of the respondents had heard of BIM software tools, but only 2.5% were fully focused on delivering all of their designs in BIM. Most of these architects and engineers have educated themselves on their own initiative for generally two commercial software tools, Graphisoft ArchiCAD and Autodesk REVIT. No openBIM tools or other alternatives were known to most of the respondents. Still, the low or nonexistent demand results in BIM trainees themselves being resistant and reluctant to learn new tools and workflows, as they perceive this as a waste of time and a hindrance to their productivity. The biggest challenge is to increase the diversity of BIM training options, enlarging the demand side, which will lead to higher penetration of BIM in the private sector, and this, from the other side, can easily lead to its implementation in the public sector and national legislation. Continuous education and training of employees are invoking real costs that could influence introduction of BIM standards. We always must have in mind that even the most proficient BIM user will need to learn continuously and evolve with their knowledge.

4.2. Discussion of the Results of Case Studies Using BIM in Achievement of Energy Efficiency BEM

4.2.1. The Scope of Realized Case Studies

Two case studies were conducted and are briefly presented below. It is important to fix the concepts that help to achieve better energy consumption through all phases of the building life cycle. Proper studies are necessary to maximize the saving while keeping energy consumption at the minimum level at the same time.

The first case study aimed at showing that proper urban planning can help to take advantage of suitable location, orientation, less shadowing (self-one or projected by neighboring buildings), and urban regulations favoring energy efficiency (number of glass openings, possibility to create balconies, terraces, flat roofs, acceptance of solar appliances and photovoltaic panels, etc.). When urban planning is fixed, the design process cannot vary these conditions, and these conditions could be easily checked through e-permit evaluation (the EUnet4DBP-European Union net for Digital Building Permits has made visible the importance of geographic information systems (GIS) and BIM connection for these parameters). This concept, called "intelligent communities life cycle" in the case of North Macedonia, is extremely useful, as one of the greatest shares in energy consumption is in public heating combined with private heating, one specific and very interesting energy mode in continental countries.

This "public energy consumption" makes it difficult to extract the exact bill for apartments and for housing blocks, and so a measure taken in this sense will be impossible to track per single use. Thus, the approach had to be changed toward developing sustainable communities through performance analysis solutions. This would have to be converted into individual building analysis that could permit there to be a counter per building instead of per user, enabling the design of new buildings as well as renovations that will lead to significantly less energy consumption, while maintaining occupant comfort. This situation means that personal behavior toward energy savings has to be extrapolated to "community" behavior, quite more difficult to track, empower, and enhance. Larger groups of buildings designed as residential homing blocks and neighborhoods will better apply these energy design concepts to entire communities and not just individual flats or single housing [42–51].

4.2.2. The Approach for Analysis of Case Study Results

TRAINEE has focused on the PV energy modeling proposal (as clean and CO_2-free energy, which works perfectly in cold and sunny countries, such as North Macedonia) and probably could have a second part for monitoring and analysis and considering how it can be integrated into a digital twin and a building digital twin for sustainable design. This approach will be more and more important, as we need SDGs (sustainable development goals) settled by the UN in 2015 to be globally achieved in 2030, net-zero buildings by 2050 in the UK and Europe, AIA 2030 Commitment in US, etc. In most cases, the communities that need to be reconfigured with PV solar systems to reach these goals are already built, and their share with respect to new buildings is significant [43–45]. TRAINEE has been focused on both fields: one to retrofit existing buildings with solar systems and one to help architects and engineers to simulate energy gains and losses in new building designs. The application of BIM in both fields is slightly different; for the first case, we have to create a BIM model using techniques such as laser-scan, plus the transformation of a point cloud into a digital BIM model, and then applying the energy consumption to the model (as in a new housing scheme) [52,53].

Case studies were performed as six-phase activities on two different concepts of buildings: block housing and single (individual) house.

- Phase 1 is based on selecting the housing block or the new design of a single house (Figure 7a).
- Phase 2 (Figure 7b) takes into account the existing conditions of the physical context (latitude, weather, solar orientation, wind, existence of rivers, mountains, and neighboring buildings).
- Phase 3 is based on the calculation of existing solar irradiation from meteorological data from Meteonorm, PVGIS, or Macedonian Geologic Institute. In the TRAINEE program, Solarius PV by ACCA Software has been used to estimate the photovoltaic

solar production from real data of the solar irradiation available, acquired from the main reference climate databases and its deployment in terraces or tilt roofs.

- Phase 4 configures the automated drawing of the single-line electrical diagram of the photovoltaic installations, having the option of customization with the addition of electrical panels (in AC and DC), electrical protections on the output or input, types of cables, etc. This single-line electrical diagram of the photovoltaic system installation is represented in a complete plan with general data and legends of graphic symbols with details of the types of components used and is ready for printing or export in PDF, DXF, DWG, etc. formats.

- Phase 5 explores the tracking of consumption per hour, day, week, month, trimester, quarter, and/or year.

- Phase 6 fixes the energy savings (Figure 8a) and money flow control (Figure 8b). Money flow control is a very important issue, as it can be the driver to make the investment or not. Normally, the investment will be made if it is self-paid or if the recovery of the investment, commonly called ROI, or return of investment, is made in a very short number of years. Long recoveries or high investments make difficult the acceptance of the change to other kinds of energies.

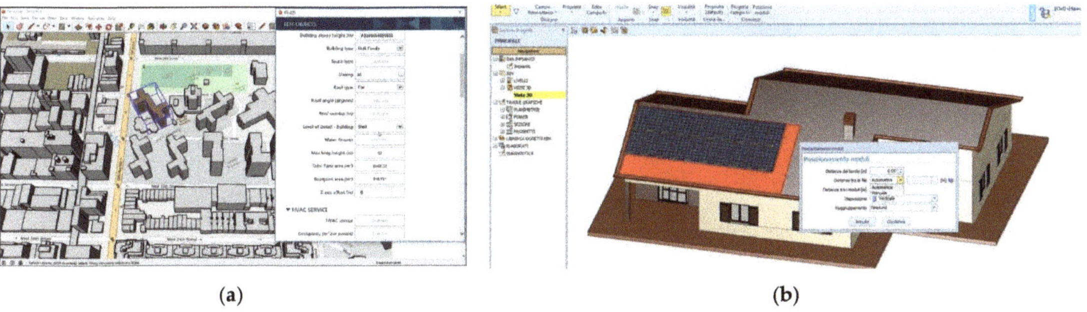

(a)   (b)

Figure 7. Procedures for realization of case studies: (**a**) Phase 1: selecting the object for BEM analysis; (**b**) Phase 2: physical context of the selected object, adapted from Ref. [41].

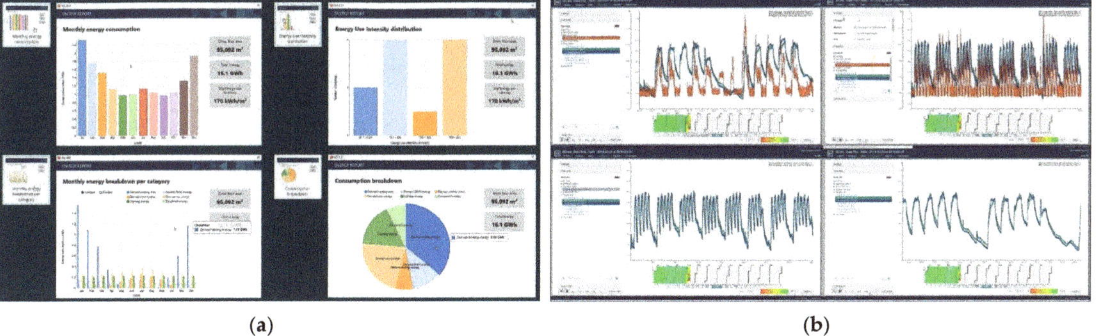

(a)   (b)

Figure 8. Phase 6 of realized case studies: (**a**) representation of energy consumption; (**b**) control over costs (money flow), adapted from Ref. [41].

4.2.3. Interpretation of the Case Study Results

The two case studies referred to different types of buildings: block housing (Figure 9) and single (individual) house, both located in Skopje, North Macedonia. The tasks involved creation of an optimized model integrated with BIM and combination of tools to enable modeling of the energy performance of the buildings.

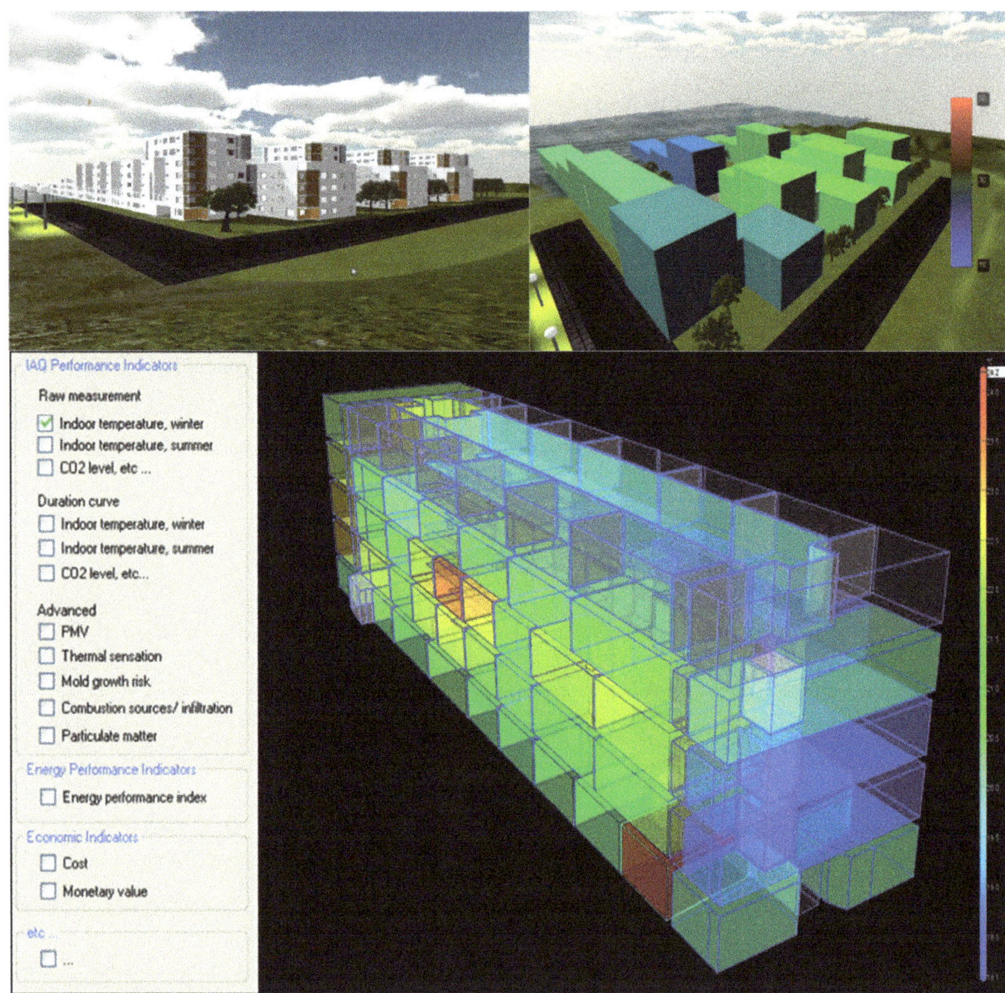

Figure 9. Case study no. 1 for expressing use of BIM in energy modeling: building block in Skopje, North Macedonia.

The central method for integrating BIM and BEM was used through a shared environment for both procedures, using specific data in IFC format. IFC provides an environment of interoperability among IFC-compliant software applications in all phases of building design, construction, and maintenance [44].

The assessment levels were comprised as optimization stages:

- BIM model (IFC/ACCA Software + Edificius)
- Energy model (SolariusPV)
- Optimization software multi-objective building optimization (TerMusPLUS)

For the second case study on the single house, the goal was to identify the best comfort, energy, and cost scenarios by using a simulation model for the purpose of building refurbishment (Figure 10). The experiment was simulated at a passive level, considering vernacular strategies of the southeastern EU regions, such as natural ventilation and the use of sun protection (comfort-economic). The model is used to find the optimum with active systems, applying cost-optimal criteria of energy and economic efficiency. A second further stage could be to introduce environmental labeling for the different grades of comfort, as in PassivHaus, BREEAM, Minergie, HQE, LEED, dgnb, VERDE, or WELL, to create an evaluation of the needs for each label, and a next phase could be the adoption of proper tracking with IoT (Internet of things) sensors, connected to the BIM and BEM model to identify which communities, blocks, or apartments are better or worse energy users.

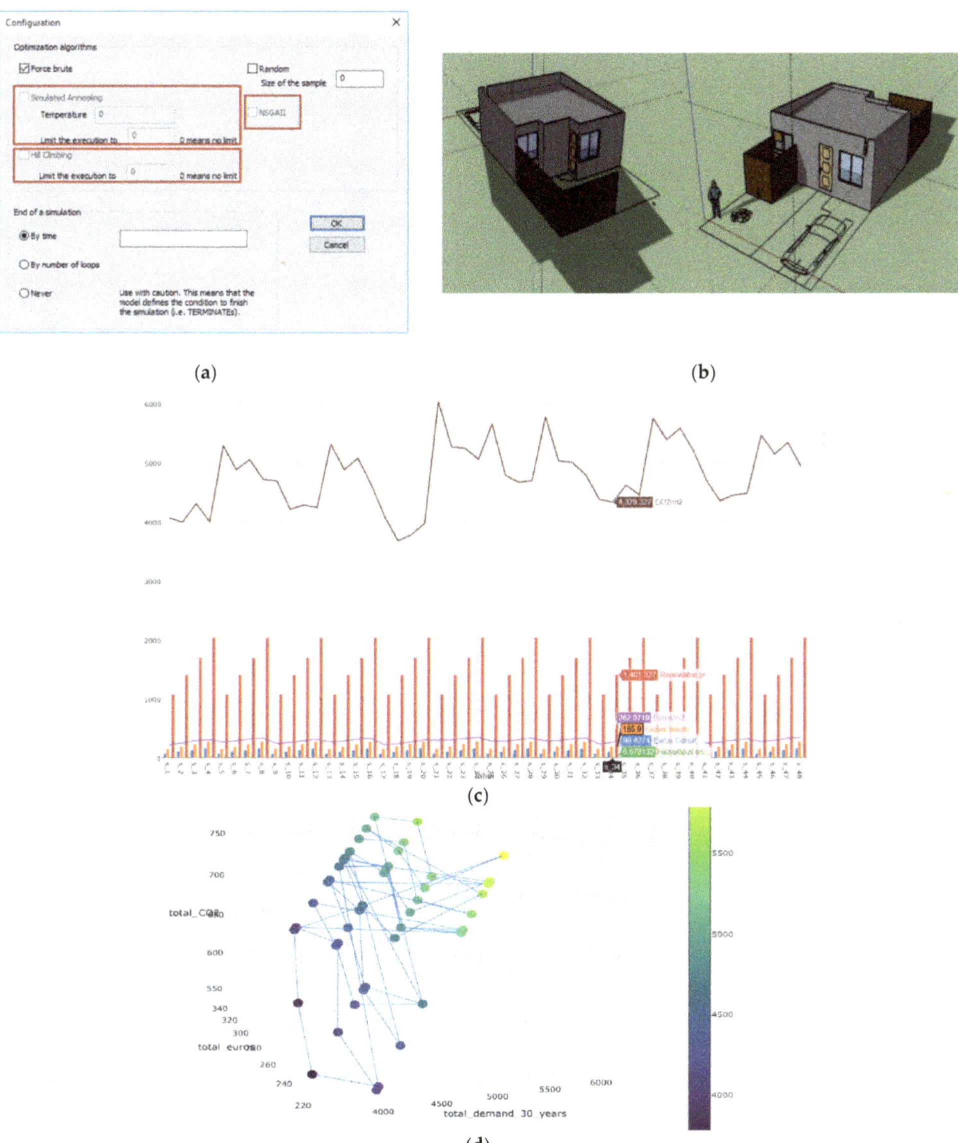

Figure 10. Case study no. 2 for expressing use of BIM in energy modeling: individual house in Skopje, North Macedonia. (**a**) Configuration of geographical location; (**b**) conceptual design from BIM model and checking for solar radiation and shadowing and first visual checker of PV and solar paneling location; (**c**) actual energy consumption and environmental impacts calculated with BEM; (**d**) estimated predictions for future building performance using BEM.

4.2.4. TRAINEE's Digital Procedure to Implement BEM Digital Models

These case studies have served the purpose of expressing the usability of BIM in performing energy modeling of existing buildings. The TRAINEE project has created a digital procedure to implement BEM digital models in housing communities to track their consumptions and savings (Figure 11). Implementation of the procedure is pending. It was shown that even when the building itself is not initially constructed with a BIM model, converting the existing building data into a digital model can serve for modeling the future building performance and selecting the retrofitting interventions that will satisfy the cost-gain ratio. The transformation of all housing communities into digital models to manage their energy consumption on a real-time basis after a PV and solar paneling intervention will be the next challenge.

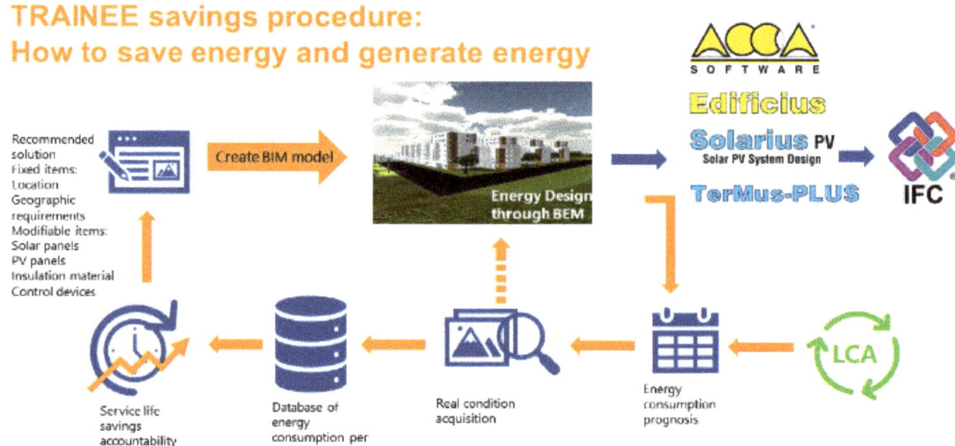

Figure 11. Digital procedure to implement BEM digital models in housing communities developed by TRAINEE project to track energy consumptions and savings.

The real savings numbers after one BEM design in a real community or block housing and its tracking and counter-comparison are pending, but the expected figure is that digitizing the housing sector by using BIM and BEM can serve to track the energy consumption trends and to identify the measures that can result in final energy savings. It is expected that significant numbers of up to 20% in energy savings for new buildings and up to 38% with retrofitting of existing building stock can be achieved through the proposed procedure.

4.3. Proposal of a National BIM Roadmap

4.3.1. Structure of the Proposal of the National BIM Roadmap

The proposal of the set of measures and actions that should lead to a more structured, governed, and coordinated approach toward BIM implementation on a national level is designed to engage all relevant stakeholders and to entitle the leading role of representative organization (body), where the government itself should take its responsibility, at least in establishing such a coordinating body (if not taking the role itself) [45].

The set of measures are chosen from the types of action previously identified and are spreading through all four fields of action (Figure 12) in order to guaranty in-depth penetration in all relevant spheres and to ensure a stable fundament for transition from traditional to digital construction. The proposed strategies and actions are targeted to overcome barriers identified in the BIM survey. The basic structure of the proposed roadmap is shown on Figure 12.

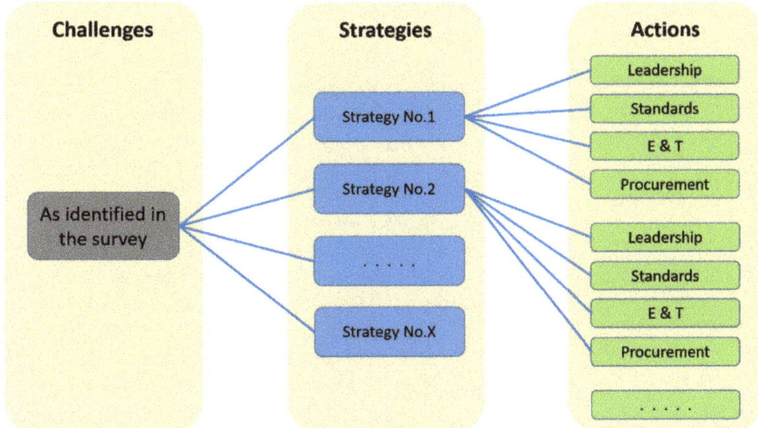

Figure 12. Basic structure and fields of action of the proposed BIM roadmap.

The detailed structure of the roadmap is then presented in Appendix A. It describes the relevant strategy proposed for meeting the identified challenges and the adequate measures set to mitigate the barriers, which are all listed in corresponding vertical columns of the graphical representation of the roadmap. Measures are to be undertaken in all four fields of action as indicated in the corresponding horizontal lines representing the listed measures. The roadmap also indicates the expected timeframe for the implementation of a certain action. It is expected that the proper implementation of well-planned measures can bring the digitization of the construction sector on a significantly higher level in the timeframe of four years.

4.3.2. General Recommendations for National BIM Adoption

The proposed roadmap for BIM adoption aims at triggering the attention of the responsible public and governmental authorities to be more proactive toward pushing the digitization of the national construction sector forward in order to keep pace with developed countries and global development. This document should act as a baseline and template document to define the step-by-step process of digital transformation and to delegate the roles of all responsible actors in the process. The document is structured by summarizing the best approaches and successful practices of developed countries and offers the structuring of a system of measures that can lead to the achievement of defined goals in collaboration with all relevant governmental and non-governmental stakeholders.

Based on the experience of developed countries that have been successful in implementation of BIM and are already harvesting the benefits that follow after the phase of "initial implementation", general recommendations that ensure success of the process of BIM adoption can be formulated as:

- introducing a BIM mandate is shown as a successful push forward incentive;
- clear framework for procurement of BIM in public sector construction projects is a positive approach;
- identifying a champion for BIM. Acting as a central figure in the process of initiating BIM adoption, the BIM champions are recognized by supporting BIM adoption and providing profound expert advice to overcome appearing challenges.

Better integration of BIM tools is necessary in order to achieve these goals and report energy savings. A European Commission study finds that the most successful countries in this area are the ones that have integrated BIM requirements into their public procurement legislation and that supported the implementation of BIM through public sector interventions in a more effective way if they are in line with private sector initiatives [31].

The regulatory environment will also define the potential for change, and, in order to support productivity growth, regulators may mandate the use of BIM to increase transparency and cooperation across the industry; publish supplier performance data; create cost transparency throughout the construction industry; transform regulations to support productivity; and consider labor interventions to ensure skills development instead of relying heavily on a cheap migrant temporary workforce [1].

5. Conclusions

BIM is definitely promoting serious changes to the construction industry. It changes its perspectives, its standard procedures, its way of traditional functioning, but BIM brings its own potential to harvest benefits in all stages of the project life cycle.

According to the findings of the survey, it had been shown that BIM is still in a premature level of implementation in North Macedonia. Although there are individual successful examples, the sector is not prepared, and the whole picture has to be created with thorough planning and activating stakeholders of all levels of the construction sector and, even more importantly, the relevant authorities and market actors. Transformation toward BIM as a new technology and methodology for executing construction in all phases requires a strategic plan. In order to achieve BIM's full adoption, enforcement by the government is needed. At the early stage of implementation, the enforcement by government toward BIM, as a driver for the required change, is necessary.

Showing successful examples, promoting benefits, and changing the culture of traditional construction are only some of the measures that guarantee change of perception and eagerness for embracing the positive changes.

6. Limitations and Further Studies

This work presents the current stage of implementation of BIM in the Macedonian construction sector. It includes an insight into the current attitude of the professionals and workers in the sector toward BIM implementation, identified through a conducted survey presented as the first pillar of the paper's actions. However, the results obtained by the survey should be considered by taking into account the limitations of this study, related to the total response rate of the target sample. As stated in Section 3.1.1, from the total of 57,000 employees in the national construction sector, the sample size of contacted respondents sent questionnaires was 1500, i.e., 2.6% of the total sample size. The obtained 312 valid responses represent a 20.8% respondent answer rate relative to the sent questionnaires, but it is an even lower answering rate relative to the total sample size. This should be considered as the frame of the survey's limitations.

Although this study is among the first dealing with this subject, it opens many different fields for further, deeper research, it raises new questions and uncertainties and requires precise analysis and evaluation of the adequacy of some internationally recognized and accepted approaches to correspond best to the national circumstances.

Some of the plans for future studies are already elaborated in the main text of this work, as they become evidently necessary through discussion of the achieved results of the performed work while identifying the uncovered area of this study. These ideas will be listed in brief, organized according to the three axes targeted by the paper:

(a) Future steps to improve the BIM survey will be to spread its scope to other respondents from the group of BIM users (not BIM professionals), to collect information on the construction phases where the use of BIM is mostly present, and, of greatest importance, to investigate in more detail the inclusion of BIM in formal education programs in order to be able to propose appropriate actions to overcome skills shortages.

(b) Regarding the use of BIM in evaluating the energy performance of buildings (both new and already constructed), there are several directions to extend this study. After developing a PV energy modeling procedure, the next part of the study could be dedicated to monitoring and analyzing how they can be integrated into a digital twin and a building digital twin for sustainable design. Then, the introduction of environmental labeling for the different grades of comfort is a useful way to express the building's value, and the third direction for further studies could be converting the existing building data into a digital model to serve for modeling the future building performance and for selecting the retrofitting interventions that will satisfy the cost-gain ratio.

(c) Considering the proposed document for the National Roadmap for implementation of BIM that elaborates strategic actions designed to support sustained use of BIM, it is necessary to continuously monitor and evaluate the effectiveness of its realization. This will ensure that the pace of development is well defined, ensuring keeping track with the objectives.

This threefold area for further studies guarantees that the interest to invest in BIM development is in the upward position and will ensure new achievements in the topic of BIM. It will also serve the sustainability of the paper's results as it opens field for further research and reveals the unanswered questions and challenges that the national adoption of BIM is facing.

Author Contributions: This is a joint research article with individual contributions of the authors as follows. Conceptualization, L.S.-G. and H.S.; formal analysis, L.S.-G., I.S., A.K., M.G., I.P. and R.I.; methodology, L.S.-G., I.P., T.C., and T.F.; resources, L.S.-G., I.S., A.K., M.G., I.P. and R.I.; software, I.P.A.; visualization, I.P.A.; writing—original draft, L.S.-G. and I.P.; writing—review and editing, L.S.-G., R.I., T.C. and T.F. All authors have read and agreed to the published version of the manuscript.

Funding: This research was funded by the project 785005-TRAINEE-H2020-EE-2016-2017/H2020-EE-2017-CSA-PPI, and the conference presentation at SP2021 was supported by the project 101033743—SEEtheSkills—H2020-LC-SC3-2018-2019-2020/H2020-LC-SC3-EE-2020-2.

Institutional Review Board Statement: At the time when the survey involving participation of respondents was conducted, there weren't any institutional requirements for requesting an approval. However, the survey was conducted in accordance with the Declaration of Helsinki, following EU GPDR rules and the ethics principles of the project TRAINEE.

Informed Consent Statement: Informed consent was obtained from all subjects involved in the study, as anonymous survey participants.

Data Availability Statement: Not applicable.

Conflicts of Interest: The authors declare no conflict of interest.

Appendix A

Table A1. The summary of the proposed National Roadmap for introduction of BIM.

Challenges	Strategies	Actions	Timeframe	Field of Action
Lack of demand	Public sector taking the lead	Incentives for BIM requirements in all public works	1st year	Leadership
		Coordinated asset information requirements across public sector	2nd year	Standards
		Training schemes for public sector	1st year	E&T
		BIM mandates in procurement of public works	2nd year	Procurement
	Establish central body	Identify a suitable individual/executive to lead the implementation program	1st year	Leadership
		Communication between BIM body and national standardization agency	1st year	Standards
		Specifying BIM competencies when accrediting BIM program	2nd year	E&T
		Make provision for developing government construction contracts	2nd year	Procurement

Table A1. *Cont.*

Challenges	Strategies	Actions	Timeframe	Field of Action
Lack of BIM skills (both training and skilled professionals)	Promoting success cases	BIM competitions	2nd year	Leadership
		Scholarships for BIM education	2nd year	E&T
		Mandate for skilled BIM professionals for government projects	3rd year	Procurement
	Building BIM capacities	Awareness seminars	1st year	Leadership
		BIM center of excellence	2nd year	Standards
		Developing self-assessment tool for BIM skills	1st year	E&T
		BIM R&D actions	3rd year	Procurement
	Developing BIM education and training	Develop a base level of learning outcomes targeted at alternative NFQ levels	1st year	Leadership
		Develop a BIM certification program to be managed by the National Standards Authority	2nd year	Standards
		BIM in Schools and universities	3rd year	E&T
	BIM certification and connection with NQF	Leveling of BIM skills with NQF levels and creation of BIM occupational standards	2nd year	Leadership
		Links with international standards for BIM certification	3rd year	Standards
		Developing different levels of BIM certification and training programs	3rd year	E&T
		BIM certification requirements for contract of works in public sector	2nd year	Procurement
Not existing BIM standards and protocols	Incorporate international standards	Enhance international collaboration to make use of structured, validated, comparable datasets	1st year	Leadership
		Support replication of international and European standards	2nd year	Standards
		BIM training based on international standards	2nd year	E&T
	Open BIM adoption	Consultation with key stakeholder groups for commitment to adopting open, internationally recognized information standards	1st year	Leadership
		"National Tools" based on open tools that help drive general conformance with standards, avoid duplication of effort, and avoid exclusion of SMEs	4th year	Standards
		BuildingSmart certification	3rd year	E&T
		Ensure that public procedures (planning, mapping, building inspections, public procurement) require information in compliance with standards	3rd year	Procurement
Missing culture for collaborative projects	Removing impediments	Set up a central platform to act as digital resources and tools bank to ease BIM introduction	2nd year	Leadership
		Establishing national protocols for using BIM	3rd year	Standards
		BIM guide workshops	2nd year	E&T
		Support through online means the clients to receive, review, manage, and assess BIM on their projects	3rd year	Procurement
	Creating platform for cross-sectional collaboration	Establishing knowledge hub for exchange of experience	1st year	Leadership
		Links with international BIM platforms to ensure standardization	2nd year	Standards
		Advise on effective BIM implementation	3rd year	E&T
		BIM fund	4th year	Procurement

Legend of fields of action: Leadership action; Standards; E&T; Procurement.

References

1. Barbosa, F.; Woetzel, J.; Srdhar, M.; Parsons, M.; Bertram, N.; Brown, S.; Mischke, J.; Ribeirinho, M. *Reinventing Construction: A Route to Higher Productivity*; McKinsey Global Institute: New York, NY, USA, 2017.
2. Alreshidi, E.; Mourshed, M.; Rezgui, Y. Requirements for cloud-based BIM governance solutions to facilitate team collaboration in construction projects. *Requir. Eng.* **2016**, *23*, 1–31. [CrossRef]
3. Doumbouya, L.; Gao, G.; Guan, C. Adoption of the Building Information Modeling (BIM) for construction project effectiveness: The review of BIM benefits. *Am. J. Civ. Eng. Archit.* **2016**, *4*, 74–79.
4. Powel, M.; Pancevski, I. Education and Employment Sector Linkages for the Building and Construction Sector, Bridging the Skills Gaps, UNDP Project. 2020. Available online: https://arhiva.skills4future.mk/en/development-of-stakeholder-mapping-analysis-education-and-employment-sector-linkages-and-gaps-for-the-building-and-construction-sector (accessed on 5 February 2022).
5. Almeida, P.R.; Solas, M.Z.; Renz, A.; Bühler, M.M.; Gerbert, P.; Castagnino, S.; Rothballer, C. *Shaping the Future of Construction: A Breakthrough in Mindset and Technology. Future of Construction*; World Economic Forum: Geneva, Switzerland, 2016. [CrossRef]
6. World Economic Forum. Shaping the Future of Construction—An Action Plan to Accelerate Building Information Modeling (BIM) Adoption. 2018. Available online: https://www3.weforum.org/docs/WEF_Accelerating_BIM_Adoption_Action_Plan.pdf (accessed on 1 February 2022).
7. EU BIM Task Group. *Handbook for the Introduction of Building Information Modeling by the European Public Sector*; EU BIM Task Group, EC: Brussels, Belgium, 2017.
8. Luciani, C.; Garagnani, S.; Mingucci, R. BIM tools and design intent. Limitations and opportunities. In *Practical BIM 2012—Management, Implementation, Coordination and Evaluation*; Kensek, K., Peng, J., Eds.; University of Southern Carolina: Los Angeles, CA, USA, 2012.
9. Smart Market Report. The Business Value of BIM for Infrastructure, Dodge Data and Analytics. 2017. Available online: https://www2.deloitte.com/content/dam/Deloitte/us/Documents/finance/us-fas-bim-infrastructure.pdf (accessed on 1 February 2022).
10. Smart Market Report. The Business Value of BIM for Owners, McGraw-Hill Construction. 2014. Available online: https://i2sl.org/elibrary/documents/Business_Value_of_BIM_for_Owners_SMR_(2014).pdf (accessed on 1 February 2022).
11. Ibrahim, H.S.; Hashim, N.; Jamal, K.A.A. The Potential Benefits of Building Information Modelling (BIM) in Construction Industry. *IOP Conf. Ser. Earth Environ. Sci.* **2019**, *385*, 012047. [CrossRef]
12. Eastman, C.M.; Eastman, C.; Teicholz, P.; Sacks, R. *BIM Handbook: A Guide to Building Information Modeling for Owners, Managers, Designers, Engineers and Contractors*; John Wiley and Sons: Hoboken, NJ, USA, 2011.
13. Patrick MacLeamy. The Future of the Building Industry (3/5): The Effort Curve, HOK. Available online: https://www.youtube.com/watch?v=9bUlBYc_Gl4&ab_channel=HOKNetwork (accessed on 24 January 2022).
14. Smith, P. BIM Implementation—Global Strategies. *Procedia Eng.* **2014**, *85*, 482–492. [CrossRef]
15. Ullah, K.; Lill, I.; Witt, E. An Overview of BIM Adoption in the Construction Industry: Benefits and Barriers. In *10th Nordic Conference on Construction Economics and Organization*; Emerald Reach Proceedings Series; Lill, I., Witt, E., Eds.; Emerald Publishing Limited: Bingley, UK, 2019; Volume 2, pp. 297–303.
16. Burgess, G.; Jones, M.; Muir, K. *BIM in the UK House Building Industry: Opportunities and Barriers to Adoption*; University of Cambridge: Cambridge, UK, 2018.
17. CitA. BIM Innovation Capability Programme (BICP), Global BIM Study: Lessons for Ireland's, BIM Programme. 2017. Available online: https://arrow.tudublin.ie/cgi/viewcontent.cgi?article=1016&context=beschrecrep (accessed on 24 January 2022).
18. McAuley, B.; Hore, A.; West, R. *BICP Global BIM Study—Lessons for Ireland's BIM Programme*; Construction IT Alliance (CitA) Limited: Dublin, Ireland, 2017.
19. NBC. Roadmap to Digital Transition for Ireland's Construction Industry 2018–2021. 2017. Available online: https://issuu.com/constructionitalliance/docs/nbc_roadmap_to_digital_transition_updated_2020 (accessed on 19 November 2021).
20. Silva, M.J.F.; Salvado, F.; Couto, P.; Vale, A.Á. Roadmap Proposal for Implementing Building Information Modelling (BIM) in Portugal. *Open J. Civ. Eng.* **2016**, *6*, 475–481. [CrossRef]
21. Yang, J.B.; Chou, H.Y. Mixed approach to government BIM implementation policy: An empirical study of Taiwan. *J. Build. Eng.* **2018**, *20*, 337–343. [CrossRef]
22. Gerges, M.; Austin, S.; Mayouf, M.; Ahiakwo, O.; Jaeger, M.; Saad, A.; El Gohary, T. An investigation into the implementation of Building Information Modeling in the Middle East. *Electron. J. Inf. Technol. Constr.* **2017**, *22*, 1–15.
23. Succar, B.; Kassem, M. Macro-BIM adoption: Conceptual structures. *Autom. Constr.* **2015**, *57*, 64–79. [CrossRef]
24. Ghaffarianhoseini, A.; Tookey, J.; Ghaffarianhoseini, A.; Naismith, N.; Azhar, S.; Efimova, O.; Raahemifar, K. Building information modelling (BIM) uptake: Clear benefits, understanding its implementation, risks and challenges. *Renew. Sustain. Energy Rev.* **2017**, *75*, 1046–1053. [CrossRef]
25. Mtya, A.; Windapo, A. Drivers and barriers to the adoption of building information modelling (BIM) by construction firms in South Africa. *Innov. Prod. Constr.* **2019**, 215–223. [CrossRef]
26. Senthilkumar, V. An exploratory study on the building information modeling adoption in United Arab Emirates municipal projects- current status and challenges. *MATEC Web Conf.* **2017**, *120*, 02015. [CrossRef]

27. Ahmed, S. Barriers to Implementation of Building Information Modeling (BIM) to the Construction Industry: A Review. *J. Civ. Eng. Constr.* **2018**, *7*, 107. [CrossRef]
28. Hamma-Adama, M.; Kouider, T. Comparative Analysis of BIM Adoption Efforts by Developed Countries as Precedent for New Adopter Countries. *Curr. J. Appl. Sci. Technol.* **2019**, *36*, 1–15. [CrossRef]
29. Damali, Y.; Kamalesh, P.; Omar, N. Challenges Involved in Adopting BIM on the Construction Jobsite. *EPiC Ser. Built Environ.* **2021**, *2*, 302–310.
30. Bargstädt, H.J. Challenges of BIM for Construction Site Operations. *Procedia Eng.* **2015**, *117*, 52–59. [CrossRef]
31. Elagiry, M.; Marino, V.; Lasarte, N.; Elguezabal, P.; Messervey, T. BIM4Ren: Barriers to BIM Implementation in Renovation Processes in the Italian Market. *Buildings* **2019**, *9*, 200. [CrossRef]
32. Lee, G.; Borrmann, A. BIM policy and management. *Constr. Manag. Econ.* **2020**, *38*, 413–419. [CrossRef]
33. Stojanovska-Georgievska, L.; Spasevska, H.; Ivanov, R. Beyond the numbers—Result oriented report of the project TRAINEE. *Zenodo* **2020**, *1*, 11. [CrossRef]
34. Panchevski, I.; Ivanov, R.; Stamenkova, Z. Survey Report on Penetration and Acceptance of BIM Practice, D3.1. Project Deliverable of EU Project TRAINEE. 2018. Available online: http://trainee-mk.eu/images/TRAINEE/Deliverables/D3.1_Survey_about_penetration_and_acceptance_of_BIM_practice_TRAINEE.pdf (accessed on 24 January 2022).
35. Stojanovska-Georgievska, L.; Sandeva, I.; Krleski, A.; Spasevska, H.; Ginovska, M.; Panchevski, I.; Ivanov, R.; Arnal, I.P.; Cerovsek, T.; Funtik, T. A Brief Report on the Actions towards the Introduction of BIM in the Macedonian Construction Sector. *Environ. Sci. Proc.* **2021**, *11*, 8. [CrossRef]
36. European Construction Sector Observatory. BIM in the EU Construction Sector. 2019. Available online: https://ec.europa.eu/growth/sectors/construction/observatory/trend-papers_en (accessed on 19 October 2021).
37. Zhong, Y. BIM-based Construction Training Program for Low-skilled Workforce: Visualization and Development of Equipment. Tools and Procedures for Construction Tasks. Master's Thesis, Bauhaus-Universität Weimar, Weimar, Germany, 2016.
38. Cerovsek, T. A review and outlook for a 'Building Information Model' (BIM): A multi-standpoint framework for technological development. *Adv. Eng. Inform.* **2011**, *25*, 224–244. [CrossRef]
39. BuildingSMART. Central Model Integration. 2016. Available online: BuildingSmart.org (accessed on 8 February 2022).
40. Daniotti, B.; Gianinetto, M.; Della Torre, S. *Digital Transformation of the Design, Construction and Management Processes of the Built Environment*; Springer: Berlin/Heidelberg, Germany, 2019.
41. BibLus BIM. BIM and Energy Certification: Here Is the BEM That Will Revolutionize the Work of Technicians. 2018. Available online: https://bim.acca.it/bim-e-certificazione-energetica-bem (accessed on 8 February 2022).
42. International Awards Program. 2019. Available online: https://www.buildingsmart.org/awards/bsi-awards-2019 (accessed on 20 December 2021).
43. Carvalho, J.P.; Almeida, M.; Bragança, L.; Mateus, R. BIM-Based Energy Analysis and Sustainability Assessment—Application to Portuguese Buildings. *Buildings* **2021**, *11*, 246. [CrossRef]
44. Mateus, R.; Silva, S.M.; de Almeida, M.G. Environmental and cost life cycle analysis of the impact of using solar systems inenergy renovation of Southern European single-family buildings. *Renew. Energy* **2019**, *137*, 82–92. [CrossRef]
45. Stojanovska-Georgievska, L.; Ivanov, R.; Panchevski, I.; Perez Arnal, I. Proposal for National Roadmap for Introduction of BIM, D3.2. Project Deliverable of EU Project TRAINEE. 2020. Available online: http://trainee-mk.eu/images/TRAINEE/Deliverables/D32_ZBKK_Roadmap.pdf (accessed on 20 December 2021).
46. Chong, H.-Y.; Lee, C.-Y.; Wang, X. A mixed review of the adoption of Building Information Modelling (BIM) for sustainability. *J. Clean. Prod.* **2017**, *142*, 4114–4126. [CrossRef]
47. Carvalho, J.P.; Bragança, L.; Mateus, R. Sustainable building design: Analysing the feasibility of BIM platforms to support practical building sustainability assessment. *Comput. Ind.* **2021**, *127*, 103400. [CrossRef]
48. Reeves, T.; Olbina, S.; Issa, R.R.A. Guidelines for using building information modeling for energy analysis of buildings. *Building* **2015**, *5*, 1361–1388. [CrossRef]
49. Ardda, N.; Mateus, R.; Bragança, L. Methodology to Identify and Prioritise the Social Aspects to Be Considered in the Design of More Sustainable Residential Buildings—Application to a Developing Country. *Buildings* **2018**, *8*, 130. [CrossRef]
50. Li, X.; Wen, J. Review of building energy modeling for control and operation. *Renew. Sustain. Energy Rev.* **2014**, *37*, 517–537. [CrossRef]
51. Zhao, H.-X.; Magoulès, F. A review on the prediction of building energy consumption. *Renew. Sustain. Energy Rev.* **2012**, *16*, 3586–3592. [CrossRef]
52. Hong, T.; Chen, Y.; Luo, X.; Luo, N.; Lee, S.H. Ten questions on urban building energy modeling. *Build. Environ.* **2019**, *168*, 106508. [CrossRef]
53. Farzaneh, A.; Monfet, D.; Forgues, D. Review of using Building Information Modeling for building energy modeling during the design process. *J. Build. Eng.* **2019**, *23*, 127–135. [CrossRef]

 *buildings*

Article

The Identification, Development, and Evaluation of BIM-ARDM: A BIM-Based AR Defect Management System for Construction Inspections

Kieran W. May *, Chandani KC, Jose Jorge Ochoa, Ning Gu, James Walsh, Ross T. Smith and Bruce H. Thomas

Australian Research Centre for Interactive and Virtual Environments, University of South Australia, Adelaide, SA 5095, Australia; chandani.kc@mymail.unisa.edu.au (C.K.); Jorge.OchoaPaniagua@unisa.edu.au (J.J.O.); Ning.Gu@unisa.edu.au (N.G.); James.Walsh@unisa.edu.au (J.W.); Ross.Smith@unisa.edu.au (R.T.S.); Bruce.Thomas@unisa.edu.au (B.H.T.)
* Correspondence: kieran.may@mymail.unisa.edu.au

Abstract: This article presents our findings from a three-stage research project, which consists of the identification, development, and evaluation of a defect management Augmented Reality (AR) prototype that incorporates Building Information Modelling (BIM) technologies. Within the first stage, we conducted a workshop with four construction-industry representatives to capture their opinions and perceptions of the potentials and barriers associated with the integration of BIM and AR in the construction industry. The workshop findings led us to the second stage, which consisted of the development of an on-site BIM-based AR defect management (BIM-ARDM) system for construction inspections. Finally, a study was conducted to evaluate BIM-ARDM in comparison to the current paper-based defect management inspection approach employed on construction sites. The findings from the study revealed BIM-ARDM significantly outperformed current approaches in terms of usability, workload, performance, completion time, identifying defects, locating building elements, and assisting the user with the inspection task.

Keywords: Augmented Reality; mixed reality; Building Information Modelling; defect management; construction inspection; quality assurance; immersive visualisation

Citation: May, K.W.; KC, C.; Ochoa, J.J; Gu, N.; Walsh, J.; Smith, R.T.; Thomas, B.H. The Identification, Development, and Evaluation of BIM-ARDM: A BIM-Based AR Defect Management System for Construction Inspections. *Buildings* **2022**, *12*, 140. https://doi.org/10.3390/buildings12020140

Academic Editor: Xianbo Zhao

Received: 24 December 2021
Accepted: 24 January 2022
Published: 28 January 2022

Publisher's Note: MDPI stays neutral with regard to jurisdictional claims in published maps and institutional affiliations.

1. Introduction

Within the last decade, Building Information Modelling (BIM) has received increased attention from the academic community with a trend towards integrating BIM within the architectural design process (Figure 1). BIM improves the quality of the documentation produced, enabling the BIM model to be leveraged throughout the construction-project life-cycle. Due to the comprehensive data provided by BIM models, advanced visualisation tools such as Augmented Reality (AR) and Virtual Reality (VR) have been recognized as effective media to visualise and interact with BIM data. AR is defined as a tool that combines the real and virtual worlds by superimposing virtual content onto the physical world. Using this concept, researchers have explored the integration of AR with BIM models in the construction-project life-cycle. Some common examples include Safety and Training [1,2], Risk Management [3,4], Architectural Design [5–7], Facilities Management [8,9], Education and Learning [10,11], and Building Performance Simulation [12,13].

In this article, we present our findings from a three-stage research project consisting of the identification, development, and evaluation of a BIM-based AR defect management (BIM-ARDM) system for conducting on-site construction inspections. In the first stage, we identified and examined the current uses, barriers, and potential uses for the integration of immersive AR and BIM within the Architecture, Engineering, and Construction (AEC) industries. This was achieved by conducting a workshop consisting of focus-group interviews with four construction-industry representatives, each with varying roles and levels of experience working in the construction industry.

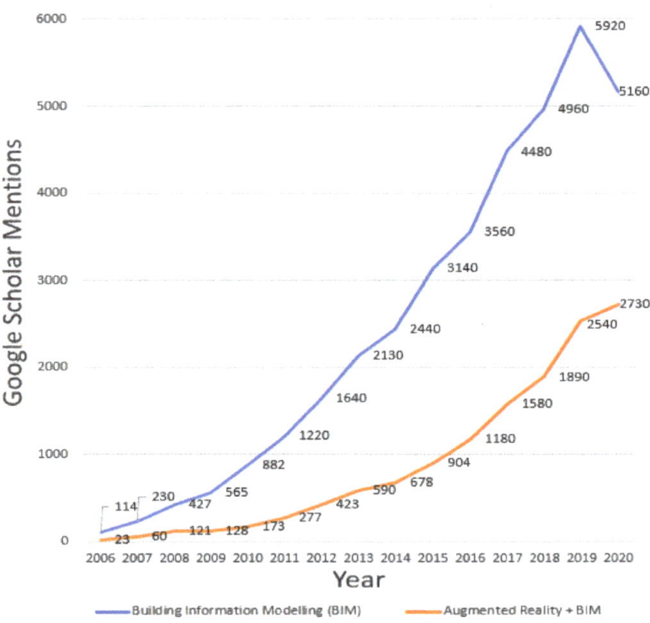

Figure 1. The growth of BIM and BIM + AR based on the yearly Google Scholar article mentions (patents and citations excluded). The following query terms were used: BIM—("Building Information Modeling" OR "Building Information Modelling"), BIM + AR—("Building Information Modeling" OR "Building Information Modelling"), AND ("Augmented Reality" OR "AR" OR "Mixed Reality").

The findings from this workshop led us to the second stage, which consisted of the development of an on-site defect management BIM-ARDM prototype. The technical implementation was designed to address the concerns and potential features highlighted by the industry representatives during the focus-group interviews. In the subsequent third stage, we conducted a pilot study with 11 participants to evaluate the experimental BIM-ARDM system by comparing it to the current defect management techniques employed in the construction industry.

Currently, digital technologies have not been adequately adopted in the construction industry, and their potential is yet to be fully exploited. Current construction processes such as conducting defect management inspections on buildings remain relatively traditional and primarily consist of extracting analogue drawings and building information data from BIM models and printing them as paper-based documents to use during the on-site construction inspection. Due to the level of detail required for construction personnel to sufficiently conduct defect management inspections, a significant amount of paper-based documents are generally required on-site. Whilst conducting the inspection, analogue drawings of the architectural model are used as a reference by the inspector to compare how elements, such as a light switch or door, compare to the architectural plans.

Identifying construction defects from printed drawings can be challenging and requires extensive training to precisely read the 2D drawings and understand their spatial location in the physical space. During the inspection, building-defect data are managed by writing inspections notes onto a paper-based or digital checklist. Due to the conventional approach for conducting on-site inspections involving a highly manual component, this process could potentially lead to errors, mistakes, or loss of data occurring before, during, and after the inspection takes place. Additionally, the conventional method for defect management inspections requires a significant amount of time allocated to set up the inspection and summarize post-inspection data. This was demonstrated in a recent study by Ma et al. [14] that illustrated the inspection process using conventional approaches that required almost double the time to plan the inspection (40 min), summarize results (50 min), and communicate (20 min) in comparison to performing the actual construction-site defect management inspection (60 min).

To improve productivity, and minimise unnecessary delays and reworks caused by misinterpretation of plans, and drawings and imprecise information transfer from three-dimensional to two-dimensional environments, an improved information-exchange medium is required.

We believe a more effective approach could be achieved by leveraging the visualisation capabilities of AR to directly link the BIM model with the physical construction site allowing the BIM model location and specification data to be shown in-situ to enhance a construction personnel's spatial awareness and understanding of the architectural design. This process also enables inspection data to be digitally recorded during the inspection and linked directly back to the BIM model, which we believe can further mitigate data loss, mistakes, and discrepancies that could potentially occur by having workers manually inputting data from paper-based documents into digital systems.

The specific contributions of this study are:

1. Identification and categorization of current uses, potentials uses, and barriers for BIM and AR integration within the construction industry.
2. An improved AR system for identifying construction defects during an on-site construction inspection.
3. A set of novel AR visualisations, and features to improve defect management inspection performance.
4. A Revit plugin to autonomously link data recorded during the construction inspection back to the original BIM model.
5. Data analysis software to evaluate and assess construction inspection performance through eye-tracking, and head-tracking data linked to a four-dimensional visualisation.

In the remainder of this article, we explore and summarize the findings from previous works that have integrated AR technologies and BIM to conduct on-site construction inspections. We also identify and discuss the current technological challenges associated with the integration of AR and BIM within the AEC industries. We present our interview results from a workshop highlighting the current uses, potential uses, and barriers associated with integrating BIM and AR within the construction industry. Next, we present a summary of an experimental BIM-ARDM system that details the implementation and design decisions involved in the development of the software. We then present our findings from a pilot study of the proposed BIM-ARDM which demonstrates significant results for the usability, workload, performance, completion time, locating building elements, identifying defects, and preferences in comparison to a traditional paper-based approach. We conclude the study with a discussion of the limitations of our BIM-ARDM prototype and some final remarks.

2. Background

In this section, we explore and summarize the implementations and findings from previous works that explore the usage of AR and BIM to conduct on-site construction inspections. Subsequently, we identify and discuss two common technological and software limitations that inhibit the current optimum usage of AR within the AEC industries.

2.1. AR Supporting BIM for Construction Inspections

Shin and Dunston [15] stated two issues needed to be addressed before AR technology could become prevalent in the AEC industries: limitations of AR technology (e.g., tracking) and identifying applicable areas within the construction industry that AR could be used for. The researchers identified eight tasks where AR could benefit the construction industry: layout, excavation, positioning, inspection, coordination, supervision, commenting, and strategising. The following year Shin and Dunston [16] presented ARCam: an AR-based prototype aimed at improving the performance of steel-column inspections. The hardware consisted of a mounted stationary video camera attached to a HiBall-3100 tracking system. Inspectors observed virtual content superimposed onto the physical steel columns through a traditional touch-screen display. A study with sixteen graduate civil engineering students was conducted to compare ARCam with conventional methods for steel-column inspections. The findings demonstrated that ARCam required less task load based on the NASA-Task-Load-Index results [17] and was a more intuitive mode of operation for conducting the inspection task. However, conventional methods outperformed ARCam in regards to inspection precision and performance. The authors concluded the graphical appearance of the virtual model, tracking, and calibration set up all required improvements to improve the feasibility of ARCam.

In recent years, researchers have utilized more modern immersive AR displays for construction-based inspections. Portalés et al. [18] developed an AR tablet-based on-site inspection tool that incorporated visualisations to superimpose virtual content onto prefabricated buildings. Similarly, García-Pereira et al. [19] presented an annotation-based tablet AR inspection tool for prefabricated buildings. The system consisted of a visualisation that allowed inspectors to change the transparency of the virtual building model by manipulating the value of a virtual slider. Various types of annotations were also incorporated into the system including text-based annotations, photographic annotations, and 2D-drawing annotations. A study with 11 participants that had previous experience in construction-based inspections was conducted to test the various functions in the system. Participants were tasked to calibrate the virtual model, change the transparency of the virtual model, revisit a pre-existing annotation, make two new annotations, revisit one of their own annotations, and delete an annotation. The results from the study showed that the usability of the system received an 81.36 (excellent) ranking based on the System Usability Scale (SUS) [20]. A Likert-based ranking (1 strongly disagree to 5 strongly agree) revealed participants believed the presented system could reduce the inspection time (mean = 4.18), improve documentation during the inspection process (mean = 4.09), be suitable for real work environments (mean = 4.45), and be valuable for documenting the geometry of elements during inspections (mean = 4.73).

Feng et al. [21] used the HoloLens to develop an AR-based inspection tool that superimposed BIM models onto a physical construction site. The researchers utilized an AR-based interface allowing inspectors to check off construction elements within a virtual holographic checklist. Park et al. [22] proposed a conceptual proactive construction defect management framework that integrated BIM, AR, and data-collection methods to improve upon the conventional manual construction-inspection processes. The proposed framework presents a solution to address each of the three stages required in a typical knowledge-management process: 1. capture, 2. retrieval, and 3. reuse. The presented solutions consisted of (1) a proposed data-collection template to improve the quality of data captured during the inspection, (2) a defect domain ontology for improved retrieval and access of defect data, and (3) an AR-based system that incorporates image-matching techniques to overcome the conventional manual construction-inspection practices. The following year Kwon and Park et al. [23] presented two AR-based prototypes to conduct defect management inspections both remotely and on-site. The remote system consisted of having an on-site worker take a picture of the physical element, which is relayed into a server. The 2D image is then run through an image-matching system to identify the corresponding virtual element to compare the differences between the physical and virtual elements. The on-site defect management system consisted of a mobile-based AR application that superimposed virtual building models onto the physical building using a marker-based tracking technique. Zhou et al. [24] developed an AR system to inspect segment displacement during tunnel construction. Th eresults from a case study comparing the proposed AR system to conventional methods demonstrated that the AR system can reduce the total duration of the inspection. Furthermore, participants claimed the AR system was more intuitive and overall a more-effective approach to diagnose segment displacement in comparison to conventional methods. Other works have proposed the usage of an IFC-based framework to improve the interoperability of equipment used to conduct self-inspections of buildings [25].

Other prototypes that do not directly support AR, but that demonstrated potential benefits for construction inspections, were presented by Ma et al. [14]. The authors proposed a non-immersive collaborative tablet-based quality-management inspection application that integrated BIM models and indoor positioning. The system was tested by having a construction manager conduct an on-site inspection using both the proposed tablet-based system and a traditional paper-based system. The results revealed that the construction manager spent approximately 30% more time using the proposed tablet-based system for conducting the actual on-site inspection. However, taking into consideration other factors associated with the entire inspection process (i.e., inspection-task planning, inspection results summarizing, and communication), the researchers concluded the system could save approximately 50% of the time for the entire inspection process. The results from this test were shown to seven construction site engineers to gather qualitative feedback on the system. The engineers stated the ability to avoid manually entering inspection data from paper records into computers, maintain the latest inspection progress from the BIM model, adhere to inspection standards, and improve stakeholder communication as the key advantages of the proposed tablet system.

2.2. Current Challenges

Based on our previous literature review of AR supporting BIM for construction inspections and the results from our workshop, we identified two common technical limitations that we believe require improvements before BIM-based AR technologies can be adopted from research into the construction industry. Firstly, before developing a BIM-based AR tool, a workflow is required to transfer the BIM model from a BIM-based CAD platform (e.g., Revit) into an AR-supported software development kit (SDK) or game engine (e.g., Unity or Unreal) without loss of BIM metadata. Once this has been achieved, a tracking technique is then required to superimpose the virtual BIM model onto the physical construction site at a one-to-one mapping whilst the user navigates through the environment.

2.2.1. Integrating BIM Models from CAD Platforms to Game Engines and SDKs

The lack of interoperability to communicate BIM data across different platforms has been one of the primary technical barriers associated with the development of BIM-based AR/VR prototypes built off SDKs or game engines. Currently, the Industry Foundation Class (IFC) remains the most widely adopted approach for transferring BIM models across platforms. IFC is defined by BuildingSMART as a data model that "can define physical components of buildings, manufactured products, mechanical/electrical systems, structural analysis models, energy analysis models, cost breakdowns, work schedules, and more" [26]. This process allows partial metadata associated with the BIM model to be saved within a single file format, which can be exported and imported across BIM-based CAD platforms that provide native IFC support. However, game engines, such as Unity, do not provide native support for IFCs, and third-party libraries [27–29] have been released to parse and import IFC models into game engines. The primary limitation associated with IFC is that only specific metadata associated with the BIM model are maintained that can result in a significant loss of BIM metadata when exporting from the native CAD format (e.g., rvt file format) to IFC.

However, recently, the Unity Development team released Unity Reflect: A software package that creates a natural uni-directional linkage between a CAD platform (Revit) and the Unity game engine (https://unity.com/products/unity-reflect, accessed on 20 December 2021). Although, future work is still required to achieve a natural bi-directional exchange of BIM data between CAD platforms and game engines.

2.2.2. Tracking and Registration

The tracking accuracy associated with the alignment of virtual holographic models onto the physical world remains one of the prominent ongoing research fields within the AR research community. Typically, AR tracking techniques [30] are camera-based, and they are defined through two categorizations: marker-based [31] and marker-less tracking [32,33]. Marker-based tracking methods rely on tracking the position of physical markers (e.g., Vuforia [34] frame targets, image targets, QR codes, or ARToolKit markers [35]) to achieve an alignment between the virtual world and the real world. Although marker-based tracking techniques yield a more-precise tracking accuracy than marker-less techniques, specific limitations associated with marker-based techniques make it difficult to employ within a real-world construction site. Firstly, the most obvious limitation is that physical markers are required for the tracking to be achieved. As a result, various physical markers would be needed to be set up on the physical construction site, which would be tedious. Furthermore, the placement of these physical markers would have to match the exact position of corresponding virtual markers on the BIM model at a one-to-one mapping. This means a point cloud or spatial map of the construction site would be required as a reference to ensure the virtual building model is precisely placed at the corresponding position of the real-world construction site.

Alternatively, marker-less tracking techniques rely strictly on the depth sensors and cameras built into the AR hardware to simultaneously map and localize the virtual world in real time. Although currently tracking performance is sacrificed when using marker-less tracking techniques, as AR hardware and tracking algorithms continue to advance, marker-less tracking techniques for on-site construction inspections will become more feasible. Research by Kopsida and Brilakis [36] explored three AR tracking techniques that would be suitable for on-site construction inspections: model-based, marker-less using monocular SLAM, and marker-less using Red-Green-Blue—Depth (RGB-D) devices. The researchers concluded that using the built-in RGB-D depth-sensors of devices such as the Microsoft Kinect and Google Project Tango was the most feasible out of the three marker-less tracking solutions for on-site construction-inspection accuracy. Recent works by Hübner et al. [37] have further validated this claim by demonstrating the accuracy of RGB-D sensors built into the Microsoft Hololens 1 Head-Mounted Display (HMD), which were capable of tracking within a two-centimetre accuracy in indoor environments.

3. Workshop

A workshop was conducted to facilitate focus-group interviews with four construction-industry representatives in attendance, each with different roles and levels of experience working within the construction industries. All participants were active professionals with multiple years of prior experience working in the construction industry. Each participant was selected from different construction companies to represent a range of practices in the construction sector including sole trader, small to medium enterprises, and larger practices. The specific roles and expertise of each participant are displayed in Table 1. Due to the data collected in the focus-group interviews being entirely qualitative, with the workshop process involving demonstrations and open discussions, the number of participants selected to take part in the workshop was purposefully kept to a smaller scale. The primary goal of this workshop was to capture the participants' opinions and perceptions regarding the potentials and barriers associated with the integration of BIM and AR within the construction industry. Specifically, the workshop explored the following research questions:

1. What is the current adoption related to the integration of BIM and AR within the construction industry?
2. What are the barriers associated with integrating BIM and AR within the construction industry?
3. What potential scenarios across the entire construction-project life-cycle could BIM and AR be integrated for?

Table 1. This table shows the experience and roles of the workshop participants.

Participant	Role	Construction-Industry Experience
(1)	General Builder and Director	33 Years
(2)	BIM Manager	33 Years
(3)	BIM Manager	10 Years
(4)	Project Administrator	4 Years

3.1. Procedure

The workshop process consisted of presentations, interactive demonstrations, focus-group interviews, and discussions. The presentations and demonstrations were conducted prior to the interviews and discussions to familiarise participants with AR and VR technologies and to showcase some of the potential scenarios the technology could be applied to within the construction industry. The VR demonstration consisted of a first-person walk-through visualisation of a virtual BIM model. Participants were able to interact with virtual elements within the building to display the BIM metadata properties associated with each element (e.g., manufacturer, cost, material, name, etc.). The AR demonstration utilized the see-through mixed reality HMD Microsoft Hololens 1, which allowed participants to visualise a virtual scaled-down BIM model. Participants were able to rescale the BIM model using the "air tap gesture" through a two-dimensional UI attached to the model. Participants were also able to interact with the BIM model by enabling and disabling specific components and rooms within the building by toggling components on and off in the UI.

Focus-group interviews were the main instrument used to collect data during the workshop. The data collected from this workshop were analysed using a content analysis [38] approach. This process enabled us to make replicable and valid inferences from texts to the contexts of their use. A three-step methodology was employed to analyse the collected data. Firstly, data were converted from an audio recording to a transcript and were organised based on time stamps. Next, the data were coded by segmenting sentences or paragraphs into categories and labelling those categories with a term associated with the actual language of the participants. This included finding frequencies of occurring ideas (or events) in the data relevant to the research questions. Subsequently, three key themes were identified from the analysis: current uses of using BIM and AR, barriers of using BIM and AR, and scenarios and future directions for using BIM and AR.

3.2. Current Uses of BIM and AR

Despite all workshop participants having limited prior experience using AR technologies, all participants had a sufficient understanding of AR. All participants stated that the technology had not been incorporated within any of their current or previous companies. However, multiple participants stated their companies were actively utilizing VR as a walk-through visualisation tool within the design phase of their projects to "bring clients in and do a bit of a walk-through".

Although participants stated there are limited current uses of AR within the construction industry, Figure 1 demonstrates that a comprehensive amount of research has previously been conducted on the integration of AR and BIM technologies and is continually rising each year. Furthermore, the literature has also demonstrated that a significant amount of research has been conducted on the integration of AR across a range of AEC disciplines [39–41]. Due to the substantial amount of AR research conducted in AEC, and the limited adoption of AR within the AEC sectors, it is evident that a clear gap exists between the research and industry. The broader impact of AR on AEC projects is not well understood. Further discussions with AEC experts to understand the current barriers surrounding BIM and AR, and to identify applicable areas for the integration of AR and BIM in the AEC industries, are required to bridge the gap between the research and practice.

3.3. Current Barriers Surrounding BIM and AR

Before AR technology can be applied to the construction industry, it is necessary to identify the current barriers inhibiting the optimum usage of AR in construction. In this section, we explore non-technical aspects from the perspective of experts within the construction industry that need to be overcome before AR and BIM can become prevalent within the construction industry. The specific barriers identified by participants that we discuss include (1) maturity and reliability, (2) accessibility, (3) standards and data exchange, and (4) stakeholder perception and organisational culture.

3.3.1. Maturity and Reliability

Participants described the lack of maturity and reliability associated with current AR technologies as one of the leading factors for why their companies had yet to adopt AR technologies. One participant stated that "we still don't see huge benefits in jumping into AR at the moment until the technology matures a little bit more". We believe that the accessibility and tracking limitations associated with AR technologies are factors that directly correlate with the advancement of the maturity and reliability of future AR technologies.

3.3.2. Accessibility and Cost

Due to the pricing and limited AR devices available, it is difficult and expensive to acquire a high-end AR display. Currently, the Hololens II is marketed at USD 3500 (https://www.microsoft.com/en-us/hololens/buy, accessed on 20 December 2021) (Table 2), and due to the cost of hardware and resources, prices of high-end AR HMDs will likely remain expensive in the future. As a result, accessibility of the technology was stated by the participants as a major barrier for the adoption of AR within the construction industry. One participant suggested using a tablet or smartphone as the primary display to make AR more accessible: "so you are not carrying a whole bunch of devices. That is where I sort of see an easy way of marketing (AR)".

Table 2. This table shows the modern AR HMDs commercialized over the last decade and the retail prices associated with the devices. ([†] Discontinued, [‡] planned, and [§] rumoured).

AR Headset	Year	Price (USD)
Meta 1 [†]	2014	750
Meta 2 [†]	2016	494
HoloLens 1 [†]	2016	3000
Magic Leap 1 [†]	2018	2295
HoloLens 2	2019	3500
Magic Leap 2 [‡]	2021	2295
Apple AR Headset [§]	2022	3000

Participants stated that the costs associated with designing and maintaining BIM models was another hurdle as to why BIM has not become fully adopted within construction companies. Participants pointed out that this issue is especially prevalent for smaller projects with a lower budget; one participant stated: "You've got to sort of work out something to try and bring smaller projects into using it".

3.3.3. Standards and Data Exchange

Interoperability and lack of support provided to exchange BIM data across different platforms is another key barrier associated with the adoption of BIM and AR technologies. One participant stated, "the models coming through are actually being dumbed down and all the data is stripped out of them". Although IFC was introduced as a standardized data model to exchange BIM data across platforms, only a limited portion of meta-data associated with the BIM are maintained when exporting to IFC. The participant further commented: "We have seen an increase in people sharing models in IFC rather than native Revit or Techno or whatever else. That's great as you get some geometry with it, but you've lost a whole heap of data".

Finally, the other challenges are related to the highly fragmented and inconsistent capabilities across hardware platforms and operating systems. There is an imbalance in the projects within the construction industry itself with architects, services, and structural industries still using basic CAD and 2D media. The level of inconsistencies in the platforms used by different industries including architects, consultants, and construction professionals affect the level of accuracy through the supply chain. One participant stated, "we're getting jobs where we still get CAD and 2D drawings".

3.3.4. Stakeholder Perception and Organisational Culture

Stakeholder perception and organisational culture also pose challenges for the integration of AR and BIM technologies within the construction industry. This was highlighted by a participant who claimed "the major holdup to the advancement and development (of AR in construction) is not the technology, it is actually the people and processes". Most of the construction industry is still very traditional and is quite reluctant to adopt new technology, especially related to the usage of BIM and AR. One participant stated "there are negative perceptions created when using new technology," and "people really have to build the confidence before they take the next step".

One participant also commented on the architectural designers' reluctance to design and develop BIM models: "The payoff for (BIM) is for the person in front. This means the driver either has to be the end user clients that are going to see a potential payoff if the entire tool chain is using BIM to its full extent". As a solution to the aforementioned issue, one participant commented: "I think the only way to push the clientele to drive BIM through the process is to make it an industry standard".

3.4. Potential and Future Directions for Using BIM and AR

AR systems provide fast and easy access to information through a three-dimensional medium that creates huge opportunities and potential for the adoption of AR in a range of construction disciplines. Some of the potential uses of AR in construction mentioned by participants were as follows: (1) quality assurance and defect management inspections, (2) safety inductions and training, (3) risk management, (4) facilities management, and (5) building performance simulation.

3.4.1. Quality Assurance and Defect Management Inspections

Using AR to conduct quality assurance or defect management inspections was one particular area of interest mentioned by multiple participants throughout the workshop. Some of the comments from the participants included the following: (1) "I see (AR) as particularly valuable at this stage in terms of reviewing things before you install ceilings before you clad walls or things where you actually should be doing your QA checks"; (2) "If you have an AR environment (where) you can go into the room so you don't necessarily need 100% accuracy of measurement, you are not actually constructing but you just have to do a building review—'Are all the elements that should be in this room actually installed? Can I sign off'?"; and (3) "Just doing QA work is a really good use (for AR) and there is no reason why you can't do that right now".

3.4.2. Safety Inductions and Training

Participants also described the possibilities of using AR to train construction workers on the safety implications associated with the construction site: "the only (way) we would potentially use (AR) is probably safety and design upfront. Walking around (a BIM) and as you're going through the safety and design review phase looking at certain areas that you potentially might have some concerns". The benefit of using AR for safety and training is that users can simulate, share, and exchange valuable information without any physical medium. A participant further stated "I think one of the uses for (AR) is safety induction because you want to familiarise some person with the safety of the site, the egress paths, where areas are restricted. There's no reason why all that information can't be contained within a model which then could be used in a training environment to familiarise someone with the building very quickly". One participant also stated the following negative aspect of using AR for safety and training: "I think it's very difficult to replace a real-life situation with a virtual situation, you might be able to visualise it but to get all other interaction that you would have on a construction site, and all the other safety factors you got to be aware of, I still think it is not quite there".

3.4.3. Risk Management

Similarly, participants identified risk management as another area of interest for the integration of AR and BIM. Participants described developing a potential AR application which "provides live visualisation of a building along with the associated high-risk areas, and exclusion zones, and the ability to manage risks and report them". Linking geolocation coordinates and the project schedule to the BIM could be further integrated to provide four-dimensional visualisations of the BIM model [42].

3.4.4. Facilities Management

Facilities management was described by participants as "a major driver for AR, it comes into great usage for AR". The ability to use AR for see-through objects and to visualise occluded elements was referenced many times throughout the workshop as a potentially valuable use of the technology. One participant described the following scenario: "if you can walk into a building and look up at the ceiling and you know what's above it, you can point at the ceiling and go: where is the air-conditioning unit for this room?" Previous research such as ARWindow has demonstrated an optimistic outlook to adopt AR technologies within facilities management (FM) from the perspective of industrial facility managers [43].

3.4.5. Building Performance Simulation and Visualisation

Using the immersive three-dimensional qualities of AR to simulate and assess the performance of building design features was also described as one of the potential uses for AR within the construction industry. One of the participants stated, "say acoustically you have to meet certain acoustic requirements, if you could integrate something that spread out certain sound levels in a room there definitely is a benefit".

3.5. Potential Features

Throughout the workshop, participants highlighted several features when describing their optimal AR applications for construction. Specifically, participants mentioned the following: (1) visualisation and sequencing, (2) UI design and user-friendliness, and (3) accessibility as important features when developing an AR-based construction application.

3.5.1. Visualisation and Sequencing

Participants described the ability to visualise sequences and the ability to "pull apart a model and put it back together" to develop a better understanding of "what you need to go in first" as a potential feature for an AR system. Similarly, another participant stated: "in terms of interaction, turning things on and off (in the BIM) so people can deconstruct and reconstruct the building as many times as they want until they get it right and say OK this is the way I want to do it". The ability to sequence components is demonstrated in commercial AR-based construction applications such as Gamma AR (https://gamma-ar.com/, accessed on 20 December 2021) and Trimble Connect (https://connect.trimble.com/, accessed on 20 December 2021). Colour-coding virtual building elements, and visualising occluded elements, was also a feature demonstrated by BIM Holoview (http://www.bimholoview.com/, accessed on 20 December 2021).

3.5.2. UI Design and User-Friendliness

The user-friendliness of the technology was a continuous theme throughout the workshop. Participants were asked to describe their optimal AR applications, and participants responded that it "has to be really simple to use", with an intuitive UI design. Furthermore, it should be a "really easy process, so you can wander around a site, point an iPad in a direction, and it knows exactly where they are and shows them what they should be seeing. You actually take the human element out of it".

3.5.3. Accessibility

Integrating AR with existing BIM-based CAD platforms can also make AR more attractive and accessible to construction companies. One participant stated "you get someone comfortable in one platform and then the next job they go to its working on a different platform and what they are able to do one they are not able to do (in the other). It just brings frustration for people". However, due to the lack of development tools, support, and flexibility provided by BIM-based CAD platforms, developers tend to utilize SDKs and game engines such as Unreal (https://www.unrealengine.com/, accessed on 20 December 2021) or Unity (https://unity.com/, accessed on 20 December 2021) for AR software development. Participants also discussed the potential of using AR to access everything needed during a construction task within a single compact device. When referring to an AR inspection application, one participant commented "if you've got everything in your iPad you can then actually start using checklists and whatever else you need in one device so you (don't need to) carry a whole bunch of devices".

3.6. Summary

A summary of the potential future applications proposed by the workshop respondents is as below:

A1. Using AR technologies to conduct on-site quality assurance and defect management inspections on a construction site.
A2. The ability to use AR technologies to visualise BIM models and go through the safety and design review phase and identify which areas on the construction site may have some potential safety concerns.
A3. An AR-based learning/educational/training risk-management application designed to familiarise a person with the safety requirements and hazards associated with the construction site.
A4. A facilities-management-focused AR application that allows facility managers to visualise obstructed or invisible objects within a physical building.
A5. An AR-based building performance simulation and analysis tool that provides the user feedback on whether a room in the building meets specific performance requirements (e.g., acoustics).

Throughout the workshop, the participants also proposed several potential features when describing their optimal AR applications. We categorized and summarized the potential features highlighted by workshop participants into four categorizations: visualisations, interface, tracking, and data exchange.
Visualisations

F1. Using AR to visualise occluded objects (e.g., wiring, pipes) on a construction site.
F2. Sequencing a BIM model into specific components to insert things, and having a UI providing visual feedback on the insertion process.
F3. Representing data using three-dimensional visualisations as opposed to traditional two-dimensional architectural analogue drawings.

F4. Design the AR application to be highly interactive. Users should have the ability to visualise the building and turn specific components on and off.

Interface

F5. Ensuring the general user-friendliness of the AR application and UI is a priority. It needs to be very simple and easy to use for a non-technical user.
F6. Having checklists and other documents required for a construction task to be all accessible within a single device.
F7. An integration of the AR system with existing BIM platforms to improve accessibility.

Tracking

F8. An improved AR tracking system that is very simple to use and setup.
F9. Linking AR to digital point survey type accuracy for improved tracking.

Data Exchange

F10. A system or workflow capable of maintaining BIM data when exchanging BIM data across platforms to avoid loss of data.

In the remainder of this article, we explore the implementation of A1. Using AR to conduct on-site construction inspections from the above-listed potential applications. We also present the implementation of several features integrated within the system highlighted by participants from the list of potential features. Specifically, we address all the listed potential features (F1–F9) in the BIM-ARDM system with the exception of F10 due to the scope of the development.

The decision-making process, which led to the selection of an AR-based defect management application from the above list of potential applications identified by workshop participants, was based on the following factors. Firstly, we narrowed down the primary potential applications that received the most optimistic responses amongst the workshop participants. We then consulted with AEC experts from both the research and industrial sectors to discuss which of the potential applications AR could have the most significant impact. Finally, a discussion amongst the researchers who have an extensive background working with AR technologies determined that implementing an AR-based defect management application would be feasible to develop as the first step for adopting the results of the focus-group interviews to advance AR development and adoption in AEC.

4. BIM-ARDM System Development

Current industry approaches for identifying defects on a construction site are still relatively manual and require construction inspectors to make estimations based on paper-based, two-dimensional plans and analogue drawings. Higher-tier AEC firms have begun integrating tablet-based application such as Aconex Field (https://help.aconex.com/aconex/our-main-application/using-aconex/field, accessed on 20 December 2021) into their project workflow. These applications are designed to improve the data-management process by having inspectors input data directly into a UI containing digital checklists as they conduct their inspections. Additionally, the quality of inspection data captured during on-site inspections still requires inspectors to interpret two-dimensional plans and analogue drawings to make estimations on whether elements were built correctly. Furthermore, these tools only provide the capability of storing inspection data into independent databases and information-management systems separate from the BIM model. As an alternative to the conventional inspection systems, researchers have proposed the usage of AR technology to superimpose the virtual BIM model onto the physical construction site. However, from the previous work highlighted in Section 2.1, we identified the following drawbacks of previous AR systems, which are specifically addressed in the BIM-ARDM system:

1. All AR implementations that incorporated HMDs used a holographic interface to input and record data (e.g., checklists). We describe in Section 4.2.1 why we believe this design consideration would not be suitable on a real-world construction site.

2. No AR systems were proven to perform more effectively than conventional approaches for conducting on-site construction inspections.
3. Limited AR features and visualisations were integrated to support the inspector with their on-site inspection performance.
4. No AR systems were tested on a real-world construction project.
5. No AR systems were capable of storing inspection data back to the original BIM model.
6. No systems were capable of quantitatively assessing or evaluating the on-site inspection performance.

Based on the aforementioned issues and potential features previously highlighted by workshop participants in Section 3, we designed and developed an on-site BIM-ARDM system for construction defect management inspections. The primary goal during the development phase was to develop a prototype that improved upon the conventional paper-based methods that utilizes two-dimensional plans and analogue drawings to conduct defect management inspections. Our approach for achieving this was to leverage the visualisation capabilities provided by AR technologies (Microsoft HoloLens 2) to visualise the three-dimensional BIM model superimposed at a one-to-one mapping over the physical construction site (Figure 2).

Our system also incorporates a handheld tablet-based interface that connects directly to the HoloLens 2 through a customized TCP/IP network architecture allowing a two-way transmission of data between the two devices. The handheld tablet-based application contains a dynamically produced checklist providing inspectors with an interface to directly input inspection data into.

In this section, we provide a complete overview of the BIM-ARDM prototype and present the four major components involved in the development of the overall BIM-ARDM system.

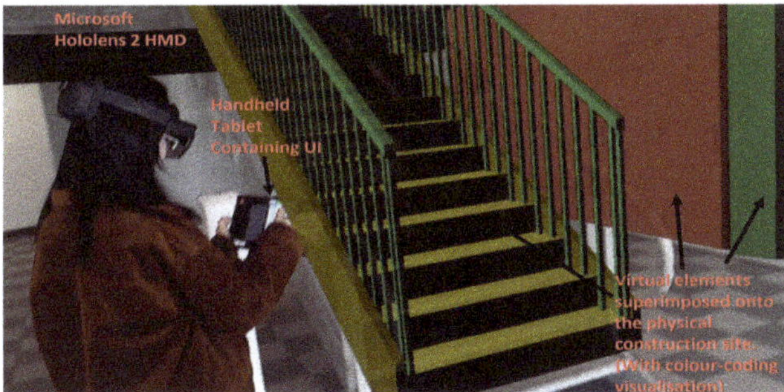

Figure 2. This figure shows a third-person perspective of the BIM-ARDM system. The BIM model is superimposed over the physical site at a one-to-one mapping. A colour-coded visualisation represents correct (green), minor (yellow), and major (red) defects.

4.1. Implementation

The BIM-ARDM system was developed to run off a specific set of hardware and software components. For the hardware, the see-through AR head-mounted display Microsoft HoloLens 2 was used to show the virtual BIM model on the physical construction site. Additionally, a Samsung Galaxy S6 Lite tablet was used in-synch with the HoloLens 2 to provide inspectors with a tablet-based UI to input inspection data and control the various visualisations and features integrated within the BIM-ARDM system.

In terms of the software, all software components in the BIM-ARDM system were developed entirely in the C# programming language. The AR and playback analysis applications were built with the Unity 3D Game Engine. The Revit Plugin was developed with the Revit API. The Unity libraries we used to develop the AR application consisted of the Mixed-Reality Toolkit (MRTK) (github.com/Microsoft/MixedRealityToolkit-Unity/releases, accessed on 20 December 2021) for HoloLens 2 integration, Vuforia (https://developer.vuforia.com/, accessed on 20 December 2021) to integrate a marker-based tracking solution for coordinate space calibration, and the 3D User Interaction Toolkit (3DUITK) (github.com/WearableComputerLab/VRInteractionToolkit, accessed on 20 December 2021) [44] to integrate a "Bendcast" visualisation [45] to aid users in locating elements. An overview of the various components used in the AR application is presented in Figure 3A.

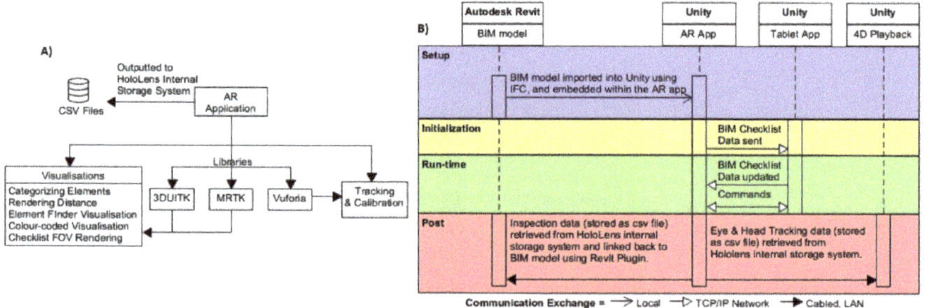

Figure 3. These diagrams present the system architecture of the BIM-ARDM system. The UML diagram (**A**) presents an overview of the primary components integrated into the AR application. The sequence diagram (**B**) demonstrates how the main components within BIM-ARDM interact.

4.2. BIM-ARDM System Overview

The BIM-ARDM system is comprised of four major components. The first component is a tablet-based application designed to handle two primary functions: (1) input and data collection, which consists of an easy-to-use checklist interface allowing inspectors to input and collect specific inspection data whilst they conduct their on-site inspection and (2) controlling the various AR visualisations and features integrated into the system. The second component is an AR-based application running off the HoloLens 2 HMD, which allows the inspector to experience the immersive three-dimensional visualisations integrated into the system. The AR-based application and tablet-based application are directly connected through a TCP/IP network protocol. The third component is a Revit plugin that synchronizes data automatically collected during the inspection back to the original Revit model. The fourth component is a post-inspection data-analysis tool designed to evaluate and assess the inspection performance of the inspector through eye-tracking and head-tracking data collected during the inspection. A diagram showing how the four major components within the BIM-ARDM system interact is presented in Figure 3B.

4.2.1. Tablet-Based Application

The development of an intuitive and easy-to-use UI for inputting checklist data was one of the primary design considerations during the development of the BIM-ARDM system. It was evident, based on participants' responses during the workshop, that practical AR-based applications in AEC must be simple, easy to use, and have an intuitive UI for construction personnel to operate (F5). Participants also commented on the importance of having checklist and other documents required for a construction inspection to be accessible within a single device (F6). As a result, we aimed to design a complete UI that contained the functional requirements required to perform a defect management inspection (F6) whilst still maintaining a simple and easy-to-use UI design (F5).

Initially, we explored using an AR-based holographic UI that utilized a combination of gestures and audio commands as user input. However, based on our informal observations, we felt a non-tactile UI would lack intuitiveness and user-friendliness when operating the UI due to the complex nature of the UI design. Furthermore, the likelihood of false positives occurring when using a gesture-based interface would likely contribute to increased errors occurring whilst the inspector conducts their inspections. We also discussed the potential of using a speech-based input system; however, due to construction sites being relatively noisy environments, we concluded that a speech-based input system would not be suitable for a real-world construction site.

We concluded that a tablet-based interface that could seamlessly communicate data with the HoloLens II was the most viable and user-friendly approach for inspectors to operate on a real-world construction site. To achieve this, we developed a TCP/IP network architecture where a local server is hosted off the Android tablet, and the HoloLens II connects to the local server as a client. This allowed us to create a bi-directional communication, allowing us to synchronously transmit commands and data between both devices.

The tablet-based application consists of three primary interfaces that allow the inspector to operate the BIM-ARDM system. The first interface, "Visualisations" (Figure 4A) enables the inspector to control (turn on/off) and configure the different visualisations integrated into the AR-based system. The second interface, "Checklist" (Figure 4B) allows the inspector to input inspection data, and the third interface, "Diagnostics and Calibration" (Figure 4C) is used to setup the calibration of the AR system and display diagnostic feedback associated with the tablet-based and AR-based applications.

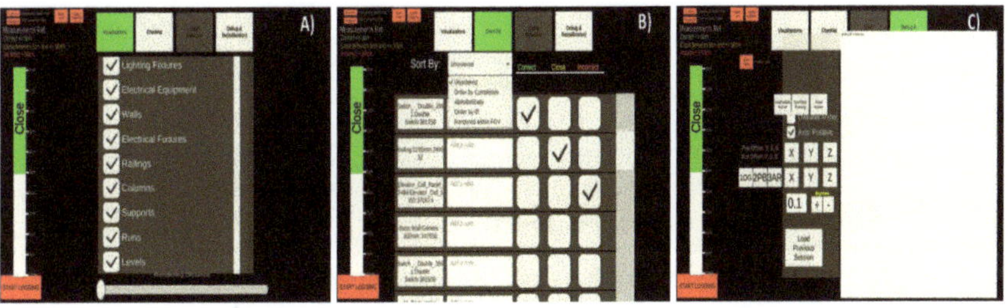

Figure 4. The left image (**A**) presents the UI design for the visualisations tab; image (**B**) displays the UI design for the checklist tab where a dynamic checklist of each element within the BIM model was generated; image (**C**) presents the UI design for the diagnostics tab.

Visualisations: The visualisation interface (Figure 4A) acts as a control panel for the visualisations, allowing the inspector to control and configure two AR visualisations, categorizing elements and rendering the distance. The interface also controls some additional AR features including toggling the visibility of the BIM model and enabling/disabling the generated AR spatial mapping mesh.

Checklist: The checklist interface (Figure 4B) provides the inspector with a UI to input inspection data into whilst they conduct their construction inspection. Upon loading the application, each element in the BIM model is generated within the checklist interface, and the inspector can assign each element to one of the following three values:

1. Correct (no defect): The construction is correct, contains no defects, and is correctly aligning with the virtual element. In the pilot study, we defined this as a less-than-5 cm drift between the virtual and physical elements.
2. Close (minor defect): The construction slightly differs from the virtual element. In the pilot study, we define this as a 5 cm-to-50 cm drift between the virtual and physical elements.
3. Incorrect (major defect): The construction significantly differs from the virtual element. In the pilot study, we defined this as a greater-than-or-equal-to-50 cm drift between the virtual and physical elements.

Upon assigning the element to one of the three above-listed values in the checklist, the corresponding holographic element within the AR environment will update its colour. Within the checklist interface, each element also contains a text-box area where inspectors can input any additional comments related to the inspection of each element.

The checklist interface also consists of a built-in sorting option feature that allows the inspector to sort elements based on the following categorizations:

1. Unordered: Displays elements in their default, randomly generated order.
2. Sort by completion: Sorts elements that are currently unassigned before elements that have been assigned.
3. Alphabetical order: Sorts elements by alphabetical order.
4. Sort by ID: Sorts elements numerically based on their Revit models unique ID.
5. Rendered within field of view (FOV): Only displays checklist elements that are actively being rendered within the inspector's FOV.

The system also provides the capability for inspectors to capture images of individual elements by selecting an element within the checklist UI and pressing the capture-image button. Images are captured through the HoloLens 2's built-in camera and stored within the HoloLens 2 internal storage system. Subsequently, checklist data inputted into the tablet application during the inspection are autonomously logged to a .csv file, which is also stored within the HoloLens 2 internal storage system. We also integrated support to load previous .csv files within the AR-based application, which allows inspectors to load their previous inspection sessions. Finally, a Revit plugin was developed to import all data recorded in the BIM-ARDM system back to the original Revit model.

Diagnostics and Calibration: The diagnostics and calibration interface (Figure 4C) was designed to support two primary functions. Firstly, if a potential issue emerges whilst operating the BIM-ARDM system, this interface allows the inspector to determine the problem that caused the issue to occur. This is achieved by having a console UI that displays the various debug statements outputted by the BIM-ARDM system. Secondly, the menu contains a set of buttons that allow the inspector to setup and configure the tracking calibration process.

4.2.2. AR-Based Application

The AR-based application is comprised of two primary functions. The first function is the visualisations and features that we developed to support inspectors with their on-site construction inspection performance. The second function is the developed tracking and calibration system that is used to superimpose the BIM model onto the real-world construction site at a one-to-one mapping.

Visualisations and Features: User-friendliness (F5) and interactivity (F4) were two features mentioned by several participants throughout the workshop. Participants also encouraged representing data through three-dimensional visualisations as opposed to traditional two-dimensional analogue drawings (F3). To achieve this, we developed a set of five three-dimensional visualisations designed to assist the inspector with conducting their inspection. The developed AR visualisations can be controlled and configured by the inspector directly through the tablet-based application. The specific visualisations were (1) the ability to visualise specific elements based on their categorizations (F2 and F4), (2) controlling the camera's rendering distance to only visualise specific elements within a specified distance from the user, (3) an element location-finder visualisation, (4) an element colour-coded visualisation, and (5) a tool that only displays elements within the checklist UI that are currently being looked at by the inspector. A more-detailed description and overview of the design motivations associated with each of the visualisations is presented below.

1. Categorizing elements: The ability to turn on and off specific elements was a feature highlighted by workshop participants (F4). Due to our system workflow consisting of an IFC-based approach to import BIM models into Unity, partial metadata associated with the BIM are maintained. As a result, unique IDs and categorizations of building elements that are automatically generated in Revit are accessible within the Unity game engine. Using this data, we developed an approach to automatically and dynamically generate a list of categorizations based on the corresponding categories associated with each building element. Within the tablet interface, inspectors can toggle on or off different categories to enable or disable the visibility of each AR building element associated with the given category; for example, electrical fixtures or signs as illustrated in Figure 5. Based on our

informal observations, we believe this visualisation to be most effective when inspecting complex and dense BIM models. The primary advantage of the visualisation is that it allows the inspector to focus on specific elements and filter out any irrelevant elements to avoid confusion and improve visibility.

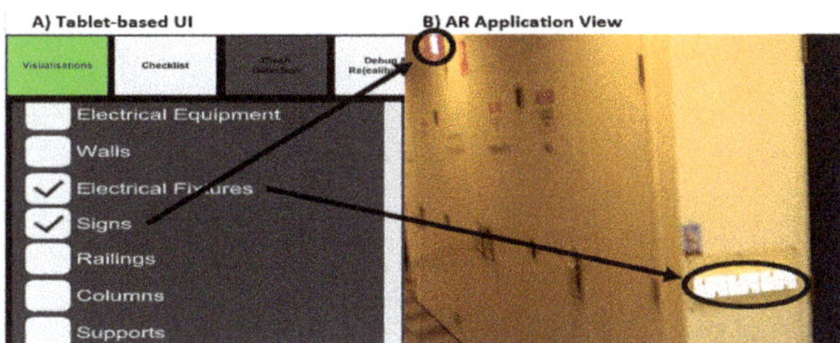

Figure 5. This figure demonstrates the categorizing elements' visualisation, which only displays specific elements in the AR view (**B**) that have been enabled by the inspector on the tablet UI (**A**).

2. Rendering distance: A visualisation was also developed that allows the inspector to control the visibility of elements by manipulating the rendering distance of the AR camera. A slider on the tablet UI can be controlled by the inspector to specify the rendering-distance threshold. A transparent holographic box is displayed in the AR environment that provides visual feedback to the inspector of the specified rendering-distance threshold. All elements between the user and the displayed holographic box are visible to the AR user; whereas, elements outside the box are not rendered by the camera (Figure 6).

The visualisation was developed based on two specific design motivations. Firstly, due to our software rendering several non-occluded objects at a time, we believed a function to mitigate the amount of virtual objects being rendered by the camera could significantly improve the visibility of elements within the AR environment. We initially tested this visualisation on a smaller-scale environment with a Revit model containing approximately thirty elements, and based on our informal observations, we did not find the visualisation particularly useful. However, when testing this visualisation on a real-world construction site with more than 2000 elements, we discovered this visualisation had the potential to become more effective as the complexity and density of the BIM model increased. This was demonstrated when it became difficult to focus on specific elements within the scaled-up BIM model due to having several non-occluded virtual elements actively being rendered by the AR camera. Therefore, using the visualisation to only display specific elements closer to the inspector significantly improved the visibility of elements.

The second motivation is that using this feature can significantly increase the frame rate of the AR application, thus improving performance. It became evident when testing the BIM-ARDM prototype on a real-world construction site that required a BIM model with over 2000 virtual elements actively rendered within the AR camera that performance was an issue. However, when using this visualisation, we discovered the performance of the AR application system significantly improved due to the visualisation mitigating the number of elements rendered by the AR camera.

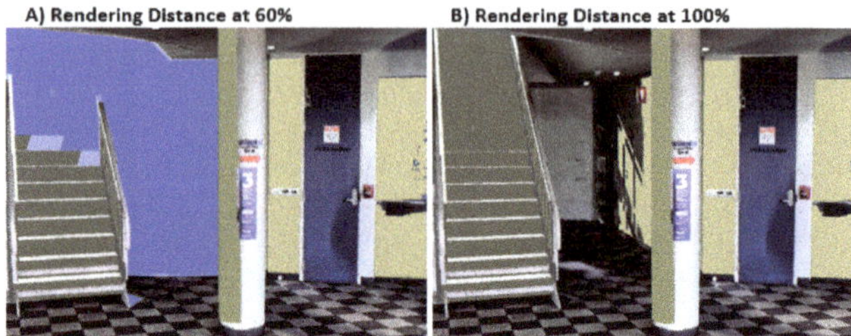

Figure 6. This figure demonstrates the rendering-distance visualisation, which mitigates the amount of elements visible by controlling the camera-rendering distance.

3. Element finder visualisation: A visualisation was also developed to make it easier for inspectors to locate virtual elements whilst conducting their inspection. This visualisation is autonomously enabled within the AR view when the inspector selects a checklist element through the checklist UI. The visualisation consists of a virtual Bezier curve that appears in front of the inspector guiding them to the location of the virtual element that was selected on the checklist (Figure 7).

When testing an early iteration of the system, we found it extremely difficult to locate and identify holographic elements based on their corresponding checklist name. Therefore, the design motivation of this visualisation was to make it easier for an inspector to locate and identify checklist elements. We initially developed a visualisation that would display a virtual arrow in front of the user and point to the direction of the selected element. However, based on our informal observations we found this approach would not be suitable to locate virtual elements within a dense environment where elements are clustered or closely confined. As a result, we re-designed the visualisation to incorporate a bendable ray that would provide a direct path from the position of the inspector to the centre of the selected element.

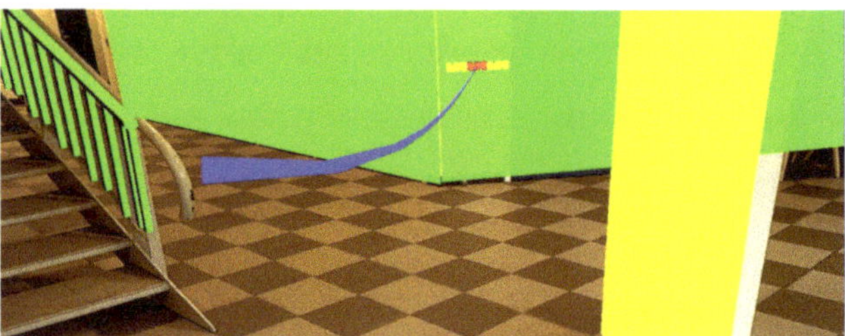

Figure 7. This figure shows the Bendray visualisation, which acts as a visual guidance to direct users to specific elements. In this example, the inspector selected a lighting switch in the tablet-based UI.

4. Colour-coded visualisation: During the construction inspection, inspectors can input into the tablet-based checklist interface whether elements were built in the correct (no defect), close (minor defect), or incorrect (major defect) positions. A colour-coding visualisation tool was integrated into the system to autonomously update the colour of the corresponding virtual element once it is checked off the checklist by the inspector. The colour of the virtual element is modified from their default material colour to a green value to represent if the element is correct (i.e., no defect), yellow if close (i.e., minor defect), or red if incorrect (i.e., major defect). The primary motivation of this visualisation is to provide inspectors with visual feedback on what elements have already been inspected within the checklist. We believe this could improve inspection performance as inspectors can easily and efficiently keep track of the elements they have previously checked off during the inspection.

5. Checklist field-of-view (FOV) rendering: The final feature integrated into our BIM-ARDM system was designed to only display elements within the checklist UI that

the inspector is currently looking at within the AR application (Figure 8). Based on the size of the BIM model, checklists can potentially produce a large and comprehensive list of elements. Therefore, we believed that as the complexity of the BIM model increased, it would become more difficult for inspectors to identify specific elements from the AR environment within the checklist UI. Furthermore, specific checklist elements may also share the same or similar names due to having multiple copies or variations of the same elements (e.g., windows) within the Revit model. As a result, the motivation of this feature was to develop an approach that would improve the inspector's ability to identify a specific element within the checklist UI. Based on our informal observations, we believe that this feature achieves this goal by making it significantly easier for an inspector to identify and locate an element within the checklist UI.

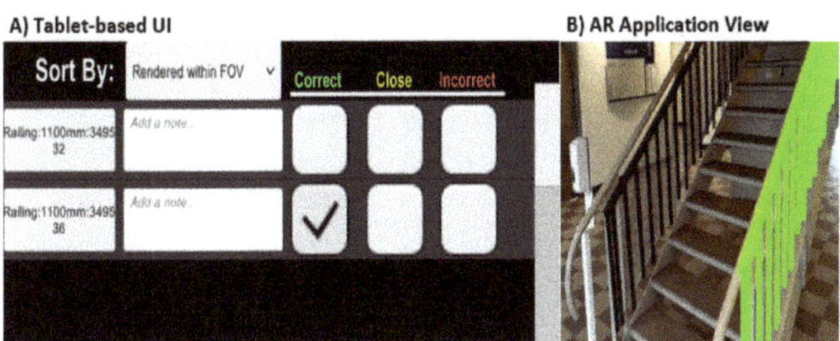

Figure 8. This figure demonstrates the checklist FOV renderer, which only displays elements in the checklist UI (**A**) that the inspector is currently looking at in the AR view (**B**). In this example, the inspector is looking at the stair rails, which appear in the tablet-based checklist UI.

Tracking and calibration process: During the workshop, participants stated they require digital surveying-type accuracy (F9) whilst still maintaining a very simple process to setup the tracking (F8). Due to inspectors requiring millimetre-precise tracking accuracy between the virtual BIM model and the physical construction site to obtain accurate results, we prioritized tracking accuracy over developing a user-friendly tracking system. However, we still integrated various features into our tracking system (e.g., spatial anchors and a re-calibration system) to improve the tracking setup process.

Previously, we discussed two common methods to superimpose BIM models onto a physical construction site: marker-based and markerless tracking. To obtain the most-accurate tracking results, we employed a hybrid tracking approach that uses a marker-based tracking technique to initially calibrate the system and subsequently a markerless approach for maintaining the tracking without markers. After the initial calibration was setup, we also assigned anchors to the BIM model to ensure minimal drift occurs whilst the user navigates through the building. Using this process, we could achieve up to a 1 cm accuracy precision, but this was heavily dependent on ensuring that the position of the physical marker was being tracked at the identical position of the corresponding virtual marker in the BIM model as depicted by Figure 9. To improve the simplicity and user-friendliness of our tracking system (F8), a set of additional features were integrated into our tracking system. This includes spatial anchors that allow inspectors to store and load previously calibrated tracking environments so that a one-time calibration is required for each construction site. However, based on our testing, we found anchors to be relatively inconsistent with the HoloLens 2, and it was highly dependent on the spatial mapping accuracy produced by the built-in depth sensors.

To make minor adjustments to the tracking, we also developed an easy-to-use tracking-adjustment system that can be controlled by the inspector directly through the tablet-based application. Simple debugging tools are also available in the tablet-based application to further assist the inspector with the tracking setup process.

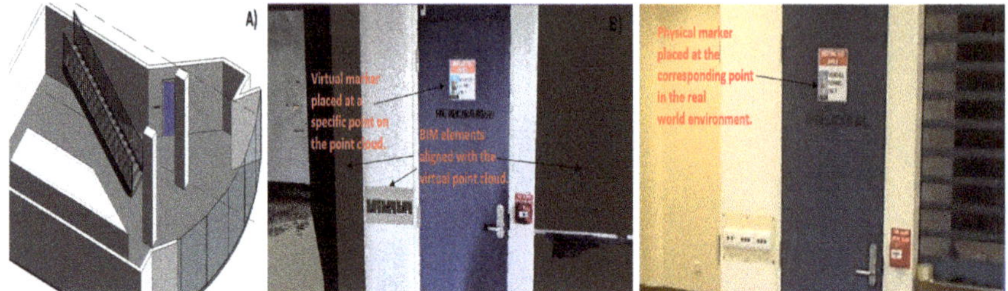

Figure 9. This diagram shows the basic calibration setup. The left image (**A**) shows the BIM model of the real-world building. Image (**B**) shows the BIM model superimposed onto the virtual point cloud. Image (**C**) shows the physical marker placed at the corresponding position of the virtual marker within the real-world environment.

4.2.3. Revit Plugin

During the workshop, participants described the importance of maintaining BIM data (F10) and integrating the AR system with existing BIM platforms (F7). Although the BIM-ARDM system was built off the Unity game engine, we developed a Revit plugin to ensure all data collected during the inspection using the BIM-ARDM system can be synchronized back to the original BIM model. The plugin can be executed directly within Revit and automatically parses a .csv file outputted by the BIM-ARDM software. Data are then linked back to the Revit model based on the unique ID of each element. The recorded data consist of two primary categories:

1. The checklist data, which contain data inputted by the user using the tablet interface. These data include whether the element is correct, close (minor defect), or incorrect (major defect), as well as any other additional comments inputted by the inspector.
2. Images captured of on-site elements, which are captured through the built-in HoloLens 2 camera and can be captured by selecting a capture image on the tablet-based interface. Running the Revit plugin links each picture taken during the inspection back to each of the corresponding Revit elements and attaches the images to the elements as a raster image parameter as illustrated in Figure 10.

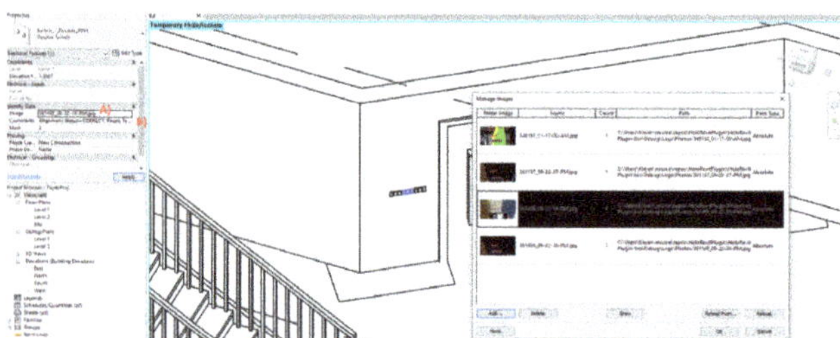

Figure 10. This figure showcases how checklist data were linked back to the original Revit model after running our developed Revit plugin. Image (**A**) shows images captured of elements that are stored as raster images in the given elements "Image" parameter. Image (**B**) shows additional data associated with checklist data that are stored in the elements "Comments" parameter.

4.2.4. 4D Playback Analysis Tool

We also present an experimental analysis tool built into our system, which aims to monitor the performance quality of the inspector's defect management inspection. During the inspection, the head position and rotation vectors, converged eye gaze position and rotation vectors of each pupil, calibration data, and associated timestamps (every 0.1 s) are logged and stored within the HoloLens as a .csv file. After the inspection, the playback analysis software then allows users to playback this data as a 4D visualisation, which displays the inspector's head and eye-gaze location at any time interval throughout the inspection. Additional options are built into the system allowing users to slow down, speed up, play in real-time, or jump between different time intervals using a 2D slider. A heatmap visualisation, which dynamically generates heatmaps at every point that the user looked at on the construction site, was incorporated into the system (Figure 11). A three-colour gradient palette (red, yellow, and green) was used to represent which areas the user was most focused on throughout the inspection. The advantages of this tool are that it allows managers to ensure the inspection has been conducted thoroughly and correctly. Using the playback analysis software, users can identify areas within the BIM model that were potentially neglected based on the inspector's eye-tracking data.

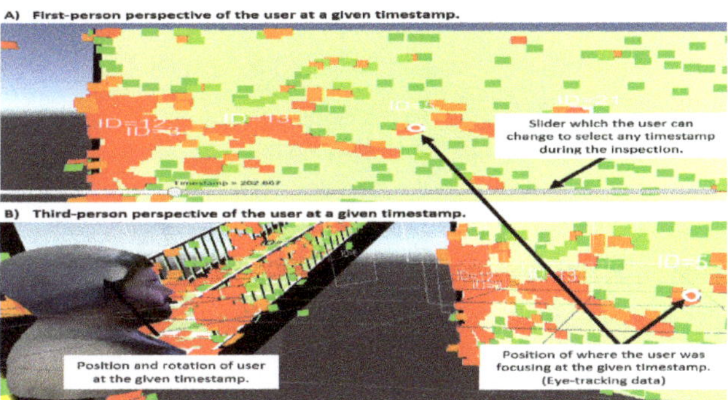

Figure 11. This figure showcases the developed 4D Playback Analysis Tool. Image (**A**) presents a playback of the inspector from a first-person perspective and Image (**B**) from a third-person perspective. The user can adjust the timestamp of the system, and it will display the exact position of where the user was in the real world at that time and where the user was looking. A heatmap is dynamically produced to keep track of where the user has and has not looked.

4.3. System Limitations

The presented BIM-ARDM system consists of two common AR limitations, tracking accuracy and performance.

4.3.1. Tracking Accuracy

Due to defect management inspections requiring millimetre-precision accuracy to ensure the accuracy of inspection data, tracking was one of the primary limitations associated with the system. Although we did not conduct any formal tests or analysis on our tracking accuracy, based on our information observations, we believe the tracking accuracy would generally produce between 1–2 cm. As a result, during our pilot study, participants were instructed to only consider virtual elements greater than 5 cm from the corresponding physical elements as a defect. It also became evident during user-testing that due to depth-perception issues associated with the HoloLens display, the depth of virtual objects would appear to change based on the position and angle of the user.

4.3.2. Performance

Throughout the development of the AR-based system, performance was the primary design consideration during the algorithmic design and implementation of the presented visualisations and features. Due to the HoloLens running as a standalone device, we had to ensure the system would be capable of maintaining a steady frame rate whilst rendering complex geometric building models in addition to several visualisations and features. As a result, our algorithmic implementation of all visualisations and features maintained an $O(1)$ time complexity.

This allowed our implementation to run on very large BIM models consisting of complex geometry. We tested and validated the performance of our BIM-ARDM system on a real-world construction site (Adelaide Festival Plaza (https://armarchitecture.com. au/projects/adelaide-festival-plaza-precinct-masterplan/, accessed on 20 December 2021) that consisted of a complex BIM model containing over 2000 virtual models. The results from this test demonstrated that the system was capable of running at a steady frame rate of 10–15 FPS. However, using the rendering-distance visualisation further improved the system's frame rate.

Despite these results demonstrating the scalability of the BIM-ARDM system, the performance-design limitation resulted in specific features being cut from the system due to not meeting performance requirements. Future iterations of AR displays will continue to improve their processing capabilities, and, as a result, more performance-demanding visualisations and features will become feasible in the future for standalone AR displays.

5. User Study

A comparability pilot study with 11 participants was conducted to assess the performance of our proposed BIM-ARDM system in comparison to the conventional paper-based defect management techniques currently employed within the construction industry. The 11 participants primarily consisted of researchers from the university. All participants who volunteered to take part in the study had some prior experience using AR technology (45% use AR 1–10 times per month, 27.3% daily, 18.2% 1–10 times per year, and 9.1% weekly). From the selection of 11 participants, 10 participants lacked prior experience in conducting on-site construction inspections within a professional capacity.

5.1. Experimental Design

The experiment comprised of two conditions. The first condition was the conventional approach for conducting defect management inspections, which utilized a three-dimensional perspective and two-dimensional orthographic paper-based drawings of the building model as a reference. The second condition was the experimental BIM-ARDM prototype, which superimposes the virtual building model onto the physical construction site and incorporates several visualisations and features as described in Section 4. The experiment consisted of a within-study design, and the order of conditions varied for each participant to counterbalance data. The different AR visualisations provided in the AR condition could be toggled on and off based on the participant's preference. However, participants were encouraged to test each visualisation at least once. During each condition, participants were instructed to complete a defect management inspection task that involved identifying and recording on-site construction defects within a building at the university.

The types of defects we focused on during the study was whether the placement (position or rotation) of a physical constructed element matched the position of the corresponding virtual building element. Participants were instructed to categorize defects into three categorizations: correct (i.e., no defect) when the virtual element was less than 5 cm of the constructed element, close (i.e., minor defect) when the virtual element was between 5 cm and 50 cm, and incorrect (i.e., major defect) when the virtual element was greater than or equal to 50 cm from the constructed element. During each condition, participants were required to inspect a total of thirty elements within the building and record whether each element was correct, close, or incorrect into a tablet-based checklist interface (Figure 4). From the thirty elements inspected by participants in each condition, eighteen of the elements were correct (no defects), eight elements were close (minor defects), and four elements were incorrect (major defects). Each condition contained a different set of defects and correctly positioned elements, and these elements were counterbalanced between the two conditions. Despite the paper-based condition relying entirely on using paper-based drawings, participants still wore a HoloLens 2 HMD with a blank display for data-collection purposes and to replicate the environmental conditions of the AR condition.

5.2. Results

A combination of objective and subjective metrics were utilized to capture specific data throughout the experiment. The collected subjective data consisted of a System Usability Scale (SUS) [20], the Paas Mental Effort ranking [46], a combination of shared and specific Likert-scale based questions, and open-ended short-answer questions as illustrated in Table 3. The objective data, which were used to evaluate the participants' performance consisted of completion time and error rates based on the experimental design.

5.2.1. Usability

The usability for both systems was measured by having participants complete the SUS questionnaire upon completion of each condition. The mean SUS results revealed participants ranked the usability of the paper-based system at 42.04 (grade = F; adjective rating: awful; SD = 19.963), whereas the AR-based system received a mean rating of 81.36 (grade = A; adjective rating: excellent; SD = 9.107). A Wilcoxon Signed rank test verified that participant SUS ratings for the AR system were significantly higher ($p < 0.01$, r = -0.87) than the paper-based approach.

5.2.2. Features and Visualisations

A series of Likert-based (not effective (1) to very effective (5)) questions (Table 3 and Figure 12) were asked to participants to capture their preferred features and visualisations that assisted them with their inspection task performance. For the AR condition, the Bendray (M = 4.909, SD = 0.301) and colour-coding elements (M = 4.545, SD = 0.934) visualisations were ranked among the two most-useful visualisations. Participants also stated the ability to visualise specific elements based on categories (M = 3.454, SD = 1.368) and sort elements (M = 3.545, SD = 1.572) within the checklist (for AR) were effective features. Lastly, the rendering distance visualisation (M = 2.454, SD = 0.934) was ranked as a relatively neutral feature by participants.

Table 3. Questionnaire results for each condition (paper = paper-based condition; AR = augmented-reality condition). A Wilcoxon Signed rank test was used to determine the significance for each Likert-based question (Blue shading represents AR condition, red represents paper-based condition).

#	Statements		Cond	Mean	SD	p
Q1	Based on the mental effort scale, please rank the cognitive-load of the task.	Very, Very Low Cognitive-Load 1 2 3 4 5 6 7 8 9 Very, Very High Cognitive-Load	Paper	6.545	1.368	0.003
			AR	3.727	1.793	
	Shared-Condition Statements					
Q2	I found this system was useful in assisting me to complete the inspection task.	Strongly Disagree 1 2 3 4 5 Strongly Agree	Paper	2.454	1.128	0.003
			AR	4.909	0.301	
Q3	I found this system was useful for locating building elements in the building.	Strongly Disagree 1 2 3 4 5 Strongly Agree	Paper	2.454	1.035	0.003
			AR	4.818	0.404	
Q4	I found this system was useful to identify construction defects.	Strongly Disagree 1 2 3 4 5 Strongly Agree	Paper	2.272	0.786	0.003
			AR	4.545	0.522	
	Condition-Specific Statements					
Q5	I found the paper-based architectural plans were useful in assisting me to complete the task.	Strongly Disagree 1 2 3 4 5 Strongly Agree	Paper	2.181	0.981	0.003
Q6	I found the visualisations were useful in assisting me to complete the task.		AR	4.818	0.404	
Q7	I found the tablet user-interface effective for inputting data into a checklist and controlling the visualisations.	Strongly Disagree 1 2 3 4 5 Strongly Agree	AR	4.272	1.009	nil
	Visualisations and Features Questions					
Q8	Sorting elements in the checklist.	Not Effective 1 2 3 4 5 Very Effective	Paper	2.3	1.337	0.01
			AR	3.545	1.572	
Q9	Bendray visualisation.	Not Effective 1 2 3 4 5 Very Effective	AR	4.909	0.301	nil
Q10	Rendering-distance visualisation.	Not Effective 1 2 3 4 5 Very Effective	AR	2.454	0.934	nil
Q11	Colour-coding visualisation.	Not Effective 1 2 3 4 5 Very Effective	AR	4.545	0.934	nil
Q12	Visualising elements based on categories.	Not Effective 1 2 3 4 5 Very Effective	AR	3.454	1.368	nil
	Post-Questions					
Q13	Please tick the condition you found MOST effective for completing the inspection task.	AR / Paper		100%		
Q14	Please tick the condition that was your personal preference.	AR / Paper		90.9%		9.1%
Q15	Please tick the condition you felt you were less likely to make mistakes/errors with.	AR / Paper		100%		

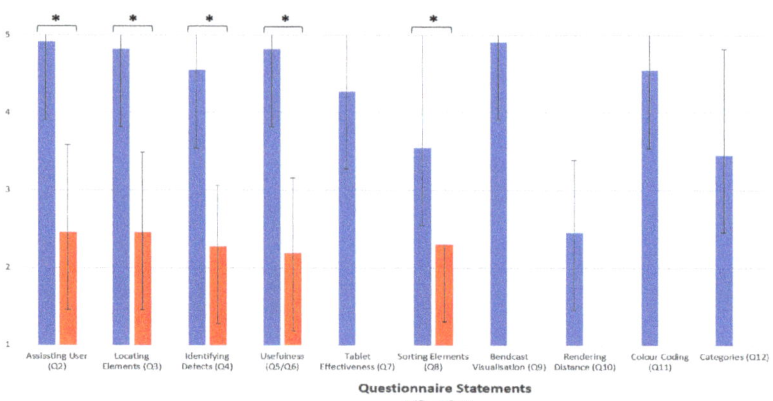

Figure 12. Questionnaire Results for each condition visualised in a bar chart with standard deviation (*: statistically significant).

5.2.3. Cognitive Load Scale

The Paas scale was used to measure the cognitive load of participants for conducting on-site inspections for both paper-based and AR conditions. A Wilcoxon Signed rank test revealed participants reported a significantly lower cognitive load ($p < 0.01$, r = -0.88) was required to complete the task when using the AR-based system (M = 3.72, SD = 1.79) in comparison to the paper-based condition (M = 6.54, SD = 1.36). A graphical comparison of the mental-effort ratings is presented in Figure 13.

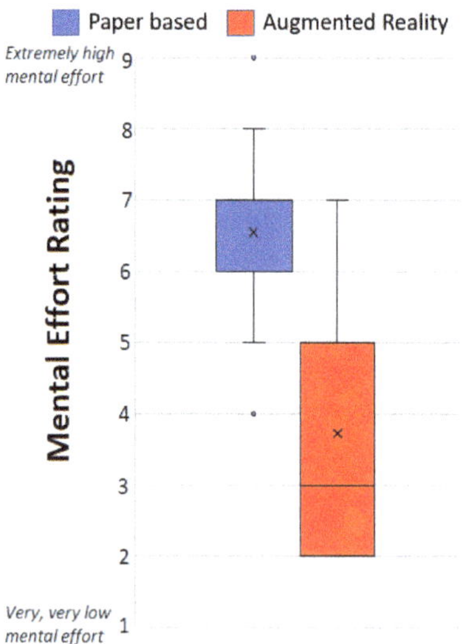

Figure 13. Paas mental effort for paper-based and AR-based conditions. (Box and whiskers plot.)

5.2.4. Participant Performance

A paired T-test determined, on average, that participants completed the inspection task significantly faster (t(10) = -3.06, $p = 0.01$, r = 0.69) when using the AR-based system (M = 639.6 s, SD = 103.4) in comparison to the paper-based approach (M = 1029.3s, SD = 457).

The overall performance of each participant's inspection for both conditions was measured by looking at whether the participant was able to correctly identify whether each element was built in the correct (no defect), close (minor defect), or incorrect (major defect) positions. A paired T-test analysis of the results from the checklist data showed, on average, that participants had a significantly lower error rate (t(10) = −3.05, p = 0.01, r = 0.69) when using the AR-based system (M = 24.7 (82% correct), SD = 2.78) in comparison to the paper-based approach (M = 17.6 (59% correct), SD = 1.9).

A closer examination of participants' error rates also showed significantly lower error rates for correctly identifying minor defects (t(7) = −10.5, p < 0.01, r = 0.97) using the AR-based system (M = 1.75) in comparison to the paper-based approach (M = 7.75). However, we did not find any significance in error rates for identifying correct (p = 0.49) and major defects (p = 0.1) between AR-based and paper-based conditions.

During the tasks, participants were also asked to input a distance approximation (in centimetres) for how far apart virtual elements that they identified as defects appeared to differ from the corresponding physical-building elements. Results from this data showed a strong correlation (R^2 = 0.978) for the participant's ability to correctly estimate distances of minor and major positional defects when using the AR system. However, results from the paper-based condition revealed that participants performed poorly (R^2 = 0.486) when estimating the distances between virtual elements on the architectural paper-based plans with their corresponding physical elements (Figure 14). We were unable to find significance for the mean error distance of the approximations made by participants for correct (p = 0.53) and major defects (p = 0.23) between AR and paper-based approaches. However, we found that participants' mean error of distance approximations for minor defects were significantly lower (p = 0.02, r = −0.8) when using the AR-based system (M = 22.5 cm, SD = 6.48) in comparison to the paper-based approach (M = 40 cm, SD = 11.16) as illustrated in Table 4.

A further examination into the most commonly misidentified elements by participants using both AR-based and paper-based methods revealed some interesting discrepancies (Table 5). Among the five most commonly misidentified elements by participants whilst using the AR system, four of the elements were correct but were misidentified by participants as minor positional defects. In contrast, the paper-based condition produced the complete opposite results where five of the six most commonly misidentified elements were minor defects but were misidentified by participants as correct elements (i.e., non-defects).

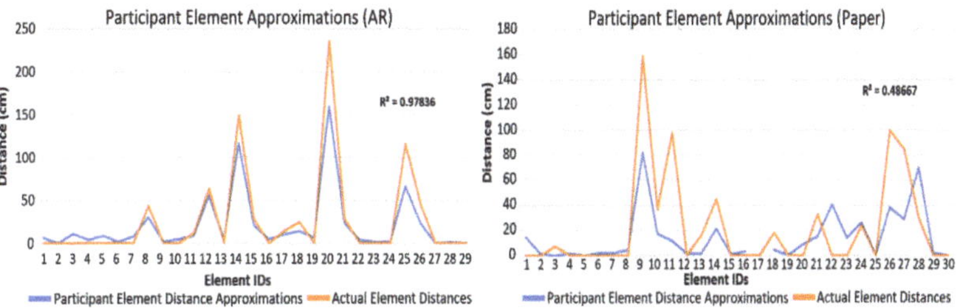

Figure 14. Participant approximations of the distances between virtual and physical elements.

Table 4. Summary of common errors made by participants using AR and paper-based conditions.

Cond	Correct		Minor		Major	
	Mean Error Rate	**Mean Error Dist**	**Mean Error Rate**	**Mean Error Dist**	**Mean Error Rate**	**Mean Error Dist**
AR	2.33	3.86 cm	1.75	22.5 cm	0.5	41.8 cm
Paper	2	5.47 cm	7.75	40 cm	6.75	70.3 cm
p-val	0.49 Paired T-test	0.53 Wilcoxon Signed Rank Test	<0.001 Paired T-test	0.02 Wilcoxon Signed Rank Test	0.1 Wilcoxon Signed Rank Test	0.23 Paired T-test

Table 5. This table presents a comparison of the most-common errors made during the inspection when using the paper-based and AR condition.

Element ID: Name	Incorrect Percentage	Participant Answers	Correct Answer
		Augmented Reality	
1: basic wall	6 (54.5%)	major (1 m) × 2, major (30 cm), major (10 cm, correct, major (rot 45)	minor defect (rot)
4: stair springer	5 (45.4%)	minor 15 cm × 2, minor (bit to the left), minor 20 cm, incorrect 60 cm	correct (0 cm)
20: light 37782	5 (45.4%)	minor 5 cm, minor 30 cm, minor 10 cm, minor 15 cm, minor 7 cm	correct (0 cm)
23: light 377731	4 (36.3%)	minor 5 cm × 2, minor 10 cm × 2	correct (0 cm)
8: railing	4 (36.3%)	minor 7 cm, minor 10 cm, major 60 cm	correct (0 cm)
		paper-based	
3: switch 361548	10 (90.9%)	correct × 10	minor (7.5 cm)
14: basic wall	9 (81.8%)	correct × 5, major × 4	minor (44.3 cm)
27: exit sign 2	9 (81.8%)	correct × 4, minor (30 cm) × 2, minor (20 cm)	major (85 cm)
28: fire sign	8 (72.7%)	major 2 m, major 1 m × 2, major 3 m, correct × 3	minor (30 cm)
18: light sign	8 (72.7%)	correct × 8	minor (18.2 cm)
13: switch 361508	8 (72.7%)	correct × 8	minor (16.4 cm)

5.2.5. Feedback

Preference: Upon completion of the study, we asked participants to pick which condition was most effective, less likely to make errors, and their overall personal preference. All participants stated the AR condition was more effective in completing the task, and that they were less likely to make mistakes during the task using AR. One participant (9.1%) stated the paper-based condition was their personal preference due to their prior experience conducting construction-site inspections with paper-based plans.

General Feedback: A variety of subjective feedback about the paper-based and AR systems was captured by asking participants what they liked and disliked about each condition. Regarding the paper-based approach, many of the dislikes were about the general difficulty of reading and interpreting plans. When participants were asked what they liked about the paper-based plans, two participants stated that the ability to see the room on the drawings without having to physically walk around it was useful. A full breakdown of participant comments is presented in Table 6.

Table 6. This table provides a summary of the subjective open-ended short-answer comments made by participants after completing both conditions.

Statements	Summary of Comments (AR)	Summary of Comments (Paper)
What do you like and/or dislike about the… *(AR Cond)* tablet-based user-interface? *(Paper Cond)* paper-based architectural plans?	Dislikes - Too much colour on the UI ($\times$2), - Elements disappearing when looking down to the tablet when using the Rendered within FOV function. ($\times$3), - Sorting checklist changing the element positions ($\times$1). Likes - Ease of use ($\times$8), - Bendray to identify elements ($\times$2), - Colour-coded visualisation ($\times$2), - Rendered within FOV (x1), - Categories visualisation to filter elements ($\times$1), - Controlling visualisations with the user-interface ($\times$2)	Dislikes - Ability to detect small differences ($\times$3) - Readability of paper plans ($\times$7) - Annoyance reading plans ($\times$1) - Annoyance of carrying tablet and paper-based plans ($\times$1) - Increased mental effort ($\times$1) Likes - The ability to see the room without having to walk around it. ($\times$2)
What did you like about this system?	- Ease of use ($\times$6), - Locating elements within Bendray ($\times$7), - Colour-coded visualisation ($\times$2), - Categories visualisation to filter elements ($\times$1), - Sort functionality ($\times$2), - Rendered within FOV function ($\times$1), - Sorted by completion & alphabetical order ($\times$1), - Inputting data into checklist ($\times$1), - Improved spatial awareness ($\times$1), - Use of tablet to control Hololens visualisations ($\times$1)	- Does not cause any motion sickness ($\times$1) - Inputting data into the tablet checklist user-interface ($\times$5) - Simple to use ($\times$4)
What did you NOT like about this system?	- Bendray coming out of eye ($\times$2), - Mild motion sickness ($\times$1) - Rendered within FOV function not keeping items on top. ($\times$2), - Holographic objects appearing to change based on the users position (depth perception) ($\times$1), - Difficult identifying where some elements should be ($\times$1), - Superimposing virtual objects on physical objects looks strange. ($\times$1)	- Interpretability of plans and ability to identify defects. ($\times$7) - Lack of spatial awareness understanding. ($\times$1) - Fatigue from holding the tablet for long period of time ($\times$1)

5.3. Discussion

An exploration into the most-common misidentified elements revealed that, for the AR condition, participants primarily misidentified correct elements as minor defects, whereas, for the paper-based condition, participants most commonly misidentified minor defects as correct elements. We believe the most-likely explanation for participants misidentifying correct elements as minor defects when using the AR system was due to centimetre drifts occurring in the tracking as participants navigated throughout the building. We also noticed that two of the five most commonly misidentified elements for the AR condition were ceiling lights, which were physically placed relatively far apart from the participants. Therefore, due to the physical distance between the participants and virtual lights, we speculate that depth perception most likely contributed to participants misinterpreting correct elements as minor defects. This was also backed up by post-questionnaire comments where one participant stated: "In some cases, the 3D object positions were a bit different in different viewpoints but not much"; another participant stated: "I found it was sometimes difficult for some elements to identify where they should be, especially the lights"; and another participant commented: "Hololens depth can be hard to read".

6. Conclusions

This article presented the findings of a three-stage research project, which consisted of the identification, development, and evaluation of the BIM-ARDM system for conducting on-site construction inspections. In summary, the research project described in this article achieved the following contributions:

1. Identification and categorization of current uses, potentials uses, and barriers for BIM and AR integration within the construction industry.
2. An improved AR system for identifying construction defects during an on-site construction inspection.
3. A set of novel AR visualisations, and features to improve defect management inspection performance.
4. A Revit plugin to autonomously link data recorded during the construction inspection back to the original BIM model.
5. Data analysis software to evaluate and assess construction inspection performance through eye-tracking, and head-tracking data linked to a four-dimensional visualisation.

The study results showed that the three-dimensional capabilities of the BIM-ARDM system significantly outperformed conventional analogue drawings in terms of the inspection-task performance, the mental workload, the completion time, locating building elements, identifying defects, and assisting the user. Results from the Likert-based questionnaires and qualitative feedback showed participants preferred using the BIM-ARDM system and ranked BIM-ARDM significantly better than conventional approaches.

We believe the current barrier that limits the adoption of our developed BIM-ARDM prototype within the construction industry is the tracking limitations associated with current AR hardware. This will limit the type and scale of defects that can be determined using the BIM-ARDM system. However, tracking is still very much an ongoing research question within the AR research community, and advancements to tracking are continuously developing. Although our BIM-ARDM system is capable of achieving up to 1 cm accuracy, we believe a more-reliable and stable tracking system that could consistently maintain less than a 1 cm accuracy would support the adoption of the BIM-ARDM system within the construction industry.

Author Contributions: Conceptualization, K.W.M., N.G., J.W., R.T.S. and B.H.T.; data curation, K.S.M.; formal analysis, K.W.M. and C.K.; investigation, K.W.M.; methodology, K.W.M.; software, K.W.M.; supervision, N.G., J.W., R.T.S. and B.H.T.; validation, K.W.M.; visualization, K.W.M.; writing—original draft, K.W.M.; writing—review and editing, K.W.M., J.J.O., N.G., J.W., R.T.S. and B.H.T. All authors have read and agreed to the published version of the manuscript.

Funding: This research was funded by he University of South Australia (UniSA) Research Themes Investment Scheme (RTIS) grant (101807).

Institutional Review Board Statement: The studies conducted in this project are approved by the University of South Australia's Human Research Ethics Protocol (protocol code 203586, 2/5/2021 and protocol code 202527, 8/26/2019).

Informed Consent Statement: Informed consent was obtained from all subjects involved in the studies.

Data Availability Statement: The data presented in this study are available in a non-personally identifiable format on request from the corresponding author.

Acknowledgments: We would like to thank Philippe Naudin from ARM Architecture (https://armarchitecture.com.au/people/philippe-naudin/, accessed on 20 December 2021) for his expertise and assistance during the design of the BIM-ARDM software.

Conflicts of Interest: The authors declare no conflict of interest.

References

1. Eiris, R.; Gheisari, M.; Esmaeili, B. PARS: Using augmented 360-degree panoramas of reality for construction safety training. *Int. J. Environ. Res. Public Health* **2018**, *15*, 2452. [CrossRef]
2. Park, C.S.; Kim, H.J. A framework for construction safety management and visualization system. *Autom. Constr.* **2013**, *33*, 95–103. [CrossRef]

3. Albert, A.; Hallowell, M.R.; Kleiner, B.; Chen, A.; Golparvar-Fard, M. Enhancing construction hazard recognition with high-fidelity augmented virtuality. *J. Constr. Eng. Manag.* **2014**, *140*, 04014024. [CrossRef]
4. Kim, K.; Kim, H.; Kim, H. Image-based construction hazard avoidance system using augmented reality in wearable device. *Autom. Constr.* **2017**, *83*, 390–403. [CrossRef]
5. Carrasco, M.D.O.; Chen, P.H. Application of mixed reality for improving architectural design comprehension effectiveness. *Autom. Constr.* **2021**, *126*, 103677. [CrossRef]
6. Ahn, K.; Ko, D.S.; Gim, S.H. A study on the architecture of mixed reality application for architectural design collaboration. In *Proceedings of the International Conference on Applied Computing and Information Technology* ; Springer: Berlin/Heidelberg, Germany, 2018; pp. 48–61. [CrossRef]
7. Walsh, J.A.; Baumeister, J.; Thomas, B.H. Spatial Augmented Reality Visibility and Line-of-Sight Cues for Building Design. In Proceedings of the 27th ACM Symposium on Virtual Reality Software and Technology, Osaka, Japan, 8–10 December 2021; Association for Computing Machinery: New York, NY, USA, 2021. [CrossRef]
8. El Ammari, K.; Hammad, A. Remote interactive collaboration in facilities management using BIM-based mixed reality. *Autom. Constr.* **2019**, *107*, 102940. [CrossRef]
9. Ammari, K.E.; Hammad, A. Collaborative BIM-based markerless mixed reality framework for facilities maintenance. In Proceedings of the 2014 International Conference on Computing in Civil and Buidling Engineering, Orlando, FL, USA, 23–25 June 2014; pp. 657–664. [CrossRef]
10. Messadi, T.; Newman, W.E.; Braham, A.; Nutter, D. Immersive learning for sustainable building design and construction practices. *J. Civ. Eng. Archit* **2017**, *11*, 841–852. [CrossRef]
11. Osello, A.; Del Giudice, M.; Guinea, A.M.; Rapetti, N.; Ronzino, A.; Ugliotti, F.; Migliarino, L. Augmented Reality and gamification approach within the DIMMER Project. In Proceedings of the INTED2015 9th International Technology, Education and Development Conference, Madrid, Spain, 2–4 March 2015; pp. 2707–2714.
12. Fukuda, T.; Yokoi, K.; Yabuki, N.; Motamedi, A. An indoor thermal environment design system for renovation using augmented reality. *J. Comput. Des. Eng.* **2019**, *6*, 179–188. [CrossRef]
13. May, K.W.; Walsh, J.; Smith, R.T.; Gu, N.; Thomas, B.H. VRGlare: A Virtual Reality Lighting Performance Simulator for real-time Three-Dimensional Glare Simulation and Analysis. ISARC. In Proceedings of the International Symposium on Automation and Robotics in Construction, Kitakyushu, Japan, 26–30 October 2020; Volume 37, pp. 32–39. [CrossRef]
14. Ma, Z.; Cai, S.; Mao, N.; Yang, Q.; Feng, J.; Wang, P. Construction quality management based on a collaborative system using BIM and indoor positioning. *Autom. Constr.* **2018**, *92*, 35–45. [CrossRef]
15. Shin, D.H.; Dunston, P.S. Identification of application areas for Augmented Reality in industrial construction based on technology suitability. *Autom. Constr.* **2008**, *17*, 882–894. [CrossRef]
16. Shin, D.H.; Dunston, P.S. Evaluation of Augmented Reality in steel column inspection. *Autom. Constr.* **2009**, *18*, 118–129. [CrossRef]
17. Hart, S.G.; Staveland, L.E. Development of NASA-TLX (Task Load Index): Results of Empirical and Theoretical Research. In *Human Mental Workload*; Hancock, P.A., Meshkati, N., Eds.; Advances in Psychology; Elsevier: Amsterdam, The Netherlands, 1988; Volume 52, pp. 139–183. [CrossRef]
18. Portalés, C.; Casas, S.; Gimeno, J.; Fernández, M.; Poza, M. From the paper to the tablet: on the design of an AR-based tool for the inspection of pre-fab buildings. Preliminary results of the SIRAE project. *Sensors* **2018**, *18*, 1262. [CrossRef]
19. García-Pereira, I.; Portalés, C.; Gimeno, J.; Casas, S. A collaborative augmented reality annotation tool for the inspection of prefabricated buildings. *Multimed. Tools Appl.* **2019**, *79*, 6483–6501. [CrossRef]
20. Brooke, J. *"SUS-A Quick and Dirty Usability Scale"*. *Usability Evaluation in Industry*; CRC Press: Boca Raton, FL, USA, 1996. ISBN 9780748404605.
21. Feng, C.W.; Chen, C.W. Using bim and Mr to improve the process of job site construction and inspection. *Wit Trans. Built Environ.* **2019**, *192*, 21–32. [CrossRef]
22. Park, C.S.; Lee, D.Y.; Kwon, O.S.; Wang, X. A framework for proactive construction defect management using BIM, augmented reality and ontology-based data collection template. *Autom. Constr.* **2013**, *33*, 61–71. [CrossRef]
23. Kwon, O.S.; Park, C.S.; Lim, C.R. A defect management system for reinforced concrete work utilizing BIM, image-matching and augmented reality. *Autom. Constr.* **2014**, *46*, 74–81. [CrossRef]
24. Zhou, Y.; Luo, H.; Yang, Y. Implementation of augmented reality for segment displacement inspection during tunneling construction. *Autom. Constr.* **2017**, *82*, 112–121. [CrossRef]
25. Hernández, J.L.; Martín Lerones, P.; Bonsma, P.; Van Delft, A.; Deighton, R.; Braun, J.D. An IFC interoperability framework for self-inspection process in buildings. *Buildings* **2018**, *8*, 32. [CrossRef]
26. Available online: https://technical.buildingsmart.org/standards/ifc/ (accessed on 20 December 2021).
27. Available online: https://www.tridify.com/ (accessed on 20 December 2021).
28. Available online: https://github.com/helpsterTee/Unity-IFCEngine (accessed on 20 December 2021).
29. Available online: https://www.pixyz-software.com/ (accessed on 20 December 2021).
30. Marchand, E.; Uchiyama, H.; Spindler, F. Pose estimation for augmented reality: A hands-on survey. *IEEE Trans. Vis. Comput. Graph.* **2015**, *22*, 2633–2651. [CrossRef]

31. Wagner, D.; Langlotz, T.; Schmalstieg, D. Robust and unobtrusive marker tracking on mobile phones. In Proceedings of the 2008 7th IEEE/ACM International Symposium on Mixed and Augmented Reality, Washington, DC, USA, 15–18 September 2008; pp. 121–124. [CrossRef]

32. Comport, A.I.; Marchand, E.; Pressigout, M.; Chaumette, F. Real-time markerless tracking for augmented reality: The virtual visual servoing framework. *IEEE Trans. Vis. Comput. Graph.* **2006**, *12*, 615–628. [CrossRef]

33. Simon, G.; Fitzgibbon, A.W.; Zisserman, A. Markerless tracking using planar structures in the scene. In Proceedings of the IEEE and ACM International Symposium on Augmented Reality (ISAR 2000), Munich, Germany, 5–6 October 2000; pp. 120–128. [CrossRef]

34. Xiao, C.; Lifeng, Z. Implementation of mobile augmented reality based on Vuforia and Rawajali. In Proceedings of the 2014 IEEE 5th International Conference on Software Engineering and Service Science, Beijing, China, 27–29 June 2014; pp. 912–915. [CrossRef]

35. Kato, H.; Billinghurst, M. Marker tracking and hmd calibration for a video-based augmented reality conferencing system. In Proceedings of the 2nd IEEE and ACM International Workshop on Augmented Reality (IWAR'99), San Francisco, CA, USA, 20–21 October 1999; pp. 85–94. [CrossRef]

36. Kopsida, M.; Brilakis, I. BIM Registration Methods for Mobile Augmented Reality-Based Inspection. In Proceedings of 11th European Conference on Product and Process Modelling eWork and eBusiness in Architecture, Engineering and Construction, (ECPPM 2016), Limassol, Cyprus, 7–9 September 2016; Christodoulou, S., Scherer, R., Eds.; CRC Press: Boca Raton, FL, USA, 2016. [CrossRef]

37. Hübner, P.; Clintworth, K.; Liu, Q.; Weinmann, M.; Wursthorn, S. Evaluation of HoloLens Tracking and Depth Sensing for Indoor Mapping Applications. *Sensors* **2020**, *20*, 1021. [CrossRef]

38. Krippendorff, K. *Content Analysis: An Introduction to Its Methodology*, 2nd ed.; Sage Publications: Thousand Oaks, CA, USA, 2004.

39. Rankohi, S.; Waugh, L. Review and analysis of augmented reality literature for construction industry. *Vis. Eng.* **2013**, *1*, 1–18. [CrossRef]

40. Ahmed, S. A review on using opportunities of augmented reality and virtual reality in construction project management. *Organ. Technol. Manag. Constr. Int. J.* **2018**, *10*, 1839–1852. [CrossRef]

41. Alizadehsalehi, S.; Hadavi, A.; Huang, J.C. From BIM to extended reality in AEC industry. *Autom. Constr.* **2020**, *116*, 103254. [CrossRef]

42. Hakkarainen, M.; Woodward, C.; Rainio, K. Software architecture for mobile mixed reality and 4D BIM interaction. In Proceedings of CIB-W78 25th International Conference on Information Technology in Construction, Santiago, Chile, 15–17 July 2008, Rischmoller, L., Eds; Universidad de Talca: Talca, Chile, 2009; pp. 1–8.

43. Gheisari, M.; Goodman, S.; Schmidt, J.; Williams, G.; Irizarry, J. Exploring BIM and mobile augmented reality use in facilities management. In Proceedings of the 2014 Construction Research Congress: Construction in a Global Network, Atlanta, GA, USA, 19–21 May 2014; Castro-Lacouture, D., Irizarry, J., Ashuri, B., Eds.; American Society of Civil Engineers: Reston, VA, USA, 2014; pp. 1941–1950. [CrossRef]

44. May, K.; Hanan, I.; Cunningham, A.; Thomas, B. 3DUITK: An Opensource Toolkit for Thirty Years of Three-Dimensional Interaction Research. In Proceedings of the 2019 IEEE International Symposium on Mixed and Augmented Reality Adjunct (ISMAR-Adjunct) Beijing, China, 10–18 October 2019; pp. 175–180. [CrossRef]

45. Riege, K.; Holtkamper, T.; Wesche, G.; Frohlich, B. The bent pick ray: An extended pointing technique for multi-user interaction. In Proceedings of IEEE Symposium on 3D User Interfaces (3DUI), Alexandria, VA, USA, 25–26 March 2006; pp. 62–65. [CrossRef]

46. Paas, F. Training Strategies for Attaining Transfer of Problem-Solving Skill in Statistics: A Cognitive-Load Approach. *J. Educ. Psychol.* **1992**, *84*, 429–434. [CrossRef]

buildings

Article

Method to Identify the Likelihood of Death in Residential Buildings during Coastal Flooding

Axel Creach [1,*], Emilio Bastidas-Arteaga [2], Sophie Pardo [3] and Denis Mercier [1]

[1] Laboratory of Physical Geography: Actual and Quaternary Environments, Geography and Planning Department, Sorbonne University, 191 Rue Saint-Jacques, 75005 Paris, France; denis.mercier@sorbonne-universite.fr
[2] Laboratory of Engineering Sciences for Environment, LaSIE UMR CNRS 7356, La Rochelle University, Avenue Michel Crépeau, CEDEX 1, 17042 La Rochelle, France; ebastida@univ-lr.fr
[3] Laboratory of Economics and Management of Nantes, LEMNA—Nantes University, Chemin de la Censive du Tertre–Bâtiment Erdre, 44322 Nantes, France; sophie.pardo@univ-nantes.fr
* Correspondence: axel.creach@sorbonne-universite.fr

Abstract: Tools exist to predict fatalities related to floods, but current models do not focus on fatalities in buildings. For example, Storm Xynthia in France in 2010 resulted in 41 drowning deaths inside buildings. Therefore, there has been increasing recognition of the risk of people becoming trapped in buildings during floods. To identify buildings which could expose their occupants to a risk of death in the case of flooding, we propose the use of the extreme vulnerability index (VIE index), which identifies which buildings are at greatest risk of trapping people during floods. In addition, the "mortality function method" is used to further estimate the expected number of fatalities based on (1) groups of vulnerable people (e.g., aged or disabled), (2) the location of buildings in relation to major watercourses, and (3) the configuration of buildings (e.g., single or multiple entries and single or multiple stories). The overall framework is derived from case studies from Storm Xynthia which give a deterministic approach for deaths inside buildings for coastal floods, which is suited for low-lying areas protected by walls or sandy barriers. This methodology provides a tool which could help make decisions for adaptation strategy implementation to preserve human life.

Keywords: coastal flooding risk; loss of life; fatality assessment; residential buildings; climate change adaptation; VIE index

Citation: Creach, A.; Bastidas-Arteaga, E.; Pardo, S.; Mercier, D. Method to Identify the Likelihood of Death in Residential Buildings during Coastal Flooding. *Buildings* **2022**, *12*, 125. https://doi.org/10.3390/buildings12020125

Academic Editor: Tiago Miguel Ferreira

Received: 16 December 2021
Accepted: 20 January 2022
Published: 26 January 2022

Publisher's Note: MDPI stays neutral with regard to jurisdictional claims in published maps and institutional affiliations.

1. Introduction

Floods killed 8 million people in the last century [1]. The conditions under which 235 fatalities occurred in relation to 13 floods events (hurricanes and storms) were analyzed. It was concluded that 68% of the deaths were caused by people drowning in cars and on foot (33 and 25%, respectively), while just 6% occurred inside buildings [1]. By contrast, 54% (out of 771 people) were drowned in buildings during Hurricane Katrina in 2005 [2], while all of the fatalities (41) caused by Storm Xynthia in France were those drowned in buildings [3,4]. Boissier [5] suggested that for high-magnitude, low-frequency events, most of the fatalities occur inside buildings.

Storm Xynthia (28 February 2010) had major impacts along the Atlantic coast of France, especially between the Loire and Gironde estuaries. The analysis of fatalities provided by Vinet et al. [3,4] for Storm Xynthia highlighted how buildings can trap people during floods. In addition to factors such as the time of day (e.g., nighttime) and the fact no warning was given [6], they showed that (1) most drownings occurred when the flood level exceeded 1 m, (2) 90% of drownings occurred inside buildings located within 400 m of the flood defenses which failed, and (3) 78% of the drownings occurred inside single-story houses.

Subsequently, Creach et al. [7] defined areas where people could be trapped inside buildings in case of flooding as zones of "extreme vulnerability".

In response to Storm Xynthia, the French government instigated the "black zone" policy, which listed 1628 buildings in areas of extreme flood risk for demolition at a cost of EUR 315.7 million [8]. This policy attracted significant criticism, however, principally on the grounds of the lack of methodological transparency [9], expenditure, and failure to engage local communities in a timely manner. However, of principal concern was the fact that the strategy did not result in any appreciable reduction in human vulnerability to flooding hazards along French coasts, as it was also limited in scope to areas flooded in 2010 [10,11]. Thus, the policy was a reactive adaptation which lacked cost-benefit analysis and time to make an enlightened choice in association with locals.

The French Atlantic Coast is particularly susceptible to flooding. For instance, it is predicted that a centennial flood could cause 354,079 ha to be submerged [12]. In this area, there are 535,500 permanent inhabitants [13] and 136,711 buildings [12], 22% of which are single-story constructions [13]. Furthermore, flooding is likely to be exacerbated further under the current trajectory of global warming and resulting sea level rise [14–16]. Moreover, other areas along the French Atlantic coast are also exposed to floods [12], and additionally, the sea level is rising, which will inevitably increase coastal flooding.

To address these problems, in this paper, we propose a global framework to evaluate potential fatalities specifically inside buildings in relation to coastal flooding. To achieve this, the novelty is to use extreme vulnerability index (VIE index) to evaluate the buildings at greatest risk of flooding and to assess the implications for human fatalities [7] in addition to the population's features. Using lessons learned from Storm Xynthia, this yields a deterministic approach which is suited for low-lying areas exposed to coastal floods with numerous single-story houses, features which are those of the French Atlantic coast. Thus, this paper will demonstrate the value of the VIE index in assessing and locating most risky houses which could, in fact, help decision makers to establish strategies to protect human life from flood risk in the future.

2. Materials and Methods

2.1. State of the Art of Fatality Assessment Methods

2.1.1. General Principles of Fatality Assessment

The assessment of fatalities due to floods is a relatively new research field [17,18], with research shifting from the hydrological process to its management [19]. Several major studies have been conducted in the UK, Netherlands, and Canada [20,21] because of greater data availability pertaining to the conditions under which flood-related fatalities occurred. These data have been crucial to the development of numerical models. In particular, these studies focused on coastal or inland floods due to breaches in flood defenses, and they differ from other work that has focused on fatalities caused by tsunamis or hurricanes [22–24].

According to Di Mauro et al. [21], fatalities are mainly influenced by the number of exposed people which could be reduced by preventive evacuation. The flood magnitude, frequency, and location, as well as the types of buildings and people's vulnerability and behavior are shown to be the main variables controlling fatality risk.

Fatalities can be evaluated in two different ways on the basis of scale [21,25]: (1) at the microscale, which pertains to "individual risk", and (2) the macroscale, which considers the overall risk to society. Microscale studies are useful for understanding individual human responses to flooding, but data are lacking, and as a result, numerical models rely on interpolation and possess limited predictive power [21,26]. Macroscale risk is simpler in terms of assessing the likelihood that deaths will occur in relation to several factors [20,21]. This probability has been estimated through examining the statistical relationships between the fatalities and characteristics of past floods [17]. For example, Klijn et al. [27] estimated that 0.3% of people could die in the Netherlands due to a flood, and according to Jonkman et al. [17], 1% of exposed people globally are expected to die in coastal floods.

2.1.2. Methods for Assessing Fatalities

Three different methods are used to assess fatalities in relation to floods [21]. First, the Life Safety Model (LSM) focuses on individual risk. It was developed in Canada to assess potential fatalities due to dam failure [28,29], and it involves modeling of the behavior of individuals during floods using an automated 2D cellular model. The health of individual people is considered in the model. The method requires specific data about the hazard characteristics (e.g., timing and magnitude), the age and health of the people, and the building type, configuration, and general accessibility for the purpose of evacuation [21]. This model is useful for simulating evacuation planning and evaluating the mitigative effects of warning systems on crisis management [21]. It can also be used for educational purposes to define the best course of action on what to do in the case of flooding.

Second, there is the Flood Risk to People (FRP) methodology—developed in the UK [30–32]—which can assess both societal and individual risk. The societal risk aims to evaluate the potential consequences of a flood. It functions by multiplying the exposed population by a rating factor defined for different sectors. This rating is based on the assessment of the characteristics of the flood (e.g., water height, water flow, and debris content), area vulnerability (e.g., type of land use and building configuration to offer shelters), and the population characteristics (e.g., age and mobility). For each of these aspects, areas with similar properties are demarcated. The number of people at risk is obtained through a census [33], which is then divided into residential buildings, and the number of potential deaths is thus estimated. This method is used to assess the consequences of different floods to derive a large-scale risk assessment.

Uncertainties exist regarding the evaluation of the number of people at risk depending on the timing of the flood [30–32]. The methodology nevertheless provides an estimate of the areas at greatest risk of flooding. It was originally designed to be used as an operational tool for decision makers in the UK, but only the flood hazard parameter was used. Priest et al. [34] adapted the FRP methodology for continental floods in Europe. They showed that the FRP methodology overestimated the number of deaths outside the UK due to the fact it did not consider population behavior and preventive evacuation. The Risk to Life Model (RLM) introduced a "mitigation" component to the methodology, which is defined by the level of awareness and the ease by which people can be evacuated. It is divided into four classes, depending on the evacuation rate, from >75% to >25%.

Third, the mortality functions method [17,25,35] focuses on fatalities related to coastal and inland floods following the failure of flood defenses. It is particularly useful for areas reclaimed from the sea (polders) with flat land protected from floods by flood defenses. The goal of the method is to estimate the fraction of fatalities from the total population exposed to flooding to derive the number of potential fatalities (Equation (1)):

$$N = F_D N_{EXP} \tag{1}$$

where N is the potential number of deaths, F_D is the mortality fraction, and N_{EXP} is the number of exposed people.

The mortality fraction (F_D) is defined by the severity of the hazard, which is controlled by the extent of the flooded area, water depth, the kinetic properties of the flood, or the proximity to flood defenses. Jonkman et al. [17] employed a 2D model to simulate different flooding scenarios. The simulated flooded area is divided into three zones ranked according to the conditions of the flood defenses: (1) the breached zone (torrential flow), (2) the zone of rapidly rising water, and (3) the zone with minimal flooding. For each zone, a mortality function is given depending on the flood characteristics (i.e., from 1 (certain death) to 0 (no risk of death)). Thus, in the breached zone (maximum of 300 m behind the breach), the mortality fraction is theoretically equal to 1 [35], though most recent works on Storm Katrina show a value of 0.053 [2].

The mortality function is then multiplied by the number of exposed people (N_{EXP}). This number depends on (1) the total number of inhabitants in the flooded area, (2) the number of people evacuated before the flood, (3) the number of people in the process of evacuating during the flood, and (4) the number of people rescued. The method is simple in that it only requires an estimation of the number of fatalities from the total number of exposed people in a specific area. However, the model parameters are based primarily on the characteristics of the flood.

In theory, the model could be modified to explicitly account for fatalities inside buildings. Consideration of a building's configuration as a control on the vulnerability of people to flooding would result in a different mortality function estimate. A simple solution would be to multiply this value by the total number of exposed people inside buildings.

2.1.3. Limits of Fatality Assessments

The advantages and limitations of the various methods are listed in Table 1. Generally, micro-tools require numerous hypotheses regarding (1) the behavior of people during floods, (2) flood wave kinetics, and (3) accurate data about the occupation of buildings. Furthermore, powerful software is then needed to simulate the spatial and temporal properties of the flood and the time needed for evacuation in different flood scenarios.

Table 1. Advantages and limitations of the three fatality assessment methods.

Type or Scale	Method	Advantages	Drawbacks
Micro	Life Safety Model (LSM)	Realistic, accurate locations of deaths	Technical, fine data needed for modulization
Macro	Flood Risk to People (FRP)	Takes into account several dimensions of vulnerability	Data not fully available in France
Macro	Mortality functions	Easy to use	Mainly driven by hazard characteristics

Macro-models try to assess the potential number of deaths through several factors pertaining to flood characteristics and the vulnerability of people. Thus, an important objective is to assess the potential for fatalities that occur inside buildings at the building scale.

The FRP methodology [30–32,34] provides a holistic approach including parameters related to the flood hazard, such as the vulnerability of a given area and the vulnerability of people living in or occupying that area. However, this approach was originally designed in relation to floods in the UK and is relevant thanks to a wide range of accurate data available there.

The mortality function method [17] is advantageous for its simplicity of use. However, as stated earlier, it relies on the characteristics of floods to define the mortality fraction. In this paper, we suppose that the configurations and locations of buildings are also crucial contributing factors to fatalities. The methodology proposed is a complementary approach to the mortality function method but with an increased emphasis on vulnerability.

The factors used in various methods are summarized in Table 2. This table shows that it is a challenging task to measure and model all the parameters which could lead to fatalities [17,20,34]. The most complete (FRP) method includes six over seven of these parameters, while the others integrate just four or five. Three factors refer to flood characteristics. Di Mauro and De Brujn [36] stated that results are largely influenced by the extent of the flooded area, which then raises questions about the simulation of flood hazards. The introduction of a new parameter based on vulnerability may considerably enhance fatality estimation.

Table 2. Parameters integrated into the main flood fatality assessment methods (adapted from [20]).

Model	Sourced	Water Depth	Water Velocity	Rate of Water Level Rise	Warning and Evacuation	Preparedness	Collapse of Buildings	Vulnerability of Individuals (Weight, Height, Gender, Clothing)	Data Obtained from Real Floods (HP) or Laboratory Research (L)
LSM	Lumbroso et al. [29]	X	X	X	X		X		
Mortality Functions	Waarts [37]	X	X	X	X				HP
	Vrouwenvelder and Steenhuis [38]	X		X	X		X		HP
	Jonkman [35,39]	X	X	X	X	X			HP/L
FRP	HR Wallingford et al. [31,32]	X	X	X	X		X	X	HP
	Priest [34]	X	X	X	X		X		HP

At present, buildings are only considered in terms of risk of collapse [20], potential shelters [31], or for their own vulnerability [40]. According to Di Mauro and De Bruijn [36], the fact that buildings are scarcely integrated in this type of study is due to the lack of knowledge of mortality inside buildings. In this respect, the studies of the fatalities due to Storm Xynthia [3,4] and Hurricane Katrina [2] are very useful.

Table 2 also shows that most of the tools are based on specific flood case studies, which could affect their applicability to other territories or flood events [17,20,34].

Estimating the number of exposed people to floods also affects the estimation of fatalities [36]. Quantifying the number of exposed people is challenging because this will vary according to the time of day, season, and the amount of time taken to evacuate particular types of buildings (e.g., residential homes and offices). According to Jonkman [35], the goal of mortality functions is not to provide an exact estimate of the number of deaths but to assess the risk level. For Di Mauro and De Bruijn [36], it is more appropriate to give the result as a percentage of fatalities rather than an exact number. To give an appropriate estimation, the best way would be to multiply flood and evacuation scenarios [36] to identify flood-prone areas in which more fatalities are likely.

In summary, the existing tools do not provide a specific assessment of fatalities inside buildings at the building scale. Therefore, we propose a framework that integrates and assesses the risk of death inside buildings at the building scale, which could be an add-on to other methods.

2.2. VIE Index: A Tool to Locate Buildings Which Could Expose Their Occupants to Death
2.2.1. Context

Due to a lack of knowledge about the death risk inside buildings at risk of flooding, there are few integrated fatality assessment methods [36]. We therefore used the VIE index method (*Vulnérabilité Intrinsèque Extrême*, i.e., extreme vulnerability assessment) recently proposed by Creach et al. [7] to integrate the role of the building type and configuration in relation to assessments of fatality risk. Following Storm Xynthia, the deficiencies linked to policies such as the "black zone" and general methods used to identify buildings that put people at serious risk [9,41] resulted in the VIE index being designed to explicitly assess the role of buildings in trapping people during floods. The VIE index does not focus on the risk of building collapse, which is related to the quality and design of buildings. Both tsunami [42–48] and inland flood hazards [34,49] are not treated by the VIE index as ends in themselves. Its main goal is to identify buildings which could trap their occupants during flooding because of their configuration and location. In this way, the VIE index focuses more on the vulnerability parameters than on hazard characteristics.

2.2.2. VIE Index Methodology

The VIE index is based on four major criteria which contributed to people having been trapped in residential buildings during Storm Xynthia [3,4] (Figure 1):

- Cr1: Potential water depth inside buildings;
- Cr2: Distance to flood defenses;
- Cr3: Architectural typology of buildings, since single-story constructions are more likely to trap people than multi-story buildings, where people could escape upstairs;
- Cr4: Proximity to a rescue point to facilitate ease of evacuation.

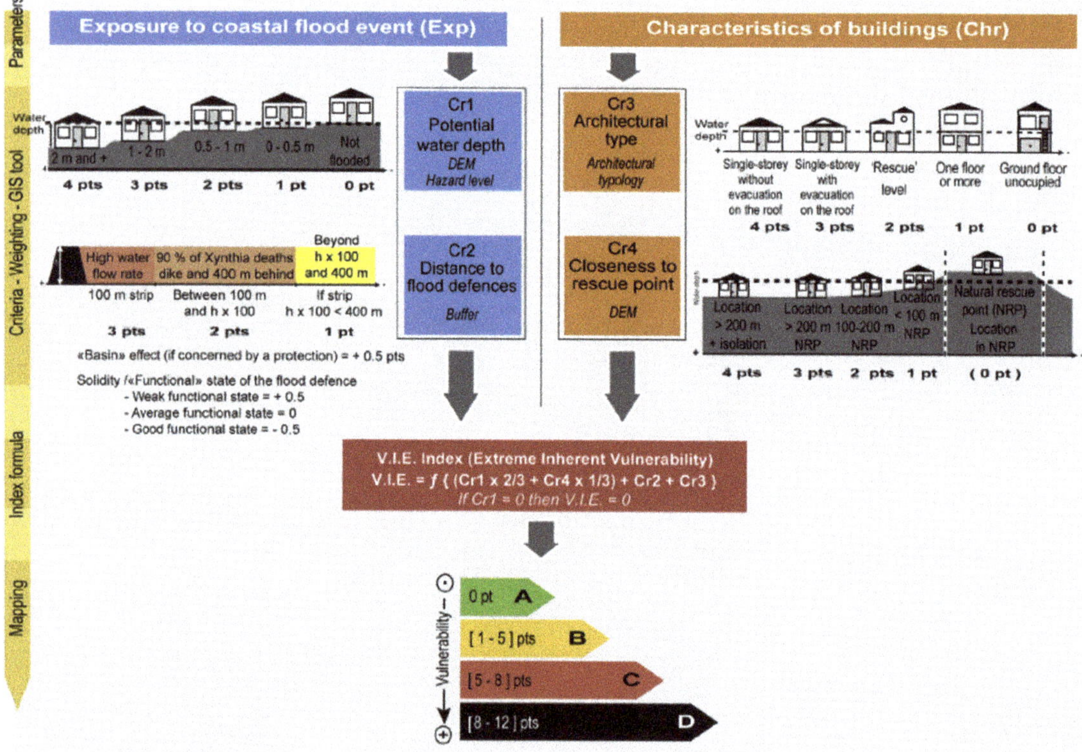

Figure 1. The VIE index methodology (source: [7]). "DEM" corresponds to Digital Elevation Model used for calculation, and "h × 100" strip corresponds to a strip behind flood defenses for which the width is 100 times the height of the dike.

Each of the criteria is rated from 0 (no vulnerability) to 4 (high vulnerability). Creach et al. [11] proposed the formula below (Equation (2)), validated through statistical analysis, to demarcate buildings that pose the greatest risk in terms of trapping people during floods (Figure 1):

$$\text{VIE} = \left(\text{Cr1} \times \frac{2}{3} + \text{Cr4} \times \frac{1}{3} \right) + \text{Cr2} + \text{Cr3} \tag{2}$$

The results range from 0 (no vulnerability) to 12 (maximum vulnerability). To map the results of the index, they are divided into four classes which represent different levels of vulnerability:

- Green class (VIE index = 0): buildings are not exposed to floods and therefore do not endanger people;
- Orange class (VIE index = 1–5): buildings are of a suitable design to reduce risk to people during floods. The level of risk for people is low;
- Red class (VIE index = 5–8): the risk for people is high but non-lethal if appropriate action is taken, except for older, younger, or disabled people;
- Black class (VIE index = 8–12): the risk for people is very high and could result in fatalities in the case of flooding.

The VIE index method has been validated through statistical analysis [11]. The first results were then validated by comparison with the locations of deaths during Storm Xynthia [11]. The calculation of the VIE index shows that 83% of fatalities occurred in buildings classified "black", while 17% of the fatalities fell within the "red" class. Thus, this shows the good ability of VIE index method to identify buildings in which death may occurs in the case of a coastal flood.

2.3. A Derived Method for Evaluating the Risk of Fatalities inside Buildings during Floods

From the methods and their limitations reviewed in Section 2, a derived method for assessing fatalities is proposed which involves the following:

- Focusing on vulnerability more than on hazards;
- Using data about fatalities that occurred inside buildings during Storm Xynthia;
- Incorporating the VIE index method to assess the vulnerability of buildings for people.

We included the FRP framework because of its holistic approach [30–32,34] and the mortality function method [17] for its simplicity in calculating a mortality fraction for areas of equal vulnerability. In contrast with Kolen et al.'s paper [6], where they applied the rule of thumb of 1% deaths during coastal floods for Storm Xynthia (see Appendix A), we calculated a specific mortality fraction for each area, which assumed that the probability of death varied according to the building's vulnerability to flooding. Though only a single event, the data for Storm Xynthia [3,4] enabled us to calculate the following in a determinist way:

- A relationship between fatalities related to Storm Xynthia and buildings which posed the greatest risk to people, as determined by the VIE index [11]. This allowed us to estimate a parameter close to the FRP framework's area vulnerability.
- A relationship between the age of deceased people and the total number fatalities. This allowed us to estimate a parameter close to the FRP framework's people vulnerability.

These two parameters are then used to estimate a mortality fraction.

Another important aspect of Storm Xynthia was the importance of secondary houses (i.e., houses which are held by people who do not live there throughout the year and which are mainly used for leisure during holiday times) in the coastal municipalities impacted by the storm, which had a major influence on the number of exposed people. This is an additional parameter to be estimated for assessing the exposed population.

In summary, we propose an add-on methodology that assesses fatality risk, taking into account the two previously mentioned aspects for the assessment of the mortality rate (Figure 2).

2.3.1. Lessons Learned from Storm Xynthia's Fatalities

The circumstances regarding deaths inside buildings during Storm Xynthia are listed in Table 3, according to previous detailed work by Vinet et al. [3,4]. This shows that 29 of the 41 deaths by drowning occurred inside buildings in the municipality of La Faute-sur-Mer.

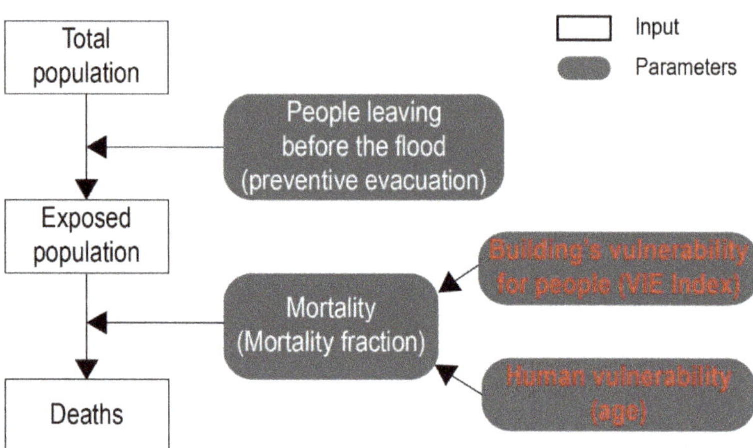

Figure 2. Proposed factors influencing the number of fatalities inside buildings due to flooding (adapted from [21]). Text in red corresponds to new parameters we propose using for calculation of the mortality fraction.

Table 3. Synthesis of death circumstances inside buildings during Storm Xynthia with regard to VIE index results.

		Number of Deaths	% of Total (29 Deaths)
Total Number of Deaths		**29**	**100%**
Deaths distribution depending on vulnerability class	Black	24	83%
	Red	5	17%
Deaths distribution depending on age	Under 15	3	10%
	15–60	4	14%
	Above 60	22	76%
Deaths distribution depending on occupation	Principal houses	20	69%
	Secondary houses	9	31%

From the VIE index, people residing in a house in a black zone are at substantially greater risk of fatality than those in a house in a red zone. According to Devaux et al. [50], 1661 out of 1996 flooded buildings in the municipality of La Faute-sur-Mer were likely to expose their occupants to a risk of death, according to the VIE index [7]. Of these 1661 buildings, 63% (1027 buildings) were "black" zone buildings and 37% (604 buildings) were "red" zone buildings, according to the definitions of the VIE index. However, 83% of the deaths were located inside "black" zone buildings according to Table 3, compared with just 17% inside "red" zone buildings. According to these results, "orange" zone buildings are not considered to be likely to expose their occupants to a risk of death, because their vulnerability level should allow people to escape (upstairs or to a shelter close to the house).

In addition, elderly or very young people are at greatest risk of being drowned. In 2006, the census reported that 45% of the inhabitants of La Faute-sur-Mer were above the age of 60 [51]. Crucially, over 76% of deaths occurred in people older than 60 (Table 3). By contrast, people aged 15–60 years represented 45% of the population in 2006. However, only 14% of the total number of deaths related to Storm Xynthia affected this age group (Table 3).

Four deaths of people aged from 15–60 were located inside black zone houses, thus confirming the risks associated with this classification of building. For the red zone buildings, 33% of the deaths of people aged under 15 and 18% of people aged above 60 occurred, which also confirms the danger of this classification of building. From these data, the probability of death relating to age and building type could be estimated.

A final point concerned the occupation of principal and secondary houses. From a total of 3737 buildings in 2006 [51], 86% were secondary houses, and only 14% were principal houses. Storm Xynthia occurred during a holiday weekend, and it is probable that several principal houses were unoccupied while several secondary houses were occupied. Table 4 shows that 69% of deaths occurred in principal houses while 31% were located inside secondary houses. This finding has important implications for estimating the exposed population.

Table 4. Results of the mortality fraction by building vulnerability class and age of population.

VIE Index Typology	Age Category	Mortality Fraction
	Under 15 $\left(F_D^{B-15} \right)$	2.61%
Black class	15–60 $\left(F_D^{B\,15-60} \right)$	1.20%
	Above 60 $\left(F_D^{B+60} \right)$	5.38%
	Under 15 $\left(F_D^{R-15} \right)$	2.22%
Red class	15–60 $\left(F_D^{R\,15-60} \right)$	0%
	Above 60 $\left(F_D^{R+60} \right)$	2.04%

Based on these different observations, we focused on the most vulnerable zone (red and black classes), and a mortality fraction was proposed for each situation using Equation (3):

$$F_D^{i,j} = \frac{N^{i,j}}{N_{EXP}^{i,j}} \tag{3}$$

where $i = \{B, R\}$ refers to the building vulnerability class (black or red class), $j = \{-15, 15-60, +60\}$ is the age of the population, $F_D^{i,j}$ is the mortality fraction according to the building vulnerability class and the age of the inhabitants, $N^{i,j}$ is the potential number of deaths for each situation, and $N_{EXP}^{i,j}$ is the corresponding exposed population.

Detailed calculations of the mortality fraction for each situation are given in Appendix B, and the results are summarized in Table 4 and Appendix C (lines 15–20). It needs to be said that black buildings could lead to death for all, whereas red class buildings are relatively safe except for the young and the elderly, who can have difficulties moving inside a flooded house.

2.3.2. Estimating a Global Mortality Fraction per Municipality ($F_D^{MUNICIPALITY}$)

Based on these mortality fractions, a specific mortality function per municipality, called $F_D^{MUNICIPALITY}$, is calculated, which includes the following:

- Age of the local population according to census data;
- Proportion of black and red houses according to the VIE index results.

As shown by Equation (4), each mortality fraction is multiplied by the proportion of each age category. Then, the proportions between the black and red houses are included:

$$\begin{aligned} F_D^{MUNICIPALITY} = &\left[\left(F_D^{B,-15} \times P^{-15} \right) + \left(F_D^{B,15-60} \times P^{15-60} \right) + \left(F_D^{B,+60} \times P^{+60} \right) \right] \times P^B \\ &+ \left[\left(F_D^{R,-15} \times P^{-15} \right) + \left(F_D^{R,15-60} \times P^{15-60} \right) + \left(F_D^{R,+60} \times P^{+60} \right) \right] \times P^R \end{aligned} \tag{4}$$

where $F_D^{B,-15}$ is the mortality fraction for people aged under 15 in black zone houses, P^{-15} is the proportion of people aged under 15 in the municipality population, $F_D^{B,15-60}$ is the mortality fraction for people aged 15–60 in black zone houses, P^{15-60} is the proportion of people aged 15–60 in the municipality population, $F_D^{B,+60}$ is the mortality fraction for people aged above 60 in black zone houses, P^{+60} is the proportion of people aged above 60 in the municipality population, P^B is the proportion of black zone houses among dangerous buildings (red and black houses), $F_D^{R,-15}$ is the mortality fraction for people aged under 15 in red zone houses, $F_D^{R,15-60}$ is the mortality fraction for people aged 15–60 in red zone houses, $F_D^{R,+60}$ is the mortality fraction for people aged above 60 in red zone houses, and P^R is the proportion of red houses among dangerous buildings (red and black zone houses).

This method enables having a more accurate assessment of the mortality fraction for each city. For example, a city with mostly red zone houses and people aged 15–60 will have a low mortality fraction.

2.3.3. Estimating the Exposed Population per Municipality (N_{EXP})

To estimate the most exposed population per municipality in the case of coastal flooding, the fatality risk in relation to building vulnerability to flooding needs to be emphasized. In particular, this concerns the occupants of both the red and black zone houses as classified by the VIE index. This number is estimated using the total number of the potentially deadliest buildings and the average number of people per household, as defined by census data.

The importance of secondary houses will be integrated into the population estimation. This is achieved using census data in which the proportion of principal and secondary houses is collected. For each of them, an average occupation rate based on data from Storm Xynthia is proposed, which is 69% for principal houses and 31% for secondary houses (Table 4). This results in a global occupation rate for the deadliest buildings. This is then multiplied by the average people per household and gives Equation (5):

$$N_{EXP}^{MUNICIPALITY} = [(P_{PH} \times T_{PH}) + (P_{SH} \times T_{SH})] \times N^{RB} \times N_{HF} \tag{5}$$

where $N_{EXP}^{MUNICIPALITY}$ is the population potentially exposed to a risk of death per municipality in the case of coastal flooding, P_{PH} is the proportion of principal houses in the municipality, T_{PH} is the estimated occupation rate of principal houses (69%), P_{SH} is the proportion of secondary houses in the municipality, T_{SH} is the estimated occupation rate of secondary houses (31%), N^{RB} is the number of black and red buildings identified by the VIE index, and N_{HF} is the average number of people per household of the municipality.

This formula allows the proportion of the deadliest houses (which could vary from one municipality to another) to be considered, as well as secondary houses which could radically increase the number of people at risk of fatality. In the case of La Faute-sur-Mer, according to the formula, a total of 1182 people were exposed to flooding during Storm Xynthia (Appendix B).

2.3.4. Assessing the Potential Number of Deaths (N)

Using Equation (4) ($F_D^{MUNICIPALITY}$) and Equation (5) ($N_{EXP}^{MUNICIPALITY}$), the potential number of deaths (N) for each municipality, depending on its specific characteristics on the basis of the formula of Jonkman et al. [17], was calculated (Equation (1)). For the specific case of La Faute-sur-Mer, the application of the formula yielded a total of 29 potential deaths, which is an accurate result in light of the data on fatalities in this municipality.

The strength of this method is that it allows for the inclusion of parameters pertaining to the vulnerability of houses, which could vary from one municipality to another (altimetry and architectural type of buildings). The method also accounts for the importance of secondary houses, which could be an important parameter to assess the exposed population in the case of some touristic municipalities exposed to coastal floods.

3. Results

The results of both the VIE index and potential fatality assessment were calculated for seven municipalities on the French Atlantic coast [7,52]. Three of them are located in Baie de l'Aiguillon, and they were directly impacted by Storm Xynthia: La Faute-sur-Mer

(29 deaths), L'Aiguillon-sur-Mer (no deaths), and Charron (3 deaths). The others are located on Noirmoutier Island, which was impacted little by the storm, but the configuration of human settlement is similar to the municipalities impacted in 2010 [52,53] and has been impacted by coastal floods in the past [54,55].

The results of these previous studies show good validation of the model for the three studied municipalities impacted by Storm Xynthia (see Section 2.2.2), as well as confirming that it was highly risky places with black houses representing levels of 48%, 28%, and 1% in La Faute-sur-Mer, l'Aiguillon-sur-Mer, and Charron, respectively [7]. This was due to recent allotments of single-story type buildings constructed in the lower parts of these municipalities. On the other hand, the municipalities of Noirmoutier Island showed a low level of black houses which was always under 4%, except for in La Guérinière (20%) [52]. In general, buildings in flood-prone areas on the island are located relatively far from the sea (500 m or more).

Here, we focus on the case of La Guérinière, since it is the most exposed municipality of Noirmoutier Island [52]. However, data and detailed calculations of potential fatalities are given for each of the seven municipalities in Appendix C.

3.1. Study Site

Noirmoutier Island is in the central part of the French Atlantic coast (Figure 3). It is a low-lying island with an area of 49 km^2 and with 68% of its territory located beneath the level of the storm surge associated with Storm Xynthia [53]. To protect it from the sea, a network of 24 km of flood defenses was established on the east coast of the island, while the west coast is protected by a sandy barrier [56]. La Guérinière is in the central part of the island, namely on its narrowest part (800 m between the west and east coasts). It is the smallest municipality of the island with an area of 7.8 km^2, and 80% of its territory could potentially be flooded [50], therefore making it the most exposed town on the island. Its position increases the risk of coastal floods. In addition, in the event of failure of the eastern flood defenses (5-m elevation flood defenses) or the sandy barrier (5.5-m elevation), the resulting damage would be catastrophic.

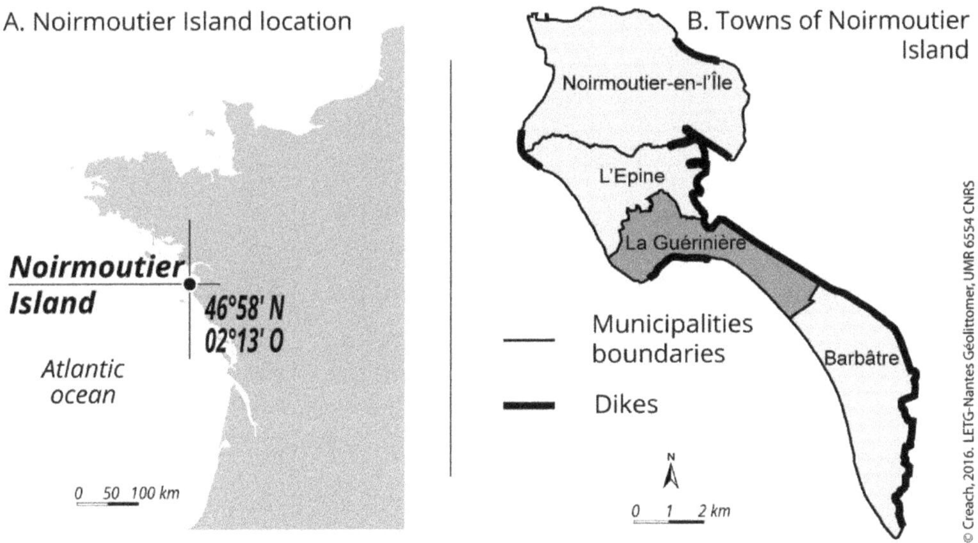

Figure 3. Location of the town of La Guérinière (source: [57]).

Today, Noirmoutier is noted for its leisure activities [58]. In the last 40 years, the number of buildings increased by 162%, and the population increased by 19% [57], with most of this urbanization occurring in low-lying areas. Thus, should a centennial flood occur, one half of the residential buildings of the island would be at extreme risk of flooding [52].

With a total of 1460 inhabitants in 2011 [51], La Guérinière is the smallest municipality of the island with 15.5% of the island's total population. However, the number of residential buildings has increased by 144% since 1968 [51], and they represent 17% of the total number of residential buildings on the island (2817 of 16,438 buildings). Of these buildings, 74% are secondary homes, while the average is 66% on the island [51]. For a centennial-scale flood, 63% of the residential buildings of La Guérinière could be flooded [52] (Table 5).

Table 5. VIE index results for La Guérinière municipality (adapted from [57]). Non-identified buildings correspond to those for which the VIE index cannot be calculated due to missing data (i.e., architectural type unknown) or which are not residential buildings.

Class	Number of Buildings	% of Buildings among "Total Identified Buildings"
Green class	1034	37%
Orange class	329	12%
Red class	885	31%
Black class	569	20%
Total identified buildings	2817	100%
Non-identified buildings	1251	ø
Total buildings	4068	ø

Although only 3% of the area of Noirmoutier was flooded during Storm Xynthia [50], and though no major storms have impacted the island since the 1950s, major floods have occurred in the past [55,59]. For example, the storm of 1937 is known to have flooded a large part of the island even though no deaths occurred, probably because of the lower population density [55]. Even though Noirmoutier Island was not recently impacted by coastal floods, it remains at risk, since low-lying areas are heavily urbanized.

3.2. VIE Index Results

The VIE index results for La Guérinière were presented in two recent publications, with one presenting the results for the whole of Noirmoutier Island for two flood scenarios [52] and the other focusing on the case of La Guérinière and presenting four different flood scenarios [57]. In this section, the "medium scenario", which corresponds to a centennial return period, is selected for analysis. As recommended by the French Ministry for the Environment, this scenario needs to be used for regulatory documents for urban planning in flood-prone areas [60]. It needs to be based on either the highest known historical sea level or on the results of a centennial flood numerical model [60]. For Noirmoutier Island, the sea level measured in the harbor of Noirmoutier-en-l'Île was used. It was measured to be 4.20 m NGF (French legal altimetric datum). It needs to be said that only the water depth was used to specify the hazard, as the VIE index focuses on vulnerability inside buildings, with the proximity to coastal defenses being a way to estimate high rising water inside.

Calculation of the VIE index requires several steps for each criterion, which are linked to different maps. Here, we refer to Creach's [61] atlas to review each map of the process. It should be noted that the results presented here refer to buildings for which the VIE index has been calculated ("total identified buildings" given in Table 5), which could defer from the total number of buildings in the territory as most of them are not residential buildings (e.g., garages or garden sheds) or sometimes data are missing (i.e., architectural typology).

La Guérinière has a total amount of 1783 residential houses located under 4.20 m NGF and which could be flooded in a centennial flood (Table 5). It represents 24% of the total number of potentially inundated buildings of the island, and 75% of them in La Guérinière could be submerged by a water depth >1 m, which is considered a level of extreme danger for people [17,30,62]. A 1-m water level would result in 28% of the buildings on the island being inundated. While 40% of the coast of the island is artificial, few buildings are close to flood defenses. Less than 23% could be directly affected by high rising water if the flood defenses failed. However, in La Guérinière, where the island is the narrowest, 56% of residential buildings are particularly vulnerable to a dike failure (Cr2). The architectural typology (Cr3) shows that 60% of the potentially inundated buildings of La Guérinière are single-story, with the average on the island being 64%. Finally, since Noirmoutier Island is a low-lying territory, there are few natural shelter areas (i.e., above the maximum water level). In La Guérinière, these correspond to the sandy barrier. Additionally, 35% of La Guérinière's buildings (Cr4) are less than 100 m from a natural shelter area (average of 25% on Noirmoutier Island), while 14% are located more than 200 m away (average of 9% on Noirmoutier Island).

The calculation of the VIE index confirmed that La Guérinière is the most exposed municipality of Noirmoutier Island. At the island scale, the exposure of residential buildings for people is not particularly high; the green class (no vulnerability of the buildings for people) is 54%, the red class is 26% (potential risk of death due to individual behavior or vulnerable people), and the black class is 5% (potential death due to the vulnerability of the building for people). In La Guérinière, according to Table 5, 37% of residential buildings are in the green class, 12% are in the orange class, and most importantly, 31% are in the red class, while 20% are in the black class. Therefore, there is a considerable risk of death for people living in 52% of the residential buildings of La Guérinière, compared with 31% for the whole island. The black class is far more representative than the other municipalities, as this class reaches a maximum of 4% on Noirmoutier-en-l'Île and less than 1% on L'Epine and Barbâtre. On Noirmoutier-en-l'Île, it comprises 291 buildings, whereas in La Guérinière, it is 569.

As shown in Figure 4, most of the black houses are relatively close to the southwest coast, where the sandy barrier is narrow, less elevated, and thus poorly protected by flood defenses. These houses are also of a single-story type, which does not allow for vertical evacuation, contrary to the surrounding red houses which are located at the same altimetry but offer multiple stories. A rapid rise of water is possible, meaning there is insufficient time for people to evacuate, except if a preventive evacuation is carried out or if it is possible to escape upstairs. On the other hand, buildings at the two ends of La Guérinière are located on the barrier (green class) or are protected by it, and therefore the rise in water level would be slower (red or orange class).

3.3. Potential Fatalities

From the VIE index results, and with reference to the census data which are given in Appendix C, it is possible to estimate the potential number of deaths in relation to floods.

3.3.1. Estimating the Mortality Fraction (F_D)

The mortality fraction was calculated from the age of the population, the proportion of potentially lethal houses (red and black VIE index classes), and the integrated mortality fraction (Appendix C). The age composition of the population for 2011 (to allow comparison with Storm Xynthia's context of 2010) and the proportion of red and black houses among the potentially lethal houses are given in Appendix C (from line 6 to 9 and lines 3 and 4, respectively).

Derived from Equation (4) and according to the data from Appendix C, we have Equation (6):

$$F_D^{LaGueriniere} = [(0.261 \times 0.138) + (0.120 \times 0.482) + (0.538 \times 0.379)] \times 0.39$$
$$+ [(0.222 \times 0.138) + (0 \times 0.482) + (0.204 \times 0.379)] \times 0.61) = 0.018 \tag{6}$$

This indicates a mortality fraction of 1.82%, which is less than that for La Faute-sur-Mer (2.45%). The difference is explained by the potentially lethal black zone houses, which are less represented in La Guérinière (39%) than in La Faute-sur-Mer (60%). Thus, in proportion, fewer buildings will expose the occupants to an extreme risk of death.

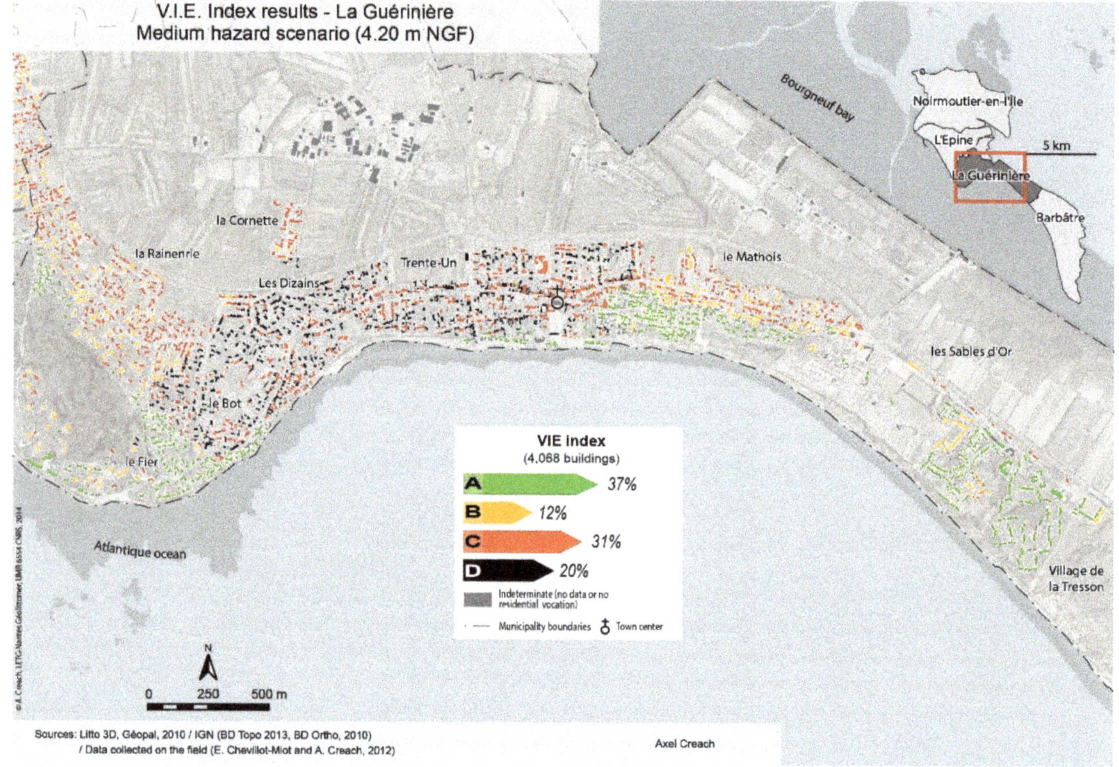

Figure 4. VIE index results for La Guérinière municipality (adapted from [57]).

Concerning the age of the population, the elderly are less represented in La Guérinière (38%) than in La Faute-sur-Mer (45%), but the representation of young people is large (14% in La Guérinière, relative to La Faute-sur-Mer (10%)).

La Guérinière is the municipality with the highest mortality fraction of Noirmoutier Island (1.5% in Noirmoutier-en-l'Île, 1.2% in Barbâtre, and 1.1% in L'Epine). This is mainly because of the importance of black buildings among potentially lethal houses; these represent 39% of houses in La Guérinière but only 19% in Noirmoutier-en-l'Île, 2% in Barbâtre, and 0.2% in L'Epine. Thus, the probability of dying inside a residential house in La Guérinière is higher than in the other municipalities of Noirmoutier Island.

3.3.2. Estimating the Exposed Population (N_{EXP})

According to Equation (5), the exposed population was estimated from the average number of people per household, which is given in line 14 of Appendix C (2.1 inhabitants per household). Appendix C also shows the proportion of principal houses (line 11; 26%) and secondary houses (line 12; 74%) and the occupation rate (Table 4 and Appendix C, lines 21 and 22, respectively, at 69% for principal houses (T_{PH}) and 31% for secondary houses (T_{SH})), and the number of potentially lethal houses (line 5; 1454 (N^{RB})):

$$N_{EXP}^{LaGueriniere} = [(0.26 \times 0.69) + (0.74 \times 0.31)] \times 1454 \times 2.10 = 2146.65 \tag{7}$$

Equation (7) gives a total of 1247 of people potentially exposed to flooding. The fact that this number is smaller than the number of potentially lethal houses is due to the importance of secondary houses, which represent 74% of the buildings of La Guérinière. This number of potentially exposed people is slightly higher than the number of people exposed to Storm Xynthia in La Faute-sur-Mer (1182 according to Equation (A3) in Appendix B). The number of potentially exposed people in La Guérinière is the highest for Noirmoutier Island but not as high as elsewhere (1118 in L'Epine, 1054 in Noirmoutier-en-l'Île, and 1022 in Barbâtre). This difference is mainly due to the slightly higher number of potentially lethal houses in La Guérinière (1454) than in the other locations (1228 in L'Epine, 1220 in Noirmoutier-en-l'Île, and 1214 in Barbâtre).

3.3.3. Number of Potential Fatalities (N)

By applying the formula for the mortality function method (Equation (1)) using (1) the mortality fraction calculated for La Guérinière ($F_D^{La\ Gueriniere} = 1.8\%$) and (2) the number of potentially exposed people ($N_{EXP}^{La\ Gueriniere} = 1247$), the potential number of deaths could be estimated using Equation (8):

$$N^{La\ Gueriniere} = F_D^{La\ Gueriniere} \times N_{EXP}^{La\ Gueriniere} = 1.8\% \times 1247 = 22.71 \tag{8}$$

In the case of a centennial flood in La Guérinière, 23 fatalities could be recorded inside houses. This is lower than the number of deaths that occurred in La Faute-sur-Mer during Storm Xynthia (29 deaths recorded). As explained previously, this is the result of the mortality fraction being lower in La Guérinière ($F_D^{La\ Gueriniere} = 1.8\%$) than in La Faute-sur-Mer ($F_D^{La\ Faute-sur-Mer} = 2.45\%$).

The other municipalities on the island may also experience fewer fatalities in the case of a centennial flood (16 in Noirmoutier-en-l'Île, 12 in L'Epine, and 12 in Barbâtre) because of a lower mortality fraction but also an exposed population (Appendix C, line 25).

This result confirms that La Guérinière is the municipality most at risk of fatalities on Noirmoutier Island, even if it is not as exposed as La Faute-sur-Mer.

4. Discussion

Regarding the VIE index, the key limitations are discussed in the work of Creach et al. [7]. Different uncertainties are responsible for overestimation of the number of flooded houses. These include (1) the use of a "static" method to estimate the potential water depth, (2) no consideration of flood defenses to reduce water depth (as recommended by the French Ministry for Environment (see [62])), and (3) estimation of the water depth in relation to the local topography. It is an evaluation of the worst scenario possible. If a building is likely to be flooded, even though the probability is low, it needs to be considered at risk of inundation. The precautionary principle needs to be considered when protecting human life.

The novelty for fatality assessment is to identify those buildings in which deaths are most likely, which allows for proposing adaptation strategies to reduce their vulnerability. However, there are several specific limitations to the approach proposed in this paper. First, it is a deterministic approach, as the mortality fractions are only estimated from Storm Xynthia. This limitation is evident when considering that the mortality fraction is 0 for people aged 15–60 in red houses (Table 4), because no deaths were recorded in this case during Storm Xynthia [3,4]. In fact, deaths are less likely to occur in red houses for "healthy" people [1,4], but there is still a degree of risk. It was similarly necessary to consider principal and secondary houses because they represent more than 50% of the residential buildings in the studied coastal municipalities [51]. Despite the availability of accurate data about the number of secondary houses, even though there are general data proposed at the municipality scale, it is difficult to estimate the population exposed to a risk of flooding. The number of people occupying secondary houses can vary because of holiday periods, which themselves are seasonal. Only very fine resolution studies would permit the elucidation of details concerning the exact location and number of occupied secondary houses, but this would require numerous interviews [63,64] or new considerations about the occupancy of residential houses throughout the year, with this type of research still being in progress for the French Atlantic coast [64]. In fact, the hypothesis we used could be seen as an upward occupancy in regard to the most frequent season for coastal floods to occur on the French Atlantic coast [59]. Winter is not the period at which residential homes are the most occupied.

Analysis of more past flood events is clearly needed to validate the methodology and to refine the mortality fractions. For instance, the 1953 storm in the Netherlands [65], Canvey Island in the UK [21,26], and Hurricane Katrina in the USA [2] are well-documented storms for which data concerning deaths inside buildings are available. Therefore, the application of the methodology presented in this paper is a first step which can be adapted and applied to the French Atlantic coastal municipality, as they exhibit similar characteristics in terms of the degree of urbanization, architectural typology, number of secondary houses, and population. In addition, the current lack of accurate data, both on fatalities related to Storm Xynthia and census information, as well as other parameters that could influence the number of fatalities (e.g., gender or health condition), have not been considered [1]. Finally, there is also a disparity between the accuracy of the outputs given by the VIE index, which can permit the identification of each residential house based on its vulnerability and ranking from fatality assessments, which provide an estimation of the potential number of deaths at the municipality scale. This is because of difficulties such as the lack of available high-resolution spatial data in France. This is particularly true for other parameters such as the building type, size, and function (i.e., principal or secondary house) as well as the number of occupants and their ages and genders.

Applying the method to other past floods could also allow for addressing another limitation of the proposed method: building collapses are not considered, as no collapse occurred during Storm Xynthia [4]. However, this is usually integrated in such studies [20], as it could amplify the risk of death [39].

Despite these limitations, the proposed methodology is suited for low-lying areas protected from the sea by walls or sandy barriers, which vast territories along the French Atlantic coast or in Netherlands and UK have. It could be proposed as an add-on to the LSM, the FRP, and the mortality functions methods. It allows one to identify buildings which are safe or unsafe, depending on the exposed population in an agent-based model, and it could add another input to the FRP method and allow one to address a specific mortality fraction, depending on the building typology, other than those currently estimated by the mortality functions method. Moreover, it should be seen as a useful experimental tool for enabling decision makers to reduce the vulnerability of buildings and to protect people from future floods, which could increase with the sea level's rising due to climate change. The achievability of this goal is enhanced by the possibility of undertaking cost–benefit analyses. Overall, it provides a complementary approach to existing fatality assessment methods by considering the risk of death inside buildings.

5. Conclusions

The methodology presented herein is useful for considering how fatalities may result from people being trapped in buildings during coastal floods. The primary example of Storm Xynthia showed how the building type, configuration, and location could considerably increase the risk of death [3,4].

The VIE index [7] provided a method of assessing which buildings posed the greatest risk to occupants during times of flooding. This index is based on four criteria, which were identified as predominant in the estimation of human vulnerability during Storm Xynthia: (1) potential water depth inside buildings, (2) distance to flood defenses, (3) the architectural typology of the buildings, and (4) proximity to a rescue point. These criteria are used to assign a rating to houses which present the greatest risks to people. The results are divided into four vulnerability classes (green to black), of which two represent a potentially lethal situation for occupants: (1) the red class, where death risk is more related to at-risk behavior or vulnerable people, and (2) the black class, where the configurations and locations of people contribute more to the risk of fatalities during periods of flooding.

From the assessment of the vulnerability of houses for people, a deterministic quantification of the death risk was proposed using the mortality function method developed by Jonkman et al. [17] and data from Storm Xynthia. This method estimates the probability of death (F_D) among an exposed population (N_{EXP}). According to Jonkman et al. [17], the probability of death occurring depends on the hazard characteristics. In this paper, the probability of death was shown to have been conditioned not just by the zone (i.e., the "black class") of the occupied houses but also the age of the occupants. Based on the conditions under which fatalities occurred during Storm Xynthia [3,4], a mortality fraction of the deaths was estimated depending on the vulnerability of the houses for people (red or black class) and the age of the population (under 15, 15–60, and above 60 years). By integrating census data and the VIE index results at the municipality scale, it is possible to estimate a specific probability of death for people. Then, the exposed population is calculated depending on the average number of households living in dangerous houses (red and black classes of the VIE index) considering the combined effects associated with the principal and secondary houses. This allows the potential number of deaths to be calculated.

The model gives good results, with the VIE index highlighting the high vulnerability of buildings in La Faute-sur-Mer and L'Aiguillon-sur-Mer—the most impacted towns during Storm Xynthia—for which the death toll was counterbalanced by numerous secondary houses unoccupied at the time of the flood. The method has also been applied to Noirmoutier Island, which was not impacted by Storm Xynthia, but with similar characteristics regarding buildings' locations and configurations. However, the expected deaths there are less important due to a less important mortality fraction. This is due to the fact that, in general, buildings are located farther from the sea, so their vulnerability level is lower.

Despite several limitations of the deterministic approach, specifically the need for more data regarding the circumstances of death, and the fact that it is impossible to assess the exact number of deaths, the estimate provided by the methodology is a first step toward a complementary approach that links with other existing methods for assessing fatalities in relation to floods and which is suited for urbanized coastal flood-prone areas protected from the sea by sandy barriers or walls. Moreover, the present methodology could be used for evaluation with cost-efficiency analysis (CEA) for building mitigation strategies. This may be a very useful tool for decision makers to save lives by identifying the best adaptation strategies to reduce vulnerability to coastal floods, the magnitude and frequency of which are expected to increase in response to rising sea levels [16,66].

6. Patents

- Different fatality assessment methods due to floods exist but do not integrate building characteristics;
- The VIE index framework allows one to assess a building's vulnerability for people;
- Coupled with census data, it allows one to evaluate a specific mortality fraction per municipality;
- This derived fatality assessment method is useful for working on a building's adaptation.

Author Contributions: Conceptualization, A.C., E.B.-A., S.P. and D.M.; data curation, A.C.; investigation, A.C., E.B.-A., S.P. and D.M.; methodology, A.C.; software, A.C.; supervision, A.C., E.B.-A., S.P. and D.M.; validation, A.C., E.B.-A., S.P. and D.M.; writing—original draft, A.C., E.B.-A., S.P. and D.M.; writing—review & editing, A.C., S.P., E.B.-A. and D.M. All authors have read and agreed to the published version of the manuscript.

Funding: This research was supported by the COSELMAR research program, funded by the Regional Council of Pays de la Loire (2012–2017): www.coselmar.fr (accessed on 19 January 2022).

Institutional Review Board Statement: Not applicable.

Informed Consent Statement: Not applicable.

Data Availability Statement: The data presented in this study are available on request from the corresponding author.

Acknowledgments: We are grateful to the following: Freddy Vinet from Université Montpellier III for providing data regarding the circumstances under which fatalities occurred during Storm Xynthia; the Departmental Service of Fire and Rescue of Vendée for their helpful discussions; and all those who helped us to collect data in the field, especially the researchers of LETG UMR 6554 CNRS and Communauté de Communes de l'île de Noirmoutier. This research was supported by the COSELMAR research program, funded by the Regional Council of Pays de la Loire (2012–2017): www.coselmar.fr (accessed on 19 January 2022). We also warmly thank Jan Bloemendal for improving the English.

Conflicts of Interest: The authors declare no conflict of interest.

Appendix A. Estimation of the 1% Value of Deaths in the Case of Storm Xynthia

Jonkman [35] and Jonkman et al. [17] proposed the generic value of 1% of deaths (F_D) among an exposed population (N_{EXP}) in case of flooding, and it was successfully applied in the case of Storm Xynthia by Kolen et al. [6]. Considering the number of deaths ($N = 29$) and an exposed population (N_{EXP}) estimated to be 3000 people, the mortality fraction (F_D) is equal to 1%, as shown below (Equation (A1)):

$$F_D = \frac{N}{N_{EXP}} = \frac{29}{3000} = 0.01 \qquad \text{(A1)}$$

However, the value of N_{EXP} can be criticized. It corresponds approximately to the combined population of La Faute-sur-Mer and L'Aiguillon-sur-Mer, which are neighboring and were both impacted by Storm Xynthia. The total population of the two municipalities was 3291 in 2006 and 3001 in 2011 [51]. However, the deaths were only recorded in La Faute-sur-Mer; its population was 1008 inhabitants in 2006. Thus, Equation (A2) yields

$$F_D = \frac{N}{N_{EXP}} = \frac{29}{1008} = 0.0288 \qquad \text{(A2)}$$

Here, the mortality fraction (F_D) is close to 3%. It could be improved by considering that not every building was flooded, and thus certain municipal areas were not affected. The exposed population was clearly lower, and if so, the mortality fraction was higher. Thus, we consider that the value of 1% was not sufficient in this case.

This contrasts with other values used in the literature. For instance, for the 1953 storm at Canvey Island (UK), the mortality fraction was estimated to be 0.4% [36], and in The Netherlands, it was proposed to have a mortality fraction of 0.3% for coastal floods [27].

Appendix B. Calculation of the Mortality Fraction Depending on Building Vulnerability and the Age of Casualties for La Faute-sur-Mer

To estimate a mortality fraction using the VIE index, which considers red or black houses and the ages of people, we used the data from Storm Xynthia (see Section 2.3.1). The results are given in Table 4. In this appendix, we give details on the calculation of each mortality fraction (F_D) according to data from Storm Xynthia for the La Faute-sur-Mer municipality (29 deaths) as derived from Equation (2). We needed to estimate the number of people impacted (N_{EXP}) to do so.

Census data used for the calculation are given in Appendix C (lines 6–14).

It should be noted that all buildings identified by the VIE index were not flooded during Storm Xynthia (due to the limitations discussed in Section 4). Table A1 shows the number of buildings effectively flooded, which needed to be used for estimating the mortality fractions from Storm Xynthia. Among the flooded buildings, 63% were black houses, and 37% were red houses.

Table A1. Differences between buildings theoretically (according to VIE index results) and effectively flooded during Storm Xynthia.

	Theoritically Flooded (According to VIE Index) for Storm Xynthia	Effectively Flooded for Storm Xynthia	% Effectively or Theoretically Flooded
Red class	885	604	68.2%
Black class	1305	1027	78.7%
Total	2190	1631	74.5%

Appendix B.1. Estimating the Impacted Population (N_{EXP})

Since we know the number of deaths (29), the context of the casualties (Table 4), and the number of effectively flooded buildings (1631) [9], it is possible to apply Equation (4) as follows (Equation (A3)):

$$N_{EXP}^{La\ Faute-sur-Mer-Xynthia} = [(P_{PH} \times T_{PH}) + (P_{SH} \times T_{SH})] \times N^{RB} \times N_{HF} = 1181.7 \quad (A3)$$

where N_{EXP} is the population potentially exposed to a risk of death in the case of a coastal flood, P_{PH} is the proportion of principal houses in the municipality (=13.4%), T_{PH} is the occupation rate of principal houses (69%), P_{SH} is the proportion of secondary houses in the municipality (85.9%), T_{SH} is the occupation rate of secondary houses (31%), N^{RB} is the number of black and red buildings identified by the VIE index and flooded during Storm Xynthia (1631), and N_{HF} is the average household of the municipality (2.02).

The result showed a total number of 1182 persons exposed to death risk during Storm Xynthia.

Appendix B.2. Mortality Fraction for People Aged under 15 Living in Black Houses

Based on the results from Equation (A3) ($N_{EXP}^{La\ Faute-sur-Mer-Xynthia} = 1181.7$), it was possible to estimate the number of those aged under 15 who were exposed to the flood during Storm Xynthia that were living in black houses. This number was driven by the proportion of people under 15 in the municipality population ($P^{-15} = 10.3\%$, line 7 from Appendix C) and the proportion of black houses effectively flooded during the storm ($P^B = 63\%$). This gives Equation (A4):

$$N_{EXP}^{B-15} = \left(N_{EXP}^{La\ Faute-sur-Mer-Xynthia} \times P^{-15} \right) \times P^B = (1181.7 \times 0.103) \times 0.63 = 76.7 \quad (A4)$$

The result gives a total number of 77 people aged under 15 who were exposed to the flood in that were black houses during Storm Xynthia (N_{EXP}^{B-15}). Among them, two deaths were recorded (N^{B-15}). The mortality fraction (F_D^{B-15}) is given by Equation (A5):

$$F_D^{B-15} = \frac{N^{B-15}}{N_{EXP}^{B-15}} = \frac{2}{76.7} = 0.0261 \quad (A5)$$

The mortality fraction for people under 15 in black houses (F_D^{B-15}) was 2.61% according to the data from Storm Xynthia in La Faute-sur-Mer.

Appendix B.3. Mortality Fraction for People Aged 15–60 Living in Black Houses

The people aged 15–60 represented 44.7% of the local population in 2006 (P^{15-60}; line 8 from Appendix C). The total number of the exposed population ($N_{EXP} = 1181.7$) and the proportion of black houses effectively flooded during the storm ($P^B = 63\%$) are given in Equation (A6):

$$N_{EXP}^{B15-60} = \left(N_{EXP}^{La\ Faute-sur-Mer-Xynthia} \times P^{15-60} \right) \times P^B = (1181.7 \times 0.447) \times 0.63 = 332.8 \quad (A6)$$

The result gives a total of 333 people aged 15–60 who were exposed to flooding during Storm Xynthia (N_{EXP}^{B15-60}). Among them, four deaths were recorded (N^{B15-60}). The mortality fraction (F_D^{B15-60}) is given by Equation (A7):

$$F_D^{B15-60} = \frac{N^{B15-60}}{N_{EXP}^{B15-60}} = \frac{4}{332.8} = 0.0120 \tag{A7}$$

The mortality fraction for people aged 15–60 who were in black houses (F_D^{B15-60}) was 1.20% according to the data from Storm Xynthia in La Faute-sur-Mer.

Appendix B.4. Mortality Fraction for People Aged above 60 Living in Black Houses

People aged above 60 represented 45% of the local population in 2006 (P^{+60}; line 9 from Appendix C). Using the total number of people exposed in the population (N_{EXP} = 1181.7) and the proportion of black houses effectively flooded during the storm (P^B = 63%) gives Equation (A8):

$$N_{EXP}^{B+60} = \left(N_{EXP}^{La\ Faute-sur-Mer-Xynthia} \times P^{+60} \right) \times P^B = (1181.7 \times 0.449) \times 0.63 = 334.3 \tag{A8}$$

This indicates that a total of 334 people aged above 60 were exposed to flooding during Storm Xynthia (N_{EXP}^{B+60}). Among them, 18 deaths were recorded (N_{EXP}^{B+60}). The mortality fraction (F_D^{B-15}) is given by Equation (A9):

$$F_D^{B+60} = \frac{N^{B+60}}{N_{EXP}^{B+60}} = \frac{18}{334.3} = 0.0538 \tag{A9}$$

The mortality fraction for people aged above 60 in black houses (F_D^{B-15}) was 5.38% according to data from Storm Xynthia in La Faute-sur-Mer.

Appendix B.5. Mortality Fraction for People Aged under 15 Living in Red Houses

People aged under 15 comprised 10.3% of the local population in 2006 (P^{-15}; line 7 from Appendix C), and those effectively flooded red houses during Storm Xynthia represented 37% (P^R). We could estimate the number of people aged under 15 effectively exposed to flooding (N_{EXP}^{R-15}) using Equation (A10):

$$N_{EXP}^{R-15} = \left(N_{EXP}^{La\ Faute-sur-Mer-Xynthia} \times P^{-15} \right) \times P^R = (1181.7 \times 0.103) \times 0.37 = 45.03 \tag{A10}$$

This gives a total of 45 people aged under 15 who were exposed to flooding in red houses during Storm Xynthia (N_{EXP}^{R-15}). Among them, one death was recorded (N^{R-15}). The mortality fraction (F_D^{R-15}) is given by Equation (A11):

$$F_D^{R-15} = \frac{N^{R-15}}{N_{EXP}^{R-15}} = \frac{1}{45.03} = 0.0222 \tag{A11}$$

The mortality fraction for people aged under 15 in red houses (F_D^{R-15}) was 2.22% according to data from Storm Xynthia in La Faute-sur-Mer.

Appendix B.6. Mortality Fraction for People Aged 15–60 Living in Red Houses

Using the total number of the exposed population (N_{EXP} = 1181.7), the proportion of people aged 15–60 (P^{15-60} = 44.7%; line 8 from Appendix C), and the proportion of red houses effectively flooded during the storm (P^R = 37%), we could estimate the exposed population of people aged 15–59 living in red houses (N_{EXP}^{R15-60}) using Equation (A12):

$$N_{EXP}^{R15-60} = \left(N_{EXP}^{La\ Faute-sur-Mer-Xynthia} \times P^{15-60} \right) \times P^R = (1181.7 \times 0.447) \times 0.37 = 195.5 \tag{A12}$$

This gives a total of 196 people aged 15–60 exposed to flooding in red houses during Storm Xynthia (N_{EXP}^{R15-60}). Among them, no deaths were recorded (N^{R15-60}). The mortality fraction (F_D^{R15-60}) is given by Equation (A13):

$$F_D^{R15-60} = \frac{N^{R15-60}}{N_{EXP}^{R15-60}} = \frac{0}{195.80} = 0.0 \tag{A13}$$

The mortality fraction for people aged 15–60 living in red houses (F_D^{R15-60}) was 0% according to data from Storm Xynthia in La Faute-sur-Mer.

Appendix B.7. Mortality Fraction for People Aged above 60 Living in Black Houses

Using the total number of the exposed population (N_{EXP} = 1181.7), the proportion of people aged above 60 (P^{+60} = 45%; line 9 from Appendix C), and the proportion of red houses effectively flooded during the storm (P^R = 37%), we could estimate the exposed population of people aged above 60 living in red houses (N_{EXP}^{R+60}) using Equation (A14):

$$N_{EXP}^{R+60} = \left(N_{EXP}^{La\ Faute-sur-Mer-Xynthia} \times P^{+60} \right) \times P^R = (1181.7 \times 0.449) \times 0.37 = 196.3 \tag{A14}$$

This gives a total of 197 people aged above 60 exposed to flooding living in red houses during Storm Xynthia (N_{EXP}^{R+60}). Among them, four deaths were recorded (N^{R+60}). The mortality fraction (F_D^{R+60}) is given by Equation (A15):

$$F_D^{R+60} = \frac{N^{R+60}}{N_{EXP}^{R+60}} = \frac{4}{196.3} = 0.0204 \tag{A15}$$

The mortality fraction for people aged above 60 living in red houses (F_D^{R+60}) was 2.04% according to data from Storm Xynthia in La Faute-sur-Mer.

Appendix C

Table A2. Data used for calculations in Section 3 (Results).

# Line		La Faute-sur-Mer		L'Aiguillon-sur-Mer		Charron		Noirmoutier-en-l'Île		L'Epine		La Guérinière		Barbâtre	
		#	%	#	%	#	%	#	%	#	%	#	%	#	%
	(i)	Building's vulnerability (calculated with VIE Index)													
1	Buildings in flood prone area	2406	X	2337	X	491	X	2227	X	1747	X	1783	X	1775	X
2	"Orange" buildings	216	X	253	X	187	X	1007	X	519	X	329	X	561	X
3	"Red" buildings (P^R)	885	40.4%	1262	60.6%	293	96.4%	929	76.1%	1225	99.8%	885	60.9%	1189	97.9%
4	"Black" buildings (P^B)	1305	59.6%	822	39.4%	11	3.6%	291	23.9%	3	0.2%	569	39.1%	25	2.1%
5	Potentially lethal buildings (N^{RB} = #3 + #4)	2190	100%	2084	100%	304	100%	1220	100%	1228	100%	1454	100%	1214	100%
		Census data (source: INSEE [48])													
	Census's year	2006		2006		2006		2011		2011		2011		2011	
	(j)	Population													
6	Total population	1008	100%	2283	100%	2140	100%	4550	100%	1713	100%	1460	100%	1786	100%
7	Under 15 (P^{-15})	104	10.3%	225	9.9%	448	20.9%	614	13.5%	204	11.9%	201	13.8%	267	14.9%
8	15-60 ($P^{15\text{-}60}$)	451	44.7%	893	39.1%	1259	58.8%	2224	48.9%	802	46.8%	704	48.2%	784	43.9%
9	Above 60 (P^{+60})	453	44.9%	1165	51.0%	433	20.2%	1713	37.6%	706	41.2%	554	37.9%	735	41.2%
		Houses													
10	Residential houses total	3737	100%	2334	100%	897	100%	6984	100%	2124	100%	2667	100%	3172	100%
11	Principal houses (P_{PH})	499	13.4%	1140	48.8%	839	93.5%	2219	31.8%	846	39.8%	695	26.1%	856	27.0%
12	Secondary houses (P_{SH})	3210	85.9%	1116	47.8%	30	3.3%	4554	65.2%	1197	56.4%	1963	73.6%	2221	70.0%
13	Unoccupied houses	28	0.7%	78	3.3%	29	3.2%	211	3.0%	81	3.8%	9	0.3%	95	3.0%
		Average people per households													
14	Average people–Principal houses (N_{HF} = #6/#10)	2.02		2.00		2.55		2.05		2.02		2.10		2.09	
		Standard values (see Section 2.3)													
		Mortality fraction values (FDV) per age and building's vulnerability													
15	F_D^{B-15}	0.0261													
16	F_D^{B15-60}	0.0120													
17	F_D^{B+60}	0.0538													
18	F_D^{R-15}	0.0222													
19	F_D^{R15-60}	0.0000													
20	F_D^{R+60}	0.0204													

Table A2. Cont.

# Line		La Faute-sur-Mer		L'Aiguillon-sur-Mer		Charron		Noirmoutier-en-l'Île		L'Epine		La Guérinière		Barbâtre	
		#	%	#	%	#	%	#	%	#	%	#	%	#	%
		Occupation rate (Tx) per type of houses													
21	Tx Principal houses (T_{PH})							69.0%							
22	Tx Secondary houses (T_{SH})							31.0%							
		Mortality fraction (F_D)													
	$F_D = [(F_D{}^{B-15} \times P^{-15}) + (F_D{}^{B15-60} \times P^{15-60}) + (F_D{}^{B+60} \times P^{+60}) \times P^B] + [(F_D{}^{R-15} \times P^{-15}) + (F_D{}^{R15-60} \times P^{15-60}) + (F_D{}^{R+60} \times P^{+60}) \times P^R]$													
23	$F_D{}^{MUNICIPALITY} = [(\#15 \times \#7) + (\#16 \times \#8) + (\#17 \times \#9) \times \#4] + [(\#18 \times \#7)+(\#19 \times \#8) + (\#20 \times \#9) \times \#3]$	2.4%		2.1%		0.9%		1.5%		1.1%		1.8%		1.2%	
		Exposed population (N_{EXP})													
	$N_{EXP} = [P_{PH} \times T_{PH}) + (P_{SH} \times T_{SH})] \times N^{RB} \times N_{HF}$													
24	$N_{EXP}{}^{MUNICIPALITY} = [(\#11 \times \#21) + (\#12 + \#22)] \times \#5 \times \#14$	1587		2025		508		1054		1118		1247		1022	
		Number of potential fatalities ($N = F_D{}^{MUNICIPALITY} \times N_{EXP}{}^{MUNICIPALITY}$)													
25	$N = \#23 \times \#24$	38		43		5		16		12		23		12	

References

1. Jonkman, S.N.; Kelman, I. An analysis of the causes and circumstances of flood disaster deaths. *Disasters* **2005**, *29*, 75–97. [CrossRef]
2. Jonkman, S.N.; Maaskant, B.; Boyd, E.; Levitan, M.L. Loss of Life Caused by the Flooding of New Orleans after Hurricane Katrina: Analysis of the Relationship between Flood Characteristics and Mortality. *Risk Anal. Off. Publ. Soc. Risk Anal.* **2009**, *29*, 676–698. [CrossRef]
3. Vinet, F.; Lumbroso, D.; Defossez, S.; Boissier, L. A Comparative Analysis of the Loss of Life during Two Recent Floods in France: The Sea Surge Caused by the Storm Xynthia and the Flash Flood in Var. *Nat. Hazards* **2012**, *61*, 1179–1201. [CrossRef]
4. Vinet, F.; Boissier, L.; Defossez, S. La Mortalité Comme Expression de La Vulnérabilité Humaine Face Aux Catastrophes Naturelles: Deux Inondations Récentes En France (Xynthia, Var, 2010). *VertigO—Rev. Électron. Sci. Environ.* **2011**, *11*, 28. [CrossRef]
5. Boissier, L. La Mortalité Liée Aux Crues Torentielles Dans le Sud de la France: Une Approche de la Vulnérabilité Humaine Face au Risque D'inondation. Ph.D. Thesis, Université Paul Valéry-Montpellier III, Montpellier, France, 2013.
6. Kolen, B.; Slomp, R.; Jonkman, S.N. The Impacts of Storm Xynthia February 27–28, 2010 in France: Lessons for Flood Risk Management. *J. Flood Risk Manag.* **2013**, *6*, 261–278. [CrossRef]
7. Creach, A.; Pardo, S.; Guillotreau, P.; Mercier, D. The Use of a Micro-Scale Index to Identify Potential Death Risk Areas Due to Coastal Flood Surges: Lessons from Storm Xynthia on the French Atlantic Coast. *Nat. Hazards* **2015**, *77*, 1679–1710. [CrossRef]
8. Cour des Comptes. *Les Enseignements des Inondations de 2010 sur le Littoral Atlantique (Xynthia) et Dans le Var*; Cours des Comptes: Paris, France, 2012; p. 299.
9. Mercier, D.; Chadenas, C. La tempête Xynthia et la cartographie des « zones noires » sur le littoral français: Analyse critique à partir de l'exemple de La Faute-sur-Mer (Vendée). *Norois* **2012**, *222*, 45–60. [CrossRef]
10. Pitié, C.; Bellec, P.; Maillot, H.; Nadeau, J.; Puech, P. *Expertise des Zones de Solidarité Xynthia en Charente-Maritime*; CGEDD/MEEDDM: Paris, France, 2011; p. 186.
11. Pitié, C.; Puech, P. *Expertise Complémentaire des Zones de Solidarité Délimitées en Vendée Suite à la Tempête Xynthia Survenue Dans la Nuit du 27 au 28 Février 2010*; CGEDD/MEEDDM: Paris, France, 2010; p. 80.
12. CETMEF; CETE Méditerranée; CETE Ouest. *Vulnérabilité du Territoire National Aux Risques Littoraux*; CETMEF/DLCE: Plouzané, France, 2009; p. 163.
13. MEDDE. *Mieux Savoir Pour Mieux Agir: Principaux Enseignements de la Première Evaluation des Risques D'inondation sur le Territoire Français-EPRI 2011*; MEDDE: Paris, France, 2012; p. 72.
14. Nicholls, R.J.; Brown, S.; Hanson, S.; Hinkel, J. *Economics of Coastal Zone Adaptation to Climate Change*; The World Bank: Washington, DC, USA, 2010; p. 62.
15. Creach, A. Coastlines with Increased Vulnerability to Sea-level Rise. In *Spatial Impacts of Climate Change*; ISTE Ltd.: London, UK; John Wiley and Sons Inc.: Hoboken, NJ, USA, 2021; pp. 71–92, ISBN 978-1-78945-009-5.
16. IPCC. *IPCC Special Report on the Ocean and Cryosphere in a Changing Climate*; IPCC: Geneva, Switzerland, 2019; p. 1170.
17. Jonkman, S.N.; Bočkarjova, M.; Kok, M.; Bernardini, P. Integrated Hydrodynamic and Economic Modelling of Flood Damage in the Netherlands. *Ecol. Econ.* **2008**, *66*, 77–90. [CrossRef]
18. Vinet, F. *Le Risque Inondation. Diagnostic et Gestion*; Sciences du risque et du danger; Tec & Doc Lavoisier: Paris, France, 2010; ISBN 978-2-7430-1263-2.
19. Yang, Q.; Zheng, X.; Jin, L.; Lei, X.; Shao, B.; Chen, Y. Research Progress of Urban Floods under Climate Change and Urbanization: A Scientometric Analysis. *Buildings* **2021**, *11*, 628. [CrossRef]
20. Brazdova, M.; Riha, J. A Simple Model for the Estimation of the Number of Fatalities Due to Floods in Central Europe. *Nat. Hazards Earth Syst. Sci.* **2014**, *14*, 1663–1676. [CrossRef]
21. Di Mauro, M.; Bruijn, K.M.D.; Meloni, M. Quantitative Methods for Estimating Flood Fatalities: Towards the Introduction of Loss-of-Life Estimation in the Assessment of Flood Risk. *Nat. Hazards* **2012**, *63*, 1083–1113. [CrossRef]
22. Koshimura, S.; Oie, T.; Yanagisawa, H.; Imamura, F. Developing Fragility Functions for Tsunami Damage Estimation Using Numerical Model and Post-Tsunami Data from Banda Aceh, Indonesia. *Coast. Eng. J.* **2009**, *51*, 243–273. [CrossRef]
23. Marchand, M.; Buurman, J.; Pribadi, A.; Kurniawan, A. Damage and Casualties Modelling as Part of a Vulnerability Assessment for Tsunami Hazards: A Case Study from Aceh, Indonesia. *J. Flood Risk Manag.* **2009**, *2*, 120–131. [CrossRef]
24. Zhai, G.; Fukuzono, T.; Ikeda, S. An Empirical Model of Fatalities and Injuries Due to Floods in Japan. *JAWRA J. Am. Water Resour. Assoc.* **2006**, *42*, 863–875. [CrossRef]
25. Jonkman, S.N.; van Gelder, P.H.A.J.M.; Vrijling, J.K. An Overview of Quantitative Risk Measures for Loss of Life and Economic Damage. *J. Hazard. Mater.* **2003**, *99*, 1–30. [CrossRef]
26. Di Mauro, M.; Lumbroso, D.M. *Hydrodynamic and Loss of Life Modelling for the 1953 Canvey Island Flood*; Keble College: Oxford, UK, 2008; pp. 1117–11126.
27. Klijn, F.; Baan, P.; De Bruijn, K.M.; Kwadijk, J. *Overstromingsrisico's in Nederland in een Veranderend Klimaat. Verwachtingen, Schattingen en Berekeningen voor het Project Nederland Later*; Delft Hydraulics: Delft, The Netherlands, 2007; p. 165.
28. Johnstone, W.M.; Sakamoto, D.; Assaf, H.; Bourban, S. Architecture, Modelling Framework and Validation of BC Hydro's Virtual Reality Life Safety Model. In Proceedings of the ISSH 2005, Nijmegen, The Netherlands, 23–24 May 2005.
29. Lumbroso, D.; Sakamoto, D.; Johnstone, W.; Tagg, A.; Lence, B. The Development of a Life Safety Model to Estimate the Risk Posed to People by Dam Failures and Floods. *Dams Reserv.* **2011**, *21*, 21. [CrossRef]

30. HR Wallingford; Middlesex University; Risk & Policy Analysts Ltd. *Flood Risks to People—Phase 2: The Flood Risks to People Methodology*; Defra/Environment Agency: London, UK, 2006; p. 103.

31. HR Wallingford; Middlesex University; Risk & Policy Analysts Ltd. *Flood Risks to People—Phase 2: Guidance Document*; Defra/Environment Agency: London, UK, 2006; p. 91.

32. HR Wallingford; Risk & Policy Analysts Ltd.; Middlesex University. *Flood Risks to People—Phase 1*; Defra/Environment Agency: London, UK, 2003; p. 123.

33. Crone, J. *Comparing Census Data with Other Geographical Data. Support Census Geography Course*; University of Edinburgh: Scoteland, UK, 2013.

34. Priest, S.; Wilson, T.; Tapsell, S.; Penning-Rowsell, E.; Viavattene, C.; Fernandez-Bilbao, A. *Building a Model to Estimate Risk to Life for European Flood Events*; Middlesex University, HR Wallingford: London, UK, 2007; p. 185.

35. Jonkman, S.N. *Loss of Life Estimation in Flood Risk Assessment*; Theory and Applications; Delft University of Technology: Delft, The Netherlands, 2007.

36. Di Mauro, M.; de Bruijn, K.M. Application and Validation of Mortality Functions to Assess the Consequences of Flooding to People. *J. Flood Risk Manag.* **2012**, *5*, 92–110. [CrossRef]

37. Waarts, P.H. *Method for Determining Loss of Life Caused by Inundation*; TU Delft Faculty of Industrial Design Engineering: Delft, The Netherlands, 1992.

38. Vrouwenvelder, A.C.W.M.; Steenhuis, C.M. *Secondary Flood Defences in the HoekscheWaard, Calculation of the Number of Fatalities for Various Flood Scenarios*; TU Delft Faculty of Industrial Design Engineering: Delft, The Netherlands, 1997.

39. Jonkman, S.N.; Vrijling, J.K.; Vrouwenvelder, A.C.W.M. Methods for the Estimation of Loss of Life Due to Floods: A Literature Review and a Proposal for a New Method. *Nat. Hazards* **2008**, *46*, 353–389. [CrossRef]

40. Nofal, O.M.; van de Lindt, J.W.; Cutler, H.; Shields, M.; Crofton, K. Modeling the Impact of Building-Level Flood Mitigation Measures Made Possible by Early Flood Warnings on Community-Level Flood Loss Reduction. *Buildings* **2021**, *11*, 475. [CrossRef]

41. Magnan, A.; Duvat, V. *Des Catastrophes... "Naturelles"?* Essais et Documents; Editions Le Pommier: Paris, France, 2014; ISBN 978-2-7465-0679-4.

42. Dall'Osso, F.; Gonella, M.; Gabbianelli, G.; Withycombe, G.; Dominey-Howes, D. Assessing the Vulnerability of Buildings to Tsunami in Sydney. *Nat. Hazards Earth Syst. Sci.* **2009**, *9*, 2015–2026. [CrossRef]

43. Dall'Osso, F.; Dominey-Howes, D. *A Method for Assessing the Vulnerability of Buildings to Catastrophic (Tsunami) Marine Flooding*; Sydney Coastal Councils Group Inc.: Sydney, Australia, 2009; p. 138.

44. Gauraz, A.L.; Valencia, N.; Koscielny, M.; Guillande, R.; Gardi, A.; Leone, F.; Salaun, T. Tsunami Damages Assessment: Vulnerability Functions on Buildings Based on Field and Earth Observation Survey. In Proceedings of the EGU General Assembly Conference Abstracts, Vienna, Austria, 19–24 April 2009; Volume 11, p. 5785.

45. Gonzalez-Riancho, P.; Aguirre-Ayerbe, I.; Garcia-Aguilar, O.; Medina, R.; Gonzalez, M.; Aniel-Quiroga, I.; Gutierrez, O.Q.; Alvarez-Gomez, J.A.; Larreynaga, J.; Gavidia, F. Integrated Tsunami Vulnerability and Risk Assessment: Application to the Coastal Area of El Salvador. *Nat. Hazards Earth Syst. Sci. NHESS* **2014**, *14*, 1223–1244. [CrossRef]

46. Leone, F.; Lavigne, F.; Paris, R.; Denain, J.-C.; Vinet, F. A Spatial Analysis of the December 26th, 2004 Tsunami-Induced Damages: Lessons Learned for a Better Risk Assessment Integrating Buildings Vulnerability. *Appl. Geogr.* **2011**, *31*, 363–375. [CrossRef]

47. Mück, M.; Taubenböck, H.; Post, J.; Wegscheider, S.; Strunz, G.; Sumaryono, S.; Ismail, F.A. Assessing Building Vulnerability to Earthquake and Tsunami Hazard Using Remotely Sensed Data. *Nat. Hazards* **2013**, *68*, 97–114. [CrossRef]

48. Reese, S.; Cousins, W.J.; Power, W.L.; Palmer, N.G.; Tejakusuma, I.G.; Nugrahadi, S. Tsunami Vulnerability of Buildings and People in South Java—Field Observations after the July 2006 Java Tsunami. *Nat. Hazards Earth Syst. Sci.* **2007**, *7*, 573–589. [CrossRef]

49. Kelman, I.; Spence, R. An overview of flood actions on buildings. *Eng. Geol.* **2004**, *73*, 297–309. [CrossRef]

50. Devaux, E.; Désiré, G.; Boura, C.; Lowenbruck, J.; Bérenger, N.; Rouxel, N.; Romain, N. *La Tempête Xynthia du 28 Février 2010—Retour D'expérience en Loire-Atlantique et Vendée-Volet Hydaulique et Ouvrages de Protection*; CETE-Ouest/DREAL Pays-de-la-Loire/DDTM Loire-Atlantique/DDTM Vendée: Nantes, France, 2012; p. 78.

51. INSEE. *Recensement Général de la Population 2011 (Publié le 26 Juin 2014)*; INSEE: Paris, France, 2014.

52. Creach, A.; Pardo, S.; Mercier, D. Diagnostic préventif de la vulnérabilité des constructions résidentielles pour leurs occupants face au risque de submersion marine appliqué à l'île de Noirmoutier (Vendée, France). *VertigO—Rev. Électron. Sci. Environ.* **2017**, *17*, 30. [CrossRef]

53. Creach, A.; Chevillot-Miot, E.; Mercier, D.; Pourinet, L. Vulnerability to Coastal Flood Hazard of Residential Buildings on Noirmoutier Island (France). *J. Maps* **2015**, *12*, 371–381. [CrossRef]

54. Eugène, J.-L. *Principales Catastrophes et Cataclysmes Dans l'Île de Noirmoutier Depuis le IIIème siècle: Répertoire–Causes–Remèdes-Vers L'établissement d'un Plan de Prévention des Risques Naturels*; Vivre l'île 12 sur 12: Noirmoutier, France, 2016; p. 23.

55. Garnier, E.; Henry, N.; Desarthe, J. Visions croisées de l'historien et du courtier en réassurance sur les submersions: Recrudescence de l'aléa ou vulnérabilisation croissante? In *Gestion des Risques Naturels—Leçons de la Tempête Xynthia*; Matière à débattre et décider; Quae: Versailles, France, 2012; p. 23, ISBN 978-2-7592-1820-2.

56. Fattal, P.; Robin, M.; Paillart, M.; Maanan, M.; Mercier, D.; Lamberts, C.; Costa, S. Effets des tempêtes sur une plage aménagée et à forte protection côtière: La plage des Éloux (côte de Noirmoutier, Vendée, France). *Norois* **2010**, *215*, 101–114. [CrossRef]

57. Creach, A.; Bastidas-Arteaga, E.; Pardo, S.; Mercier, D. Vulnerability and costs of adaptation strategies for housing subjected to flood risks: Application to La Guérinière France. *Mar. Policy* **2020**, *117*, 103438. [CrossRef]
58. Chauvet, A.; Renard, J. *La Vendée. Le Pays. Les Hommes*; Editions du Cercle d'or: Les Sables-d'Olonne, France, 1978.
59. Garnier, E.; Surville, F. *La Tempête Xynthia Face à L'histoire; Submersions et Tsunamis sur les Littoraux Français du Moyen Âge à nos Jours*; Le Croît Vif: Saintes, France, 2010; ISBN 978-2-36199-009-1.
60. Chadenas, C.; Creach, A.; Mercier, D. The Impact of Storm Xynthia in 2010 on Coastal Flood Prevention Policy in France. *J. Coast. Conserv.* **2013**, *18*, 529–538. [CrossRef]
61. Creach, A. Cartographie et Analyse Economique de la Vulnérabilité du Littoral Atlantique Français Face au Risque de Submersion Marine (2 Volumes). Ph.D. Thesis, Université de Nantes, Nantes, France, 2015.
62. MEDDTL. Circulaire du 27 Juillet 2011 Relative à la Prise en Compte du Risque de Submersion Marine Dans Les Plans de Prévention des Risques Naturels Littoraux. Available online: https://www.bulletin-officiel.developpement-durable.gouv.fr/notice?id=Bulletinofficiel-0025182&reqId=9497407c-3c16-4990-b1e9-687cb5e0ba7a&pos=8 (accessed on 19 January 2022).
63. Bontet, C.; Blondy, C.; Donnat, S.; Plumejeaud, C.; Riollet, J.-P.; Vacher, L.; Vermandé, M.; Vye, D. *Propriétaires et Usages des Résidences Secondaires en Charente-Maritime*; UMR LIENSs CNRS-Université de La Rochelle, Charente-Maritime Tourisme, CCI La Rochelle & CCI Rochefort et Saintonge: La Rochelle, France, 2016; p. 65.
64. Bontet, C.; Plumejeaud-Perreau, C.; Blondy, C.; Vacher, L.; Vye, D. Does Inhabitants Having a Second Home Play Only a Second Role in Coastal Territories? A Case-Study in Charente-Maritime (France). In Proceedings of the New Perspectives on Second Homes, Tourism, Leisure and Global Change Commission, Stockholm, Sweden, 9–11 June 2014.
65. Gerritsen, H. What Happened in 1953? The Big Flood in the Netherlands in Retrospect. *Philos. Transact. A Math. Phys. Eng. Sci.* **2005**, *363*, 1271–1291. [CrossRef]
66. Nicholls, R. Planning for the Impacts of Sea Level Rise. *Oceanography* **2011**, *24*, 144–157. [CrossRef]

MDPI

Review

Causal Effects between Criteria That Establish the End of Service Life of Buildings and Components

Ana Silva [1,*], Jorge de Brito [2], André Thomsen [3], Ad Straub [3], Andrés J. Prieto [4] and Michael A. Lacasse [5]

[1] CERIS, Department of Civil Engineering, Architecture and Georesources, IST-University of Lisbon, Av. Rovisco Pais, 1049-001 Lisbon, Portugal

[2] Department of Civil Engineering, Architecture and Georesources, IST-University of Lisbon, Av. Rovisco Pais, 1049-001 Lisbon, Portugal; jb@civil.ist.utl.pt

[3] Department Management in the Built Environment (MBE), Faculty of Architecture and the Built Environment, Delft University of Technology, Julianalaan 34, 2628 BL Delft, The Netherlands; A.F.Thomsen@tudelft.nl (A.T.); A.Straub@tudelft.nl (A.S.)

[4] School of Engineering, Pontificia Universidad Católica de Chile, Avenida Vicuña Mackenna, Santiago 4860, Chile; andres.prieto@ing.puc.cl

[5] National Research Council Canada, Construction Research Centre, Ottawa, ON K1A 0R6, Canada; michael.lacasse@nrc-cnrc.gc.ca

* Correspondence: ana.ferreira.silva@tecnico.ulisboa.pt

Abstract: In the last decades, considerable work has been done regarding service life prediction of buildings and building components. Academics and members of the CIB W080 commission, as well as of ISO TC 59/SC14, have made several efforts in this area and created a general terminology for the concept of service life, which is extremely relevant for property management, life cycle assessment (LCA) and life cycle costs (LCC) analyses. Various definitions can be found in the literature that share common ideas. In fact, there are different criteria that trigger the end of a building's service life, but the trap that building practitioners too often fall into and that should be avoided is dividing a problem into separate boxes, labels, and specializations without the mutual cohesion and interaction, and ignoring human behavior. Some definitions of service life are discussed in this review paper, in which the cause-effect processes underlying aging and decay are described. These descriptions highlight the continuous interrelation between different criteria for the end of a building's service life, considering too often neglected and misunderstood causes of the end of life.

Keywords: service life; buildings' components; property management

Citation: Silva, A.; de Brito, J.; Thomsen, A.; Straub, A.; Prieto, A.J.; Lacasse, M.A. Causal Effects between Criteria That Establish the End of Service Life of Buildings and Components. *Buildings* **2022**, *12*, 88. https://doi.org/10.3390/buildings12020088

Academic Editor: David Arditi

Received: 20 December 2021
Accepted: 14 January 2022
Published: 18 January 2022

Publisher's Note: MDPI stays neutral with regard to jurisdictional claims in published maps and institutional affiliations.

1. Introduction

Like humans, buildings and their components also are "born", "get older" and "die" [1]. The "birth" of a building is easy to identify, i.e., the beginning of its service life starts as soon as the building is put into use [2]. However, the end of a building's service life is difficult to predict, as this depends on both objective and subjective demands. The "end of service life" can be seen as a normative concept, since without human intervention (e.g., maintenance actions) the building may physically endure until collapse.

How then can the concept of service life be defined? Perhaps the first definition of service life can be found in the Hammurabi Code (c.1950 to 1910 BC) as "a house should not collapse and kill anybody" [3]. In the last decades, several studies and methodologies have been developed regarding service life prediction of buildings and components. These studies have arisen from the need for reliable knowledge about the durability of building elements so that more sustainable strategies can be adopted regarding maintenance activities and the management of a building's life cycle. Service life data have mostly been used in research fields that include property management [4], life cycle analysis (LCA), and life cycle cost analysis (LCC). LCA and LCC results are strongly dependent on assumptions regarding the service life of buildings and their components. These assumptions are often limited to standardized values of 30 to 50 years but, in reality, the buildings and their components' lifetime vary considerably, first because each building (as a whole) is a unique prototype and then because some components present shorter service lives. Therefore, the adoption of standard assumptions tends to lead to incorrect results [5]. A critical literature review performed by Silva and de Brito [6] revealed that guidelines and standards propose a standardized average value for all the cladding solutions, usually adopting a too optimistic value in less durable claddings and a too pessimistic in more durable claddings. This study [6] reveals that the estimations provided by sampling seem to be more reliable and adjusted to the actual durability of building components.

The concepts and terminology in service life prediction present some variations, with different meanings and connotations, depending on the scope of the different studies carried out. In this sense, in the literature, different terms can be used with the same meaning, and the same terminology can be used to describe different concepts.

Different standardization documents [7–10] have been developed, which were intended to harmonize the concept of service life. Awano [11] separates the life of a building into two categories: "real life", where the life of a building is conditioned by its physical condition; and "service life", which is conditioned by the capability of a building to fulfill its function and other performance criteria. Brand [12] mentioned that buildings lose their capability to meet minimum performance requirements at two different rates: a slower rate related to the natural aging of materials, referring to the physical deterioration that occurs over time; and a more variable but faster rate, related to changes dictated by aesthetic (e.g., new materials or construction trends) or legal motives (e.g., fire safety or the presence of materials that are not allowed anymore, as asbestos), or due to variations in the social context (e.g., changes in users' demands). Moreover, Brand [12] divided the building into durability layers, i.e., construction subsystems whose degradation occurs at different paces, among which are the "structure", the "skin", the "systems" and the "interior lay-out and finishes" [1]. Therefore, instead of an end of service life, one can speak of different ends of service life of the various parts of the building, according to the performance criteria considered relevant. The average service life results from the average of the values of the remaining service times of each element considering the relative importance of each one. For example, some elements can easily be replaced, e.g., ceramic wall cladding tile, while the degradation of structural elements can jeopardize the building's use, leading to its rehabilitation or demolition.

There are several definitions in the literature that describe the loss of a building's ability to meet minimum performance requirements. Service life estimates should consider the various reasons for establishing the end of service life of a building and its components but there are various interrelating parameters. In this study, an exhaustive literature review is provided, based on the collection of information over several years, and based on the authors' experience on service life prediction, who are active members of the CIB W080 and ISO TC 59/SC14 commission. This review presents different perspectives regarding the concept of service life and obsolescence and the criteria that trigger the need for maintenance, repair, or replacement of a building and its components. Section 2 describes the concepts of service life and obsolescence; Section 3 confronts the concepts of serviceability and functionality; Section 4 considers the cause-effect dimensions of obsolescence; Section 5 discusses the causal effects between criteria to establish the end of service life of a building and its components.

2. Service Life and Obsolescence

In the ISO 15686-1 [13] standard, service life is defined as the period after installation during which a building or its components meets or exceeds the performance requirements. Another relevant concept is the "design service life" [9,13], also referred to as "planned service life" [7] or "design life" [8] or "design working life" [10] and defined as the *service life that the designer intends an item (product, component, assembly or construction) to achieve when subjected to the expected service conditions and maintained according to a prescribed maintenance management plan*. Haapio and Viitaniemi [14] refer to service life as the period during which a building or component is able to fulfill the users' objective and subjective needs, without unacceptable maintenance costs or losses to third parties.

A building's safety and functionality are usually seen as objective needs [15–19]. Aesthetic and social reasons, as well as the users' comfort and well-being, are usually seen as subjective demands, varying from user to user, and changing over time [20,21]. In some studies [12,22], it is revealed that the building components are often replaced before the end of their service life, and this is mainly due to subjective reasons, leading to "obsolescence-based" maintenance actions.

The term obsolescence is defined, according to ISO 15686-1 [13], as the loss of the ability of a building element to perform satisfactorily due to changes in performance requirements. In dictionaries, obsolescence is described as the process of becoming old-fashioned or out-of-date [23]. Various authors [24–29] relate to a building's obsolescence to their loss of performance or utility, which can be caused by their physical deterioration or due to economic and social motivations, technological or political changes, or even fluctuations in users' needs. In this sense, an obsolete building element may not be damaged or dysfunctional, but instead, it can no longer fulfill the users' needs considering more recent and up-to-date standards [30].

Therefore, the service life and the obsolescence of building elements can be described as the incapability of fulfilling performance requirements and the user's needs, but these requirements are constantly changing over time. In this sense, the meaning of "utility" is a key concept in this definition [31]. Nevertheless, it is not possible to measure objectively the concept of "utility", which makes it extremely difficult to measure the limit that establishes the end of service life or the instant after which the constructive element becomes obsolete. Usually, a conventional limit is adopted to establish the end of service life of a building element, considering various acceptance criteria.

3. Serviceability and Functionality

The first contribution in the scope of the analysis of a building's functionality appeared at the beginning of the 70s; Markus et al. [32] highlighted the relevance of understanding the humans' role in the definition of the criteria to evaluate a building's performance. In the 80s, CIB [33] introduced the adoption of test methods to quantify physical properties (e.g., thermal conditions, through the assessment of the heating/cooling demands) to ascertain a building's performance. Rush [34] put forward the idea of analyzing a building's performance as a whole, considering building physics, social and psychological aspects, as well as economic factors. The concept of *serviceability* in construction was initially introduced by Master and Brandt [35], referring to *serviceability* as the capability of a building or component to perform the function for which it has been specifically designed and used. Andersen and Brandt [36] related the concept of service life with those of serviceability and functionality, referring that a building's service life can be described as the period of time during which at least the basic performance properties are maintained at an acceptable level. Davis and Szigeti [37] described *serviceability* as the ability of the building to support the activities performed by the users and owners, being able to perform the functions for which they were designed.

The users' demands are considered to be increasingly dynamic [38]. In this sense, buildings and components must be able to constantly adapt in order to fulfill the users' demands and to comply with the performance-based building legislation, which itself is constantly changing [39,40].

In recent decades, various methodologies have been put forward to appraise buildings' functionality. Lützkendorf et al. [41] evaluated a building's functional performance, considering a building's use, its accessibility, and capability to adapt to the users' changing needs. Preiser and Vischer [42] indicated that building performance in a post-occupancy stage can be defined as the boundary between criteria and design [43,44]. The study of Blok and Teuffel [45] also relates the term functionality with flexibility, describing it as the capability of the building to adapt to new needs. This study highlights that the "functional service life" of a building is more a question of a stakeholder's decision rather than a physical or technical service life. Various authors [46–49] have created a new digital tool to estimate the functional performance and service life of heritage buildings, taking into account external risks (e.g., static-structural, atmospheric, and anthropic conditions) and variables related to building vulnerability (e.g., geological location, constructive system, roof design, conservation state, amongst other variables). Augenbroe [50] proposed an innovative virtual experiment to measure how well a technical solution fulfills the users' requirements, focusing on the impact of building performance simulation in a building's design. More recently, climate change has emerged as a new challenge to attaining the expected life of a building. Hence, the impact of climate change on the functionality of a building and its elements has been discussed [51,52], where the need to change a building's components, elements, or systems must be evaluated so that a building can accommodate new requirements to adapt to the changing environmental conditions.

The two terms are interrelated, but they refer to different concepts, as discussed in more detail in Section 5.2.

4. Cause-Effect Dimensions of Obsolescence

The concept of obsolescence is difficult to define. Buildings become obsolete due to a variety of reasons, which are typically interdependent and correlated with each other [53]. Common causes can easily be found between different types of obsolescence.

Once considered 'obsolete', a building is typically subjected to maintenance, rehabilitation, or replacement/demolition. The reasons why the building is subjected to maintenance are commonly categorized as being either predictable or unpredictable. Marteinsson [54] has denoted that, whilst the physical degradation is a measurable parameter, social and legal reasons are more challenging or even impossible to predict.

Iselin and Lemer [55] suggested that building obsolescence tends to occur due to endogenous (e.g., inadequate use of materials) and exogenous causes (e.g., climate degradation agents). In another approach, Nutt et al. [56] suggested that the obsolescence of buildings can occur due to either physical deterioration or behavioral factors (e.g., users' needs).

Various authors [57,58] differentiate "physical deterioration" from "physical obsolescence", suggesting that whereas "physical deterioration" occurs due to expected phenomena (e.g., wear and tear), "physical obsolescence" occurs due to sudden events, linked to the users' behavior. Other authors [55,59,60] applied the two terms arbitrarily, considering that the physical deterioration of a building is an underlying form of physical obsolescence [26].

In a more recent approach, Thomsen and Van der Flier [28,29] suggested that the conceptualization of obsolescence could be described by the quadrant-matrix presented in Figure 1. This correlation matrix considers four parameters, thereby establishing a conceptual model to describe a building's obsolescence and considers the correlation between them. Quadrant A represents the "building obsolescence" as initially described by Baum [61]. This type of depreciation occurs due to internal or endogenous characteristics of buildings, namely inadequate design, lack of maintenance, and the natural wear and aging process that leads to a building's physical deterioration. Quadrant B represents the "physical location obsolescence", in which a building's obsolescence occurs due to exogenous factors related to changes in local conditions, e.g., changes in the environment by adjacent buildings or by traffic, changes in local laws, emerging standards, and rising functional demands. Quadrant C represents behavioral conditions related to the proprietor's or users' attitudes, as misuse or changes in the required functions. Quadrant D is related to behavioral effects due to changes in local conditions, e.g., increase of the criminality and social depreciation of the neighborhood, loss of market status and value, availability of better options.

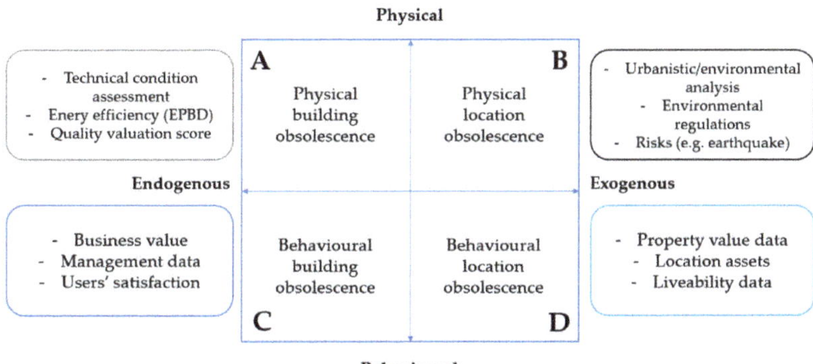

Figure 1. Extended diagram of the two main cause-effect dimensions of obsolescence (adapted from Thomsen and Van der Flier [28,29]). Copyright 2022 Copyright Informa UK Limited; Copyright 2011 Copyright Thomsen, A. and Van der Flier, K.

The analytical model proposed in Figure 1 is based on the premise that obsolescence occurs due to a sequence of complex, frequent, and interconnected cause-effect processes at different scales within and between the four quadrants of the model, leading to the depreciation of the building.

These seem to be the principal cause-effect dimensions of obsolescence since even if the physical service life can be prolonged indefinitely, the end of service life of a given element will occur anyway, since it is rooted in the users' behavior and can be caused by aesthetic, legal, technological or functional, economic, or social reasons [61–63].

5. Discussion of the Criteria That Trigger the End of a Building's Service Life

Analyzing the criteria that trigger the end of a building's service life in separate layers seems an impossible task, since the end of a building's service life occurs, most of the time, due to a set of factors that are interrelated to each other. Notwithstanding the evident challenges, various authors, and as is found in different standards documents [1,2,7,8,13,64–67], in a simplified way, have split the concept of service life into different types of service life. In the following sections, the criteria that trigger the end of a building's service life are discussed, in accordance with various types of service life or obsolescence as referenced in the literature.

5.1. End of Service Life Triggered by Physical Degradation

Contrarily to the building owner's common beliefs, the physical deterioration is not the dominant factor for the end of a building's service life [68,69]. In real world situations, the end of service life due to physical deterioration can have distinct connotations, according to the type of building or component analyzed [12,70].

For structural elements, due to safety requirements, the end of service life is usually conditioned by economic or functional criteria, rarely reaching the end of their physical service life [71].

Concerning the building envelope, the end of service life depends on the type of use of the building. According to Itard and Meijer [72], fundamental differences can be observed in commercial compared to residential, rented compared to owned, and single compared to joint ownership buildings. The performance of maintenance activities is also highly dependent on the tenure profile. Usually, a planned preventive maintenance strategy tends to be adopted in the rented sector (which has revenue intents). On the other hand, in the residential owner-occupied sector, maintenance activities are essentially reactive, in response to existing anomalies due to physical deterioration. Therefore, the end of a building's physical service life is usually attained or exceeded [28,29]. On the other hand, a regular maintenance plan is usually adopted for buildings with revenue intents; thus, the façades are subjected to major interventions even before they have reached the end of their physical service life (considering a conventional limit state according to the owner's requirements).

The end of service life of infrastructure networks is rarely triggered by their physical deterioration since installations are regularly subjected to regular maintenance actions [55]. Therefore, the replacement of equipment and installations is usually conditioned by technical or legal reasons. In this case, "obsolescence" cannot be recognized as the inability of infrastructural networks to meet performance requirements, but rather, as the incompatibility between the performance of the infrastructural network, and the increasing requirements of the users over time [73].

For interior finishes, the end of service life is usually conditioned by fashion or aesthetic criteria, rooted in changes in users' focus on and interest in aesthetics.

Various codes and standards provide an average standardized value for the design life of a building or component (which can vary between 50 to 120 years) [10,12,13]. However, it has been shown through various studies that, when considered as a whole, buildings seldom reach their maximum value, conventionally assumed for the end of a building's service life. According to Thomsen and van der Flier [74], defining the instant after which a building fails to comply with the essential performance requirements is a very challenging task. In practice, the end of a building's physical service life can be limitless, when regular and adequate maintenance policies are in place, and a building can, in contrast, be demolished long before reaching its physical or functional limit due to social or legal motives [75].

In real world scenarios, the relevance of physical deterioration to trigger the end of a building's service life is linked to the type of use, construction methods, ownership (and tenure), and culture. Iizuka [76] shows that, in 74% of the circumstances, physical deterioration is responsible for the demolition of bridges, whereas this reason only motivates 26% of the refurbishment of the current building stock [22]. Horst et al. [77] suggested that, in the USA, the service life of housing ranges from 11 to 32 years, whereas schools only undergo general renovations after 42 years, with a tendency to be discarded after 60 years, mainly due to functional motivations. In this study, it was revealed that the end of service life of tertiary office buildings in Japan tended to be limited by economic reasons, with expected service lives ranging between 23 and 41 years. Wuyts et al. [78] refer that several Japanese houses are used for less than a generation prior to being demolished or abandoned and that the average service life of residential buildings is of only 25 years. Pinder and Wilkinson [23] and Gann and Barlow [79] suggested that the end of service life of an office building is usually triggered by the mutable and growing number of requirements over time. For example, in the United Kingdom, an office building's service life decreased to 40–50 years in the 1950s and 1960s, and to 20–25 years in the 1990s.

5.2. End of Service Life Triggered by Functional and Technological Demands

The concept of functional service life [13] is related to three principal parameters that are interrelated (Figure 2); these include:

(i) Functionality, which is usually related to the users' demands, considering the relation between the users' needs and the ability of the building to fulfill these needs or to perform to a given level of demand;
(ii) Serviceability, which is related to the extent to which the building is appropriate or valuable in addressing users' requirements;
(iii) Suitability, which is defined as a building's capability to support the functions or activities required by users [43].

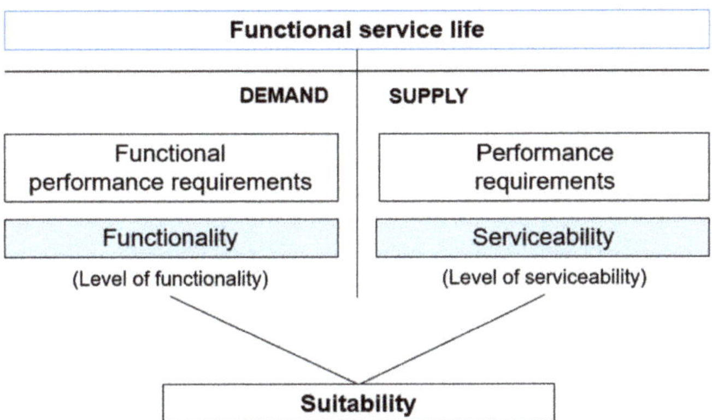

Figure 2. Connection between functionality, serviceability, and suitability.

For example, the same place or room can have different functions (e.g., office work, dining room, or a medical office). When a given place is not suitable for a given use, there is a disparity between the level of functionality required and the level of serviceability offered for that use by the place [13].

Various authors [80,81] have indicated that the end of service life is commonly related to technical criteria, whether conveyed as safety, functional or performance demands. The "functional service life" is described as the period during which the building fulfills adequate functional, or operational performance levels [82]. Under specific conditions, a building's functional performance can be linked to its physical properties [83] (e.g., when the building's physical deterioration compromises its normal use). According to Landman [84], a building's functional service life is intrinsically related to its flexibility and adaptability.

Blok and Teuffel [45] proposed a framework to estimate the end of service life, suggesting values ranging between 30 and 80 years. This huge variation is due to the variability of demands of new stakeholders and advances in the building technology, which tend to trigger the end of a building's functional service life [50,85–87]. For example, for health care facilities, technological criteria seem to be the crucial reason to establish the end of a building's service life, related to upgrading towards modern medical and health care technology; this can lead to a reduction of about 50% of a building's design service life [88,89]. In reality, the functional service life is based on a demand and supply chain, which is shared between functionality and serviceability, i.e., between the users' needs and a building's functions.

5.3. End of Service Life Triggered by Legal Requirements

A building's service life is often conditioned by legal, political, regulatory, or statutory requirements [90–93]. Buildings are designed to comply with current standards; thus, the end of their legal service life is usually triggered as new building legislation is proposed [53,94]. Some authors [56,95] refer to legal obsolescence as control obsolescence, since the end of service life or the need for rehabilitation is triggered by modifications to the legal instruments that control the design and maintenance of buildings.

Wilkinson et al. [96] suggested another term to designate the legal obsolescence, which is political obsolescence, this occurring when trigged by alterations in public or community benefits. Pourebrahimi et al. [26] highlight that the concept of political obsolescence overlaps that of social obsolescence, but the first is ruled by measurable regulatory principles, whereas the second is rooted in social behavior and subjective expectations. According to Williams [97], legal obsolescence can be defined as statutory obsolescence, but the meaning is similar, considering that the building reaches the end of its statutory or legal service life when it is unable to satisfy statutory obligations.

The performance requirements referred to in the definitions provided in ISO 15,686 [13] are usually related to national building codes or stakeholders' expectations. These legal regulations are frequently related to safety concerns, and economic and cultural circumstances [98]. Once again, it appears that the end of a building's service life is established by criteria that are not mutually exclusive, i.e., there are several motivations that are interconnected, and the end of service life is difficult to split into distinct or isolated concepts.

New policies and guidelines, as well as new urban development plans, can trigger the end of a building's service life (e.g., new thermal comfort codes, fire safety codes, environmental hazards, among others) [99–101].

Various reasons can promote the extension of a building's service life, requiring substantial maintenance costs [102]. In some instances, the service life is prolonged due to political or compulsory regulatory alterations, e.g., when a heritage classification is awarded to a building [103]. In other circumstances, demolishing and reconstructing might impose new and stricter regulations, thereby causing the stakeholders to maintain the building in unsuitable conditions for prolonged periods of time [100,101].

5.4. End of Service Life Triggered by Social or Community Motivations

Various authors [104,105] have defined social obsolescence as the decline in suitability or serviceability of a building due to social changes centered on individual or collective perceptions. A building can be in perfect condition but still be abandoned, often long before it reaches undesirable levels of degradation, simply because more advanced, and apparently superior solutions are available [23,106].

A building faces a declining capacity to fulfill the shifting and rising expectations of users over time, as social and political alterations occur [107,108]. These rising expectations are rooted in behavioral aspects, as the users' personal experiences and desires.

Social obsolescence is linked to the social variability of a building's cultural framework, varying at global, regional, and local scales. In this sense, social obsolescence is referred to, in some instances, as cultural obsolescence [109], since this depends on local customs, cultural principles, the users' standard of living and working circumstances. In particular, the acceptance criteria or the tolerance for the deterioration of a building varies by country, within the same country, and even within the same neighborhood. For example, Power [110] has suggested that large-scale social housing commonly represents reduced levels of design and construction quality, minimized construction costs, thereby jeopardizing the safety, comfort, and durability of buildings. Nevertheless, the majority of the tenants reveal reduced levels of demand, thus prolonging a building's service life, whilst in neighborhoods having high economic power, the end of service life can be precipitated due to the users' high level of demand.

Williams [97] describes social obsolescence in a more general way, adopting the term community obsolescence, referring to a building's inability to meet the users' needs due to local conflicts of interest arising from the use of a building. The author highlighted that obsolescence is a subjective idea, since "one man's trash is another man's treasure", and an obsolete building according to a given user or for a given use, may still be very useful for a different user or for other uses.

A building's obsolescence due to social, cultural, or community reasons is rooted in social trends that occur due to alterations in [26,111]: individual and collective preferences and social perceptions; users' lifestyle and the social role of housing and the underlying affairs with family, amenities, schools, transport, and jobs; closeness to familiar attractions; neighborhood identity and local culture; among others.

Preiser and Vischer [42] mentioned that the reasons underlying the performance of rehabilitation actions are, in increasing order of importance: aesthetics, social motivations, psychological aspects; functionality; safety; and health. The users recognize the importance of mitigating the evolution of a building's physical deterioration, but still do not adopt specific measures in this direction. Frequently, the users only complain about a building's conditions, when severe degradation levels have already been attained, compromising a building's safety [112].

For decades, the stakeholders and users believed that a building can last forever, but several practical studies demonstrated that the real service life can be abruptly ended, long before the design service life [113]. The example of the Quattrograna West District at Avellino (in Italy) shows how social motives can impose the demolition and reconstruction of an entire neighborhood, after an earthquake, disregarding a building's physical deterioration. The district was rebuilt to promote the users' comfort and well-being, and to stimulate social cohesion [114,115]. Another example is the Gailanxi region of Chongqing city, where the early demolition of buildings due to urban renewal requirements leads to significantly short buildings' service lives [116].

The demolition of buildings presents several drawbacks. From an environmental point of view, a building's demolition produces wastes, whose disposal, management, and reuse have an environmental and financial impact. A building's demolition also encompasses social problems, due to the need for ensuring the compensation for financial support of the demolition or alternative housing for those dislocated [110].

Frequently, economic, or financial reasons supersede criteria for physical degradation or social motivations [117]. Different stakeholders have different points of view; investors are predominantly concerned with economic targets, disregarding the users' social value or a building's cultural worth. A study performed for housing organizations, and concerning demolition strategies, raised a suspicion that secret motivations can influence the decision to intervene or demolish. In some situations, the owners may want to demolish or rehabilitate their buildings, even if they comply with functional requirements or provide lower levels of physical deterioration, in order to force the disposal of unwelcome tenants and to revitalize and improve the attractiveness of their assets [118].

5.5. End of Service Life Triggered by Aesthetics or Architectural Trends

The end of service life due to aesthetics or architectural trends is almost impossible to model, since this is related to a high degree of subjectivity and personal opinion. Different authors [96,119,120] have introduced the concept of aesthetic or visual obsolescence, indicating that a building becomes obsolete from an aesthetic point of view when it is considered outdated or old-fashioned and no longer fulfills the currently prevalent architectural trends [96].

Aesthetic obsolescence can also be connected to alterations in architectural styles [121]. Other authors [27,57,58,122] have indicated that a building becomes aesthetically obsolete due to changes in its aesthetic attributes (image, fashion trends, among others). Pourebrahimi et al. [26] have indicated that these alterations in style or architecture, or in aesthetic ideals, are unavoidable, and an attractive building today will necessarily become an objectionable building, from an aesthetic point of view, sometime in the future. Marteinsson [123] highlighted that the changes in architectural trends tend to promote the adoption of new requirements concerning the aesthetic appearance of the building, new designs, or the choice of new or different construction materials.

More than would be desirable, aesthetic motivations lead to the replacement of building components, necessarily neglecting their physical deterioration or their technical or economic performance. According to Alaimo and Accurso [124], undertaking rehabilitation actions in buildings is often conditioned by their aesthetic appearance, as well as by their social and urban context.

The Gillender Building is a good example in which aesthetic concerns and social criteria led to a sudden end of the building's service life. This building was constructed in 1897 and demolished in 1910, with a view to being replaced by an even taller building. Abramson [125] listed various examples of "speedy obsolescence" in the United States and particularly in New York (e.g., the Plaza Hotel, the Western Union Building, and the Grand Central Terminal). In the 1910s, this kind of obsolescence became endemic to the city center of America's urban agglomerations, given real estate (economic factors), architectural and cultural reasons. Nevertheless, other examples can be found worldwide, in which fashion trends or new demands from an aesthetic point of view helped bring about the end of a building's service life. In the U.K., 46% of buildings on local authority estates are dismantled after 11 to 32 years [126]. In Japan, office buildings frequently have service lives ranging between 23 and 41 years [127].

For non-structural buildings elements, as interior finishing, the end of service life is commonly triggered by variations in the demographics, i.e., the users' requirements change over their life, and it is simply a matter of time before fashion and style decide the fate of a building. In this sense, the same building element may fulfill an elder owner's needs, but fail to comply with the younger generation's expectations (e.g., the use of rugs as a floor covering system).

In hotels, the replacement and renovation of building components usually occur long before the end of their physical and functional service life, being triggered by aesthetic and fashion motivations, which tend to occur every 5 to 7 years [128].

6. Conclusions

The preparation of this paper is a result of efforts from members of the CIB W080 working commission for the conceptualization of building service life. In this study, a theoretical discussion is presented regarding the concept of service life and obsolescence, collected from different perspectives and points of view on an entire complex concept that is service life.

A valuable review has been provided for which the causal effects between the different criteria that dictate the end of buildings' service life are discussed. The impact of endogenous and exogenous factors affecting a building's service life is described, also considering the influence of human behavior on the underlying cause-effect processes that contribute to the end of service life of buildings and components.

In the literature, the different criteria that trigger the end of a building's and building components' service life are described as being separate and unique. From this study, it is now apparent that the different criteria are interrelated and that the end of service life is conditioned by a combination of motives, in cause-and-effect chains, related to e.g., building type, function, location, and ownership.

Due to the multitude of different factors, general conclusions about the lifespan and life expectancy of buildings are overall not possible. More is known and possible at the level of specific building types and parts, particularly in the residential sector, such as applications in LCA. Because this knowledge is essential for meeting the goals of the Paris Agreement (e.g., footprint reduction, lifespan extension, circular construction, and waste minimization), continued research should be focused on relevant specific parts of the building stock.

The literature analyzed in this paper revealed that obsolescence appears to occur, in the majority of the situations, before the physical deterioration of buildings and their components. This knowledge suggests that the current focus on the durability or longevity of the buildings should be rethought. In future studies, it may be interesting to evaluate, e.g., through surveys and case studies analysis, the criteria that most often lead to the end of the service life of buildings and components, evaluating different demand levels, as well as use and ownership profiles.

Author Contributions: Conceptualization, A.S. (Ana Silva), J.d.B.; formal analysis, A.S. (Ana Silva), J.d.B., A.T., A.S. (Ad Straub), A.J.P., M.A.L.; writing—original draft preparation, A.S. (Ana Silva), J.d.B., A.T., A.S. (Ad Straub), A.J.P.; writing—review and editing, A.S. (Ana Silva), J.d.B., A.T., A.S. (Ad Straub), A.J.P., M.A.L. All authors have read and agreed to the published version of the manuscript.

Funding: This research was partially funded by FCT (Portuguese Foundation for Science and Technology) through the project BestMaintenance-LowerRisks (PTDC/ECI-CON/29286/2017) and by the Agencia Nacional de Investigación y Desarrollo (ANID) of Chile through the project ANID FONDECYT 11190554.

Institutional Review Board Statement: Not applicable.

Informed Consent Statement: Not applicable.

Data Availability Statement: The data presented in this study are available on request from the corresponding author.

Acknowledgments: The authors gratefully acknowledge the support of the CIB W080 commission, the CERIS Research Centre (Instituto Superior Técnico-University of Lisbon), the Pontificia Universidad Católica de Chile, TU Delft (The Netherlands) and Canada National Research Council (NRC).

Conflicts of Interest: The authors declare no conflict of interest.

References

1. Silva, A.; de Brito, J.; Gaspar, P.L. *Methodologies for Service Life Prediction of Buildings: With a Focus on Façade Claddings*; 1st ed.; Springer: Geneva, Switzerland, 2016.
2. Rudbeck, C. *Methods for Designing Building Envelope Components Prepared for Repair and Maintenance*; Report R-035; Department of Buildings and Energy, Technical University of Denmark: Kongens Lyngby, Denmark, 1999; ISBN 87-7877-037-8.
3. Harper, R.F. *The Code of Hammurabi, King of Babylon, About 2250 BC: Autographed Text, Transliteration, Translation, Glossary, INDEX OF Subjects, Lists of Proper Names, Signs, Numerals, Corrections and Erasures, with Map, Frontispiece and Photograph of Text*; University of Chicago Press: Callaghan, VA, USA, 1904.
4. Straub, A. Estimating the service lives of building products in use. *J. Civ. Eng. Arch.* **2015**, *9*, 331–340. [CrossRef]
5. Grant, A.; Ries, R.; Kilbert, C. Life cycle assessment and service life prediction; A case study of building envelope materials. *Res. Anal.* **2014**, *18*, 187–200.
6. Silva, A.; de Brito, J. Service life of building envelopes: A critical literature review. *J. Build. Eng.* **2021**, *44*, 102646. [CrossRef]
7. Architectural Institute of Japan (AIJ). *The English Edition of the Principal Guide for Service Life Planning of Buildings*; Architectural Institute of Japan: Tokyo, Japan, 1993.
8. *BS 7543*; Guide to Durability of Buildings and Building Elements, Products and Components; BSI (British Standards Institution): London, UK, 1992.
9. *CSA S478-1995*; Guideline on Durability in Buildings; CSA (Canadian Standards Association): Mississauga, ON, Canada, 1995; pp. 9–17.
10. *Eurocode 0*; Basis of Structural Design; British Standards: London, UK, 2002.
11. Awano, H. *Towards Sustainable Use of the Building Stock*; OECD—Organisation for Economic Cooperation and Development: Paris, France, 2006.
12. Brand, S. *How Buildings Learn—What Happens after They're Built*; Orion Books: London, UK, 1997.
13. *ISO 15686-1 Buildings and Constructed Assets—Service Life Planning—Part 1: General Principles and Framework*; International Organization for Standardization: Geneva, Switzerland, 2011.
14. Haapio, A.; Viitaniemi, P. How workmanship should be taken into account in service life planning. In Proceedings of the 11th International Conference on Durability of Building Materials and Components, Istanbul, Turkey, 11–14 May 2008; pp. 1466–1482.
15. Bordass, W.; Cohen, R.; Staneven, M.; Leaman, A. Assessing building performance in use 2: Technical performance of the Probe buildings. *Build. Res. Inf.* **2001**, *29*, 103–113. [CrossRef]
16. Bordass, W.; Cohen, R.; Staneven, M. Assessing building performance in use 3: Energy performance. *Build. Res. Inf.* **2001**, *29*, 114–128. [CrossRef]
17. Bordass, W.; Leaman, A.; Ruyssevelt, P. Assessing building performance in use 5: Conclusions and implications. *Build. Res. Inf.* **2001**, *29*, 144–157. [CrossRef]
18. Humphreys, M. Quantifying occupant comfort: Are combined indices of the indoor environment practicable? *Build. Res. Inf.* **2005**, *33*, 317–325. [CrossRef]
19. Nicol, F.; Roaf, S. Post-occupancy evaluation and field studies of thermal comfort. *Build. Res. Inf.* **2005**, *33*, 338–346. [CrossRef]
20. Rybczynski, W. *The Look of Architecture*; Oxford University Press: New York, NY, USA, 2001.
21. Leaman, A.; Bordass, W. Assessing building performance in use 4: The Probe occupant surveys and their implications. *Build. Res. Inf.* **2001**, *29*, 129–143. [CrossRef]
22. Aikivuori, A.M. Critical loss of performance—What fails before durability. In Proceedings of the 8th International Conference on Durability of Building Materials and Components, Vancouver, BC, Canada, 30 May 30–3 June 1999; pp. 1369–1376.
23. Pinder, J.; Wilkinson, S.J. *Measuring the Gap: A User Based Study of Building Obsolescence in Office Property*; RICS Research Foundation: London, UK, 2001.
24. Mansfield, J.R. Much Discussed, Much Misunderstood: A Critical Evaluation of the Term "Obsolescence". In Proceedings of the RICS Cutting Edge Conference, London, UK, 6–8 September 2000.
25. Mansfield, J.R.; Pinder, J.A. 'Economic' and 'functional' obsolescence: Their characteristics and impacts on valuation practice". *Prop. Manag.* **2008**, *26*, 191–206. [CrossRef]
26. Pourebrahimi, M.; Eghbali, S.R.; Roders, A.P. Identifying building obsolescence: Towards increasing buildings' service life. *Int. J. Build. Pathol. Adapt.* **2020**, *38*, 635–652. [CrossRef]

27. Grover, R.; Grover, C. Obsolescence—A cause for concern? *J. Prop. Invest. Financ.* **2015**, *33*, 299–314. [CrossRef]
28. Thomsen, A.; Van der Flier, K. Understanding obsolescence: A conceptual model for buildings. *Build. Res. Inf.* **2011**, *39*, 352–362. [CrossRef]
29. Thomsen, A.; Van der Flier, K. Obsolescence, conceptual model and proposal for case studies. In Proceedings of the 23rd Conference of the European Network of Housing Research: ENHR 2011, Toulouse, France, 5–8 July 2011.
30. Lemer, A.C. Infrastructure obsolescence and design service life. *J. Infrastruct. Syst.* **1996**, *2*, 153–161. [CrossRef]
31. Smith, M.; Whitelegg, J.; Williams, N. *Materials Intensity in the Built Environment*; Greening the Built Environment, Earthscan Publications Ltd.: London, UK, 1998.
32. Markus, T.; Whyman, P.; Morgan, J.; Whitton, D.; Maver, T.; Canter, D.; Fleming, J. *Building Performance*; Applied Science Publishers: London, UK, 1972.
33. CIB. *Working Commission W060 CIB Report 64: Working with the Performance Approach to Building*; CIB: Rotterdam, The Netherlands, 1982.
34. Rush, R.D. *The Building Systems Integration Handbook*; Architectural Press, The American Institute of Architects: New York, NY, USA, 1986.
35. Masters, L.W.; Brandt, E. Systematic methodology for service life prediction of building materials and components. *Mater. Struct.* **1989**, *22*, 385–392. [CrossRef]
36. Andersen, T.; Brandt, E. The use of performance and durability data in assessment of life time serviceability—Performance and durability data. In Proceedings of the 8th International Conference on Durability of Building Materials and Components, Ottawa, QC, Canada, 30 May–3 June 1999; pp. 1813–1820.
37. Davis, G.; Szigeti, F. Are facilities measuring up? Matching building capabilities to functional needs—Using the serviceability tools and methods. In Proceedings of the 8th International Conference on Durability of Building Materials and Components, Ottawa, QC, Canada, 30 May–3 June 1999; pp. 1856–1866.
38. Blok, R.; Herwijnen, F.V.; Kozlowski, A.; Wolinski, S. *Service Life and Life Cycle of Building Structures*; COST C-12; Improving Building Structural Quality by New Technologies: Lisbon, Portugal, 2002.
39. Lacasse, M.A.; Vanier, D.J.; Kyle, B.R. Towards integration of service life and asset management tools for building envelope systems. In Proceedings of the 7th Conference on Building Science and Technology: Durability of Buildings Design, Maintenance, Codes and Practice, Toronto, ON, Canada, 20–21 March 1997; pp. 153–167.
40. Foliente, C. Developments in performance-based building codes and standards. *For. Product. J.* **2000**, *50*, 12–21.
41. Lützkendorf, T.; Speer, T.; Szigeti, F.; Davis, G.; Le Roux, P.; Kato, A.; Tsunekawa, K. *A Comparison of International Classifications for Performance Requirements and Building Performance Categories Used in Evaluation Methods*; Performance Based Building, Technical Research Centre of Finland (VTT)/Association of Finnish Civil Engineers (RIL): Helsinki, Finland, 2005; pp. 61–80.
42. Preiser, W.; Vischer, J. (Eds.) *Assessing Building Performance*; Routledge, Taylor and Francis: London, UK, 2006.
43. Hammond, D.; Dempsey, J.J.; Szigeti, F.; Davis, G. Integrating a performance-based approach into practice: A case study. *Build. Res. Inf.* **2005**, *33*, 128–141. [CrossRef]
44. Grussing, M.N.; Uzarski, D.R.; Marrano, L.R. Building infrastructure functional capacity measurement framework. *J. Infrastruct. Syst.* **2009**, *15*, 371–377. [CrossRef]
45. Blok, R.; Teuffel, P. Indicators for functional service life of building structures. In Proceedings of the International Conference Civil Engineering for Sustainability and Resilience (CESARE'14), Amman, Jordan, 24–27 April 2014; p. 52.
46. Macías-Bernal, J.M.; Calama-Rodríguez, J.M.; Chávez-de Diego, M.J. Modelo de predicción de la vida útil de la edificación patrimonial a partir de la lógica difusa. *Inf. Constr.* **2014**, *66*, e006. [CrossRef]
47. Prieto, A.J.; Silva, A.; de Brito, J.; Alejandre, F.J. Functional and physical service life of natural stone claddings. *J. Mater. Civ. Eng.* **2016**, *28*, 04016150. [CrossRef]
48. Ibáñez, A.J.P.; Bernal, J.M.M.; de Diego, M.J.C.; Sánchez, F.J.A. Expert system for predicting buildings service life under ISO 31000 standard; Application in architectural heritage. *J. Cultural Herit.* **2016**, *18*, 209–218. [CrossRef]
49. Prieto, A.J.; Silva, A.; de Brito, J.; Macías-Bernal, J.M. Serviceability of facade claddings. *Build. Res. Inf.* **2018**, *46*, 179–190. [CrossRef]
50. Augenbroe, G. *The Role of Simulation in Performance-Based Building*; Building Performance Simulation for Design and Operation; Routledge, Taylor and Francis: London, UK, 2019.
51. Lacasse, M.A.; Gaur, A.; Moore, T.V. Durability and climate change-implications for service life prediction and the maintainability of buildings. *Buildings* **2020**, *10*, 53. [CrossRef]
52. Prieto, A.J.; Verichev, K.; Silva, A.; de Brito, J. On the impacts of climate change on the functional deterioration of heritage buildings in South Chile. *Build. Environ.* **2020**, *183*, 107138. [CrossRef]
53. Reed, R.; Warren-Myers, G. Is sustainability the 4th form of obsolescence? In Proceedings of the 16th Pacific Rim Real Estate Society, (PRRES) Conference, Wellington, New Zealand, 24–27 January 2010.
54. Marteinsson, B. Homes in Iceland-flexibility and service life fulfilment of functional needs. In Proceedings of the 11th International Conference on Durability of Building Materials and Components, Istanbul, Turkey, 11–14 May 2008; Volume 2010, pp. 1485–1490.
55. Iselin, D.G.; Lemer, A.C. (Eds.) *Fourth Dimension in Building: Strategies for Avoiding Obsolescence*; National Research Council, The National Academy Press: Washington, DC, USA, 1993. [CrossRef]

56. Nutt, B.; Walker, B.; Holliday, S.; Sears, D. *Obsolescence in Housing: Theory and Applications*; Saxon House and Lexington Books: Farnborough, UK, 1976.

57. Flanagan, R.; Norman, G.; Meadows, J.; Robinson, G. *Life Cycle Costing Theory and Practice*; BSP Professional Books: Oxford, MS, USA, 1989.

58. Ashworth, A. Estimating the life expectancies of building components in life-cycle costing calculations. *Struct. Surv.* **1996**, *14*, 4–8. [CrossRef]

59. API (Australian Property Institute). API Definitions. 2021. Available online: https://www.api.org.au/standards/definitions/ (accessed on 23 March 2021).

60. RICS (Royal Institution of Chartered Surveyors). RICS Valuation-Global Standards. 2017. Available online: https://www.rics.org/globalassets/rics-website/media/upholding-professional-standards/sector-standards/valuation/red-book-2017-global-edition-rics.pdf (accessed on 23 March 2021).

61. Baum, A. *Property Investment Depreciation and Obsolescence*; Routledge Publishing: London, UK, 1991.

62. Kyle, B.R. Toward Effective Decision Making for Building Management. In Proceedings of the APWA International Public Works Congress, NRCC/CPWA/IPWEA Seminar Series "Innovations in Urban Infrastructure", Philadelphia, PA, USA, 9–12 September 2001; pp. 51–69.

63. Geissdoerfer, M.; Savaget, P.; Bocken, N.M.; Hultink, E.J. The Circular Economy—A new sustainability paradigm? *J. Clean. Product.* **2017**, *143*, 757–768. [CrossRef]

64. Frohnsdorff, G.; Nireki, T. Integration of standards for prediction of service life of building materials and components and for assessment of design life of buildings. In Proceedings of the EUREKA Conference of Building Technologies, Lillehammer, Norway, 16 June 1994.

65. Nireki, T. Service life design. *Constr. Build. Mater.* **1996**, *10*, 403–406. [CrossRef]

66. Soronis, G. Standards for design life of buildings: Utilization in the design process. *Constr. Build. Mater.* **1996**, *10*, 487–490. [CrossRef]

67. Marteinsson, B. Durability and the factor method of ISO 15686-1. *Build. Res. Inf.* **2003**, *31*, 416–426. [CrossRef]

68. Sarja, A. Generic limit state design of structures. In Proceedings of the 10th International Conference on Durability of Building Materials and Components, Lyon, France, 17–20 April 2005; p. TT3-161.

69. Balaras, C.; Droutsa, K.; Dascalaki, E.; Kontoyiannidis, S. Service life of building elements & installations in European apartment buildings. In Proceedings of the 10th International Conference on Durability of Building Materials and Components, Lyon, France, 17–20 April 2005; pp. 718–725.

70. Duffy, F. *The New Office*; Conran Octopus: London, UK, 1997.

71. Gosav, I. Field studies concerning the life service life prediction. In Proceedings of the 8th International Conference on Durability of Building Materials and Components, Ottawa, QC, Canada, 30 May–3 June 1999; pp. 1392–1402.

72. Itard, L.; Meijer, F. *Towards a Sustainable Northern European Housing Stock: Figures, Facts and Future*; Sustainable Urban Areas; IOS Press: Amsterdam, The Netherlands, 2008; Volume 2.

73. Gaspar, P.L. Service Life of Constructions; Development of a Methodology to Estimate the Durability of Construction Elements, Application to Renders in Current Buildings. Ph.D. Thesis, Engineering Sciences at Instituto Superior Técnico, Lisbon, Portugal, 2009. (In Portuguese)

74. Thomsen, A.; van der Flier, K. Replacement or renovation of dwellings: The relevance of a more sustainable approach. *Build. Res. Inf.* **2009**, *37*, 649–659. [CrossRef]

75. Kohler, N.; Hassler, U. The building stock as a research object. *Build. Res. Inf.* **2002**, *30*, 226–236. [CrossRef]

76. Iizuka, H. A statistical study of life time of bridges. *Struct. Eng. Earthq. Eng. Jap. Soc. Civ. Eng.* **1988**, *5*, 51–60. [CrossRef]

77. Horst, S.; O'Connor, J.; Argeles, C. Survey on actual service lives of north American buildings. In Proceedings of the 10th International Conference on Durability of Building Materials and Components, Lyon, France, 17–20 April 2005; pp. 851–858.

78. Wuyts, W.; Miatto, A.; Sedlitzky, R.; Tanikawa, H. Extending or ending the life of residential buildings in Japan: A social circular economy approach to the problem of short-lived constructions. *J. Clean. Product.* **2019**, *231*, 660–670. [CrossRef]

79. Gann, D.M.; Barlow, J. Flexibility in building use: The technical feasibility of converting redundant offices into flats. *Construct. Manage. Econ.* **1996**, *14*, 55–66. [CrossRef]

80. Bakens, W.; Foliente, G.; Jasuja, M. Engaging stakeholders in performance-based building: Lessons from the Performance-Based Building (PeBBu) Network. *Build. Res. Inf.* **2005**, *33*, 149–158. [CrossRef]

81. Butt, T.E.; Giddings, B.D.; Cooper, J.C.; Umeadi, B.B.N.; Jones, K.G. *"Advent of Climate Change and Resultant Energy Related Obsolescence in the Built Environment", Sustainability in Energy and Buildings, Results of the Second International Conference on Sustainability in Energy and Buildings (SEB'10), Smart Innovation, Systems and Technologies*; Springer: Berlin/Heidelberg, Germany, 2011; Volume 3, pp. 211–224.

82. Fitch, M.G.; Weyers, R.E.; Johnson, S.D. Determination of end of functional service life for concrete bridge decks. *Transp. Res. Rec.* **1995**, *1490*, 60–66.

83. Ford, K.M.; Arman, M.; Labi, S.; Sinha, K.C.; Shirole, A.; Thompson, P.; Li, Z. *Methodology for Estimating Life Expectancies of Highway Assets*; National Cooperative Highway Research Program: Washington, DC, USA, 2011.

84. Landman, M. *Technical Building Properties with the Probability of Elongating the Functional Service Life*; Eindhoven University of Technology: Eindhoven, The Netherlands, 2016.

85. Slaughter, E.S. Design strategies to increase building flexibility. *Build. Res. Inf.* **2001**, *29*, 208–217. [CrossRef]
86. Meacham, B.; Bowen, R.; Traw, J.; Moore, A. Performance-based building regulation: Current situation and future needs. *Build. Res. Inf.* **2005**, *33*, 91–106. [CrossRef]
87. Trinius, W.; Sjöström, C. Service life planning and performance requirements. *Build. Res. Inf.* **2005**, *33*, 173–181. [CrossRef]
88. Mc Duling, J.; Horak, E.; Cloete, C. Service life prediction beyond the factor method. In Proceedings of the 11th International Conference on Durability of Building Materials and Components, Istanbul, Turkey, 11–14 May 2008; pp. 1515–1522.
89. Mc Duling, J.; Abbott, G. Service life and sustainable design methods: A case study. In Proceedings of the 11th International Conference on Durability of Building Materials and Components, Istanbul, Turkey, 11–14 May 2008; pp. 1663–1669.
90. Conejos, S.; Langston, C.; Smith, J. Designing for better building adaptability: A comparison of adaptSTAR and ARP models. *Habitat Int.* **2014**, *41*, 85–91. [CrossRef]
91. Langston, C. Estimating the useful life of buildings. In Proceedings of the 36th Australasian University Building Educators Association (AUBEA) Conference, Gold Coast, Australia, 27–29 April 2011.
92. Langston, C. Validation of the adaptive reuse potential (ARP) model using iconCUR. *Facilities* **2012**, *30*, 105–123. [CrossRef]
93. Chaplin, R.I. The threat of obsolescence to police precincts on the heritage 'beat'. *Int. J. Herit. Stud.* **2003**, *9*, 117–133. [CrossRef]
94. Tan, Y.; Shen, L.Y.; Langston, C. A fuzzy approach for adaptive reuse selection of industrial building in Hong Kong. *Int. J. Strateg. Prop. Manag.* **2014**, *18*, 66–76. [CrossRef]
95. Raftery, J. *Principles of Building Economics*; BSP Professional: Oxford, UK, 1991.
96. Wilkinson, S.J.; Remøy, H.; Langston, C. *Sustainable Building Adaptation: Innovations in Decision-Making*; John Wiley & Sons: West Sussex, UK, 2014.
97. Williams, A. Remedying industrial building obsolescence: The options. *Prop. Manage.* **1986**, *4*, 5–14. [CrossRef]
98. Hed, G. Service Life Planning in Building Design. In Proceedings of the 14th CIB World Building Congress, Gavle, Sweden, 22–24 February 1998; pp. 201–209.
99. Fu, F.; Pan, L.; Ma, L.; Li, Z. A simplified method to estimate the energy-saving potentials of frequent construction and demolition process in China. *Energy* **2013**, *49*, 316–322. [CrossRef]
100. Dias, W.P.S. *Useful Life of Buildings*; University of Moratuwa: Moratuwa, Sri Lanka, 2003.
101. Dias, W.P.S. Factors influencing the service life of buildings. *J. Inst. Eng.* **2013**, *46*, 1–7. [CrossRef]
102. Rauf, A.; Crawford, R.H. Building service life and its effect on the life cycle embodied energy of buildings. *Energy* **2015**, *79*, 140–148. [CrossRef]
103. Kincaid, D. Adaptability potentials for buildings and infrastructure in sustainable cities. *Facilities* **2000**, *18*, 155–161. [CrossRef]
104. Forster-Kraus, S.; Reed, R.; Wilkinson, S. Affordable housing in the context of social sustainability. In Proceedings of the ISA International Housing Conference, Glasgow, UK, 21–26 April 2009.
105. Rodi, W.N.W.; Hwa, T.K.; Said, A.S.; Mahamood, N.M.; Abdullah, M.I.; Abd Rasam, A.R. Obsolescence of green office building: A literature review. *Proc. Econ. Financ.* **2015**, *31*, 651–660. [CrossRef]
106. BRB—Building Research Board. *The Fourth Dimension in Building: Strategies for Minimising Obsolescence*; National Academy Press: Washington, DC, USA, 1993.
107. Ohemeng, F.A.; Mole, T. Value-focused approach to built asset renewal and replacement decisions. In Proceedings of the Cutting-Edge Conference, Birmingham, UK, 20–21 September 1996; RICS—Royal Institution of Chartered Surveyors: Birmingham, UK, 1996.
108. Bryson, J.R. Obsolescence and the process of creative reconstruction. *Urb. Stud.* **1997**, *34*, 1439–1458. [CrossRef]
109. Sarja, A. *Predictive and Optimised Life Cycle Management*; Taylor and Francis: London, UK, 2006.
110. Power, A. Does demolition or refurbishment of old and inefficient homes help to increase our environmental, social and economic viability? *Energy Policy* **2008**, *36*, 4487–4501. [CrossRef]
111. Sarja, A. Reliability principles, methodology and methods for lifetime design. *Mater. Struct.* **2010**, *43*, 261–271. [CrossRef]
112. Flores-Colen, I.; de Brito, J.; Freitas, V. Discussion of criteria for prioritization of predictive maintenance of building façades-survey of 30 experts. *J. Perform. Constr. Facil.* **2010**, *24*, 337–344. [CrossRef]
113. O'Connor, J. Survey on actual service lives for North American buildings. In Proceedings of the Woodframe Housing Durability and Disaster Issues Conference, Las Vegas, NV, USA, 4–6 October 2004.
114. Marino, F.P.R. The pathologies of the industrialized systems: 1.023 flats realized in Avellino (Italy) in the '80 s. In Proceedings of the Construction in the XXI Century: Local and Global Challenges, joint 2006 CIB W065/W055/W086 International Symposium, Rome, Italy, 18–20 October 2006.
115. Marino, F.P.R. Pathologies of the industrialized systems: 192 flats built in Avellino (Italy) at Quattrograna west district. In Proceedings of the 11th International Conference on Durability of Building Materials and Components, Istanbul, Turkey, 11–14 May 2008; p. T71.
116. Shen, L.Y.; Yuan, H.; Kong, X. Paradoxical phenomenon in urban renewal practices: Promotion of sustainable construction versus buildings' short lifespan. *Int. J. Strateg. Prop. Manag.* **2013**, *17*, 377–389. [CrossRef]
117. Farahani, A.; Wallbaum, H.; Dalenbäck, J.-O. The importance of life-cycle based planning in maintenance and energy renovation of multifamily buildings. *Sust. Cit. Soc.* **2019**, *44*, 715–725. [CrossRef]
118. Van der Flier, K.; Thomsen, A. *The Life Cycle of Dwellings and Demolition by Dutch Housing Associations*; Gruis, V., Visscher, H., Kleinhans, R., Eds.; Sustainable Neighbourhood Transformation; IOS Press/DUP: Amsterdam, The Netherlands, 2006.

119. Remøy, H. Out of Office: A Study on the Cause of Office Vacancy and Transformation as a Means to Cope and Prevent. Ph.D. Thesis, Delft University of Technology, Delft, The Netherlands, 2010.
120. Dunse, N.; Jones, C. Rental depreciation, obsolescence and location: The case of industrial properties. *J. Prop. Res.* **2005**, *22*, 205–223. [CrossRef]
121. Douglas, J. *Building Adaptation*, 2nd ed.; Spon Press: Abington, MA, USA, 2006.
122. Johnston, K. Obsolescence and Renewal: Transformation of Post War Concrete Buildings. Master's Thesis, Architecture, University of Maryland, College Park, MD, USA, 2016.
123. Marteinsson, B. Service Life Estimation in the Design of Buildings; A Development of the Factor Method. Ph.D. Thesis, KTH Research School Department of Technology and Built, Centre for Built Environment, University of Gävle, Gävle, Sweden, 2005.
124. Alaimo, G.; Accurso, F. The assessment of durability of discontinuous roofing: An experiment on sandwich panels. In Proceedings of the 11th International Conference on Durability of Building Materials and Components, Istanbul, Turkey, 11–14 May 2008; p. T22.
125. Abramson, D.M. *Obsolescence: An Architectural History*; The University of Chicago Press: Chicago, IL, USA, 2016.
126. Pieda, D.T.Z. *Consulting Demolition and New Building on Local Authority Estates*; U.K. Department of the Environment, Transport and the Regions: London, UK, 2000.
127. Yashiro, T.; Kato, H.; Yoshida, T.; Komatsu, Y. Survey on real life span of office buildings in Japan. In Proceedings of the CIB 90 W55/65 Symposium, Sydney, Australia, 14–22 March 1990; pp. 215–226
128. International Society of Hospitality Consultants (Ed.) *CapEx2014: A Study of Capital Expenditures in the Hotel Industry*; American Hotel & Lodging Educational Institute: Lansing, MI, USA, 2015.

Article

An Integrated Safety, Health and Environmental Management Capability Maturity Model for Construction Organisations: A Case Study in Ghana

Millicent Asah-Kissiedu [1], Patrick Manu [2], Colin Anthony Booth [3,*], Abdul-Majeed Mahamadu [3] and Kofi Agyekum [4]

[1] Department of Environmental Management and Technology, Koforidua Technical University, Koforidua KF981, Ghana; mokumih@ktu.edu.gh
[2] Department of Mechanical, Aerospace and Civil Engineering, The University of Manchester, Manchester M13 9PL, UK; patrick.manu@manchester.ac.uk
[3] Department of Architecture and the Built Environment, University of the West of England, Bristol BS16 1QY, UK; Abdul.Mahamadu@uwe.ac.uk
[4] Department of Construction Technology and Management, Faculty of Built Environment, College of Art and Built Environment, Kwame Nkrumah University of Science and Technology, Kumasi AK384, Ghana; kagyekum.cap@knust.edu.gh
*   Correspondence: Colin.Booth@uwe.ac.uk; Tel.: +44-117-328-3998

Citation: Asah-Kissiedu, M.; Manu, P.; Booth, C.A.; Mahamadu, A.-M.; Agyekum, K. An Integrated Safety, Health and Environmental Management Capability Maturity Model for Construction Organisations: A Case Study in Ghana. *Buildings* **2021**, *11*, 645. https://doi.org/10.3390/buildings 11120645

Academic Editor: David Arditi

Received: 4 November 2021
Accepted: 6 December 2021
Published: 13 December 2021

Publisher's Note: MDPI stays neutral with regard to jurisdictional claims in published maps and institutional affiliations.

Abstract: Safety, health and environmental (SHE) management is becoming a priority as construction companies (i.e., contractors) strive to reduce construction accidents and negative environmental impacts, conform to regulatory requirements, and sustain their competitiveness. Consequently, construction firms are expected to adopt and implement innovative SHE management systems to mitigate SHE risks effectively and efficiently. For construction firms to effectively do this, they need to have the adequate capability in respect of integrated SHE management. However, there is limited empirical insight regarding the integrated SHE management capabilities of construction companies. Furthermore, there is limited insight regarding the mechanisms for ascertaining the integrated SHE management capability of construction companies to guide such organisations towards SHE management excellence in their operations. Drawing on the capability maturity model integration (CMMI) concept, this study, by applying expert reviews (i.e., Delphi technique and the design methodology for capability maturity grids), developed an integrated Safety, Health and Environmental Management Maturity Model (iSHEM-CMM). The model offers capability maturity assessment on a five-level scale within five thematic categories and 20 integrated SHE management capability attributes. Based on an industrial validation by construction professionals, it is concluded that the maturity model is a useful assessment framework or tool for industry stakeholders, particularly construction firms, to evaluate the status of their current SHE management capabilities, identify strengths and improvement areas, and accordingly prioritise strategies/actions for improving their SHE management. Furthermore, clients who appoint construction companies could use the model as part of prequalification arrangements in selecting construction companies with an adequate SHE management capability.

Keywords: capability attribute; capability maturity model; construction; integrated safety; health and environmental management

1. Introduction

The construction sector remains one of the key generators of adverse environmental impacts and is among one of the highest contributors of work-related accidents, resulting in injuries, fatalities, and illnesses [1–3]. For instance, in the USA, the construction sector accounted for over 800 worker-related deaths in 2019 [4]. Moreover, the UK construction sector recorded the highest number of fatal injuries in 2020/21 [5]. The construction sector in India, having only 7.5% of the total world labour force, also contributes 16.4% of fatal occupational accidents worldwide [6].

Furthermore, the sector accounts for a significant consumption of natural resources and energy. Estimates indicate that buildings and construction together account for 36% of final energy use, 16% of natural water, 39% of CO_2 emissions, 40% of the waste produced and 50% of all raw materials extracted [7,8]. With the volume of construction output projected to grow by more than 85% globally by 2030 [9], the impact of construction operations on the environment and workers' safety and health would be far-reaching if nothing were to be done about it. The socio-economic impacts arising from these negative environmental impacts, injuries, illnesses, and fatalities [3,10] have triggered several efforts to address the poor status of SHE management in construction. One of the prominent initiatives to address the SHE situation in construction is the implementation of management systems, particularly environmental management systems (EMSs) and safety and health management systems (SHMSs) in construction to manage SHE risks with maximum effectiveness and minimum bureaucracy [11]. This could be beneficial in reducing the number of fatalities, injuries, illnesses and potentially negative environmental impacts, leading to better SHE performance outcomes within the construction sector.

Like other countries [4–6], in Ghana, the construction industry accounts for a high number of occupational accidents and deaths as well as work-related illnesses [12]. The construction industry in Ghana is also noted for its constant degradation, pollution, substantial raw materials and energy consumption, which negatively impact the development of the country [12]. Despite these negative impacts, Agyekum et al. [13] reported that the high-risk nature of the industry, the weak institutional structures for implementing SHE standards, and laxity in the enforcement of safety and environmental legislations on construction sites have impeded the implementation of SHE standards. Due to these lapses, there has been a need to implement proactive and systematic methods that have the potential to prevent accidents and negative environmental impacts on construction sites, and that will further assist construction companies to effectively improve SHE performance outcomes in the industry. Unfortunately, the uptake of a prominent approach like the implementation of SHE management systems in the Ghanaian construction industry is low [12].

While several authors and industry stakeholders have advocated for integrated management systems for the construction industry [14], there is no single integrated SHE management framework for construction organisations to use. Therefore, there is a general lack of a robust systematic mechanism that enables construction companies to ascertain the maturity of their SHE management practices. A process improvement tool, like a capability maturity model, can offer such a mechanism. Though maturity models have been proven valuable for assessing organizational processes or practices in delivering performance for various domains, there are just a few related examples of its application to integrated SHE management in the construction industry. For instance, Hamid et al. [15] developed the integrated management system for safety, health and environmental quality (SHEQ-MS) in the construction industry. Rebelo et al. [16] also developed the integrated management system-quality, environment and safety (IMS-QES). Though closely related to SHEM-CMM, these two systems/models do not enable SHE management capability maturity assessment. In addition to these two, there have been single stand-alone maturity models (i.e., maturity models that do not integrate multiples domains) developed for safety management [17] and environmental management [18], which still do not incorporate the environmental aspects and safety aspects, respectively.

Drawing on the afore-mentioned gap, the question that arises is what integrated safety, health and environmental capability maturity model can best work for the construction industry. To answer this question, this study thus examines integrated SHE management capability. It adopts the capability maturity modelling concept for the development of an integrated safety, health and environmental management capability maturity model (iSHEM-CMM) to enable the assessment of iSHEM capability of construction firms, effective management of components of an iSHEM system, and thereby improve construction iSHEM capability maturity levels and management practices. The capability maturity model developed is a useful assessment framework or tool for industry stakeholders, particularly construction firms, to evaluate the status of their current SHE management capability, identify strengths and improvement areas, and accordingly prioritise strategies/actions for improving their SHE management. Furthermore, clients who appoint construction companies could use the model as part of prequalification arrangements in selecting construction companies with an adequate SHE management capability.

The next section presents a brief overview of integrated SHE management capability and capability maturity modelling concepts to provide the foundation for developing the iSHEM-CMM. The research method applied, including the design decisions involved in developing the iSHEM-CMM, the maturity model and validation of the model, are subsequently presented. The implications stemming from the developed capability maturity model and concluding remarks are also presented.

2. Literature Review

This section conducts a critical comparative review of the related literature. Literature reviewed is presented under two sub-sections, i.e., safety, health and environmental management capability in construction; and capability maturity models. A systematic review through content analysis of literature related to the theme under investigation was conducted. Multiple queries were conducted on online databases like Google Scholar, Web of Science (WoS), Scopus and the like. Literature relevant to the current study and which spanned 1990 to 2019 were covered. Initial reviews limited to titles and abstracts of papers were accessed to ensure relevance to the theme under investigation. The search for the relevant literature was carried out using a combination of words like 'safety and health in the construction industry', health and safety management in construction', safety and health capability in construction', 'safety, health and environment', and 'capability maturity models'. Sections 2.1 and 2.2 summarise the significant literature retrieved from the search.

2.1. Safety, Health and Environmental Management Capability in Construction

The increasing concern regarding environmental, safety and health issues and their efficient management in construction have become of utmost importance for construction organisations worldwide. Some of these organisations are complying with SHE legislation and standards by deploying systematic and proactive initiatives such as SHE management systems [19–21] to address SHE issues and their associated undesirable outcomes. Though the adoption and implementation of SHE management systems are minimal in the construction sector, several studies have highlighted their implementation in construction as an innovative approach that offers substantial improvements in operation efficiency, standard compliance, as well as in SHE performance [22–24]. A study by Yoon et al. [24] revealed that in Korea the implementation of SHE standards saw safety performance increasing by more than 30%, with fatal accidents decreasing by 10.3%. Zeng et al. [22] reported that construction companies in China were able to enter international markets, reduce waste and noise control and improve safety and health at workplaces by implementing Environmental Management Systems (EMS). These studies show that the implementation of EMSs and SHMSs is an important innovative, systematic and proactive approach in reducing construction accidents and in minimising detrimental environmental impacts of construction operations [22,24]. However, the parallel implementation of both management systems (i.e., EMS and SHMS) have been criticised for being bureaucratic, costly, paper-driven and arduous [14,19,25], hence the need for an integrated management of SHE issues in construction through a single system (i.e., an integrated SHE management system).

Integrated SHE management in construction involves identifying, assessing and managing SHE risks rightly to minimise injuries, illness, fatalities and negative environmental impacts. It requires construction companies to take into account SHE considerations in addition to cost, time and quality considerations in all phases of building and construction projects. Though the compliance to SHE regulations often leads to a reduction in work-related tragedies and adverse environmental impacts, the efficient and effective management of SHE problems in construction based on an integrated SHE management framework makes it critical for construction companies to have an appropriate organisational capability which encompasses the policies, systems, resources, information, infrastructure and personnel of the company. However, empirical work into integrated SHE management capability is missing in the growing body of construction SHE management literature. For instance, within the last few decades, several studies on SHE management systems in the construction industry have focused on: (1) awareness, motivators, costs, benefits and barriers of management systems [2,26–28]; (2) effectiveness of SHE management systems in addressing occupational accidents, SHE performance, pollution and waste reduction [24,29]; (3) integration of environment, quality, safety and health management systems and benefits [14,19,25]; and the elements of stand-alone management systems and integrated management systems [29,30]. In terms of integrated SHE management capability, there is inadequate empirical research for insights into what constitutes integrated SHE management capability and mechanisms by which it can be reliably assessed to pave the way for continuous process improvement. Given the increasing concerns over SHE performance for sustainable construction and the lack of existing frameworks for integrated SHE management in construction, the development of a simple and implementable iSHEM framework that involves capabilities or practices relevant to the efficient implementation of an iSHEM system in a construction firm is crucial. The efficient management of iSHEM capabilities or practices could lead to better SHE performance. According to the capability maturity modelling concept, the degree of process effectiveness and efficiency reflects the capability of firms and organisations to implement processes successfully, thereby showing the maturity of organisational practices [31]. Therefore, capability maturity models (CMMs) serve as a good reference framework for developing iSHEM-CMM.

2.2. Capability Maturity Models

To respond to the highly competitive external environment, organisations continuously search for effective new approaches for assessing performance and organisational capability, as well as enhancing management capabilities [32] such as business excellence models, balanced scorecards, maturity models, total quality management, business process reengineering amongst others. However, amongst these approaches, maturity models have been designed to provide organisations with guidance on how to effectively measure and improve the maturity of functional domains within these organisations [33]. Moreover, they have assisted organisations in overcoming challenges of the need for cost reduction or quality improvement in the face of competitive pressure [34].

The principal idea of maturity models is that they describe the characteristics of organisational processes or activity at different levels of maturity [35]. They have their roots in quality management and continuous process improvement [36]. In particular, the Quality Management Maturity Grid by Crosby [37] describes the behaviour exhibited by a company at five maturity levels for a set of aspects of quality management [38]. According to Van Looy et al. [36], the best-known derivative of the quality management maturity concept is the capability maturity model (CMM). CMM was developed by the Software Engineering Institute (SEI) at Carnegie Mellon University as a reference model for assessing, evaluating and improving software process maturity [39,40]. Capability maturity models (CMMs) focus on improving organisational processes and identifying several levels of maturity ranging from low to high and each maturity level details the behaviour exhibited by organisations [41,42]. The CMM framework describes the maturity of organisations according to five levels (i.e., initial, repeatable, defined, managed and optimising) and determines these levels based on key process areas or capabilities [38,39]. However, the number of maturity levels can differ, depending on the domain and the concerns motivating the model.

The CMM Integration (CMMI), which is an extension of the CMM, is a single and comprehensive framework that is appropriate for organisations of any structure and focused on guiding organisation-wide process improvement [36,41]. It has two different representations of maturity, namely staged and continuous representations [43]. The staged presentation includes five levels similar to those of the original CMM. Each maturity level consists of several process areas that are specifically demarcated to that stage. Organisations get assessed against their process areas' existence or absence and produce an overall maturity level rating [44]. This presentation is useful for organisations that are looking at improving their overall process capability. With the continuous representation, maturity within each process area is analysed separately and improvements are made accordingly [36]. This presentation offers a more flexible approach to process improvements and is suitable for organisations looking to improve specific process areas and desiring to choose areas of implementation [45].

Currently, CMM/CMMI is one of the most widely accepted frameworks for assessing organisational capability in a domain as part of continuous process improvement [39,42]. Increasingly, CMMI has become a tool used to assess and improve organisational processes, systems, products, and competencies on the evolutionary path towards excellence and attaining desired outcomes [46]. Although originally developed for process improvement within the software industry, the CMM/CMMI represent a generic framework for continuous process improvement and hence has been applied in varied domains in several industrial sectors, including construction in areas such as supply chain management, risks management, disability management, change management, Building Information Modelling, and e-business [44,47–50]. In the area of SHE in construction, CMM frameworks have also been applied, although not specifically to integrated SHE management. For instance, there is the safety culture maturity model by Fleming [51] to access safety culture maturity, the AC2E performance matrix by Carillon Plc [52] to assess construction site safety management, the health and safety maturity model by Goggin [53] to assess the maturity of safety management practices of a given construction company at the organisational level, the Environmental management maturity model of construction programs by Bai et al. [54], and also the Design for occupational safety and health capability model (DfOSH) by Manu et al. [55] to access the DfOSH capability maturity of design firms in the construction sector. Other than Goggin's model, which focuses on safety and health management practices in construction, the extant literature does not reveal any other maturity models and systematic approaches for evaluating integrated SHE management in the construction industry, thus highlighting the significance of this paper. The application of CMMI/CMM in several areas in construction, including occupational safety, health and the environment, as a useful and robust tool for assessment and continuous process improvement, therefore, supports its application to integrated SHE management to produce an iSHEM-CMM.

3. Materials and Methods

In developing the iSHEM-CMM, the approach of Maier et al. [56] on how to develop maturity grids based on organisational capability assessments was as follows. Maier's et al. [56] procedural approach consists of four steps: (1) Planning: identification of target audience, aim, scope and success areas; (2) Development: defining the various parts of the maturity model which are the process areas, maturity levels, the cell descriptors and administration mechanisms; (3) Evaluation: model verification, refinement and validation; (4) Maintenance: documentation and communication of development processes, results and changes in process areas and cell descriptors. The main design decisions in this approach are the establishment of (1) key process areas (i.e., integrated SHE management capability attributes) and (2) the capability maturity levels.

In maturity model literature, maturity models (MMs) have received recurrent criticisms, particularly its lack of theoretical framework or methodology and traceability [57]. There is a dearth of literature on how to theoretically develop a maturity model [58].

However, the development process is not demonstrated in most of the documentation of maturity models and grids. Notwithstanding, recent studies have sought to introduce a structured approach to previous work done [59].

Compared to other traditional methods, Maier et al.'s [56] was followed in this study because it provides rigorous and consistent development procedure, and also looks similar to some of the common steps in the approaches developed by other authors like SEI [60], De Bruin et al. [58], Poghosyan et al. [61], and Asah-Kissiedu et al. [12]. Sub-Section 3.1 expounds the various methods.

3.1. Maturity Model Development

Like most existing CMMI- or CMM-based models, the iSHEM-CMM follows the continuous-structure [60] since it provides a generic measurement of capability maturity level for each integrated SHE management capability attribute. The model is represented in a grid format and has two main components: capability maturity levels and integrated iSHEM capability attributes. Levels of capability maturity are allocated against the attributes, thereby creating a series of cells. Each cell contains a brief text description (i.e., descriptor) for each activity at each capability maturity level. The following subsections present the steps taken to develop the maturity model.

3.1.1. Design Decisions for Developing the iSHEM-CMM

In this section, the main design decisions outlined in Maier et al. [56] for developing maturity grids are elaborated.

Planning

Step 1: Specifying the Audience. The iSHEM-CMM is intended to assist construction firms to improve their SHE management. The expected audience of the model is thus construction firms.

Step 2: Defining the Aim. The purpose of the iSHEM-CMM is to assist construction companies to improve SHE performance in the construction sector. The aim of the maturity model is, therefore, to assist construction firms to assess their current SHE management maturity to facilitate continuous improvement.

Step 3: Clarifying the Scope. While some maturity models are designed for generic purposes, others are designed for a specific domain. The iSHEM-CMM, as the name indicates, is designed to support a particular domain, which is SHE management in the construction industry.

Step 4: Defining the Success Criteria. The development of the iSHEM capability maturity model is motivated by the need for improved guidance on SHE management processes and practices in the construction industry. The most important success criteria were, therefore: (1) Usefulness for the construction industry, determined by the relevance of the domain's components, and the ability of the model to support improvement effort within SHE management; (2) Usability determined by the clarity and the syntactic quality of the model; and (3) Coverage of key iSHEM capability attributes determined by how well the maturity model covers the areas important to focus on for ensuring effective management of SHE issues in construction companies.

Development

Step 5: Selecting the Process Areas. A key element of developing a maturity model is the identification of capability areas/attributes [56,58,61]. Therefore, the development of the iSHEM-CMM involved identifying the relevant key iSHEM capability attributes and the definitions of the levels of maturity. According to Maier et al. [56], the key process areas used in developing a maturity grid can be derived from (1) the experiences in the field of the originator and by reference to established knowledge in a particular domain; and (2) a panel of experts in the domain, especially where there is limited prior literature concerning the domain.

Considering the lack of empirical work on construction SHE management capability, this study used a panel of construction industry experts after a comprehensive systematic literature review to identify potential capability attributes for achieving effective integrated SHE management in construction. This was applied as a three-pronged sequential research approach comprising: (1) a systematic literature review to identify potential integrated SHE management capability attributes and a preliminary expert verification process to ascertain the appropriateness and comprehensiveness of the identified attributes; (2) application of expert Delphi technique to generate consensus regarding the importance of the attributes; and (3) application of voting analytical hierarchy process (VAHP) to generate weights of importance based on the outcomes of the Delphi technique. Detailed description of the application of the three-pronged approach is given in Asah-Kissiedu et al. [12].

To select suitable and qualified experts for the preliminary verification of capability attributes, the guidance suggested by Hallowell and Gambatese [62] in selecting experts was followed. This included at least five years of professional experience in the construction industry, a minimum of five years' experience in SHE management, an advanced degree in construction management or other related fields (minimum of BSc.), an affiliation with a professional body and have researched areas of environmental, health and safety management in construction. In line with the criteria, twelve (12) experts were engaged for the preliminary verification process and a total of 30 experts for the three-round Delphi survey. The experts for the preliminary verification were academics with industry experience and expertise in SHE management in construction. Such people are likely to have an up to date understanding of the subject matter of the study. Therefore, they were considered useful to engage with in the preliminary verification exercise prior to the Delphi survey. The experts involved in the Delphi survey were industry professionals.

Each of the Delphi rounds took three weeks, spanning a three-month duration. From the verification process and the Delphi rounds, the views of the experts regarding the capability attributes were collated and analysed. An agreement on 20 iSHEM capability attributes was then obtained (refer to Section 4). A detailed account of the derivation of the attributes from the aforementioned methods is reported in Asah-Kissiedu et al. [12]

Step 6: Formulating the Maturity Levels and Descriptors. The literature shows that CMMs commonly used five maturity levels [56,61,63], which aligns with the original CMM by Paulk et al. [39] Similarly, in this study, five capability maturity levels (i.e., Level 1 being the lowest maturity level and Level 5 being the highest maturity) were adopted as shown in Table 1. Capability maturity level definitions and characteristics were abstracted from the literature review and refined through expert review. In line with the guidelines by Maier et al. [56], the maturity level descriptors at the extreme ends (i.e., Level 1 being the lowest maturity level, and Level 5, being the highest maturity) were formulated based on the underlying notion of what represents maturity for each attribute. In capability maturity modelling, lower levels of maturity are used as the basis for achieving higher levels of maturity. For instance, for a construction firm to reach capability Level 5 or full maturation in a capability attribute, it should have met the requirements for the lower levels. As a result, each level is defined and characterised clearly, thus allowing companies to self-evaluate their level of maturity. It is therefore important to understand what these capability maturity levels represent in practice, as they are fundamental to assessing the capability maturity of a company. Shown below in Table 1 are the capability maturity levels and their definitions.

Step 7: Formulating the Cell Texts (i.e., maturity level descriptors). This decision point represents the intersection of the key process area (i.e., the capability attributes) and the capability maturity levels. Attribute characteristics, thus, need to be described at each level of maturity. This decision point is recognised as a significant step in developing a maturity model assessment [56]. To be able to formulate cell descriptors that are precise, concise, and clear, three considerations are described by Maier et al. [56]: (1) using a top-down or bottom approach; (2) consideration of the information source; (3) consideration of the formulation mechanism. The top-down approach involves the writing of definitions before measures or a set of practices are developed to fit the definitions, while the bottom approach involves the determination of measures before definitions are written to reflect the measures [56]. Since integrated SHE management in construction is a relatively new field in maturity model applications, not much evidence is available for what is thought to represent maturity. Consequently, a top-down approach was deemed appropriate for formulating the cell texts since this approach places emphasis first on what maturity is before how it can be measured [56]. Again, this approach was used because of the lack of empirical work on integrated SHE management capability.

Table 1. Capability maturity levels and definitions.

Capability Level	Definition
Level 1	There are no structured processes and procedures in place. Performance is consistently poor.
Level 2	Organisational processes and procedures may exist but are usually ad-hoc and unstructured. Procedures and processes are not defined. Performance is fair.
Level 3	Organisational processes and procedures are formal and defined. Process and procedure are reactive. Performance is mostly good.
Level 4	Organisational procedures and processes are planned, well-defined, proactive and generally conform to best practices. Performance is very good and consistently repeated.
Level 5	Organisational processes and procedures are standardised, fully integrated throughout the organisation, and continually monitored, reviewed for continuous improvement. Performance is exemplary and comparable to best in the industry.

In establishing what represents maturity in each of the key process areas (i.e., SHE management capability attribute) in this study, the underlying notion of maturity was obtained by reviewing various sources, including extant literature relating to the key process areas, feedback from future recipients of the model (through an expert verification), existing capability maturity models and best practice guides on subjects related to SHE management capability attributes. Therefore, existing capability maturity models like the UK Coal Journey Model by Foster and Hoult [17] and Risk Management Maturity Model (RM3) by the Office of Road and Rail, and Health and Safety Maturity [63] were reviewed to obtain the underlying notion of maturity for each of the SHE capability attributes. In summary, the cell texts were formulated using:(1) The underlying rationale of maturity of each capability attributes; and (2) The identification and the descriptions of the best and worst practices at the extreme ends of the scale (i.e., Level 1 and Level 5), such that Level 1 represented no or very low maturity and Level 5 represented the highest level of maturity which is also presented by reviews within the capability maturity model literature to ensure continuous improvement. Secondly, the other cell descriptors in between (i.e., levels 2, 3 and 4) were also deduced from the underlying notion and formulated accordingly. In the end, the model was developed with a fraction full version presented in the results section and the full version in the Appendix A, Table A1.

Step 8: Defining the Administration Mechanisms. The iSHEM-CMM was developed as a stand-alone model and targeted for application in several construction firms. Following the formulation of cell texts, the developed model and an evaluation questionnaire were sent to selected experts to further verify the model.

Evaluation

Step 9: Validating the Model. Once the iSHEM-CMM was populated, it was evaluated by construction professionals to ensure the practical utility of the model. A detailed description of the model validation is presented in Section 5.2.

Maintenance

Step 10: Documenting, communicating and maintaining the model. The purpose of the maintenance phase is to keep the final maturity model and, its elements or attributes current. Continued accuracy and relevance of the model can be ensured by its end-users during this phase. For the iSHEM-CMM, communication is in part secured through this paper.

4. Results and Discussion

The experts for the preliminary verification and Delphi survey were experts who had knowledge and experience in SHE management in the construction industry. Each of the experts is affiliated with at least one professional body, which includes: Chartered Institute of Building, Institute of Environmental Management and Assessment, Institution of Occupational Safety and Health, International Institute of Risk and Safety Management, Association of Project management, Ghana Institution of Construction, Ghana Institute of Safety and Environmental Professionals and Ghana Institute of Surveyors. The years of experience in SHE management in construction are between 5 and 17 years. The experts engaged in the study were suitable as their experience and roles relate to SHE management in construction.

From a systematic literature review, twenty-seven (27) potential capability attributes were identified. At the end of the verification process and the subsequent three-round Delphi survey, 20 integrated SHE management capability attributes were finally obtained and subsequently categorised, based on their relatedness, into five thematic areas of SHE management capability. The five thematic categories are strategy, people, process, resources, and information. The categorisation of the capability attributes is consistent with the concept of organisational capability maturity, although specific to integrated SHE management [39,64–66]. Detailed descriptions of the thematic categories and the various attributes within are presented in Table 2. The emergent iSHEM capability attributes were similar to some of the key process areas/capabilities/criteria used in existing capability maturity grids and models. For example, Goggin's [53] Health and safety maturity model for health and safety in construction proposed attributes such as management commitment, safety policy, hazard identification, resources, reporting and control, and worker involvement and commitment. The UK coal journey maturity model by Foster and Hoult [17] also included attributes such as policy and commitment, training and competence, communication and consultation, documents and operations control, incident investigations, and monitoring and auditing.

Furthermore, the design safety capability maturity model for the offshore sector by Strutt et al. [46] included attributes like education and training, research and development, organisational learning, and managing of safety in the supply chain, while the safety culture maturity model by HSE [67] also included attributes including 'training', 'management commitment and visibility', 'learning organisation', and 'safety resources'. Furthermore, the design for occupational safety and health (DfOSH) maturity grid by Manu et al. [66] proposed attributes such as DfOSH competence and training, management commitment, design risks management, physical work and ICT resources. In the building information modelling (BIM) domain in construction, Succar [68] proposed a BIM maturity matrix, which was comprised of capability attributes such as leadership, physical infrastructure, technology (encompassing software and hardware) and human resources (comprising of knowledge, resources and skills). The iSHEM capability attributes (e.g., senior management commitment to SHE; SHE policy objectives and targets; SHE management programme; SHE risk management; Management of outsourced SHE services; Physical and financial resources; SHE incidents investigation; SHE system auditing; SHE training and processes for learning lessons and knowledge management) share similarities with the above-mentioned attributes in models by HSE [67], Strutt et al. [46], Goggin's [53], Filho et al. [69], Foster and Hoult [17] and Poghosyan et al. [61], although the iSHEM capabilities have specific relevance or focus on the implementation of an iSHEM system in construction firms.

Table 2. Verified integrated SHE management capability attributes.

Thematic Category	Attributes
Strategy, i.e., the organisation's vision and top management commitment to SHE management	Senior management commitment to safety, health and environment (SHE) management
	An integrated SHE policy that serves as the foundation for a company's SHE development and implementation
	SHE objectives and targets for a company, in line with SHE policy
	SHE management programme, i.e., company's action plans for achieving SHE objectives and targets
Processes, i.e., the organisation's procedures, processes and systems for SHE management	SHE risks management, i.e., systems, processes and procedures for SHE hazards identification, risks assessment and identification risks control strategies
	Management of outsourced services, i.e., processes and mechanisms for assessing the competence of outsourced personnel, subcontractors and suppliers with regards to the management of SHE
	SHE operational control, i.e., processes, procedures and measures for controlling SHE risks, to ensure SHE regulatory compliance in operational functions and to achieve the overall SHE objectives
	SHE emergency preparedness and responses, i.e., emergency procedures and measures to minimise the impact of uncontrolled events and unexpected incidents
	SHE performance monitoring and measurement, i.e., systems, processes and procedures to monitor and measure SHE performance to ensure compliance with SHE regulations
	SHE incidents investigation, i.e., processes and procedures for investigating the causes of SHE incidents
	SHE system auditing, i.e., processes and procedures to conduct SHE audits to assess compliance and SHE management system effectiveness
People, i.e., organisation's human capital, their roles, responsibilities, and involvement in SHE management	SHE roles and responsibilities, i.e., availability of dedicated SHE roles and responsibilities within an organisational hierarchy
	SHE Training, i.e., provision of suitable SHE training for personnel
	Employee involvement and consultation at all levels in SHE management and operations
	SHE competence, i.e., the skills, knowledge and experience of personnel to undertake responsibilities and perform SHE activities
Resources, i.e., organisation's physical and financial resources required for SHE management	Physical SHE resources, i.e., provision of physical resources for SHE implementation
	Financial resources for SHE, i.e., Provision of financial resources for SHE implementation
Information, i.e., SHE related documents, data, lessons, records and their communication across an organisation	Communications, i.e., communication of relevant SHE information and requirements to personnel and other relevant stakeholders
	SHE documentation and control, i.e., provision and maintenance of adequate SHE documentation and records
	SHE lessons and knowledge management, i.e., capturing lessons learned and knowledge acquired from historical incidents and management of SHE
	Communications, i.e., Communication of relevant SHE information and requirements to personnel and other relevant stakeholders

4.1. The iSHEM Capability Maturity Model

After the capability attributes and the capability, maturity levels were obtained and an initial iSHEM-CMM was developed. The model is a multilevel framework and offers maturity assessment on a five-level scale, within five thematic categories consisting of 20 integrated SHE management capability attributes. Table 3 shows an excerpt of the iSHEM capability maturity grid, with two iSHEM capability attributes—Senior management commitment to SHE and SHE policy and maturity levels from 1–5. Due to its large size, the full version of the iSHEM-CMM is presented in the Appendix A, Table A1 of this paper.

4.2. Validating the iSHEM Capability Maturity Model

In this section, validation of the maturity model by industry experts is presented. De Brium et al. [58] recommended that the evaluation process of a maturity model should mainly focus on the model's constructs (i.e., relevance and coverage of the domains components) and the model instruments (i.e., the reference model, performance scale and assessments procedure). In view of this, the validation process involved: (1) the use of the iSHEM by construction professionals to assess the SHE management capability of construction companies; and subsequently (2) the completion of a validation survey by the professionals. The validation survey was used to appraise both content of the maturity model (i.e., the relevance and appropriateness of the capability attributes and levels) and its usability (i.e., understandability, ease of use and practicality).

4.2.1. Selection of Companies for Validation

To ensure a broad validation of the maturity model, construction professionals, including SHE experts from 70 construction firms operating in Ghana, were invited to participate in the validation process. Fifty-nine (59) construction firms consented. The construction professionals involved included Health and Safety managers (15.3%), Project managers and construction managers (13.6%), Environmental Managers (13.6%), and Site Managers, Safety, Health or Environmental Consultants and Health and Safety Officers (11.9%). A majority of the respondents (67.8%) had over five years of professional experience. This is indicative of an experienced and knowledgeable group of construction professionals.

4.2.2. Questionnaire for Validation

To validate the capability maturity model (i.e., SHEM-CMM), an evaluation questionnaire was used as the instrument (see Appendix A-Table A2). It consisted of two sections. The first section solicited information on the respondent background details. In the second section, respondents were asked to evaluate the model based on six criteria (i.e., relevance of attributes, comprehensiveness of attributes, appropriateness, adequacy of capability maturity levels, ease of understanding, ease of use and level of usefulness and practicality). These validation criteria were similar to the survey developed by Salah et al. [70] (2014).

The validation exercise required that the construction professionals were to assess their company's SHE management capability maturity by using the developed maturity model and thereafter evaluate the capability maturity model as a whole by the six criteria on the five-point Likert scale using the levels (5) Strongly agree, (4) agree, (3) Neither agree nor disagree, (2) disagree, (1) Strongly disagree.

4.2.3. Validation Results

In total, responses were obtained from 59 construction firms operating in Ghana. The results of the survey are presented in Table 4. From the validation results, it is evident that the iSHEM-CMM is comprehensive and suitable for assessing the SHE management capability maturity of construction companies. The high rating indicates a convincing level of approval of the developed capability maturity model. Regarding the relevance and comprehensiveness of integrated SHE management capability attributes, the results confirm that the capability attributes are relevant and did cover all aspects of integrated SHE management capability in construction. Concerning the correct assignment of attributes to their respective capability levels and sufficient maturation of attributes, the validation results indicated that the construction professionals were satisfied with the accuracy of the capability attributes and its correct assignment to their respective maturity levels in the developed model.

Table 3. The iSHEM-CMM (excerpt).

She Capability Attributes	Underlying Notion of Maturity	Capability Maturity Levels				
		Level 1	Level 2	Level 3	Level 4	Level 5
Senior management Commitment	As maturity increases, senior management commitment to safety, health and environmental (SHE) management becomes unwavering, visible and well-articulated across the company	• Lack of senior management commitment to SHE management • There is no resource commitment (financial and human resources) for SHE related issues	• Limited commitment by company's senior management to SHE implementation • Limited resource commitment for SHE related issues	• Partial commitment by company's senior management to SHE implementation • Show of senior management commitment is reactive (e.g., when significant risks are anticipated or response to a major environmental impacts) • An ad hoc implementation committee is established • SHE champion is identified • There is resources commitment for SHE-related issues	• Firm commitment by company's senior management to SHE implementation. • Senior management commitment is aligned to company's policy on SHE management. • Senior management are amongst the SHE champions within the organisation. • Management commitment is well articulated across the company • Sufficient resources commitment for SHE-related issues	• There is a full, unwavering and clearly visible commitment of company's senior management to SHE implementation • Senior management continuously and visibly demonstrate their commitment to SHE and show shared values directed at continually meeting SHE objectives safely • A cross functional SHE implementation committee is established including a SHE champions and members from all key management functions of the company. • There is a ring-fenced resource commitment for SHE implementation and maintenance • Company senior manager(s) are amongst SHE management champions within the industry and are recognised as industry thought-leaders in respect of SHE management

Table 3. *Cont.*

She Capability Attributes	Underlying Notion of Maturity	Capability Maturity Levels				
		Level 1	Level 2	Level 3	Level 4	Level 5
She Policy	As maturity increases, company SHE policy becomes explicitly stated, well-communicated within the organisation, and interpreted and applied consistently by all managers/supervisors and staff.	• No policy statement on SHE management	• SHE policy statement is outdated and vaguely worded • SHE policy does not meet legal requirements and employees are rarely involved in its development • Policy has not been communicated within the company and documented	• SHE policy statement is clear, setting out the intention(s) on how SHE is managed, tracked and reported • Policy meets majority of legal requirement with some employees actively involved in its development • Policy is communicated across different levels of the company, but management or supervisors and employees have inconsistent interpretations and applications of the policy • 3Policy statements are poorly documented and not displayed at workplace	• SHE policy is clear, comprehensive and well-defined, setting out the intention on SHE • SHE policy presents a clear approach to managing SHE including the required accountability and responsibility for managing SHE • SHE policy meets all the legal requirements and other requirements the company subscribes to • More relevant employees are actively involved in SHE policy formation and strategy formulation • SHE policy is actively communicated within the company and to other stakeholders • Policy is accepted, understood and consistently interpreted and applied in the same way by all manager's or supervisors and employees • SHE policy is formally documented, displayed at the workplace and is available to all stakeholders	• There is a clear policy on SHE management, setting out intention(s) on SHE management and recognising that SHE implementation is not a separate task but an integral part of the organisation SHE activities • All relevant people are engaged in SHE policy formation as wells as SHE strategy formulation, with clear actions, and accountabilities and targets • Documented policy is in place, consistent with other best-performing organisation's policies, communicated and readily available to all stakeholders • SHE policy is periodically reviewed to ensure that it remains relevant to the company, reflect industry best practices and demonstrate effectiveness and continuous improvement

Table 4. Summary of responses feedback for maturity model evaluation.

Assessment Criteria	Evaluation Response (%) ($n = 59$)						
	Strongly Agree	Agree	Neither Agree nor Disagree	Disagree	Strongly Disagree	Total (%)	Median/Mean/Standard Deviation
Attributes used in the SHEM-CMM worksheet							
Attributes are relevant to SHE management capability.	35.6	62.7	1.7	0	0	100	4.00/4.34/0.51
Attributes cover all aspects of SHE management capability.	20.3	62.7	16.9	0	0	100	4.00/4.03/0.62
Attributes are correctly assigned to their respective capability level.	15.6	71.2	13.6	0	0	100	4.00/4.02/0.54
Attributes are clearly distinct.	40.7	50.8	8.5	0	0	100	4.00/4.32/0.63
Capability maturity levels							
The capability levels sufficiently represent maturation in the attributes.	18.6	69.5	8.5	3.4	0	100	4.00/4.03/0.64
There is no overlap detected between descriptions of maturity levels.	6.8	52.5	27.1	13.6	0	100	4.00/3.53/0.82
Ease of understanding							
The capability levels are understandable	33.9	61	5.1	0	0	100	4.00/4.29/0.56
The documentations (i.e., assessment instructions) are easy to understand	13.6	71.2	11.9	3.4	0	100	4.00/3.95/0.63
The results are understandable	13.6	79.7	6.8	0	0	100	4.00/4.07/0.45
Ease of use							
The scoring scheme [i.e., drop-down options for maturity levels (1–5)] is easy to use	39	57.6	1.7	1.7	0	100	4.00/4.39/0.61
The SHEM-CMM is easy to use	18.6	71.2	8.5	1.7	0	100	4.00/4.07/0.58
Usefulness sand practicality							
SHEM-CMM is useful for assessing SHE management capability	49.2	47.5	3.4	0	0	100	4.00/4.46/0.57
SHEM-CMM is practical for use in industry	28.8	64.4	6.8	0	0	100	4.00/4.22/0.56

Furthermore, the results indicated that the majority of the construction professionals were of the opinion that capability levels, supporting documentations and the results were easy to understand. Additionally, the iSHEM-CMM was found to be easy to use, useful for assessing SHE management capability and practical for use in the construction industry by the majority of the construction professionals, particularly the ease of using the Microsoft Excel format of the maturity model and the user-friendly nature of the scoring scheme (i.e., drop-down options for capability levels) during the assessment. Based on the overall results of the validation exercise, the developed integrated SHEM-CMM was generally well-received by practitioners in the industry.

5. Conclusions

Efficient management of SHE issues has become of utmost importance for construction firms globally. Construction firms need to have the appropriate capability in terms of SHE to effectively minimise injuries, illness and negative environmental impacts through an integrated SHE management framework. Construction firms would have differing iSHEM capabilities, and it is important they understand their current iSHEM capability depth so that they can continuously improve. Similarly, it is vital that construction clients, consultants or other institutions engaging the services of construction firms are also able to evaluate the iSHEM capability of those organisations. This study adopted the capability maturity modelling concept to develop an integrated safety, health and environmental management capability maturity model for construction firms. This study addressed a significant research gap relating to iSHEM capability by identifying 20 distinct capability attributes and presenting empirical work on developing an iSHEM capability maturity model to facilitate assessments and improvement of integrated SHE management practices. The maturity model shows five maturation levels in distinct iSHEM capability attributes drawn from literature review and a Delphi survey with a panel of SHE experts. To ensure the usefulness of the maturity model in practice, the model was further improved by SHE expert verification and refinement and validated by construction professionals, including SHE experts. The findings reveal that the developed model is fit and is capable of assessing the iSHEM capability of construction firms, managing components of an iSHEM system effectively, and improving construction iSHEM capability maturity levels and management practices. The developed maturity model considers the main tasks of an iSHEM system, the underlying processes, strategies, resources, information and the people using the iSHEM system. It is anticipated that the developed iSHEM-CMM would be beneficial to construction firms and other industry stakeholders by undertaking self-assessment of their iSHEM capability to better understand their iSHEM and implement actions needed to improve it.

5.1. Implications of the Research

This study's main implication is that the developed model provides a means for construction companies to evaluate their SHE management capability systematically. This would enable them to ascertain the areas of strength and deficiency in respect of their SHE capability. On the basis of SHE management capability self-assessment, construction companies could prioritise their investments and target efforts at addressing any identified areas of capability deficiency to ensure continuous improvements and avoid sub-optimisation. Moreover, this model serves as a management tool that allows SHE management consultants to evaluate their construction clients firm's current SHE capability maturity and provide guidance on how they can improve their SHE management practices and processes. Additionally, the identified iSHEM capability attributes could be used by construction clients (including government agencies) as part of the SHE management criteria for selecting companies to undertake projects.

Several maturity models have been developed in relation to safety culture in many domains; however, until now, no maturity model has been developed that has attempted to take the maturity perspective into integrated SHE management practices in construction. The output of this study (i.e., the iSHEM-CMM) therefore offers a basis for similar capability maturity-based research with a focus on integrated SHE management capability in other industries.

5.2. Limitations and Recommendation for Further Studies

The research has limitations which are highlighted in this section. The study was based on professional views of SHE management experts and other practitioners within the Ghanaian construction industry; therefore, findings may be peculiar to SHE management in the Ghanaian construction industry. Further studies could be conducted among SHE management experts and practitioners in the construction industry of other countries to enable an appropriate comparison to be done. Another limitation identified lies in the fact that the development of the integrated SHEM-CMM focused on the construction industry alone. This may tamper with its immediate applicability to other industrial sectors. Future studies could be conducted to develop a similar model for other industries other than the construction industry. This can improve the SHEM of such industries as well.

Furthermore, another potential limitation relates to the sample size used to validate the maturity model. Available guidance for testing CMM using an expert evaluation approach [70] does not specify the minimum number of experts. Nonetheless, for expert group techniques, the recommended number of experts range from 8–12 (e.g., Delphi Technique [66]). This is because in an expert group technique, the focus tends to be on the depth of knowledge of the experts rather than the breadth of participation, i.e., the number of experts [66]. Therefore, in this study, the number of experts that were involved in the CMM can be deemed to be adequate. Notwithstanding, future studies could adopt alternative methods, e.g., large cross-sectional surveys to test the capability model.

Author Contributions: Conceptualization—M.A.-K., P.M., C.A.B. and A.-M.M.; methodology—M.A.-K., P.M., C.A.B. and A.-M.M.; validation—M.A.-K., P.M., C.A.B. and A.-M.M.; formal analysis—M.A.-K.; investigation—M.A.-K.; writing—original draft preparation—M.A.-K. and K.A.; writing—review and editing—M.A.-K., K.A., P.M., C.A.B. and A.-M.M.; supervision—P.M., C.A.B. and A.-M.M.; project administration—M.A.-K., P.M., C.A.B. and A.-M.M.; funding acquisition—M.A.-K. and P.M. All authors have read and agreed to the published version of the manuscript.

Funding: Appreciation is extended to the Commonwealth Scholarship Commission (Reference No: GHCS-2016-147) for funding this research.

Institutional Review Board Statement: This study was granted ethical approval by the research ethics committee of the Faculty of Environment and Technology, University of the West of England (UWE), Bristol, UK, (reference # FET.17.11.013; date of approval 12 January 20218). Further information about UWE Bristol's research governance and policies is available at https://www.uwe.ac.uk/research/policies-and-standards, accessed on 4 October 2021.

Informed Consent Statement: Informed consent was obtained from all subjects involved in the study.

Data Availability Statement: All data are a part of the manuscript.

Conflicts of Interest: The authors declare no conflict of interest.

Appendix A

Table A1. Integrated safety, health and environmental management capability maturity model (iSHEM-CMM).

She Capability Attributes	Integrated Safety, Health and Environmental Management Capability Maturity Model (iSHEM-CMM)				
	Capability Levels				
	Level 1	Level 2	Level 3	Level 4	Level 5
Senior management Commitment	• Lack of senior management commitment to SHE management • There is no resource commitment (financial and human resources) for SHE related issues	• Limited commitment by company's senior management to SHE implementation • Limited resource commitment for SHE related issues	• Partial commitment by company's senior management to SHE implementation • Show of senior management commitment is reactive (e.g., when significant risks are anticipated or response to a major environmental impacts) • An ad hoc implementation committee is established • SHE champion is identified • There is resources commitment for SHE related issues	• Firm commitment by company's senior management to SHE implementation • Senior management commitment is aligned to company's policy on SHE management • Senior management are amongst the SHE champions within the organisation • Management commitment is well articulated across the company • Sufficient resources commitment for SHE-related issues	• There is a full, unwavering and clearly visible commitment of company's senior management to SHE implementation • Senior management continuously and visibly demonstrates their commitment to SHE and show shared values directed at continually meeting SHE objectives safely • A cross functional SHE implementation committee is established including a SHE champions and members from all key management functions of the company • There is a ring-fenced resource commitment for SHE implementation and maintenance • Company senior manager(s) are amongst SHE management champions within the industry and are recognised as industry thought-leaders in respect of SHE management
She Policy	• No policy statement on SHE management	• SHE policy statement is outdated and vaguely worded • SHE policy does not meet legal requirements and employees are rarely involved in its development • Policy has not been communicated within the company and documented	• SHE policy statement is clear, setting out the intention(s) on how SHE is managed, tracked and reported. • Policy meets majority of legal requirement with some employees actively involved in its development • Policy is communicated across different levels of the company, but management or supervisors and employees have inconsistent interpretations and applications of the policy. • 3Policy statements are poorly documented and not displayed at workplace	• SHE policy is clear, comprehensive and well-defined, setting out the intention on SHE • SHE policy presents a clear approach to managing SHE including the required accountability and responsibility for managing SHE • SHE policy meets all the legal requirements and other requirements the company subscribes to • More relevant employees are actively involved in SHE policy formation and strategy formulation • SHE policy is actively communicated within the company and to other stakeholders • Policy is accepted, understood and consistently interpreted and applied in the same way by all managers or supervisors and employees • SHE policy is formally documented, displayed at the workplace and is available to all stakeholders	• There is a clear policy on SHE management, setting out intention(s) on SHE management and recognising that SHE implementation is not a separate task but an integral part of the organisation SHE activities • All relevant people are engaged in SHE policy formation as well as SHE strategy formulation, with clear actions, and accountabilities and targets • Documented policy is in place, consistent with other best-performing organisation's policies and is communicated and readily available to all stakeholders • SHE policy is periodically reviewed to ensure that it remains relevant to the company, reflects industry best practices and demonstrates effectiveness and continuous improvement

Table A1. *Cont.*

	Integrated Safety, Health and Environmental Management Capability Maturity Model (iSHEM-CMM)				
She Objectives and Targets	• No formal SHE objectives and targets identified and documented	• SHE objectives and targets are vaguely worded and not based on any baseline review of the company's SHE operations. They are not 'specific, measurable, attainable, relevant and timely (SMART) and prioritised. • People in relevant functional area(s)are not involved in setting SHE objectives and targets • Objectives and targets not included in critical tasks or role descriptions of employees • SHE objectives and targets are poorly documented and not communicated to employees and other stakeholders	• SHE objectives and targets are defined, formal, based on a baseline review and consistent with SHE policy and applicable legal and other regulatory requirements • Some SHE objectives and targets may be SMART and prioritised. • Some people in relevant functional areas(s) are involved in setting objectives and targets • Objectives and targets are rarely included role descriptions of employees • SHE objectives and targets are somewhat documented and informally communicated to employees and relevant stakeholders within the company	• SHE objectives and targets are formal, well defined, mostly SMART, and consistent with SHE policy and applicable legal and other regulatory requirements • More people in relevant functional areas (s)are involved in setting SHE objectives and targets • Objectives and targets are included role descriptions of employees • Objectives and targets are properly documented and formally communicated to all relevant functions across the company	• SHE objectives and targets are clear, SMART, prioritised and aligned to the overall SHE policy and focused towards continually improving SHE performance. • All relevant people are involved in setting SHE objectives and targets • Objectives and target are included in critical tasks or role descriptions of employees • SHE objectives and targets are adequately documented, monitored, routinely reviewed and updated to ensure continuous improvement.
She Management Programme	• There are no clearer or well defined SHE management programme(s) for achieving objectives and targets.	• SHE plans and programme(s) are available but without a clear definition of specific responsibilities and the time frame. • Little involvement of employees in establishing SHE plans and programme(s)	• Formal and detailed management plans and programme(s) are available • Key responsibilities, tactical steps, resources needed and schedules are clearly defined to achieve SHE objectives and targets • More involvement of employees in establishing SHE programmes	• SHE management plans and programme(s) are adequate, more detailed and integrated with company objectives, strategies and budgets • Greater number of employees' involvement in establishing SHE programmes • SHE plans and programme(s) are clearly communicated to all who need to know	• SHE management plans and programmes are dynamic and integrated with company's SHE planning strategies • Full involvement of employees and other stakeholders in establishing SHE programmes • SHE management programmes are continuously reviewed and modified to address changes to company's operations for continuous improvement of SHE programmes
She Risk Management	• No processes and procedures for SHE hazards identification, risk assessment and control	• Informal processes and procedures for SHE hazards identification and risk assessments are in place • Risk control measures are poorly defined, understood and have limited application • SHE risks assessments and management are poorly documented	• Formal processes and procedures for SHE hazards identification and risk assessment are in place • Processes and procedures for identification and management of SHE risks, focuses on the most significant and obvious SHE risks • SHE risks assessments are carried out in isolation • Risk control measures are somewhat defined and used to reactively managed identified SHE risks • Most important SHE risks assessment activities and plans are documented	• Formal, more detailed and proactive processes and procedures for SHE hazards identification and risk assessment • Processes and procedures for identification and management focusses on specific, hazards and risks, including less obvious and immediate risks • Processes and procedures are consistently applied to identify and manage SHE risks • SHE risks control measures are well defined, understood and implemented in a consistent manner • All levels of SHE employees and other stakeholders can contribute to risks assessments • Appropriate SHE risks assessment records are accurately documented and maintained	• Well-defined processes and procedures for SHE risks management are in place and practicable • SHE risk management processes and procedures are embedded into company's SHE planning activities and considered as a core measure of operational excellence • The approach to SHE risks assessment are routinely applied consistently throughout the company in a pragmatic manner to drive continual improvement in the SHE risks profile of the company

Table A1. *Cont.*

	Integrated Safety, Health and Environmental Management Capability Maturity Model (iSHEM-CMM)				
She Capability Attributes	**Capability Levels**				
	Level 1	**Level 2**	**Level 3**	**Level 4**	**Level 5**
				• Processes and plans for SHE risks management are modelled on best practice risks assessment standards, e.g., ISO 31000	• SHE risks management processes, procedures and control measures are monitored, reviewed and improved on a regular basis to address changing circumstances and to ensure continuing success
Management of Outsourced Personnel	• No structured procedure is used in appointing competent outsourced employees, subcontractors and suppliers with regards to the management of SHE • No structured monitoring and assessment of the performance of outsourced employees, subcontractors and suppliers	• Informal procedure in place but rarely used in appointing competent outsourced SHE employees, subcontractors and suppliers • Rare monitoring and assessment of the performance of outsourced employees, subcontractors and suppliers in respect of SHE management • Procedures are poorly documented and maintained	• Formal procedures in place and used occasionally and reactively appointing competent outsource employees, subcontractors and suppliers. • Occasional and reactive assessment of the performance of outsourced employees, subcontractors and suppliers in respect of SHE management • Procedures are adequately documented and maintained	• Regular and proactive procedures are in place for appointing competent outsource employees, subcontractors in a consistent manner • Regular and proactive assessment of the performance of outsourced employees, subcontractors and suppliers in respect of SHE management • All competency definitions are explicitly defined and include industry recognised best practice • Procedures are accurately documented and maintained	• There is a well-structured procedure for appointing, monitoring and assessing the performance of outsourced personnel, subcontractors and suppliers • The well-structured and clear competence management system is integrated within the company's performance of SHE management. Competence and performance assessment procedures are reviewed regularly to ensure their current suitability and continuous improvement.
She Operational Control	• No procedures for identification of SHE operations that need to be controlled to ensure risk associated with them are minimised or eliminated • SHE risks control measures are not in place	• Informal procedures are in place for identification of SHE operations and activities that need to be controlled to ensure risk associated with them are minimised or eliminated • SHE controls measures, are unclear and poorly documented	• Formal procedures are in place for the identification of SHE operations and activities that need to be controlled • Control measures for identified SHE risks are more detailed and clearly stated • Operation control procedures and measures are adequately documented	• Formal and comprehensive procedures are in place for the identification of SHE operations and activities that need to be controlled • Control measures for identified SHE risks are comprehensive and well defined • Identified SHE operations that needs to be controlled and their associated control measures are appropriately documented and well communicated to relevant employees (e.g., suppliers, contractors and other interested parties)	• Well-structured procedures are in place for identification of SHE operations and activities that need to be controlled to ensure compliance, and to achieve objectives • Documented SHE control procedures and measures are continually reviewed and improved
She Emergency Preparedness and Response	• No emergency preparedness and response (EPAR) procedures • No measures for identification of possible emergencies and SHE accidents, and how to respond if they arise	• Undefined and inappropriate EPAR procedures and measures for identification of possible emergencies and SHE accidents, and how to respond if they arise • EPAR procedures and measures are poorly documented and are not accessible • Employees are rarely trained in emergency responses	• Defined procedures and measures are available for identification of possible emergencies and SHE accidents, and how to respond if they arise • EPAR procedures and measures are adequately documented but are not easily accessible • Employees are trained in formal emergency responses	• Well-defined and sufficient EPAR procedures and measures for identification of possible emergencies with a focus on specific emergency situations • EPAR procedures and measures are appropriately and accurately documented • EPAR procedures and measures are communicated and accessible to all employees involved • Employees are adequately trained in emergency responses	• Appropriate and comprehensive EPAR plans, procedures and measures are in place to effectively respond to emergency situations • EPAR plans and procedures are fully integrated with other control measures and benchmarked consistently against best practices • EPAR plans are periodically tested for the adequacy of the plan and the results reviewed to improve its effectiveness for continuous improvement

Table A1. *Cont.*

Integrated Safety, Health and Environmental Management Capability Maturity Model (iSHEM-CMM)

She Capability Attributes	Capability Levels				
	Level 1	Level 2	Level 3	Level 4	Level 5
She Performance Monitoring and Measurement	• No performance measuring and monitoring system in place • SHE procedures for performance monitoring and measurement (MaM) are not well developed • SHE performance indicators and measures are not established • SHE system performance is poor	• There are vague procedures for MaM of SHE performance • Some SHE performance indicators and measures are in place but are not well defined • Performance MaM are rarely undertaken • Some employees are aware of the SHE performance measures in their areas of responsibilities • SHE system performance is fair	• SHE performance MaM procedures and performance indicators and other measures are in place and defined • Performance MaM are undertaken occasionally. • Monitoring is reactive • More employees are aware of the SHE performance measures in the areas of responsibilities • SHE system performance is mostly good	• Well-defined and appropriate performance procedures, key SHE performance indicators and other measures are in place to monitor SHE performance • Performance monitoring and measurement are undertaken regularly with the purpose of improving the SHE system • Performance MaM procedures and measures are compliance led and used to track SHE performance • MaM procedures and measures are adequately documented and communicated to all employees • Employees at all levels are aware of the critical SHE performance measures in their areas of responsibility. • SHE system performance is very good and is constantly repeated	• Well-designed and defined proactive procedures and measures for monitoring, measuring and recording of SHE performance on a regular basis is in place and is institutionalised within the company, focusing on operational excellence and continuous improvement • Results of SHE performance MaM are documented and effectively communicated throughout the company, to facilitate subsequent corrective and preventive actions analysis • SHE performance MaM procedures and measures are continuously used to improve the SHE management system. Best practice is shared across the entire company. • SHE performance of the MaM system is periodically reviewed and improved to make sure they remain relevant to the company's risk profile • SHE system performance is exemplary and comparable to the best in the industry
She Incidents Investigations	• No structured processes and procedures for SHE incidents investigations • No organised evidence of SHE investigations	• Vague processes and procedures for SHE incident investigations are in place • The range of incidents investigated is limited to immediate causes of accidents and environmental aspects • Limited employees' involvement • SHE investigations processes and procedures are not documented	• Formal processes and procedures for SHE incident investigations are in place • Investigations tend to focus on the immediate and root causes of SHE incidents, near misses and environmental aspects and their impacts • Incident investigations tend to be reactive • More employees' involvement in SHE investigations. • SHE incident investigations processes and procedures are somewhat documented	• Formal comprehensive and standard processes and procedures for SHE incident investigations • Incident investigations are proactive and probe more deeply to identify direct and indirect causes of SHE incidents and environmental aspects that result in significant SHE risks • Greater employees' involvement in SHE incidents investigations • SHE incidents investigations procedures are communicated to relevant committees for appropriate recommendations and actions • SHE investigation processes and procedures are well documented and corrective actions well communicated to best utilise any lessons to be learned.	• There are documented structured processes and procedures in place for consistently high quality SHE incident investigations • SHE incident investigation procedures are linked to SHE hazard identification and risk a mitigation process and are institutionalised within the company • Outcomes of SHE incidents investigations are seen as opportunities for improvement, and are documented, monitored and shared with industry. SHE incident trends are used to identify and help manage SHE risks • Lessons learned from incidents investigations are shared and implemented across the company. • Corrective and preventive actions are reviewed regularly and updated to ensure actions taken are effective. • SHE incidents investigations procedures are routinely reviewed and updated to drive continuous improvement

Table A1. *Cont.*

	Integrated Safety, Health and Environmental Management Capability Maturity Model (iSHEM-CMM)				
She Capability Attributes	Capability Levels				
	Level 1	Level 2	Level 3	Level 4	Level 5
She System Audits	• No auditing of SHE system • No clear SHE audits processes and procedures	• Company rarely undertake planned SHE system audits. Ad hoc audit with no follow up. • SHE audits processes and procedures are not defined and may not be documented. • Procedures for assessing SHE compliance is limited • Legal and regulatory obligations noncompliance	• Company occasionally undertakes planned SHE system audits • SHE audit processes and procedures are somewhat defined and poorly documented • Most aspects of the SHE system are audited with some follow-up • Minimal legal and regulatory compliance • SHE audits processes and procedures are focused on achieving compliance with legal and regulatory obligations	• Company regularly undertakes planned SHE audits • SHE audits processes and procedures are well defined and designed, and modelled on best practice of audits • All aspects of SHE system audited with some follow-up • Total legal and regulatory obligations compliance Written recommendations, (e.g., non-compliances) are well documented and communicated to form the basis of SHE improvement and innovation. • SHE audits processes and procedures are modelled on best practice standards for auditing management system, e.g., ISO 19011:2018 guidelines for auditing management systems, OHSAS 18001:2007	• There is a company-wide standardised audit system in place and institutionalised within the company, with best practice shared internally with other functions of the company • SHE audits are undertaken regularly by competent employees to demonstrate compliance with required standards, legal and regulatory obligations. • SHE audits processes and procedures are planned and prioritised, and covers all aspects of the SHE system. • SHE audits process and procedures are reviewed periodically to ensure they are current and consistent with leading internal audit practice and standard requirements in order to ensure continuous improvement in audit processes
Roles and Responsibilities for She	• No clear SHE roles, and responsibilities (i.e., there are no roles, tasks and objectives given to people and teams to meet the organisation's SHE objectives)	• SHE roles and responsibilities are unclear with some specific responsibilities and authorities somewhat defined and developed. • SHE roles and responsibilities are not recorded in job descriptions	• SHE roles and responsibilities are mostly defined and assigned to employees • SHE roles and responsibilities are inconsistently recorded in job descriptions	• SHE roles and responsibilities are well defined, sufficiently comprehensive and well communicated to designated employees at all levels • All SHE roles and responsibilities are consistently recorded in key documentation (e.g., job descriptions) and appropriate communication media	• Clearly defined SHE roles, responsibilities and authorities at all levels of the company • SHE roles and responsibilities are unambiguous, clearly understood and accurately documented • SHE roles, responsibilities and authorities are continuously reviewed, realigned to effort and tracked to ensure proper distribution and continuous improvement
She Training	• No provision of SHE-related training for employees • No formal training needs analysis undertaken	• Provision of SHE-related training for employees is very low and unplanned. Provision of SHE training is rarely informed by a formal training needs analysis • Training needs are not well defined and documented	• Provision of SHE-related training is reactive • Provision of SHE training is occasionally informed by a formal training needs analysis • Identified training needs are somewhat defined and based on the wider competency and performance objectives • Training needs adequately documented	• Regular provision of adequate SHE-related training for employees, informed by a formal and objective training needs analysis undertaken on a regular basis • Training is typically based on employees SHE roles and respective competency objectives • Training needs are well defined and accurately documented (e.g., in the employees' personal files)	• Appropriate and timely SHE training is in place and is integral to company's human resource strategy to improve SHE performance • SHE training strategies are incorporated into the company's overall, SHE management strategies and policies • SHE-related training programmes or plans are reviewed for their effectiveness and are periodically reviewed to ensure their current suitability. • SHE-related training programmes and training are continuously assessed and updated to reflect organisational, regulatory changes and any other changes in technology and techniques, to allow continuous learning and improvement

Table A1. *Cont.*

261

She Capability Attributes	Integrated Safety, Health and Environmental Management Capability Maturity Model (iSHEM-CMM)				
	Capability Levels				
	Level 1	Level 2	Level 3	Level 4	Level 5
				• Training is usually proactive, tracked and evaluated to be improved upon	• The various training methods are incorporated into the knowledge and communication channels of the company • Training needs analysis procedures are regularly reviewed
Employee Involvement In She	• No consultation of employees on SHE-related issues • Employees are not involved and have no interested in participating in SHE related issues	• Limited consultation on SHE-related issues, but not carried out in a systematic way. • Minority of the employees are involved and interested in participating in SHE-related issues	• More consultation on SHE issues is carried out in a systematic way • Majority of the employees are involved and interested in participating in SHE-related issues	• All employees are regularly consulted on SHE-related issues and are carried out in a range of ways (e.g., surveys, workshops, site meetings and committees) • Overwhelming majority of the employees are involved and interested in participating in SHE-related issues • Employee involvement and consultation arrangements are documented and interested parties are informed	• All employees are fully consulted and actively engaged in SHE related issues at all company's levels. • All employees are interested in participating SHE related issues • Company use employee involvement to gather ideas for improvement on SHE issues • Company makes full use of employees' potential to develop shared values and a culture of trust, openness and empowerment
She Competence	• Company's employees do not have the skills, knowledge and the experience necessary for SHE management	• An overwhelming majority of company's employees have basic SHE knowledge and skills, with no employees having advanced or expert skills and knowledge • Company's employees have limited experience in SHE management tasks	• A majority of company's SHE employees have intermediate SHE skills and knowledge with very few having advanced and/or expert skills and knowledge • Company's employees have some experience in SHE management tasks	• A majority of company's employees have sufficient and advanced SHE skills, and knowledge with very few having basic or no SHE skills and knowledge • Company's employees have appropriate experience in SHE management tasks	• An overwhelming majority of company's employees have expert SHE skills and knowledge with very few or none having basic or no SHE skills and knowledge • Company's employees have vast and experience in SHE management tasks • Company's employees feel competent and capable to perform their SHE tasks.
Physical She Resources	• No physical resources available to enable SHE employees to perform SHE-related tasks.	• Company is ill-equipped with physical resources for employees to perform SHE-related tasks. • Physical SHE resources are limited • Resource provision is not or rarely informed by any strategic resource plan	• Company is equipped with adequate physical SHE resources to enable employees to perform SHE-related tasks. • Resource provision is usually reactive and occasionally informed by a strategic resource plan	• Company is well equipped with sufficient physical resources for employees to perform SHE-related tasks. • A strategic resource plan is available to inform timely provision of physical resources to enable employees to perform SHE-related tasks	• Company is fully equipped with sufficient resources in quality and quantity for employees to perform SHE-related tasks • Company's SHE physical resources are considered to be integral to SHE performance and competitiveness • Physical resources are continuously tested, upgraded and deployed • Resource plans for provision of physical resources are documented and integrated into company's processes and systems to improve effectiveness and efficiency • Resource plans are regularly reviewed to ensure the provision of adequate and current resources to meet planned and agreed targets and objectives

Table A1. Cont.

	Integrated Safety, Health and Environmental Management Capability Maturity Model (iSHEM-CMM)				
Financial Resources For She	• No financial resources for SHE implementation. • Unstable or uncertain funding	• Limited financial resources for SHE implementation and rarely informed by a strategic resource plan • No established sources of funding	• Company has adequate financial resources for SHE implementation • Provision of financial resources is occasionally informed by strategic resource plan • Established source of funding	• Company has sufficient and well organised funding lines for SHE implementation. • A strategic resource plan is available to inform timely provision of financial resources for effective SHE management • Stable sources of funding	• Dedicated and adequate financial resources in place for effective SHE implementation and considered to be an integral part of the company's finance plan • Highly stable funding. Resource plans are regularly reviewed to ensure the provision of adequate and current resources to meet planned and agreed targets and objectives
She Communications	• No formal communication of any SHE-related issues to employees • No formal communication channels for effective flow of SHE information internally and externally in the company	• Limited communication of SHE information to employees. • Communication is ad hoc and restricted to those involved in specific incidents. • Company's employees are unaware of important SHE information • Some informal and formal communication channels are established for information flow internally to all employees.	• Some communication of SHE information to employees on a need-to-know basis • There is a communication strategy for SHE information flow internally and externally occasionally to all employees. • Employees are aware of pertinent SHE information. • Specific informal and formal communication channels are in place for communicating SHE issues to employees	• Adequate SHE information is routinely and regularly communicated to all employees. Employees are aware of critical SHE information. • There are established, good and appropriate informal and formal communication channels for communicating critical SHE information and resultant actions • All levels of employees are involved, and there are robust mechanisms for them to feedback	• There is an open, proactive and effective SHE communication between the company and its employees and stakeholders. • SHE communication is a strong, and consistent two-way process. Good practice is communicated both externally and internally • The company communicates to its employees on all the SHE-related issues and aspects of the company. Established communication channels and methods are fully adopted throughout the supply chain in the company and are consistently used for efficient coordination of SHE activities. • All pertinent SHE information and resultant actions are well communicated to all employees across the company. • Communication methods for SHE information flow internally and externally are continuously monitored and regularly reviewed against identified best practices in other sectors for potential continuous improvement.
She Documentation and Control	• No organised documentations (e.g., SHE policy, SHE manual, emergency plans and work instructions etc.) and records that describe company's SHE system elements and their interrelationships	• Documentations of some elements of a company's SHE system and other related SHE records are available to employees • SHE documentations and records are not organised and are not easily traceable and accessible	• Documentations and records of more elements of a company's SHE system and other related SHE records are available to employees • SHE documentations and records are compiled and organised in a format that is somewhat traceable and accessible	• Documentations and records of all elements of the company's SHE system and other related SHE records are available to all employees • All SHE documentations are compiled and mostly organised in an appropriate format, traceable and accessible.	• SHE documentations including other related SHE records are compiled and well organised in a clear, concise and functional format, traceable and readily accessible to all. • SHE documentations and records are integrated with other organisational documentations (such as human resource plans) for continuous improvement of company's functions. • SHE reports and SHE documentations are systematically maintained regularly reviewed and updated with appropriate version control in place, based on system improvements, to drive efficiency and effectiveness of the management system.

Table A1. *Cont.*

Integrated Safety, Health and Environmental Management Capability Maturity Model (iSHEM-CMM)

She Capability Attributes	Capability Levels				
	Level 1	Level 2	Level 3	Level 4	Level 5
Lessons learned and knowledge Management	• Company has no structured system for capturing lessons in order to facilitate future improvement of the SHE management system • No promotion of knowledge sharing and lessons learned across the company • No records of lessons learned. There is great reliance on individual memory	• Company's processes and procedures for capturing and disseminating lessons learned are characterised by poor or unstructured records keeping and inconsistent data • Limited promotion of knowledge sharing and lessons learned across the company • Reliance on manual record keeping of lessons • Lessons learned are rarely used for SHE management system continuous improvement and innovation	• Company's processes and procedures for capturing and disseminating lessons learned are characterised by well-structured record keeping and good information • Knowledge sharing and lessons learned is promoted across the company • Little reliance on manual record keeping and greater usage of digital technologies for record keeping • Records of lessons learned are sometimes relied on for SHE management system continuous improvement and innovation	• Company's processes and procedures for capturing and disseminating lessons learned are characterised by routinely well-structured record keeping and consistent high-quality information • Knowledge sharing and lesson learned is promoted systematically across the company • Reliance on advanced digital technologies for capturing and disseminating lessons • Records of lessons are consistently relied upon for SHE decision making, continuous improvement and innovation • Processes and procedures for capturing and disseminating lessons learned are modelled on best practice knowledge management standards e.g., ISO 30401-2018, ISO 9001: 2015.	• There is well structured system for capturing and disseminating lessons learned and knowledge gained across the whole company. Heavy reliance on technological innovations for capturing and disseminating lessons • The processes are institutionalised within the company and are considered a key measure of operational excellence. • Knowledge and lessons learned are continuously shared and consistently relied upon across the company to continuously improve SHE • Processes and procedures for capturing and disseminating lessons learned are routinely reviewed and updated to drive continuous improvement and innovation.

Table A2. Evaluation Questionnaire.

Assessment Criteria	Level of Agreement				
	Strongly Agree 5	Agree 4	Neither Agree nor Disagree 3	Disagree 2	Strongly Disagree 1
Attributes used in the SHEM-CMM Worksheet					
Attributes are relevant to SHE management capability.	☐	☐	☐	☐	☐
Attributes cover all aspects of SHE management capability.	☐	☐	☐	☐	☐
Attributes are correctly assigned to their respective maturity level.	☐	☐	☐	☐	☐
Attributes are clearly distinct.	☐	☐	☐	☐	☐
Capability Maturity Levels					
The maturity levels sufficiently represent maturation in the attributes.	☐	☐	☐	☐	☐
There is no overlap detected between descriptions of maturity levels.	☐	☐	☐	☐	☐
Ease of Understanding					
The maturity levels are understandable	☐	☐	☐	☐	☐
The documentations (i.e., assessment instructions) are easy to understand	☐	☐	☐	☐	☐
The results are understandable	☐	☐	☐	☐	☐
Ease of Use					
The scoring scheme (i.e., drop-down options for maturity levels (1–5)) is easy to use	☐	☐	☐	☐	☐
The SHEM-CMM is easy to use	☐	☐	☐	☐	☐
Usefulness and Practicality					
SHEM-CMM is useful for assessing SHE management capability	☐	☐	☐	☐	☐
SHEM-CMM is practical for use in industry	☐	☐	☐	☐	☐

References

1. Adinyira, E.; Manu, P.; Agyekum, K.; Mahamadu, A.-M.; Olomolaiye, P. Violent behaviour on construction sites: Structural equation modelling of its impact on unsafe behaviour using partial least squares. *Eng. Constr. Archit. Manag.* **2020**, *27*, 3363–3393. [CrossRef]
2. Schmidt, J.S.; Osebold, R. Environmental management systems as a driver for sustainability: State of implementation, benefits and barriers in German construction companies. *J. Civ. Eng. Manag.* **2017**, *23*, 150–162. [CrossRef]
3. HSE. Annual Injury and Ill-Health Statistics on Work-Related Health and Safety in Great Britain. Health and Safety Executive: Bootle, Merseyside, UK, 2018. Available online: https://www.acornsafety.co.uk/hse-health-safety-statistics-released-2018/37 (accessed on 23 July 2018).
4. Bureau of Labor Statistic. National Census of Fatal Occupational Injuries in 2019. Bureau of Labor Statistics, 2020. Available online: https://www.bls.gov/news.release/pdf/cfoi.pdf (accessed on 4 October 2021).
5. Health and Safety Executive (HSE). Workplace fatal injuries in Great Britain, 2021. 2021. Available online: https://www.hse.gov.uk/statistics/pdf/fatalinjuries.pdf (accessed on 25 October 2021).
6. Kanchana, S.; Sivaprakash, P.; Joseph, S. Studies on labour safety in construction sites. *Sci. World J.* **2015**, *2015*, 590810. [CrossRef]
7. Gupta, A.R.; Deshmukh, S.K. Energy Efficient Construction Materials. *Eng. Mater.* **2016**, *678*, 35–49. [CrossRef]
8. UN Environment and International Energy Agency. Towards a Zero-Emission, Efficient, and Resilient Buildings and Construction Sector: Global Status Report. 2017. Available online: https://www.worldgbc.org/sites/default/files/UNEP%20188_GABC_en%20(web).pdf (accessed on 3 November 2018).
9. Global Construction Perspectives and Oxford Economics. Global Construction 2030: A Global Forecast for the Construction Industry to 2030. 2015. Available online: http://constructingexcellence.org.uk/wp-content/uploads/2015/09/GCP_Conference_ENR_2030_WEB.pdf (accessed on 20 February 2018).
10. Workplace Safety and Health Institute. Economic Cost of Work-Related Injuries and Ill Health in Singapore. 2013. Available online: https://www.wshinstitute.sg/~{}/media/wshi/past%20publications/2013/economic%20cost%20of%20work-related%20injuries%20and%20ill-health%20in%20singapore.pdf?la=en (accessed on 10 November 2018).
11. Sze, N.N.; Chan, D.W.M.; Chan, A.P.C.; Yiu, N. Implementation of safety management system in managing construction projects: Benefits and obstacles. *Saf. Sci.* **2019**, *117*, 23–32.
12. Asah-Kissiedu, M.; Manu, P.; Booth, C.; Mahamadu, A.-M.; Agyekum, K. Integrated safety, health and environmental management in the construction industry: Key organisational capability attributes. *J. Eng. Des. Technol.* **2021**. ahead-of-print. [CrossRef]
13. Agyekum, K.; Simons, B.; Botchway, S.Y. Factors influencing the performance of safety programmes in the Ghanaian construction industry. *Acta Structilia* **2018**, *25*, 39–68. [CrossRef]
14. Zaira, M.M.; Hadikusumo, B.H.W. Structural equation model of integrated safety intervention practices affecting the safety behaviour of workers in the construction industry. *Saf. Sci.* **2017**, *98*, 124–135. [CrossRef]
15. Hamid, A.R.A.; Singh, B.; Yusof, W.Z.W.; Yang, A.K.T. Integration of safety, health, environment and quality (SHEQ) management system in construction—A review. *Mal. J. Civ. Eng.* **2004**, *16*, 24–37.
16. Rebelo, M.F.; Santos, G.; Silva, R. Conception of a flexible Conception of a flexible integrator and lean model for integrated management systems. *Total Qual. Manag.* **2014**, *25*, 683–701. [CrossRef]
17. Foster, P.; Hoult, S. The safety journey: Using a safety maturity model for safety planning and assurance in the UK coal mining industry. *Minerals* **2013**, *3*, 59–72. [CrossRef]
18. Ormazabal, M.; Sarriegi, J.M.; Rich, E.; Viles, E.; Gonzalez, J.J. Environmental management maturity: The role of dynamic validation. *Org. Env.* **2021**, *34*, 145–170. [CrossRef]
19. Hossain, M.A.; Abbott, E.L.S.; Chua, D.K.H.; Qui, N.T.; Goh, Y.M. Design-for-safety knowledge library for BIM-integrated safety risk reviews. *Autom. Constr.* **2018**, *94*, 290–302. [CrossRef]
20. Gangolells, M.; Casals, M.; Forcada, N.; Fuertes, A.; Roca, X. Model for enhancing integrated identification, assessment, and operational control of on-site environmental impacts and health and safety risks in construction firms. *J. Constr. Eng. Manag.* **2013**, *139*, 138–147. [CrossRef]
21. Poh, C.Q.X.; Ubeynarayana, C.U.; Goh, Y.M. Safety leading indicators for construction sites: A machine learning approach. *Autom. Constr.* **2018**, *93*, 375–386. [CrossRef]
22. Ikram, M.; Zhou, P.; Shah, S.A.A.; Liu, G.Q. Do environmental management systems help improve corporate sustainable development? Evidence from manufacturing companies in Pakistan. *J. Clean. Prod.* **2019**, *226*, 628–641. [CrossRef]
23. Gianni, M.; Gotzamani, K.; Tsiotras, G. Multiple perspectives on integrated management systems and corporate sustainability performance. *J. Clean. Prod.* **2017**, *168*, 1297–1311. [CrossRef]
24. Li, Y.; Guldenmund, F.W. Safety management systems: A broad overview of the literature. *Saf. Sci.* **2018**, *103*, 94–123. [CrossRef]
25. Rebelo, M.; Santos, G.; Silva, R. Integration of standardised management systems: A dilemma of systems. *Systems* **2015**, *3*, 45–59. [CrossRef]
26. Zhang, Q.; Oo, B.L.; Lim, B.T.H. Drivers, motivations and barriers to the implementation of corporate social responsibility practices by construction enterprises: A review. *J. Clean. Prod.* **2019**, *210*, 563–584. [CrossRef]
27. Kamble, S.S.; Gunasekaran, A.; Sharma, R. Analysis of the driving and dependence power of barriers to adopt industry 4.0 in Indian manufacturing industry. *Comput. Ind.* **2018**, *101*, 107–119. [CrossRef]

28. Campos, L.; Trierweiller, A.C.; Spenassato, D.C.; Bornia, A.C.; Selih, J. Barriers for implementation of EMS: A study in the Construction Industry of Brazil and Slovenia. In Proceedings of the Production and Operations Management Society 25 Annual Conference, Atlanta, GA, USA, 9–12 May 2014; pp. 1–10.
29. Yiu, N.S.N.; Sze, N.N.; Chan, D.W.M. Implementation of safety management systems in Hong Kong construction industry—A safety practitioner's perspective. *J. Saf. Res.* **2018**, *64*, 1–9. [CrossRef] [PubMed]
30. Ghaffar, S.H.; Burman, M.; Braimah, N. Pathways to circular construction: An integrated management of construction and demolition waste for resource recovery. *J. Clean. Prod.* **2020**, *244*, 118710. [CrossRef]
31. Lacerda, T.C.; von Wangenheim, C.G. Systematic literature review of usability capability/maturity models. *Comput. Stand. Interfaces* **2018**, *55*, 95–105. [CrossRef]
32. Henning, L. Developing a Framework to Assess Day Hospital Maturity. Ph.D. Thesis, Stellenbosch University, Stellenbosch, South Africa, 2017.
33. Van Steenbergen, M.; Bos, R.; Brinkkemper, S.; de Weerd, I.V.; Bekkers, W. The design of focus area maturity models. In *Global Perspective on Design Science Research, Proceedings of the 5th International Conference, DESRIST, 2010, St. Gallen, Switzerland, 4–5 June 2010*; Springer: Berlin/Heidelberg, Germany, 2010; pp. 317–332.
34. Lahrmann, G.; Marx, F.; Winter, R.; Wortmann, F. Business Intelligence Maturity: Development and Evaluation of a Theoretical Model. In Proceedings of the 44th Hawaii International Conference on System Sciences, Kauai, HI, USA, 4–7 January 2011.
35. Lahti, M.; Shamsuzzoha, A.H.M.; Helo, P. Developing a maturity model for Supply Chain Management. *Int. J. Logist. Syst. Manag.* **2009**, *5*, 654–678. [CrossRef]
36. Van Looy, A.; De Backer, M.; Poels, G. Defining business process maturity. A journey towards excellence. *Total Qual. Manag. Bus. Excell.* **2011**, *22*, 1119–1137. [CrossRef]
37. Crosby, P.B. *Quality Is Free*; McGraw-Hill: New York, NY, USA, 1979.
38. Jokela, T.; Siponen, M.; Hirasawa, N.; Earthy, J. A survey of usability capability maturity models: Implications for practice and research. *Behav. Inf. Technol.* **2006**, *25*, 263–282. [CrossRef]
39. Paulk, M.C.; Curtis, B.; Chrissis, M.B.; Weber, C.V. Capability maturity model, version 1.1. *Softw. IEEE* **1993**, *10*, 18–27. [CrossRef]
40. Srai, J.S.; Alinaghian, L.S.; Kirkwood, D.A. Understanding sustainable supply network capabilities of multinationals: A capability maturity model approach. *Proc. Inst. Mech. Eng. Part B J. Eng. Manuf.* **2013**, *227*, 595–615. [CrossRef]
41. Software Engineering Institute (SEI) CMMI Product Team. Capability Maturity Model Integration (CMMI) Version 1.1: Improving Processes for Developing Better Products and Services. 2002. Available online: https://resources.sei.cmu.edu/asset_files/TechnicalReport/2002_005_001_14072.pdf (accessed on 23 January 2019).
42. Ngai, E.W.T.; Chau, D.C.K.; Poon, J.K.L.; To, C.K.M. Energy and utility management maturity model for sustainable manufacturing process. *Int. J. Prod. Econ.* **2013**, *146*, 453–464. [CrossRef]
43. Software Engineering Institute (SEI). CMMI for Development, Version 1.3: Improving Processes for Developing Better Products and Services. 2010. Available online: www.sei.cmu.edu/reports/10tr033.pdf (accessed on 28 July 2018).
44. Meng, X.; Sun, M.; Jones, M. Maturity model for supply chain relationships in construction. *J. Manag. Eng.* **2011**, *27*, 97–105. [CrossRef]
45. Antunes, P.; Carreira, P.; Silva, M.M. Towards an energy management maturity model. *Energy Policy* **2014**, *73*, 803–814. [CrossRef]
46. Strutt, J.E.; Sharp, J.V.; Terry, E.; Miles, R. Capability maturity models for offshore organisational management. *Environ. Int.* **2006**, *32*, 1094–1105. [CrossRef] [PubMed]
47. Sun, M.; Vidalakis, C.; Oza, T. A change management maturity model for construction projects. In *Proceedings 25th Annual ARCOM Conference, Nottingham, UK, 7–9 September 2009*; Dainty, A., Ed.; Association of Researchers in Construction Management: Reading, UK, 2009; pp. 803–812.
48. Zou, P.X.W.; Ying, C.; Tsz-Ying, C. Understanding and improving your risk management capability: Assessment model for construction organisations. *J. Constr. Eng. Manag.* **2010**, *136*, 854–863. [CrossRef]
49. Siebelink, S.; Voordijk, J.T.; Adriaanse, A. Developing and testing a tool to evaluate BIM maturity: Sectoral analysis in the Dutch construction industry. *J. Constr. Eng. Manag.* **2018**, *14*, 05018007. [CrossRef]
50. Quaigrain, R.A. Evaluating Disability Management in Construction Using Maturity Modelling and Metrics and Its Relation to Safety Performance. Ph.D. Thesis, University of Manitoba, Winnipeg, MB, Canada, 2019.
51. Fleming, M. *Safety Culture Maturity Model. Report 2000/049*; Health and Safety Executive: Norwich, UK, 2001. Available online: www.hse.gov.uk/research/otopdf/2000/oto00049.pdf (accessed on 26 June 2018).
52. Carillion Plc. Sustainability Report 2012. Carillion, 2013. Available online: http://sustainability2012.carillionplc.com/downloads/carillion-sr2012.pdf (accessed on 25 October 2016).
53. Goggin, A. Health and Safety Maturity Model for New Brunswick Construction Industry. Master's Thesis, University of New Brunswick, Saint John, NB, Canada, 2010.
54. Bai, L.; Wang, H.; Huang, N.; Du, Q.; Huang, Y. An environmental management maturity model of construction programs using the AHP-entropy approach. *Int. J. Environ. Res. Public Health* **2018**, *15*, 1317. [CrossRef]
55. Manu, P.; Poghosyan, A.; Mahamadu, A.-M.; Mahdjoubi, L.; Gibb, A.; Behm, M.; Akinade, O. Development of a design for occupational safety and health capability maturity model. In Proceedings of the Joint CIB W099 TG59 International Safety, Health and People Construction Conference, Salvador, Brazil, 1–3 August 2018; pp. 93–102.

56. Maier, A.M.; Moultrie, J.; Clarkson, P.J. Assessing organisational capabilities: Reviewing and guiding the development of maturity grids. *IEEE Trans. Eng. Manag.* **2012**, *59*, 138–159. [CrossRef]
57. Asdecker, B.; Felch, V. Development of industry 4.0 maturity model for the delivery process in supply chains. *J. Model. Manag.* **2018**, *13*, 840–883. [CrossRef]
58. Correia, E.; Carvalho, H.; Azevedo, S.G.; Govindan, K. Maturity models in supply chain sustainability: A systematic literature review. *Sustainability* **2017**, *9*, 64. [CrossRef]
59. Storbjerg, S.H.; Brunoe, T.D.; Nielsen, K. Towards an engineering change management maturity grid. *J. Eng. Des.* **2016**, *26*, 361–389. [CrossRef]
60. Software Engineering Institute (SEI). *CMMI for Development: Improving Processes for Better Products, Version 1.2*; Carnegie Mellon University: Pittsburgh, PA, USA, 2006.
61. Poghosyan, A.; Manu, P.; Mahamadu, A.-M.; Akinade, O.; Mahdjoubi, L.; Gibb, A.; Behm, M. A Web-based Design for Occupational Safety and Health Capability Maturity Indicator. *Saf. Sci.* **2020**, *122*, 104516. [CrossRef]
62. Hallowell, M.R.; Gambatese, J.A. Activity-Based safety risk quantification for concrete formwork construction. *J. Constr. Eng. Manag.* **2009**, *135*, 990–998. [CrossRef]
63. Office of Road and Rail and Health and Safety Maturity. RM3 the Risk Management Maturity Model. 2017. Available online: https://www.orr.gov.uk/guidance-compliance/rail/health-safety/strategy/rm3 (accessed on 2 November 2021).
64. Carvalho, J.V.; Rocha, A.; van de Wetering, R.; Abreu, A. A maturity model for hospital information systems. *J. Bus. Res.* **2019**, *94*, 388–399. [CrossRef]
65. Wagire, A.A.; Joshi, R.; Rathore, A.P.S.; Jain, R. Development of maturity model for assessing the implementation of industry 4.0: Learning from theory and practice. *Prod. Plan. Control.* **2021**, *32*, 603–662. [CrossRef]
66. Manu, P.; Poghosyan, A.; Mahamadu, A.-M.; Mahdjoubi, L.; Gibb, A.; Behm, M.; Akinade, O. Design for occupational safety and health: Key attributes for organisational capability. *Eng. Constr. Archit. Manag.* **2019**, *26*, 2614–2636. [CrossRef]
67. HSE. *Keil Centre Offshore Technology Report 2000-049: Safety Culture Maturity Model*; HSE Books: London, UK, 2000.
68. Succar, B. Building information modelling framework: A research and delivery foundation for industry stakeholders. *Autom. Constr.* **2009**, *18*, 357–375. [CrossRef]
69. Filho, A.P.G.; Andrade, J.C.S.; de Oliveira Marinho, M.M. A safety culture maturity model for petrochemical companies in Brazil. *Saf. Sci.* **2010**, *48*, 615–624. [CrossRef]
70. Salah, D.; Paige, R.; Cairns, P. An evaluation template for expert review of maturity models. In *International Conference on Product-Focused Software Process Improvement*; Springer International Publishing: Berlin/Heidelberg, Germany, 2014; pp. 318–321.

Article

Champions of Social Procurement in the Australian Construction Industry: Evolving Roles and Motivations

Martin Loosemore [1,*], Robyn Keast [2], Jo Barraket [3] and George Denny-Smith [4]

[1] School of Built Environment, University of Technology Sydney, Sydney, NSW 2007, Australia
[2] Faculty of Business, Southern Cross University, Lismore, NSW 2480, Australia; robyn.keast@scu.edu.au
[3] Centre for Social Impact, Swinburne University of Technology, Melbourne, VIC 3122, Australia; jbarraket@swin.edu.au
[4] School of the Built Environment, University of New South Wales, Sydney, NSW 2052, Australia; g.denny-smith@unsw.edu.au
* Correspondence: martin.loosemore@uts.edu.au

Citation: Loosemore, M.; Keast, R.; Barraket, J.; Denny-Smith, G. Champions of Social Procurement in the Australian Construction Industry: Evolving Roles and Motivations. *Buildings* **2021**, *11*, 641. https://doi.org/10.3390/buildings11120641

Academic Editor: Carlos Oliveira Cruz

Received: 31 October 2021
Accepted: 7 December 2021
Published: 11 December 2021

Abstract: There has been a recent proliferation of social procurement policies in Australia that target the construction industry. This is mirrored in many other countries, and the nascent research in this area shows that these policies are being implemented by an emerging group of largely undefined professionals who are often forced to create their own roles in institutional vacuums with little organisational legitimacy and support. By mobilising theories of how organisational champions diffuse innovations in other fields of practice, this paper contributes new insights into the evolving nature of these newly emerging roles and the motivations which drive these professionals to overcome the institutional inertia they invariably face. The results of semi-structured interviews, with fifteen social procurement champions working in the Australian construction industry, indicate that social procurement champions come from a wide range of professional backgrounds and bring diverse social capital to their roles. Linked by a shared sense of social consciousness, these champions challenge traditional institutional norms, practices, supply chain relationships, and traditional narratives about the concepts of value in construction. We conclude that, until normative standards develop around social procurement in the construction industry, its successful implementation will depend on external institutional pressures and the practical demonstration of what is possible in practice within the performative constraints of traditional project objectives.

Keywords: construction industry; champions; innovation; social innovation; social procurement; social value

1. Introduction

Social procurement involves the deliberate creation of social value by purchasing goods and services [1]. While this idea is not new [2], McNeill [3] conceptualised social procurement as a social innovation because it strategically repositions the procurement function as a tool for addressing an organisation's social objectives through the creation of new hybrid assemblages between organisations in the public, private and third sectors.

In contrast to the broader body of research on public procurement, sustainable procurement and green procurement, which has increased markedly since 2017, social procurement research is relatively new, especially in the field of construction [4,5]. The nascent body of social procurement research in the field of construction indicates that the implementation of social procurement in construction is challenging. For example, Murphy and Eadie [6] argue that social procurement is challenging to implement in the Irish construction industry because it requires the adaption of performance targets and accounting models to measure social value, and the creation of new linkages between disparate organisational functions such as procurement, sustainability and human resources. Loosemore et al. [7] showed that subcontractors in the Australian construction industries perceive social procurement as a risk to safety, productivity, cost and quality. In Sweden, Troje and Andersson [8] argue that social procurement pushes the construction sector into a state of 'institutional instability.' This is because it disturbs incumbent institutional logics, relationships and practices. Troje and Kadefors [9], and Troje and Gluch [10,11], portray an uncertain and emerging institutional field in Sweden, which is being shaped by lone organisational actors (called Employment Requirement Professionals) who are engaging in ongoing 'institutional work' to create and legitimise social procurement in their organisations. While they did not use the term champion, the roles, activities and attributes that Troje and Gluch [10,11] describe, closely reflect the definitions of organisational champions in wider organisational studies research. In making this observation, we draw on organisational studies research which defines organisational champions as informal role holders who, through insightful, diplomatic, and skilled organisational lobbying, sell new ideas to others and protect them from inertia and attack by visibly, emotionally, and materially supporting the development process and enlisting the political support of other influential decision-makers [12–14].

Although many construction researchers have recognised the vital role of champions in driving construction innovation, this research has been largely confined to the implementation of economic, technological, and environmental innovations [15–19]. While both Barraket et al. [1] and McNeill [3] recognise the critical role of organisational champions in driving social procurement, they do not elaborate on the functions of such roles. Aside from the work of Troje and Gluch [10,11] and Troje and Andersson [8], there has been little research into the role of social innovation champions in implementing social innovations, such as social procurement, in the construction industry or in other industry contexts.

The aim of this paper is to contribute to this gap in the research by mobilising the organisational champion theory as a new conceptual lens to explore what these important actors do to promote social procurement in the construction industry. More specifically, two key research questions are addressed:

1. What motivates people to become champions of social procurement in the construction industry?
2. What is the nature of these evolving roles in the construction industry?

The objective of this paper is to advance, both empirically and conceptually, the emergence of social procurement as a distinct field of practice [1]. This research is important because, as Troje and Gluch [10] note, social procurement professionals are important carriers of social sustainability practices within the construction industry. Therefore, a better understanding the nature of these practices, how they promote that agenda in the face of significant institutional resistance and what motivates them to do so, is important for selecting the appropriate people for these critical roles and facilitating inter-organisational learning to advance sectoral practices in this area.

This article proceeds with an examination of organisational champion research to explore the strategies that champions employ to bring about change, highlighting the deficit concerning its application to social procurement in construction. Next, the research approach is outlined, and findings are distilled and discussed to address the above research questions.

2. Social Procurement Champions in Construction: Motivations and Roles

In exploring the two research questions posed above, laterally relevant research into the motivations and roles of organisational champions' areas of social enterprise, social innovation, and environmental sustainability can offer potentially helpful insights.

Concerning research question one, Loosemore and Higgon [20] argued that social intrapreneurs are motivationally distinguishable from their economic counterparts by a unique sense of accountability for what they do; a powerful sense of right and wrong; a strong sense of empathy towards those who are less fortunate; a sense of profound responsibility to their environment and communities; and a sense of moral outrage against injustice and inequity in society. Lewis and Cassells [21] found that sustainability champions in construction were motivated by a strong sense of responsibility to the community; commercial opportunities for cost reduction/financial benefits from implementing sustainability initiatives; and compliance with sustainability legislation. Wood et al. [22] identified three types of sustainability champions: *saviours*; *nurturers*; and *strugglers*. Saviours believe that they are solving or resolving pressing environmental issues and use objective, positivist evidence to build their arguments. Nurturers focus on nurturing, growth and building empathetic relationships with colleagues to advocate and pursue organisational change. Strugglers are champions whose experiences are characterised by vulnerability, and thus they struggle to persuade others to their cause. More recently, Troje and Gluch [10] classified social procurement actors into three groups (idealists, problem-solvers and pragmatists), classifying their work as operational, co-creating and educational in nature. Idealists are altruistic and caring society builders who feel compelled to help others. Problem-solvers are driven by the intellectual challenge of finding a recipe that makes social procurement commercially viable. Pragmatists are driven by performative goals and compliance with political decisions, formal policies and completing a task with little sentiment.

Concerning research question two, research has traditionally depicted organisational champions as heroic, charismatic, and transformational leaders prepared to adopt and fight for a new idea by modifying and fitting it into a relevant organisational context [23–27]. Hargreaves [28] used social practice theory to conceptualise sustainability champions as 'carriers' of new ideas across organisations and functions, allowing them to co-design and refine hybrid practices over time, through trial-and-error and experimentation with new ideas. Reflecting this collaborative theme, Crosby and Bryson [29] describe social innovation champions as "monomaniacs with a vision" (p. 167) who persistently and creatively connect stakeholders and resourcing pools with opportunities to create social change. Later work by Crosby and Bryson [30] describes champions as strategic builders of cross-sector and multi-level relationships through self-sustaining 'mutual gain regimes'. Huxham and Vangen [31] describe champions as politically savvy, tireless organisers and promoters of change who are multilingual translators across different organisational and practice cultures, as well as factions. Similarly, Woolcott et al. [32] describe social innovation champions as 'value-adding connectors', and Molloy et al. [13] employ the concept of bricolage to show how social innovation champions 'make-do' with the limited resources at hand by mixing, recombining and reusing them in new ways to support an idea. According to Molloy et al. [13], social innovation champions engage in 'sensemaking' to persuade others to support their ideas. They also argue that the instability of prevailing organisational logics orientates champion sensemaking efforts in specific directions. For example, when opposing institutional logics are strong and stable as they are in construction [8], champions must direct their efforts externally to exert outside pressure to change organisations. In construction, this may involve lobbying construction clients to implement and enforce social procurement policies in projects. Molloy et al. [13] argue that hybrid solutions are more likely in this context, supporting recent references to the importance of cross-sector collaboration in implementing social procurement in construction [33]. Most recently, Troje and Andersson [8] argued that social procurement actors engaged in 'institution building work' at both operational and strategic levels. At the operational level, this work relies on references to hard facts and figures, doing the right thing, laws and regulations and the social impact of social procurement. With strategic-level arguments, work involves leveraging legitimate power to emphasise social procurement's commercial and socio-economic benefits. Earlier work by Troje [34] found that in the face of unsympathetic institutional norms, social procurement actors had to rely heavily on rhetorical strategies (ethos, pathos and logos) and market-based logic underpinned by a commercial, sales-related discourse, to persuade industry incumbents to implement social procurement. Building on this research, Troje and Gluch [10] found that social procurement champions in construction lacked a distinct "domain of practice" and relied on "self-made adjustable roles" (p. 67), which in turn relied on operational, educational and collaborative work and an organisational identity based on personal engagement rather than a formally defined role. Troje and Gluch [10] described their role as "gardeners and teachers" (p. 66), planting seeds of change and growing other people into champions of social procurement.

Table 1 summarises the core literature in the field regarding the two research questions. The contribution of Table 1 is twofold. First, it synthesises the fragmented and limited research on organisational champions in the research questions. Second, in a more immediate contribution, it provides a framework to guide our methodological design, which is described in the following section.

Table 1. Analytical framework: drivers, roles and attributes of social procurement champions.

Research Question	Factors	Key References
1. What motivates people to become champions of social procurement in construction companies?	Altruism. Opportunities for addressing social problems in a new and more efficient way. Responsibility to the community. Making change for the better. Idealism.	[10,21]
	Inspiration from others. Personal experience of disadvantage and injustice. Social networks/encouragement from others.	[35]
	Personality (determined to bring about change). Cognition (deep understanding of social problems and causes, as well as potential solutions). Feasibility (a realistic opportunity to bring about change). Desirability (need for change to happen and confidence that others support change). Skills available (confidence can bring about change). Personal experience/confidence in bringing about change successfully (having seen injustice). Social networks (social circles for encouragement). Conducive environment (social, economic, political).	[36]
	Creating new positions, roles, influences, organisational freedoms, and power bases aligning with personal goals, interests, and values.	[37]
	Career opportunities and rewards.	[38]
	Spiritual leadership (vision, altruism, compassion, social responsibility).	[39]
	Relationships between social procurement field actors which motivate and enable the sharing of information and practices. Resources that provide finance, support, and the best practice guidance to enable inter-organisational learning.	[1]
	Create change at a systems level. Moral outrage against injustice. Opportunity to create community benefit/social value.	[20]
2: What is the nature of these evolving roles in construction companies?	Sell and protect ideas. Provide emotional and material support. Channel resources to a project. Enlisting and securing support. Motivating others to come onboard. Supply ideas, energy and determination. Different levels of champions working together.	[12,13,40]
	Form supporting intra- and inter-organisational cross-sector collaborations/relationships.	[41]
	Encourage, promote, and facilitate open discussion. Protect ideas from threat.	[19]
	Engage in extracurricular tasks (outside normal job description). Self-made adjustable roles which rely on operational, education and collaborative work. Engage in institutional work (to change existing norms and practices and legitimise new social procurement roles, relationships and authority in existing structures). Adopt rhetorical strategies to bring about change.	[8,10,11,34,42]
	Value-adding connector—adding social value along the interaction chain.	[32]
	Mesh people, processes technologies, cultures and systems. Leadership (visionary, political, ethical).	[30]
	Research—collect and present evidence to back-up arguments/case. Working outside existing rules and procedures, defined roles and lines of responsibility.	[31]
	Educate stakeholders.	[41]

Table 1. *Cont.*

Research Question	Factors	Key References
2: What is the nature of these evolving roles in construction companies?	Sensemaking, sense giving and sense breaking. Connecting actors in new ways. Influencing others through combinations of rhetorical, evidence-based and rational strategies. Unlearning (breaking existing logics and institutions). Bricolage (making do, combining existing resources and redefining roles in new ways).	[13]
	Creating new proto-institutions.	[8]
	Praxis (drive change through developing new practices, trial and error and experimentation, working in different, careers, fields and business units).	[28]
	Role changes over phases of the innovation process: Design for change; develop business case; internalise change; implement change; evaluate change. Knowledge stage; persuasion stage; decision-making; implementation; confirmation.	[43,44]
	Create and leverage new cross-sector partnerships between private, public and third sector organisations.	[33]

3. Methodology

Ontologically, given the collaborative, multi-field and relational underpinnings of organisational champion motivations and roles, this research was guided by a social constructivist lens and an interpretivist epistemology. Informed by reflective practitioner theory and the Aristotelian concept of praxis [23], it employed qualitative data collection and analyses to produce in-depth accounts of the day-to-day emerging roles, practices and experiences of social procurement champions working in the construction industry.

3.1. Data Collection

Data were collected using semi-structured interviews with social procurement champions from major contracting firms in the Australian construction industry. Semi-structured interviews allowed the research team to make sense of social procurement champions' multidimensional and evolving roles. As Blackstone [45] notes, in such contexts, data validity can be undermined by using methods such as surveys or highly structured interviews. This is because the standardised manner in which questions are posed in surveys makes it difficult for respondents to articulate the uncertainties and evolving experiences and lessons surrounding them. Furthermore, given the uncertain nature of the roles we were exploring, semi-structured interviews allowed respondents the freedom to express their views on their own terms and for researchers to engage in a two-way dialogue, following unexpected leads not anticipated in the original interview questions [46]. By allowing the co-production of narratives between interviewer and participants, semi-structured interviews provided the research team with deeply reflective stories about our champions' experiences in advocating for social procurement in their organisations.

Semi-structured interviews were also well-suited to the lack of clarity in many construction organisations regarding what social procurement is and who is responsible for it [11,46]. Using a survey approach to identify champions in this environment is also unreliable and problematic due to the high risk of a survey being sent to the wrong person or misunderstood [47]. Furthermore, since there are relatively few recognised social procurement champions in the Australian construction industry, the sample size for a survey would be too small for reliable statistical analysis. Finally, and crucially, given the potential for social desirability bias in this area [48], semi-structured interviews allowed the researchers to probe respondents if they suspected they were saying what they perceived to be the 'right thing'.

Our sampling approach was purposeful and involved approaching tier-one contracting and consulting organisations in the Australian construction industry to nominate a champion of social procurement in their organisation. Tier-one firms are organisations that work on the largest construction projects (typically valued at over AUD 500 million) and were chosen because social procurement is currently restricted to major public sector projects in Australia, which these firms tend to work on [33].

Following Molloy et al. [13], our sampling criteria were based on a literature-based description of a champion (as defined above) to help organisations nominate appropriate respondents irrespective of hierarchical level. Before any interview, each respondent was also asked to confirm their role as a champion of social procurement in their organisation. Additionally, following Molloy et al. [13], potential attributional bias was minimised by using neutral terminology (e.g., non-heroic) and ensuring the anonymity of all respondents in the analysis and research reporting process. Snowball sampling was employed to access an undefined and emerging community of social procurement practitioners by asking respondents to nominate who they considered champions in other qualifying tier-one organisations.

Interviews were conducted remotely over a two-month period (March–May 2020) using Zoom due to the face-to-face restrictions imposed by the COVID-19 pandemic. The semi-structured interview questions were deliberately open-ended to enable the respondents to describe their roles on their own terms without any preconceived answers from the researchers [49]. Interview questions included:

1. What is driving SP outside and within the business?
2. What is your main motivation in doing this job?
3. Can you describe your job in terms of SP and what you actually do?
4. How important are relationships to your job, and how much time do you spend nurturing and building them?
5. Is your SP role formally established and recognised within the business; where does it sit in the organisational structure and how has it evolved over time?
6. How do you persuade people to get onboard? What methods do you use to build support for SP in the business?
7. Who are the most important people to get onboard; why and how do you do that?
8. How do you build a community of support within the organisation and across organisations?

Employing the concept of 'theoretical saturation' [49], we continued interviewing respondents until the data collection process no longer offered any new or relevant insights in relation to the research questions and core analytical categories in Table 1. This rolling approach meant that data analysis and data collection had to occur in parallel. This process resulted in a total of fifteen interviews with champions from ten different tier-one contracting and consulting companies (Table 2), which typically lasted between one and two hours. To minimise researcher bias and maximise reflexivity [50], these interviews were conducted by two researchers (an 'insider' with experience of the construction industry and an 'outsider' without any experience of construction but an expert in social policy and public management). After each interview, the researchers met to debrief and compare notes around their interpretation of how respondents answered each question. These analytical memos became an important part of the data, and were analysed following the process described below.

3.2. Analysis

All interviews were audio recorded, transcribed verbatim and analysed using thematic analysis in five stages following the protocols of Guest et al. [51] and Gioia et al. [52].

Our research questions were our analytic starting point, and the first step involved researchers repeatedly reading the interview transcripts to obtain a high level of familiarity with the data. Second, researchers conducted open (inductive) and directed (deductive) interviews, coding, organising, and generating an initial list of items/codes (first-order coding) from the dataset that had a reoccurring pattern. The analytical framework in Table 1 was used for deductive coding. Third, researchers searched for recurring patterns, linkages, categories, and subcategories within the first-order codes relating to each research question. Fourth, researchers examined how codes combined to form over-reaching themes relating to the research questions. In the fifth and final stage, emergent themes were further refined by continued searches for data that supported or refuted the initial themes, allowing for further expansion and connections between overlapping themes. This process continued in parallel with data collection until theoretical saturation occurred and no further themes emerged. Any instances of disagreement were resolved through discussion, a process which continued until 100% inter-rater agreement was achieved, providing a high level of 'fit' with the data and a confidence in the theoretical validity of the emergent themes.

This process was applied to all the interview transcripts and tabulated in Table 3, which shows some selected examples of interview quotations relating to each research question for illustrative purposes.

Table 2. Sample structure.

Respondent	Position	Gender	Organization Description
R1	Social Procurement Manager	Female	Major international construction and infrastructure contractor. Revenue: AUD 4.2 billion
R2	Social Inclusion Manager	Male	Major international construction and infrastructure contractor. Revenue: AUD 4.2 billion
R3	General Manager	Female	Major international building construction, infrastructure, investment and development company. Revenue USD 5.19 billion
R4	Head of Sustainability	Male	Major international building construction, infrastructure, investment and development company. Revenue USD 5.19 billion
R5	Technical Director Social Outcomes	Female	Major project management, engineering and consulting services firm operating in 150 countries. Revenue AUD 2.72 billion
R6	Social Programme Manager	Female	An international construction contractor which specialises in commercial high-rise buildings. AUD 3.78 billion
R7	Employee Relations Manager	Male	An international construction contractor which specialises in commercial high-rise buildings. Revenue AUD 3.78 billion
R8	Director, Communication and Stakeholder Engagement	Female	An engineering, management, design, planning, project management, consulting and advisory company Revenue AUD 1.06 billion
R9	Senior Project Manager	Male	Project management consultancy, project manages major projects, 22 employees operate across Australia
R10	Stakeholder Engagement Manager and Training Project Officer	Female	Multinational and publically listed construction, property and infrastructure company. Revenue AUD 11.1 billion
R11	Social Diversity Supply Chain Manager	Male	Multinational and publically listed construction, property and infrastructure company. Revenue AUD 11.1 billion
R12	People and Engagement Director	Male	International construction, tunnelling, rail, building and services provider Revenue: AUD 4.2 billion
R13	Workforce Development and Industry Participation Manager	Female	An international construction contractor which specialises in commercial high-rise buildings. Revenue AUD 3.78 billion
R14	Development and Services Manager	Female	Construction contractor specialising in Metro, Freight and Heavy Haul and Light Rail infrastructure. Revenue AUD 55.1 million
R15	Managing Director	Male	Major subcontractor specialising in electrical contracting

Table 3. Coding and thematic analysis process (examples).

Evidence	Codes	Analytic Categories	Emergent Themes
RQ1: What motivates people to become champions of social procurement?			
The demand is just simply going to outstrip the supply. So we've spent the last couple of years trying to– . . . we'll teach you how to do it, we'll teach you how to be cost comparative and how to actually stand on your own two feet'. (R14) *It's sometimes just the basic ability to facilitate linkages into socially primed or mission based businesses or community organizations in a way that's respectful and mutually beneficial. (R5)*	Supply outstripping demand Teaching and supporting suppliers Ability to facilitate linkages Mutually beneficial Respectful	Third-sector capacity building [43] Social capital (bonding capital, bridging capital, linking capital) [53]	Third-sector capacity building
RQ2: What is the nature of these evolving roles in the construction industry?			
Is it the language we are using, is it the culturally appropriate practice that is informing our processes. (R5) *The language of a project director, of a site engineer, of a cost planner, how they price services and products. (R11)*	Language Culture Profession Supportive Appropriate	Field of collaborative practices and the enabling role of micro process such as language [54]	Developing culturally appropriate workforces, narratives, languages and practices

Following good qualitative research practice, and to minimise any researcher bias, the above process was undertaken by a team of researchers from social sciences, public administration, and construction backgrounds to provide different perspectives on the data. This insider/outsider approach is widely used in psychology and social sciences research to provide different perspectives on data [50]. Comparing and cross-checking codes, categories, and themes between the research team further helped our positionality and reflexivity and minimised any potential disciplinary bias in our results.

Following Hennink [55], results are presented below in simultaneous analysis and interpretation to facilitate deeper reflection on the research findings and help manage researcher positionality to minimise potential subjectivity in the interpretation of results. In line with the traditions of thematic research, the results are supported by selected quotes that, through the coding process, were related to the emergent themes under each research question. Given the limits of the literature on innovation champions, especially in a construction social procurement context, the results below draw on 'laterally relevant' literature from other relevant fields to develop the discussion introducing new avenues of multidisciplinary research and theory, which can be used to enrich the nascent, empirical and theoretical foundations of social procurement in the field of construction.

4. Results and Discussion

4.1. Research Question One: What Motivates People to Become Champions of Social Procurement in Construction Companies?

When asked about their main motivation in championing social procurement within their organisations, our data yielded four main themes which could be broadly clustered under the following headings: opportunity to make a difference; opportunity to leverage the power of business to make change; supporting third sector capacity building in the sector, and political motives. Each of these categories is explored below.

4.1.1. Opportunity to Make a Difference

In responding to this question, all respondents articulated a strong desire to 'make a difference' to the community and change negative stereotypes regarding the disadvantaged people that social procurement policies were designed to help. This finding supports Lewis and Cassell's [21] references to altruism in Table 1 and provides further empirical evidence to support the benevolent attributes and 'idealist norms' that Loosemore and Higgon [20] and Troje and Gluch [34] attribute to social procurement actors that they interviewed in the construction industry:

> *I don't think the fame and the fortune are necessarily going to come. But I do think being able to play some role in making a difference to the lives of a lot of people who find it a little bit hard that's probably the most important thing (R7)*

> *I am fortunate enough to now be in a position that I can challenge people's preconceived ideas of what somebody who needs a handout looks like. It's a personal thing (R14).*

Adding to Table 1 and reflecting insights from wider cognitive career theory that highlights the importance of 'calling' and 'self-efficacy' in career choice decisions [29], our findings indicated that this sense of altruism was often based on personal exposure to disadvantages and involvement in social activism in their previous lives [17].

> *I come from a bit of a working-class background and have been exposed to a few things growing up that sort of motivates me personally. (R2)*

> *I've always had a strong appetite and active participation in social justice issues. (R5)*

4.1.2. Opportunity to Leverage the Power of Business to Cause Change

The findings also contribute new insights into career pathways into social procurement. The data showed that respondents achieved their roles through a diverse range of pathways that were mostly not associated with construction. These backgrounds included involvement in social enterprises, not-for-profits, government, charities, social purpose intermediaries or unions. Reflecting the references in Table 1, which refer to opportunities to bring about change and develop new career opportunities by leveraging their unique relationships to offer new solutions [1,36], most respondents recounted their frustrations of lacking the influence and resources to affect systemic change in their previous professional lives. All were attracted to social procurement by the opportunity to leverage the power of business to solve social problems in new and innovative ways:

> *Mission-based businesses were still not at the main table, here there's the right resourcing, there's the right opportunity for market share. (R5)*

> *Improving the system. I mean part and parcel of that delivery is actually coming up with a better way to do things (R7)*

Notably, in contrast to Loosemore and Higgon [20], as shown in Table 1, only one respondent aspired to create change at a wider systems level, based on an in-depth analysis of the systemic causes of social problems such as unemployment. Therefore, while some champions recognised that the industry needed to change, most champions were quite conservative in their ambitions and were mostly limited to bringing about change at an organisational level.

However, adding to Table 1, our findings also support Troje and Gluch's [10] and Allen's [56] insights into the 'pragmatic problem-solving' aspects of social procurement roles. While respondents all aspired to see social procurement given the same legitimacy as other key project roles, they were also realistic about the challenges they faced in normalising their practices. Interestingly, many considered that their non-construction backgrounds compromised their ability to bring about change because they were often seen as an 'outsider' to the industry. Highlighting the importance of 'social identity' to the acting-out of social procurement roles, these findings support Phua's [57] proposition that individuals define their self-concepts through the organisations with which they identify, and that these identity-based forces are the basis for the development of co-operative behaviour.

Respondents expressed various frustrations in 'normalising social procurement' especially at a project level. These included: the relative lack of priority given to social issues in construction [58]; the need to resolve competing institutional logics and goals [8,59]; and the difficulties in measuring and reporting the impact of their work in a relevant way [46]. Reflecting Grob and Benn's [60] early institutional analysis of how sustainability became imbedded into other industries, all respondents noted the importance of external regulatory and contractual requirements as a coercive force to normalise social procurement in their organisations and supply chains. Therefore, this history may offer relevant insights into how social procurement could be normalised in the construction industry:

> *It's hard for people to think about social value when they have to deliver projects under incredibly demanding time and cost constraints social value is just another distraction.* (R7)

The findings also added an interesting, gendered dimension to Kruse et al's [36] notion of conducive environments for social change in Table 1. Several female respondents agreed that gender equality advances the industry and offers new opportunities to make a difference through these new social procurement roles:

> *I took a step out of construction because I just got really disillusioned about the industry, that it was just rough, it was brutal, and lots of men and just the same. I saw that there was some change happening in this space.* (R2)

4.1.3. Supporting Third-Sector Capacity Buildings in the Third Sector

Adding to the literature in Table 1, many respondents were motivated by the prospect of supporting third-sector capacity buildings in this area. Reflecting the immaturity of the third construction sector [43], most respondents raised concerns about the inability of social enterprises and indigenous businesses (the primary focus of social procurement in Australia) to grasp the opportunities afforded by these new policies. Once again, raising the insider/outsider argument respondents described the misunderstanding of these organisations in the industry and the negative perceptions of their ability to compete for work against industry incumbents. Given the centrality of these organisations to social procurement goals [46], these findings raise critical questions for researchers, policy makers and practitioners regarding how the industry best supports third-sector capacity building to ensure the sustainability of social procurement requirements:

> *The demand is just simply going to outstrip the supply. So we've spent the last couple of years trying to build capacity in social enterprises and Indigenous businesses. We'll teach you how to do it, we'll teach you how to be cost comparative and how to actually stand on your own two feet.* (R14)

The findings also indicate that social procurement champions retain strong relational ties and loyalties in their previous professional lives. Although this is referenced in Table 1 [1], it was also found that their ability to coordinate these background relationships strategically shaped the approach each respondent took to implementing social procurement in their organisations. The link between a respondent's professional background, their coordinating potential and the approach they take to social procurement provide new insights into the drivers of social innovation in this area. This also adds new insights to the nature of cross-sector collaboration [54,61] which underpins social procurement in construction by suggesting that this is likely to be driven from a particular perspective linked to a champion's professional background. In particular, drawing on the recent work of Kyne and Aldrich [53], the roles of bonding capital (the relationships a person has with friends); bridging capital (the ability to connect people); and linking capital (the relationship between a person and formal leaders) appear to be especially important. Our findings, therefore, indicate that, if social procurement continues to become an increasing source of competitive advantage in construction, as Raiden et al. [46] predict, the choice of social procurement champion will be critical in a firms' differentiation strategy and should be considered very carefully:

It's sometimes just the basic ability to facilitate linkages into socially primed or mission based businesses or community organizations in a way that's respectful and mutually beneficial. (R5)

In its simplest sense, social procurement is a form of organizing. (R7)

4.1.4. Political Motives

Prior research has highlighted the political underpinnings of social procurement [2] and questioned its political neutrality [62]. However, adding to the literature in Table 1, only one respondent identified their politics as a motive for implementing social procurement. Instead, reflecting Troje and Gluch's [10] references to pragmatism in social procurement roles, social procurement was generally portrayed as a politically sterile, uncritical, and instrumental mechanism to bring about social change:

I was going to say my motivation is not political, but it is political, I suppose. Yeah if you come to the nub of it my motivation really is quite political, part of my motivation is to respond to the fragmentation that I know is being created deliberately. (R7)

Indeed, the politicisation of social procurement was portrayed by most respondents as risky and potentially counterproductive to their goals. This reflects De Pieri and Teasdale's [62] recent findings regarding people's tendency to ignore the underlying political motives, which inevitably underpin social policy innovations such as social procurement. It also perhaps partially explains the lack of wider ambition to bring about wider systematic change outside the confines of their organisations. As Loosemore and Phua [63] note, the construction industry's relationship with corporate social responsibility is complex, multidimensional and characterised by conflicting institutional logics. To those at project level, the politics of social procurement are largely irrelevant in delivering a project within tight budget and time constraints:

The minute we're seen as the expert or the lefty or the social justice person we don't have clout in the business anymore (R5).

4.2. Research Question Two: What Is the Nature of These Evolving Roles in Construction Companies?

When asked to describe their job in terms of social procurement, our data yielded eight main themes, which can be broadly clustered under the following headings: inspiring people; cross-functional working and linking to communities; developing culturally appropriate workforces, narratives, languages, and practices; challenging institutions and existing incumbent relationships; changing perceptions of value; managing risk; learning, educating, experimenting and innovating; and building trusting relationships. Each of these is explored below.

4.2.1. Inspiring People

Reflecting Schon's [14] portrayal of champions as heroic, inspirational and transformational leaders, most respondents agreed that, in the absence of clearly defined roles, organisational legitimacy and formal power, as well as in the face of significant institutional resistance, there was a large motivational and inspirational element to their job. Adding to Table 1 and Troje's [34] references to 'emotional work' (pathos-based strategies), respondents relied heavily on their personal commitment, determination and passion for change, as well as the power of stories to emotionally engage others in the need for change. However, reflecting the enterprise culture of construction [64] Troje and Gluch's [11] references to pragmatism, respondents universally agreed that this must be coupled with concrete examples and a practical demonstration of what was achievable in practice:

It's about inspiring people a lot of inspirational thought leadership, giving some really fun examples. (R5)

We engaged the hearts of these crusty hardnosed commercially driven supervisors and engineers, that whilst they might not be getting the cheapest subcontract out of a social enterprise they were transforming lives. (R2)

4.2.2. Cross-Functional Working and Linking to Communities

Adding to Woolcott et al.'s [32] notion of value-adding connectors, respondents also articulated the importance of working across organisational boundaries within their organisations at operational, tactical, and strategic levels. Project management, human resources, corporate social responsibility, sustainability, procurement and bidding were generally identified as key internal stakeholders in promoting social procurement. The results indicate that social procurement is not just about external inter-organisational cross-sector collaboration [1,33] but also requires internal intra-organisational collaboration. As Murphy and Eddie [6] noted, social procurement is challenging because it creates new connections between organisational functions in construction organisations that are traditionally disconnected:

I worked with our HR and our diversity inclusion team to come up with some innovative responses that could meet the government requirements (R8).

How you can get sometimes quick tactical wins, but also how do you embed strategic planning into supply chain for long term enduring outcomes with community, it's informing our bids or proposals. (R5)

4.2.3. Developing Culturally Appropriate Workforces, Narratives, Languages and Practices

Research in the field of collaborative practice highlights the important enabling role of language in allowing collaboration to occur and be sustained [42]. Reflecting this, several respondents noted that their roles also involved developing appropriate "narratives" and "culturally appropriate practices", which could connect them to the various stakeholders that they worked with. Once again, findings indicated that organisational story-telling played an especially powerful role in developing these narratives:

So if you can create a narrative where you are slowly but surely educating those people around you and providing opportunities for them to actually engage in that narrative, you're creating other champions (R5).

Sharing stories is critical (R2)

4.2.4. Challenging Institutions and Existing Incumbent Relationships

Reflecting upon Table 1, and particularly the work of Troje [42], respondents described a very traditional industry with strong institutional norms, practices, and notions of value that needed to be challenged. Adding a new industrial relations dimension to this research, surprisingly, construction unions were widely described as a major institutional barrier to implementation. Although we did not interview unions to explore this finding further, respondents argued that unions opposed these policies because the people targeted were typically non-union members and employed on contracts outside of union-negotiated enterprise agreements. Our findings therefore are not in alignment with research which argues that unions help 'all' workers [65] and raise questions for union leaders about their apparent representation of marginalised and disadvantaged groups in society:

It's a high risk environment, they don't want to do anything that's going to potentially pose risks, particularly with unions. (R2)

4.2.5. Changing Perceptions of Value

Adding to emerging debates regarding the need to widen traditional concepts of 'value' in construction [66] and institutional logics and norms in construction [8,59], respondents widely articulated the need to challenge deeply rooted norms around concepts of value and to widen narrow, professionally bounded perceptions of the construction industry's raison d'être. For the respondents, bringing about change involved moving other organisational actors beyond the measurable economic benefits of social procurement to understanding and participating in improving the lives of others through their decision-making agency and personal experience of contributing to the change:

> *Engineers and builders see construction as a technical exercise and rarely if ever ask what buildings are for social procurement is about recognising that buildings are ultimately tools for social change. This is huge change in thinking for procurement professionals who generally are taught to think in one dimension and seek the lowest price. (R3)*

> *Processes around redefining what value is they won't compromise you also need to be realistic around what that cost is, be articulate in telling someone like [company name] this is the cost and this is why it costs, and the outcomes. (R11)*

4.2.6. Managing Risk

Risk management was a recurring theme in the data, although it is largely missing from the extant literature in Table 1. This supports Loosemore et al's [33] research, which shows that social procurement is perceived to be a significant risk to budgets, programs, quality and safety in the construction industry. The risk of delivering social procurement targets was routinely transferred down the contractual chain to subcontractors, raising new questions about their willingness and capacity to deliver. As Loosemore et al. [33] note, it is also where the most significant resistance to change exists:

> *In my head it's all around risk mitigation. So if I have an engineer saying we don't need to procure from Aboriginal businesses or we don't need that person onsite, I'm already thinking about what are their perceived risks, and how can I sell it to them in a way that I'm already solving their problem so that's giving them an opportunity. (R14)*

> *Cost, cost is super important, there needs to be acceptance, endorsement, and a level of confidence and comfort that the team that they're going to be introducing on this project is going to execute it and design and deliver. (R4)*

4.2.7. Learning, Educating, Experimenting and Innovating

All respondents highlighted a significant cognitive and educational component to their roles (in both their organisations and supply chains). This reflects Molloy et al.'s [13] references to 'sensemaking' and Troje and Gluch's [10] references to 'educational work' (Table 1). However, adding to this, respondents concurred that any education needed to be experiential rather than abstract. In line with Kolb's [67] experiential learning model, which shows that this requires both 'concrete experience' and 'abstract conceptualization', respondents noted the critical role of demonstration projects to illustrate the practical feasibility and value of implementing social procurement in their organisations:

> *It's about education, it's about us–without wanting to sound arrogant, it's about us educating our subcontractors as to the potential. (R7)*

> *It's about constantly educating those around you and providing opportunity for those around you to learn and experience. Because I think this is very experiential if never actually experienced it it's very hard to understand or walk in somebody else's shoes. (R6)*

4.2.8. Building Trust and Relationships

Relationship and trust building for mutual gain were universally described as critical to social procurement implementation and the importance of Woolcott et al's [32] concept of 'value-adding connectors' referenced in Table 1 was evident in our results:

> *To make sure that we have infrastructure and the connections that enable us to–I suppose ultimately to outperform. I mean if you spend a lot of time joining the dots, you start to understand just what benefits can be derived by putting people in touch with each other. (R7)*

While the use of demonstration projects was central to building trust through credibility in delivering on project targets, respondents described the challenges of building trust in a project-based context where both internal and external stakeholders are constantly changing. This required large amounts of time to be invested in building and then re-building relationships, often outside formal work hours. Interestingly, while this raises new questions about the work–life balance of social procurement champions, most saw their roles as a cause rather than a job and considered it a worthwhile investment in developing this new field of practice:

> *You take time, you take a lot of time, you can't build a relationship in a minute, it does take time, and it gets back to all of those things about being genuine, about your transparency, about those levels of trust.* (R6)

> *I mean my phone's always on, I never turn my phone off and there are drawbacks to that I know.* (R7).

5. Conclusions

This research was set within the emerging context of social procurement in construction. In addressing the general lack of theory and empirical research in this area, the aim was to contribute to the nascent research on the organisational professional who are shaping this new field of practice. Specifically, by mobilising the long tradition of research into organisational champions in the broader field of organisation studies, the motives driving these professionals and the nature of their evolving roles were explored. Mobilizing the theoretical perspectives offered by organisational champion research in Table 1 has helped to contribute numerous new conceptual and practical insights into the function of social procurement champions in the Australian construction industry for researchers, policy makers and practitioners working in this emerging field.

Concerning research question one, our findings indicate that the champions on which the success of social procurement depends come from a wide range of professional backgrounds and bring a variety of new social capital to the construction industry which can enhance the social value it creates. Strategically, as companies innovate to gain a competitive advantage in this increasingly important area of construction performance, these relationships are critical in differentiating an organisation's approach to social procurement. It is found that these people are linked by a shared sense of altruism often rooted in personal exposure to disadvantages and social activism. Typically frustrated by their previous experiences of working in the third-sector to bring about meaningful social change, these champions put aside political judgements and ideologies and see the potential power of leveraging construction procurement to make a difference to the communities in which the industry builds. This is not easy and they achieve this by challenging institutional norms, practices and supply chains relationships, and traditional discourses around value and competitiveness in an increasingly socially conscious business environment.

For research question two, the results highlight a significant inspirational, emotive and educational element to these new roles. However, in the long-term, the results indicate that sustainable changes in industry norms and practices will depend on evidence-based demonstrations of what is possible at a project level within the constraints of other project objectives. This involves de-risking social procurement and changing deeply protected and entrenched procurement and recruitment practices, supply chain relationships and biases and stereotypes about the 'ideal' type of person and supplier working in the construction industry. This requires a strong focus on experiential learning, educating, experimenting, innovating and developing new cross-disciplinary narratives (and language) around the importance of social value. The findings show that this requires enormous investments of time and energy, often outside formal day-to-day roles. It also depends on the development of new relational and collaborative capital, knowledge and competencies to build the trusting inter-organisational and intra-organisational relationships, on which the successful implementation of social procurement depends.

The implications for practice point to the importance of providing champions with the freedom and flexibility to build their evolving social procurement roles. It also highlights the need for organisations to support their social procurement champions in challenging deeply embedded institutional norms and practices, which currently portray social procurement as a threat to effective project delivery. Education and training are key to changing these norms and practices, as is the use of demonstration projects to illustrate the impact that these policies can have. It is clear that incremental, rather than radical, innovation is most likely to succeed, and that organisations must support their champions by adequately resourcing and legitimizing their roles with a clear identity and position within an organisational structure. Developing methodologies to measure social impact to provide an evidential basis for change is also crucial.

While these results provide important, new, conceptual and practical insights to inform social procurement research, as well as practice and policy in construction, the limitation of the findings reported is the Australian-centric, exploratory and qualitative nature of the research. Although the sample size is well within the norms of qualitative research and was determined by the concepts of theoretical saturation in analysis, more research is needed to cross-validate our results. In particular, given that the sample was mainly focused on head-office-type roles and external consultants, more research is needed into how social procurement champions operate at a project level and in client organisations (public and private). This is important because the literature indicates that a variety of types of champions are needed to implement innovations into organisations. It is important to understand the different types of champions that are needed to implement these policies. Further research is also required in different international social procurement policy contexts. As noted above, social procurement policies are emerging in many countries and will inevitably differ in the scope, focus and degree to which they mandate social value outputs. The extent to which this influences the practices of social procurement champions in advancing this field of practice needs to be explored. Other important avenues of research which emerge from these findings include: the role of gender in influencing change in such a highly masculinised industry; the emotional dimensions of social procurement roles; and the role of social capital, linking capital and relational capital in facilitating cross-sector collaboration. The findings also highlight the potential value of theories of social capital, social networks, new institutionalism and cross-sector collaboration as new conceptual lenses to understand these issues. In particular, the use of social network analysis offers the potential for new quantitative insights into the types of social structures and social capital (bonding, bridging, linking), which best facilitate the inter and intra-organisational collaborations which appear to be so central to champion roles.

Author Contributions: Conceptualization, M.L., R.K., J.B. and G.D.-S.; methodology, M.L., R.K., J.B. and G.D.-S.; software, M.L., R.K., J.B. and G.D.-S.; validation, M.L., R.K., J.B. and G.D.-S.; formal analysis, M.L., R.K., J.B. and G.D.-S.; investigation, M.L., R.K., J.B. and G.D.-S.; resources, M.L., R.K., J.B. and G.D.-S.; data curation, M.L., R.K., J.B. and G.D.-S.; writing—original draft preparation, M.L., R.K., J.B. and G.D.-S.; writing—review and editing, M.L., R.K., J.B. and G.D.-S.; visualization, M.L., R.K., J.B. and G.D.-S.; supervision, M.L., R.K. and J.B.; project administration, M.L., R.K. and J.B.; funding acquisition, M.L., R.K. and J.B. All authors have read and agreed to the published version of the manuscript.

Funding: This research was funded by the Australian Research Council, grant number LP170100670.

Institutional Review Board Statement: The study was conducted according to the guidelines of the Declaration of Helsinki, and approved by the Human Research Ethics Committee of University of Technology Sydney (protocol code ETH 20 4767 approved on 19/5/2020).

Informed Consent Statement: Informed consent was obtained from all subjects involved in the study.

Conflicts of Interest: The authors declare no conflict of interest.

References

1. Barraket, J.; Keast, R.; Furneaux, C. *Social Procurement and New Public Governance*, 1st ed.; Routledge Critical Studies in Public Management; Routledge: New York, NY, USA, 2016; ISBN 978-0-415-85855-7.
2. McCrudden, C. Using Public Procurement to Achieve Social Outcomes. *Nat. Resour. Forum* **2004**, *28*, 257–267. [CrossRef]

3.   McNeill, J. *Enabling Social Innovation Assemblages: Strengthening Public Sector Involvement*; Western Sydney University: Sydney, Australia, 2017.
4.   Loosemore, M. Social Procurement in UK Construction Projects. *Int. J. Proj. Manag.* **2016**, *34*, 133–144. [CrossRef]
5.   Torres-Pruñonosa, J.; Plaza-Navas, M.A.; Díez-Martín, F.; Beltran-Cangrós, A. The Intellectual Structure of Social and Sustainable Public Procurement Research: A Co-Citation Analysis. *Sustainability* **2021**, *13*, 774. [CrossRef]
6.   Murphy, M.; Eadie, R. Socially Responsible Procurement: A Service Innovation for Generating Employment in Construction. *Built Environ. Proj. Asset Manag.* **2019**, *9*, 138–152. [CrossRef]
7.   Loosemore, M.; Alkilani, S.; Mathenge, R. The Risks of and Barriers to Social Procurement in Construction: A Supply Chain Perspective. *Constr. Manag. Econ.* **2020**, *38*, 552–569. [CrossRef]
8.   Troje, D.; Andersson, T. As above, Not so below: Developing Social Procurement Practices on Strategic and Operative Levels. *Equal. Divers. Incl. Int. J.* **2020**, *40*, 242–258. [CrossRef]
9.   Troje, D.; Kadefors, A. Employment Requirements in Swedish Construction Procurement—Institutional Perspectives. *J. Facil. Manag.* **2018**, *16*, 284–298. [CrossRef]
10. Troje, D.; Gluch, P. Populating the Social Realm: New Roles Arising from Social Procurement. *Constr. Manag. Econ.* **2020**, *38*, 55–70. [CrossRef]
11. Troje, D.; Gluch, P. Beyond Policies and Social Washing: How Social Procurement Unfolds in Practice. *Sustainability* **2020**, *12*, 4956. [CrossRef]
12. Markham, S.K. Corporate Championing and Antagonism as Forms of Political Behavior: An R&D Perspective. *Organ. Sci.* **2000**, *11*, 429–447. [CrossRef]
13. Molloy, C.; Bankins, S.; Kriz, A.; Barnes, L. Making Sense of an Interconnected World: How Innovation Champions Drive Social Innovation in the Not-for-Profit Context. *J. Prod. Innov. Manag.* **2020**, *37*, 274–296. [CrossRef]
14. Schon, D.A. Champions for Radical New Inventions—ScienceOpen. *Harv. Bus. Rev.* **1963**, *41*, 77–86.
15. Ernstsen, S.N.; Whyte, J.; Thuesen, C.; Maier, A. How Innovation Champions Frame the Future: Three Visions for Digital Transformation of Construction. *J. Constr. Eng. Manag.* **2021**, *147*. [CrossRef]
16. Sergeeva, N. What Makes an "Innovation Champion"? *Eur. J. Innov. Manag.* **2016**, *19*, 72–89. [CrossRef]
17. Shibeika, A.; Harty, C. Digital Innovation in Construction: Exploring the Firm-Projects Interface. In Proceedings of the 31st Annual ARCOM Conference, Lincoln, UK, 7–9 September 2015; Raiden, A., Aboagye-Nimo, E., Eds.; Association of Researchers in Construction Management: Lincoln, UK, 2015; pp. 1167–1176.
18. Tatum, C.B. (Bob) Construction Engineering Research: Integration and Innovation. *J. Constr. Eng. Manag.* **2018**, *144*, 04018005. [CrossRef]
19. Xue, X.; Zhang, R.; Wang, L.; Fan, H.; Yang, R.J.; Dai, J. Collaborative Innovation in Construction Project: A Social Network Perspective. *KSCE J. Civ. Eng.* **2018**, *22*, 417–427. [CrossRef]
20. Loosemore, M.; Higgon, D. *Social Enterprise in the Construction Industry: Building Better Communities*, 1st ed.; Routledge: Florence, SC, USA, 2016; ISBN 978-1-138-82406-5.
21. Lewis, K.; Cassells, S. Barriers and Drivers for Environmental Practice Uptake in SMEs: A New Zealand Perspective. *Int. J. Bus. Stud.* **2010**, *18*, 7–21. [CrossRef]
22. Wood, B.E.; Cornforth, S.; Beals, F.; Taylor, M.; Tallon, R. Sustainability Champions? Academic Identities and Sustainability Curricula in Higher Education. *Int. J. Sustain. High. Educ.* **2016**, *17*, 342–360. [CrossRef]
23. Day, D.L. Raising Radicals: Different Processes for Championing Innovative Corporate Ventures. *Organ. Sci.* **1994**, *5*, 148–172. [CrossRef]
24. Howell, J.M.; Higgins, C.A. Champions of Change: Identifying, Understanding, and Supporting Champions of Technological Innovations. *Organ. Dyn.* **1990**, *19*, 40–56. [CrossRef]
25. Rogers, E.M. *Diffusion of Innovations*, 5th ed.; Free Press: New York, NY, USA, 2003; ISBN 978-0-7432-2209-9.
26. Rost, K.; Hölzle, K.; Gemünden, H.-G. Promotors or Champions? Pros and Cons of Role Specialisation for Economic Process. *Schmalenbach Bus. Rev.* **2007**, *59*, 340–363. [CrossRef]
27. Schumpeter, J. On the Concept of Social Value. *Q. J. Econ.* **1909**, *23*, 213–232. [CrossRef]
28. Hargreaves, T. Practice-Ing Behaviour Change: Applying Social Practice Theory to pro-Environmental Behaviour Change. *J. Consum. Cult.* **2011**, *11*, 79–99. [CrossRef]
29. Crosby, B.C.; Bryson, J.M. A Leadership Framework for Cross-Sector Collaboration. *Public Manag. Rev.* **2005**, *7*, 177–201. [CrossRef]
30. Crosby, B.C.; Bryson, J.M. Integrative Leadership and the Creation and Maintenance of Cross-Sector Collaborations. *Leadersh. Q.* **2010**, *21*, 211–230. [CrossRef]
31. Huxham, C.; Vangen, S. *Managing to Collaborate: The Theory and Practice of Collaborative Advantage*; Taylor & Francis Group: Florence, SC, USA, 2005; ISBN 978-0-203-01016-7.
32. Woolcott, G.; Keast, R.; Pickernell, D. Deep Impact: Re-Conceptualising University Research Impact Using Human Cultural Accumulation Theory. *Stud. High. Educ.* **2020**, *45*, 1197–1216. [CrossRef]
33. Loosemore, M.; Denny-Smith, G.; Barraket, J.; Keast, R.; Chamberlain, D.; Muir, K.; Powell, A.; Higgon, D.; Osborne, J. Optimising Social Procurement Policy Outcomes through Cross-Sector Collaboration in the Australian Construction Industry. *Eng. Constr. Archit. Manag.* **2020**, *28*, 1908–1928. [CrossRef]

34. Gorse, C.; Neilson, C.J. Rhetorical Strategies to Diffuse Social Procurement in Construction. In Proceedings of the 34th Annual ARCOM Conference, Belfast, UK, 3–5 September 2018; Association of Researchers in Construction Management: Belfast, UK, 2018; pp. 500–514.
35. Ghalwash, S.; Tolba, A.; Ismail, A. What Motivates Social Entrepreneurs to Start Social Ventures? An Exploratory Study in the Context of a Developing Economy. *Soc. Enterp. J.* **2017**, *13*, 268–298. [CrossRef]
36. Kruse, P.; Wach, D.; Wegge, J. What Motivates Social Entrepreneurs? A Meta-Analysis on Predictors of the Intention to Found a Social Enterprise. *J. Small Bus. Manag.* **2021**, *59*, 477–508. [CrossRef]
37. Wunderer, R. Employees as "Co-intrapreneurs"—A Transformation Concept. *Leadersh. Organ. Dev. J.* **2001**, *22*, 193–211. [CrossRef]
38. de Villiers-Scheepers, M.J. Motivating Intrapreneurs: The Relevance of Rewards. *Ind. High. Educ.* **2011**, *25*, 249–263. [CrossRef]
39. Usman, M.; Ali, M.; Ogbonnaya, C.; Babalola, M.T. Fueling the Intrapreneurial Spirit: A Closer Look at How Spiritual Leadership Motivates Employee Intrapreneurial Behaviors. *Tour. Manag.* **2021**, *83*, 104227. [CrossRef]
40. *Commonwealth Indigenous Procurement Policy: 1 July 2015*; Australian Government: Canberra, Australia, 2015.
41. Visser, W.; Crane, A. *Corporate Sustainability and the Individual: Understanding What Drives Sustainability Professionals as Change Agents*; Social Science Research Network: Rochester, NY, USA, 2010.
42. Troje, D. Inter-Project Support for Creating Social Value Through Social Procurement. In Proceedings of the 36th Annual ARCOM Conference, Leeds, UK, 7 September 2020; Scott, L., Neilson, C.J., Eds.; Association of Researchers in Construction Managemen: Leeds, UK, 7 September 2020; pp. 105–114.
43. Loosemore, M. Building a New Third Construction Sector through Social Enterprise. *Constr. Manag. Econ.* **2015**, *33*, 724–739. [CrossRef]
44. Wiesner, R.; Chadee, D.; Best, P. Managing Change Toward Environmental Sustainability: A Conceptual Model in Small and Medium Enterprises. *Organ. Environ.* **2018**, *31*, 152–177. [CrossRef]
45. Blackstone, A. *Sociological Inquiry Principles: Qualitative and Quantitative Methods*; Flat World Knowledge: Irvington, NY, USA, 2012.
46. Raiden, A.; Loosemore, M.; King, A.; Gorse, C. *Social Value in Construction*; Routledge: London, UK, 2019; ISBN 978-1-351-58720-4.
47. Punch, K.F. author *Introduction to Social Research: Quantitative and Qualitative Approaches*, 2nd ed.; Sage Publications: Thousand Oaks, CA, USA, 2005; ISBN 978-0-7619-4417-1.
48. Loosemore, M.; Lim, B.T.H. Linking Corporate Social Responsibility and Organizational Performance in the Construction Industry. *Constr. Manag. Econ.* **2017**, *35*, 90–105. [CrossRef]
49. Saunders, B.; Sim, J.; Kingstone, T.; Baker, S.; Waterfield, J.; Bartlam, B.; Burroughs, H.; Jinks, C. Saturation in Qualitative Research: Exploring Its Conceptualization and Operationalization. *Qual. Quant.* **2018**, *52*, 1893–1907. [CrossRef]
50. Hayfield, N.; Huxley, C. Insider and Outsider Perspectives: Reflections on Researcher Identities in Research with Lesbian and Bisexual Women. *Qual. Res. Psychol.* **2015**, *12*, 91–106. [CrossRef]
51. Guest, G.; MacQueen, K.; Namey, E. *Applied Thematic Analysis*; SAGE Publications, Inc.: Thousand Oaks, CA, USA, 2012; ISBN 978-1-4129-7167-6.
52. Gioia, D.A.; Corley, K.G.; Hamilton, A.L. Seeking Qualitative Rigor in Inductive Research: Notes on the Gioia Methodology. *Organ. Res. Methods* **2013**, *16*, 15–31. [CrossRef]
53. Kyne, D.; Aldrich, D.P. Capturing Bonding, Bridging, and Linking Social Capital through Publicly Available Data. *Risk Hazards Crisis Public Policy* **2020**, *11*, 61–86. [CrossRef]
54. Stout, M.; Bartels, K.; Love, J.M. Clarifying Collaborative Dynamics in Governance Networks. In *From Austerity to Abundance?: Creative Approaches to Coordinating the Common Good*; Stout, M., Ed.; Emerald Group Publishing: Bingley, UK, 2018; pp. 91–115.
55. Monique, M. *Hennink Focus Group Discussions*; Understanding Qualitative Research; Oxford University Press: New York, NY, USA, 2014; ISBN 978-0-19-985616-9.
56. Allen, B. Broader Outcomes in Procurement Policy—A Case of New Zealand Pragmatism. *J. Public Procure.* **2021**, *21*, 318–341. [CrossRef]
57. Phua, F.T.T. The Antecedents of Co-operative Behaviour among Project Team Members: An Alternative Perspective on an Old Issue. *Constr. Manag. Econ.* **2004**, *22*, 1033–1045. [CrossRef]
58. Kordi, N.E.; Belayutham, S.; Zulkifli, A.R. The Attributes of Social Sustainability in Construction: A Theoretical Exploration. In Proceedings of the 35th Annual ARCOM Conference, Leeds, UK, 2–4 September 2019; Association of Researchers in Construction Managemen: Leeds, UK, 2 September 2019; pp. 629–638.
59. Jia, A.Y.; Rowlinson, S.; Loosemore, M.; Gilbert, D.; Ciccarelli, M. Institutional Logics of Processing Safety in Production: The Case of Heat Stress Management in a Megaproject in Australia. *Saf. Sci.* **2019**, *120*, 388–401. [CrossRef]
60. Grob, S.; Benn, S. Conceptualising the Adoption of Sustainable Procurement: An Institutional Theory Perspective. *Australas. J. Environ. Manag.* **2014**, *21*, 11–21. [CrossRef]
61. Keast, R.; Charles, M.; Modzelewski, P. The Cost of Collaboration–More than You Bargained For? *Power Persuade 2017*. Available online: http://www.powertopersuade.org.au/blog/the-cost-of-collaboration-more-than-budgeted-for/13/4/2017 (accessed on 10 October 2021).
62. De Pieri, B.; Teasdale, S. Radical Futures? Exploring the Policy Relevance of Social Innovation. *Soc. Enterp. J.* **2021**, *17*, 94–110. [CrossRef]

63. Loosemore, M. Responsible Corporate Strategy in Construction and Engineering: "Doing the Right Thing?". Spon Press: London, UK; New York, NY, USA, 2010; ISBN 978-0-415-45909-9.
64. Green, S.D.; Harty, C.; Elmualim, A.A.; Larsen, G.D.; Kao, C.C. On the Discourse of Construction Competitiveness. *Build. Res. Inf.* **2008**, *36*, 426–435. [CrossRef]
65. Walters, M.; Mishel, L. *How Unions Help All Workers*; Economic Policy Institute: Washington, DC, USA, 2003.
66. Green, S.D.; Sergeeva, N. Value Creation in Projects: Towards a Narrative Perspective. *Int. J. Proj. Manag.* **2019**, *37*, 636–651. [CrossRef]
67. Kolb, D. *Experiential Learning: Experience as the Source of Learning and Development*, 2nd ed.; Pearson Education: Upper Saddle River, NJ, USA, 2015.

Article

The Impact of Temporary Means of Access on Buildings Envelope's Maintenance Costs

Cláudia Ferreira [1,*], Ilídio S. Dias [1], Ana Silva [1], Jorge de Brito [1,2] and Inês Flores-Colen [1,2]

[1] CERIS, Instituto Superior Técnico, Universidade de Lisboa, Av. Rovisco Pais, 1049-001 Lisboa, Portugal; ilidio.dias@tecnico.ulisboa.pt (I.S.D.); ana.ferreira.silva@tecnico.ulisboa.pt (A.S.); jb@civil.ist.utl.pt (J.d.B.); ines.flores.colen@tecnico.ulisboa.pt (I.F.-C.)

[2] Department of Civil Engineering, Architecture and Georesources, Instituto Superior Técnico, Universidade de Lisboa, Av. Rovisco Pais, 1049-001 Lisboa, Portugal

* Correspondence: claudiaarferreira@tecnico.ulisboa.pt

Citation: Ferreira, C.; Dias, I.S.; Silva, A.; de Brito, J.; Flores-Colen, I. The Impact of Temporary Means of Access on Buildings Envelope's Maintenance Costs. *Buildings* **2021**, *11*, 601. https://doi.org/10.3390/buildings11120601

Academic Editor: David Arditi

Received: 27 October 2021
Accepted: 26 November 2021
Published: 1 December 2021

Abstract: Accessibility to buildings' envelope depends on efficient inspection and other maintenance actions of their components. When access to these components is not planned, special means of access are required to carry out the maintenance work. Means of access, besides having a fundamental role on the quality of maintenance works of building envelope components, also represents a considerable part of the maintenance costs. Thus, to optimize costs and resources in maintenance plans, assessment of the impact of the means of access on maintenance costs is crucial. For works in height, there are several alternative means of access. The choice of the most adequate solution is strongly linked to the characteristics (e.g., architecture, height) and constraints (e.g., users, surrounding space) of each building, the maintenance needs of the envelope, and the time and funds available for the intervention. Therefore, in this study, a sensitivity analysis to understand how the cost of means of access can influence the maintenance costs is carried out. Moreover, the optimisation of maintenance activities in façade claddings is also analysed. This study intends to assess whether it is advantageous to consider permanent means of access during the design phase or opt for temporary means of access. In a first stage, the impact of six temporary means of access (supported and suspended scaffolds; articulated booms; telescopic booms; scissor lifts; and rope access) on the maintenance plans developed for the six types of claddings (ceramic tiling systems—CTS, natural stone claddings—NSC, rendered façades—RF, painted surfaces—PS, external thermal insulation composite systems—ETICS, and architectural concrete façades—ACF) is examined. The impact is estimated through a stochastic maintenance model based on Petri nets. After that, a sensitivity analysis and a multi-criteria decision analysis are performed. Based on the results, general recommendations are presented concerning the maintenance strategies to adopt in the cladding solutions analysed. The results reveal that planning the means of access during the design stage can be economically beneficial for all buildings' envelope components.

Keywords: degradation; maintenance; façade claddings; means of access

1. Introduction

In the building management field, an adequate degradation condition can be easily achieved but maintaining this degradation condition over time is another matter. Regular inspections and maintenance interventions are fundamental to preserve the degradation state of the building components, whatever their size, complexity, or function [1]. Accessibility is one of the main barriers to maintenance. While there are building envelope components that are easily accessible (e.g., windows frames or even roofs), external façades usually require special means of access to carry out the maintenance works [2]. Furthermore, these are not commonly designed to provide safe access to maintenance workers. In the construction sector, works in height are probably the activities that more frequently require special means of access [3].

In buildings, accessibility to all components, for maintenance activities, should be a concern of stakeholders at the design phase [4]. For example, the external façades of buildings should be designed by considering safe and easy access of the maintenance staff. The shape of the surface and number of features influence how the façade can be accessed [1]. Not planning assets with adequate means of access (permanent and/or temporary) is usually translated into higher maintenance costs, since temporary means of access need to be used to overcome this limitation and are not usually predicted in maintenance costs' planning [5]. In the literature, there are no research studies that quantify the economic impact of temporary means of access during the service life of building components. Moreover, globally, the number of studies in the literature that deal with maintenance costs [6], the cost of the different types of inspections/maintenance of buildings [7], or the difference between what is planned and what is spent [8] is quite low. Concerning accessibility, the studies only suggest that this impact is real and significant to put into practice maintenance strategies during the life cycle. For example, according to the Construction Industry Institute (CII), temporary facilities for access are one of the main categories that influence the indirect construction costs [9]. El-Haram and Horner [10], in a study about the factors that affect maintenance costs, refer that, to reduce the maintenance costs, the number and/or the duration of the maintenance activities must be reduced. The authors also assert that the duration of the maintenance activities can be reduced by increasing the accessibility and planning maintenance resources. Moreover, Ferreira et al. [11] suggest that the fixed costs, namely the means of access, have a considerable influence on maintenance costs and, consequently, on the choice of maintenance strategies. In terms of safety, from a study about the implementation of design solutions to improve safety on the construction site during the construction stage in roofs, it is understood that permanent solutions are more expensive than temporary solutions [12]. However, in this study, the exploration stage is not considered. Still, from the findings of Rajendran and Gambatese [12], it is assumed that, when maintenance is really implemented in buildings during the life cycle, there are advantages in choosing the permanent solutions. The high initial costs of these solutions can be amortized over the years because there are no expenses with the non-predicted temporary means of access.

Methodology

In this study, a sensitivity analysis to understand how the cost of the means of access can influence the maintenance costs is carried out. The optimisation of maintenance activities in façade claddings is also analysed. This research intends to assess whether it is advantageous to consider permanent means of access during the design phase or to opt for predicted temporary means of access. To do that, in a first step, a stochastic maintenance model based on Petri nets is used to assess the impact of different temporary means of access in the maintenance plans developed for different types of claddings is examined. Six temporary means of access (supported and suspended scaffolds, articulated booms, telescopic booms, scissor lifts, and rope access) and six façade claddings (ceramic tiling systems—CTS, natural stone claddings—NSC, rendered façades—RF, painted surfaces—PS, external thermal insulation composite systems—ETICS, and architectural concrete façades—ACF) are analysed. These equipment types are chosen because they are the most common in Portugal. In a second phase, a sensitivity analysis is performed to analyse how the different façade claddings' maintenance strategy costs are influenced by the cost of the means of access, and a multi-criteria decision analysis is carried out to analyse the more beneficial alternatives. Furthermore, based on the results obtained in the multi-criteria analysis, general recommendations are presented for the maintenance strategies. In this study, only the cumulative average costs of maintenance plans are considered. As a simplification, it is assumed that both the impact of the different maintenance activities on the degradation state and the time intervals between inspections are not influenced by the means of access selected.

2. Means of Access

Means of access are assemblies/structures or equipment used to provide support to people and material during the construction and/or maintenance of the building elements, at heights above the ground, allowing activities to be carried out under safe conditions [13]. According to its nature, the means of access are classified in two types [14]: permanent and temporary. Although permanent access systems can be implemented during the operational stage of the buildings, these systems are usually planned during the design stage and implemented at the construction stage [14]. Since the main purpose of these systems is aiding the periodic maintenance of building, they should be designed to facilitate access of the maintenance personnel to the building envelope, as well as to cover the largest possible area of the façade without causing damage to it, when the system is being used [15]. These systems are generally applied in buildings where maintenance requirements are high (for example, glass façade). There are several solutions for permanent access systems [14], depending on the buildings' characteristics (architecture, height, among others) and constraints of each building (users, surrounding space, among others), the maintenance needs of the façade, and the time and funds available for the intervention [16].

On the other hand, temporary access systems are only installed before inspections or maintenance interventions take place and are removed after completion of the actions. This type of access system is used in two situations: when buildings do not have permanent means of access or when have permanent means of access, but these means are not able to cover all the singular points that need maintenance. As for permanent access systems, there are a variety of solutions for temporary access systems and its choice must take into account the characteristics and constraints of each building [16]. Among them are supported and suspended scaffolds, aerial work platforms (articulated booms, telescopic booms, and scissors lifts), and rope access (Figure 1). The temporary access systems represented in Figure 1 illustrate the means of access most often used in maintenance activities in current buildings.

2.1. Supported Scaffolds

Supported scaffolds are considered the means of access most used in building maintenance [5], as seen in Figure 1a. Scaffolding consists of a set of horizontal platforms, at different levels, linked by ladders and shoring props. In terms of disadvantages, to guarantee the safety, the temporary structure needs anchorage points. If the façade does not contemplate these points, the surface can be damaged. In addition, supported scaffolds require space and suitable ground to be installed [5]. Finally, the assembly and disassembly have a considered impact on the labour cost and construction schedule [9].

2.2. Suspended Scaffolds

Suspended scaffolds, as seen in Figure 1b, consist of a platform, rigging to suspend the platform, a hoist to move it up and down, and an anchor for the suspension ropes [17,18]. This equipment is provisionally suspended from the roof, and it is normally used in maintenance operations in high-rise buildings [18]. The choice of this equipment must consider the weather conditions (namely the action of the wind) and the characteristics of the roof. By comparison with the supported scaffolds, the assembly and disassembly times are shorter, it has a good carrying capacity for people and materials, and it allows access to façades with restrictions at the soil level [19], but it has limitations in complex features and protrusions from the façade [2].

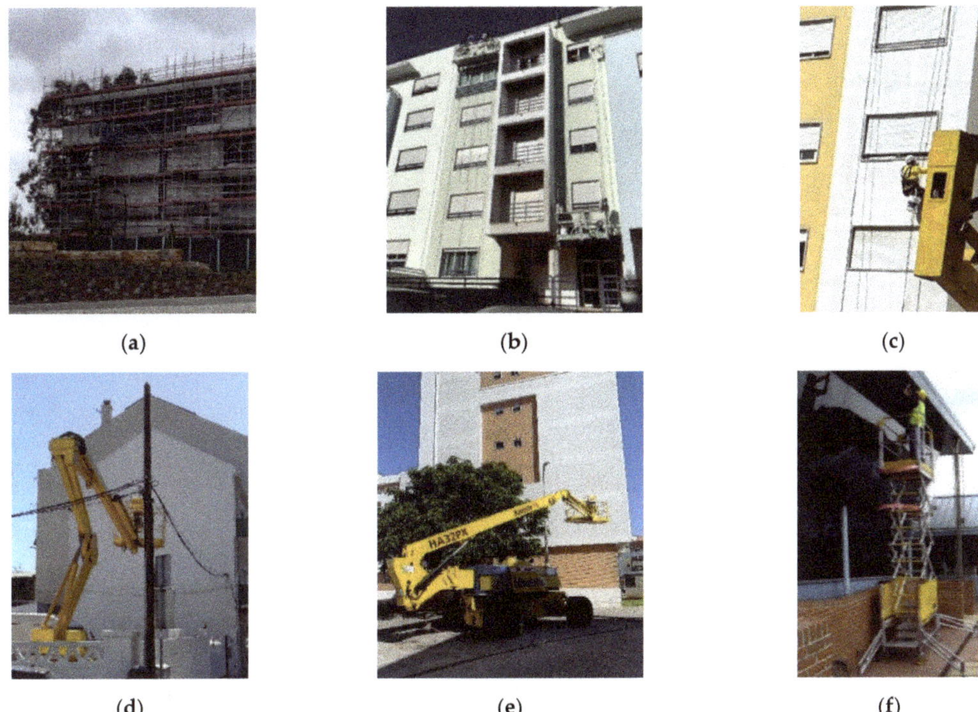

Figure 1. Illustrative examples of the different temporary means of access: (**a**) supported scaffolds; (**b**) suspended scaffolds; (**c**) rope access; (**d**) articulated booms; (**e**) telescopic booms; (**f**) scissor lifts.

2.3. Rope Access

Compared to other means of access [5,20], rope access (Figure 1c) provides easy access to all areas of the façade, ensures quick installation, and requires little equipment. Furthermore, the visual impact on the building is reduced and it does not need to damage the cladding to work on the façade [20]. On the other hand, the workers need specific training [20], and it is uncomfortable for long-term works [2,16]. The application of rope access ensures that the building meets adequate conditions for the installation of anchorage points. Moreover, good weather conditions should be assured during the intervention on the façade. The use of this technique can also jeopardize the quality of the maintenance work, thus this technique is mainly used for light maintenance actions and surveying [5,21].

2.4. Aerial Work Platforms

In terms of aerial work platforms, articulated booms (Figure 1d), telescopic booms (Figure 1e), and scissor lifts (Figure 1f) are the most common. Aerial work platforms are motorised vehicles and have reduced installation times. However, the weather conditions can also influence the performance and safety of this means of access, particularly the incidence of wind. Moreover, the ground surrounding the façade should have sufficient strength to support the equipment [14]. The work platform of the articulated and telescopic booms extends (vertical and horizontal) beyond the wheelbase of the supporting structure while scissor lifts are elevating platforms (only with vertical movement) [22]. These means of access are normally used in occasional and short-term maintenance operations in low-to-medium-height buildings. The reach of these equipment types is reduced when compared with the other means of access (approximately 20 m to articulated and telescopic booms, and 10 m to scissor lifts) [17,22].

3. Materials and Methods

To understand the influence of the means of access on the maintenance costs of building envelope elements, six temporary access systems are analysed: (i) supported scaffolds; (ii) suspended scaffolds; (iii) articulated booms; (iv) telescopic booms; (v) scissor lifts; and (vi) rope access.

3.1. Costs of Means of Access to Maintain the Building Envelope

Table 1 presents the costs of the different means of access analysed. These are average costs in Portugal, which were obtained through consultation with companies in this area of expertise from May to July 2020. As a case study, it was assumed that the equipment would be needed to intervene a 250 m^2 façade cladding.

Table 1. Unit cost, at year 0, for the different means of access.

Means of Access	Supported Scaffolds	Aerial Work Platform			Suspended Scaffolds	Rope Access
		Articulated Boom	Telescopic Booms	Scissor Lifts		
Costs (€/m^2)	10.20	8.13	11.84	10.58	4.68	2.12

Rope access is the means of access with the lowest cost (2.12 €/m^2). This means of access is suitable for cleaning operations on façades and possibly for more occasional minor interventions, considering the type of coating to be intervened. For example, this technique has been used in total replacement of the coating in painted surfaces. After rope access, suspended scaffolds are the most economical means of access, with an estimated value of around 4.68 €/m^2. Compared to rope access, this equipment has a larger working area as well as greater load capacity. Regarding aerial work platforms, articulated booms are the most economical, with an average cost of 8.13 €/m^2, which is lower than supported scaffolds (10.20 €/m^2). Based on that and considering the constraints of supported scaffolds, such as the times of assembly and disassembly, visual impact on the façade, among others, articulated booms can be a good alternative, particularly in minor interventions.

3.2. Maintenance Model

This analysis is supported by a stochastic maintenance model based on Petri nets. This model was previously developed by the authors [23], as a useful tool to analyse the trade-off between different alternatives. The maintenance model developed is a condition-based model that, in addition to the degradation process, also includes the inspection and maintenance processes.

The classification system implemented is composed of five degradation conditions, varying between A (no visual degradation) and E (generalized degradation). The degradation condition is defined based on the severity of degradation index, S_w-Equation (1).

$$S_w = \frac{\sum (A_n \times k_n \times k_{a,n})}{A \times \sum k},$$ (1)

where S_w, the severity of degradation (in %), is the ratio between the area affected by the anomalies observed in the building component, weighted according to their severity, and a reference area is equivalent to the total area with the highest possible degradation level. A_n—the area affected by anomaly n (in m^2); k_n—the multiplication factor for the anomaly n; $k_{a,n}$—the weighting coefficient according to the relative weight of the anomaly n; A—the total area of the constructive solution (in m^2); and k—the multiplying factor corresponding to the highest degradation condition of the area A. These parameters vary according to the constructive solution (for CTS, see [24]; for NSC, see [25]; for RF, see [26]; for PS, see [27]; for ETICS, see [28]; and, for ACF, see [29]).

In the maintenance model, interventions are divided in four levels: (i) inspections; (ii) cleaning operations; (iii) minor interventions; and (iv) total replacement. In the inspection, a visual assessment of the building component is carried out. In the cleaning operation, the removal of the visual anomalies (soiling, stains, and other deposits) is proceeded. In the minor intervention, in addition of the cleaning operations, localized repair and/or partial replacement are also considered. Finally, in the total replacement, the complete replacement of the building component is carried out. Since the maintenance model is condition-based, a decision of maintenance activities is taken based on the inspection diagnosis. In this case study, it is assumed that, if the degradation condition of the building component is A, no maintenance is carried out; if it is B, a cleaning operation is needed; if it is C, a minor intervention is required; and, finally, if it is D or E, total replacement must be carried out. The outputs of the maintenance model (service life, life-cycle costs, efficiency index, and number of interventions) are used to assess the different alternatives.

4. Sensitivity Analysis

To analyse how the different façade claddings' maintenance strategy costs are influenced by the cost of the means of access, a sensitivity analysis is carried out. This analysis has only impact on the maintenance costs. The impact of different maintenance activities on the degradation condition of claddings as well as the time intervals between inspections are the same regardless of the means of access selected. Furthermore, in the composition of the price of maintenance activities (cleaning operations, minor interventions, and total replacement), there is only a change in the cost of the means of access. The costs of materials, equipment, and labour remain the same. Results regarding the service life, efficiency index, and number of interventions can be consulted in Ferreira et al. [30]. In this study, there are six façade claddings (CTS, NSC, RF, PS, ETICS, and ACF), and four maintenance strategies, which are analysed below.

- Maintenance strategy 1 (MS1): includes only the total replacement of the façade. This maintenance strategy is used to characterize the strategy most adopted by the owners or managers of the buildings. The façade is only intervened when the end of their service life is reached. MS1 is applied to the six claddings.
- Maintenance strategy 2 (MS2): combination of total replacement and minor interventions. The minor intervention is added to delay or mitigate the degradation process, showing that localized repairs or replacements can be carried out to increase the service life of the claddings and, consequently, to prevent unnecessary interruptions without compromising important characteristics of the claddings. MS2 is not applied to PS due to its short service life.
- Maintenance strategy 3 (MS3): combination of total replacement, minor interventions, and cleaning operations. The introduction of cleaning operations represents the maintenance activity most applied (easier and more economical) in claddings. MS3 is not applied to PS.
- Maintenance strategy 4 (MS4): combination of total replacement and cleaning operations. It is a simplification of MS3. Since PS are not subject to minor interventions, MS4 is only applied to them.

The maintenance costs over time (C) include the costs of inspection ($C_{inspection}$) and other maintenance activities ($C_{maintenance}$) (cleaning operations, minor interventions, and total replacement) (Equation (2)).

$$C = C_{inspection} + C_{maintenance}, \tag{2}$$

To consider future investments in this analysis, the net present value of the inspection and maintenance activities is computed through Equations (3) and (4) [31].

$$C_{inspection} = \sum_{t=0}^{t_h} \frac{C_{inspection,0}}{(1+v)^t}, \tag{3}$$

$$C_{maintenance} = \sum_{t=0}^{t_h} \frac{C_{maintenance,0}}{(1+v)^t},\tag{4}$$

where $C_{inspection,0}$ and $C_{maintenance,0}$ are the inspection and other maintenance costs at year 0 (Tables 1 and 2), respectively; v is the real discount rate; and t_h is the time horizon. In this study, a private sector environment is considered. Therefore, a real discount rate of 6% is adopted [32,33]. Finally, a time horizon of 150 years is considered as the period of study. A long-term horizon is used to guarantee that a considerable percentage of the Monte Carlo simulation samples reach the end of the service life, in the more durable constructive solutions, allowing decreasing uncertainty in the analysis.

Table 2. Unit costs, at year 0, for the different maintenance activities without the means of access costs.

	CTS	NSC	RF	PS	ETICS	ACF
Inspection ($€/m^2$)			1.03			
Cleaning operations ($€/m^2$)	17.28	21.17	16.68	16.68	16.68	16.98
Minor interventions ($€/m^2$)	55.57	58.60	24.48	-	47.93	74.55
Total replacement ($€/m^2$)	58.65	139.31	26.48	20.18	85.78	96.28

The costs of the different maintenance activities (Table 2) are adapted from the literature [34]. Since the cost of the maintenance activities varies according to the materials applied, the following assumptions are adopted: (i) CTS-sandstone tiles; (ii) NSC-*Alpinina* stone; (iii) RF-current cementitious-based render; (iv) PS-paint based on acrylic polymers; (v) ETICS-expanded polystyrene boards as thermal insulation material and acrylic-based coat as finishing material; and (vi) ACF-façade cast in situ with a thickness of 25 cm and a smooth texture without paint. Only visual inspections are considered in this study.

According to the remarks presented in Section 3, for the six façade claddings, the following means of access are analysed: supported scaffold (the most commonly used means of access), suspended scaffolds, aerial work platforms (articulated booms, telescopic booms, and scissors lifts), and rope access. Finally, the situation in which maintenance activities can be carried out safely without the need of additional means of access is also analysed. The comparison of this situation with the different means of access helps to assess the influence of the means of access on the maintenance costs.

Regarding the maintenance strategies, for CTS, NSC, RF, ETICS, and ACF, four maintenance strategies are analysed: MS1, MS2, MS3, and MS3*. The main difference between MS3 and MS3* is the means of access used to perform the cleaning operations. Since for these five façade claddings the rope access is only suitable for cleaning operations; in MS3, the three maintenance activities (cleaning operations, minor interventions, and total replacement) are carried out using the same type of means of access. In MS3*, it is assumed that cleaning operations are always carried out by rope access and, because of that, for the remaining maintenance activities, the type of means of access used varies. On the other hand, for PS, rope access is also suitable for total replacement of the cladding. Therefore, only two maintenance strategies are analysed for PS: MS1 and MS4.

In Figures 2–7, maintenance costs (including inspection costs), at year 0, obtained for the different maintenance strategies and means of access, for the six façade claddings, are compared.

For a 150-year time horizon and a real discount rate of 6%, the inspection costs are 8.33 $€/m^2$ (time interval between inspection of 2 years) for PS, 5.39 $€/m^2$ (interval of 3 years) for RF and ETICS, 3.92 $€/m^2$ (interval of 4 years) for CTS and ACT, and 3.04 $€/m^2$ (interval of 5 years) for NSC.

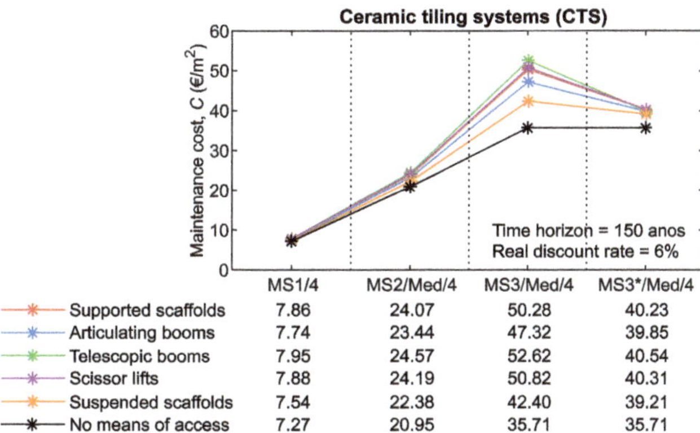

	MS1/4	MS2/Med/4	MS3/Med/4	MS3*/Med/4
Supported scaffolds	7.86	24.07	50.28	40.23
Articulating booms	7.74	23.44	47.32	39.85
Telescopic booms	7.95	24.57	52.62	40.54
Scissor lifts	7.88	24.19	50.82	40.31
Suspended scaffolds	7.54	22.38	42.40	39.21
No means of access	7.27	20.95	35.71	35.71

Figure 2. Sensitivity analysis of the different means of access in CTS maintenance strategy costs for a time horizon of 150 years.

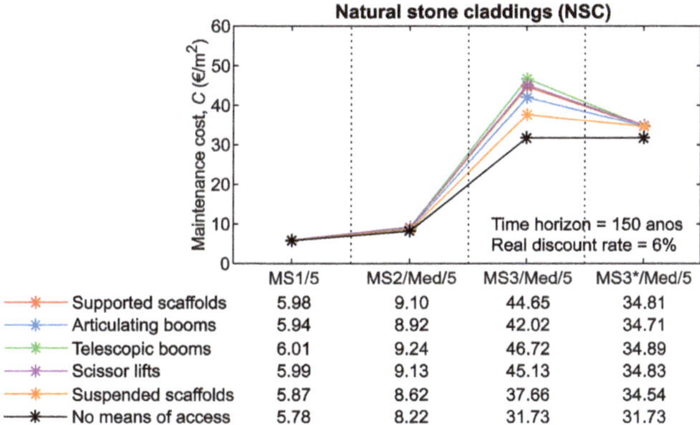

	MS1/5	MS2/Med/5	MS3/Med/5	MS3*/Med/5
Supported scaffolds	5.98	9.10	44.65	34.81
Articulating booms	5.94	8.92	42.02	34.71
Telescopic booms	6.01	9.24	46.72	34.89
Scissor lifts	5.99	9.13	45.13	34.83
Suspended scaffolds	5.87	8.62	37.66	34.54
No means of access	5.78	8.22	31.73	31.73

Figure 3. Sensitivity analysis of the different means of access in NSC maintenance strategy costs for a time horizon of 150 years.

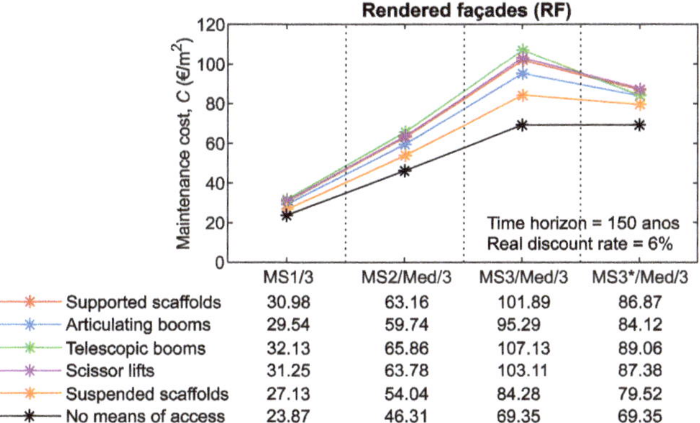

	MS1/3	MS2/Med/3	MS3/Med/3	MS3*/Med/3
Supported scaffolds	30.98	63.16	101.89	86.87
Articulating booms	29.54	59.74	95.29	84.12
Telescopic booms	32.13	65.86	107.13	89.06
Scissor lifts	31.25	63.78	103.11	87.38
Suspended scaffolds	27.13	54.04	84.28	79.52
No means of access	23.87	46.31	69.35	69.35

Figure 4. Sensitivity analysis of the different means of access in RF maintenance strategy costs for a time horizon of 150 years.

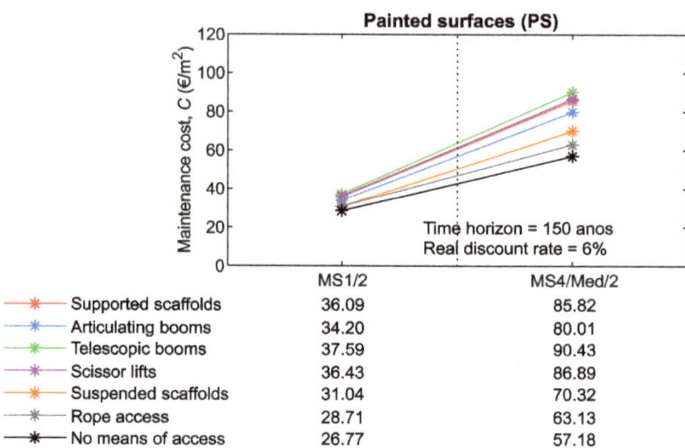

	MS1/2	MS4/Med/2
Supported scaffolds	36.09	85.82
Articulating booms	34.20	80.01
Telescopic booms	37.59	90.43
Scissor lifts	36.43	86.89
Suspended scaffolds	31.04	70.32
Rope access	28.71	63.13
No means of access	26.77	57.18

Figure 5. Sensitivity analysis of the different means of access in PS maintenance strategy costs for a time horizon of 150 years.

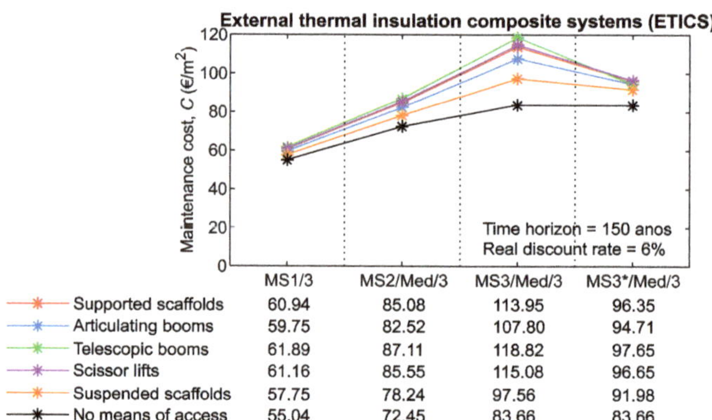

	MS1/3	MS2/Med/3	MS3/Med/3	MS3*/Med/3
Supported scaffolds	60.94	85.08	113.95	96.35
Articulating booms	59.75	82.52	107.80	94.71
Telescopic booms	61.89	87.11	118.82	97.65
Scissor lifts	61.16	85.55	115.08	96.65
Suspended scaffolds	57.75	78.24	97.56	91.98
No means of access	55.04	72.45	83.66	83.66

Figure 6. Sensitivity analysis of the different means of access in ETICS maintenance strategy costs for a time horizon of 150 years.

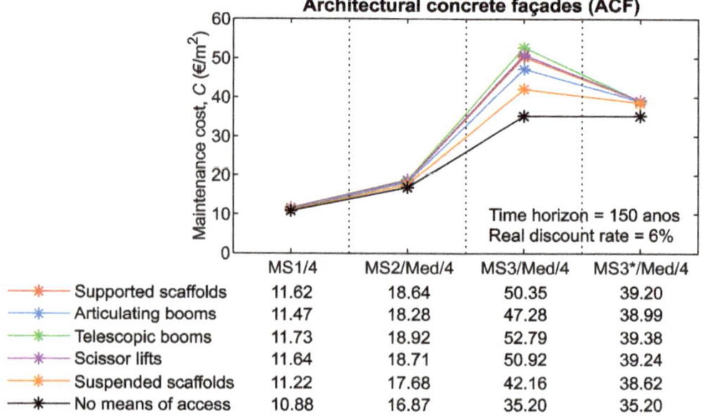

	MS1/4	MS2/Med/4	MS3/Med/4	MS3*/Med/4
Supported scaffolds	11.62	18.64	50.35	39.20
Articulating booms	11.47	18.28	47.28	38.99
Telescopic booms	11.73	18.92	52.79	39.38
Scissor lifts	11.64	18.71	50.92	39.24
Suspended scaffolds	11.22	17.68	42.16	38.62
No means of access	10.88	16.87	35.20	35.20

Figure 7. Sensitivity analysis of the different means of access in ACF maintenance strategy costs for a time horizon of 150 years.

Generally, the results obtained in the sensitivity analysis are in line with what would be expected. Firstly, as maintenance strategies become more complete (higher number of different types of interventions), there is an increase in maintenance costs. Secondly, the maintenance costs and the unit cost of the means of access (Table 1) have a direct relationship. Maintenance strategies combined with the most economical means of access correspond to the most economical options. However, the usefulness of this analysis is to quantify the savings, over the time horizon, when specific access equipment is not needed. Comparing with the different means of access and claddings, maintenance costs present an average reduction between 10% (for NSC) and 26% (for PS). When crossing this information with the service life and number of intervention values [30], the results reveal that the reduction on maintenance costs is greater for claddings and maintenance strategies subjected to a higher number of interventions during the time horizon.

For example, the impact on maintenance costs on MS3 is higher than on MS2 or MS1 (the number of interventions increases with the complexity of the maintenance strategy). In addition, for RF and PS (claddings with reduced service life and, consequently, higher number of interventions during the time horizon), the impact on maintenance costs is higher than for the remaining claddings (CTS, NSC, ETICS, and ACF). For RF/MS3 and PS/MS4, the costs of the means of access can represent a third of the maintenance costs. Furthermore, it is also found that the use of rope access to carry out cleaning operations has a significant impact on maintenance costs. For the 150-year time horizon, there is an average reduction of 20% in maintenance costs for CTS, NSC, and ACF, and 15% for RF and ETICS.

5. Multi-Criteria Analysis

In the same way as choosing the materials to be implemented in the façades, the types of maintenance strategies to be adopted, as well as the level of interventions and the time interval between inspections, the owners or managers always have the last word when it comes to deciding on the means of access to be used. Although the decision process is subjective, this decision must include different parameters, such as budget limitations, aesthetic appearance, target performance level, local constraints, the social and economic context of the building, among other criteria. Multi-criteria analysis is a useful tool to aid decision makers to coherently manage the available information. It can be used to identify a single option, to select a shortlist of options for future detailed appraisal, or to differentiate between acceptable and unacceptable possibilities [35]. The multi-criteria analysis allows ranking the different alternatives according to the criteria defined. In this study, an additive aggregation approach with compensatory rationality is implemented [36], as seen in Equation (5).

$$X_i = \sum_{j=1}^{m} \lambda_j \times x_{ij}, \text{ with } \sum_{j=1}^{m} \lambda_j = 1, \tag{5}$$

where X_i is the global ranking of the alternative i, λ_j is the weight of the criterion j, and x_{ij} is the standardized classification of alternative i according to criterion j. The different scales in the different criteria are standardized by Equation (6) for an increased order of preference and by Equation (7) for a decreased order of preference [37].

$$x_{ij} = \frac{X_{ij} - \min X_{ij}}{\max X_{ij} - \min X_{ij}}, \tag{6}$$

$$x_{ij} = \frac{\max X_{ij} - X_{ij}}{\max X_{ij} - \min X_{ij}}, \tag{7}$$

where X_{ij} is the classification of alternative i according to criterion j, and $\min X_{ij}$ and $\max X_{ij}$ are the minimum and maximum value of criterion j, respectively.

From the individual analysis and comparison of criteria, it is difficult to identify the most and least advantageous alternative from all points of view. The strengths and weaknesses differ according to the criterion analysed. In this sense, for the different façade claddings, the different maintenance strategies and means of access (options) are globally classified in terms of durability (service life), performance (efficiency index), and maintenance costs. Consequently, four scenarios, considering three criteria (efficiency index, maintenance costs, and total replacements), are analysed. The four scenarios are: (1) the same importance is given to the three criteria; (2) a higher emphasis is given to the efficiency index; (3) a higher emphasis is given to the maintenance costs; and (4) a higher emphasis is given to the number of total replacements over the time horizon. Data referring to the efficiency index and total number of replacements can be consulted in Ferreira et al. [30]. In Tables 3 and 4, the results obtained in the multi-criteria analysis are presented.

Table 3. Multi-criteria analysis.

Façade Claddings		CTS	NSC	RF	PS	ETICS	ACF	CTS	NSC	RF	PS	ETICS	ACF
Scenario		1						2					
Criteria		Weight, λ_i											
Efficiency index		33.3%						50.0%					
Maintenance costs		33.3%						25.0%					
Total replacements		33.3%						25.0%					
MS	Means of access	Standardized global rating, x_i											
MS1	SS	0.00	0.00	0.01	0.01	0.01	0.00	0.00	0.00	0.01	0.01	0.01	0.00
	AB	0.00	0.00	0.02	0.03	0.02	0.00	0.00	0.00	0.01	0.02	0.01	0.00
	TB	0.00	0.00	0.00	0.00	0.00	0.00	0.00	0.00	0.00	0.00	0.00	0.00
	SL	0.00	0.00	0.01	0.01	0.01	0.00	0.00	0.00	0.00	0.01	0.00	0.00
	SpS	0.01	0.00	0.04	0.06	0.04	0.01	0.00	0.00	0.02	0.04	0.02	0.01
	RA	-	-	-	0.08	-	-	-	-	-	0.05	-	-
	NMA	0.01	0.00	0.06	0.10	0.06	0.01	0.01	0.00	0.04	0.06	0.04	0.01
MS2	SS	0.75	0.99	0.71	-	0.59	0.97	0.69	0.92	0.69	-	0.58	0.93
	AB	0.76	0.99	0.74	-	0.61	0.98	0.69	0.92	0.70	-	0.60	0.93
	TB	0.74	0.98	0.69	-	0.57	0.97	0.68	0.92	0.67	-	0.57	0.92
	SL	0.75	0.98	0.71	-	0.58	0.97	0.69	0.92	0.68	-	0.58	0.92
	SpS	0.77	0.99	0.78	-	0.65	0.99	0.70	0.92	0.73	-	0.62	0.93
	RA	-	-	-	-	-	-	-	-	-	-	-	-
	NMA	0.80	1.00	0.84	-	0.71	1.00	0.72	0.93	0.77	-	0.66	0.94
MS3	SS	0.77	0.71	0.75	-	0.71	0.72	0.87	0.87	0.85	-	0.82	0.85
	AB	0.82	0.76	0.80	-	0.77	0.77	0.89	0.89	0.88	-	0.86	0.88
	TB	0.73	0.68	0.71	-	0.67	0.68	0.84	0.85	0.82	-	0.79	0.83
	SL	0.76	0.71	0.74	-	0.70	0.71	0.86	0.86	0.84	-	0.81	0.85
	SpS	0.89	0.83	0.88	-	0.87	0.85	0.94	0.94	0.93	-	0.92	0.93
	RA	-	-	-	-	-	-	-	-	-	-	-	-
	NMA	1.00	0.93	1.00	-	1.00	0.96	1.00	1.00	1.00	-	1.00	1.00
MS3*	SS	0.93	0.88	0.86	-	0.88	0.89	0.96	0.97	0.92	-	0.93	0.96
	AB	0.93	0.88	0.89	-	0.90	0.90	0.96	0.97	0.93	-	0.93	0.96
	TB	0.92	0.87	0.85	-	0.87	0.89	0.96	0.97	0.91	-	0.92	0.96
	SL	0.93	0.88	0.86	-	0.88	0.89	0.96	0.97	0.92	-	0.92	0.96
	SpS	0.94	0.88	0.92	-	0.92	0.90	0.97	0.97	0.95	-	0.95	0.97
	RA	-	-	-	-	-	-	-	-	-	-	-	-
	NMA	1.00	0.93	1.00	-	1.00	0.96	1.00	1.00	1.00	-	1.00	1.00
MS4	SS	-	-	-	0.73	-	-	-	-	-	0.83	-	-
	AB	-	-	-	0.79	-	-	-	-	-	0.87	-	-
	TB	-	-	-	0.69	-	-	-	-	-	0.81	-	-
	SL	-	-	-	0.72	-	-	-	-	-	0.83	-	-
	SpS	-	-	-	0.88	-	-	-	-	-	0.92	-	-
	RA	-	-	-	0.94	-	-	-	-	-	0.97	-	-
	NMA	-	-	-	1.00	-	-	-	-	-	1.00	-	-

Legend: SS—supported scaffolds; AR—articulated booms; TB—telescopic booms; SL—scissor lifts; SpS—suspended scaffolds; RA—rope access; NMA—no means of access.

Table 4. Multi-criteria analysis continuation.

Façade Claddings		CTS	NSC	RF	PS	ETICS	ACF	CTS	NSC	RF	PS	ETICS	ACF
	Scenario			3						4			
	Criteria	Weight, λ_i											
	Efficiency index	25.0%						25.0%					
	Maintenance costs	50.0%						25.0%					
	Total replacements	25.0%						50.0%					
MS	Means of access	Standardized global rating, x_i											
MS1	SS	0.01	0.00	0.02	0.03	0.02	0.00	0.00	0.00	0.01	0.01	0.01	0.00
	AB	0.01	0.00	0.05	0.08	0.05	0.01	0.00	0.00	0.01	0.02	0.01	0.00
	TB	0.00	0.00	0.00	0.00	0.00	0.00	0.00	0.00	0.00	0.00	0.00	0.00
	SL	0.00	0.00	0.02	0.03	0.02	0.00	0.00	0.00	0.00	0.01	0.00	0.00
	SpS	0.02	0.00	0.11	0.15	0.10	0.02	0.00	0.00	0.02	0.04	0.02	0.01
	RA	-	-		0.20	-	-	-	-	-	0.05	-	-
	NMA	0.04	0.01	0.17	0.25	0.16	0.03	0.01	0.00	0.04	0.06	0.04	0.01
MS2	SS	0.83	0.97	0.64	-	0.46	0.94	0.76	0.96	0.76	-	0.65	0.94
	AB	0.87	0.98	0.72	-	0.53	0.95	0.77	0.96	0.77	-	0.67	0.94
	TB	0.81	0.96	0.59	-	0.42	0.93	0.76	0.96	0.74	-	0.64	0.94
	SL	0.83	0.97	0.63	-	0.45	0.94	0.76	0.96	0.75	-	0.65	0.94
	SpS	0.92	0.99	0.84	-	0.63	0.97	0.78	0.97	0.80	-	0.70	0.95
	RA	-	-	-	-	-	-	-	-	-	-	-	-
	NMA	1.00	1.00	1.00	-	0.77	1.00	0.79	0.97	0.84	-	0.73	0.96
MS3	SS	0.16	0.08	0.29	-	0.28	0.11	0.87	0.87	0.85	-	0.82	0.85
	AB	0.32	0.17	0.43	-	0.43	0.22	0.89	0.89	0.88	-	0.86	0.88
	TB	0.04	0.01	0.17	-	0.16	0.03	0.84	0.85	0.82	-	0.79	0.83
	SL	0.13	0.06	0.26	-	0.25	0.09	0.86	0.86	0.84	-	0.81	0.85
	SpS	0.59	0.32	0.66	-	0.67	0.40	0.94	0.94	0.93	-	0.92	0.93
	RA	-	-	-	-	-	-	-	-	-	-	-	-
	NMA	0.95	0.52	0.97	-	1.00	0.64	1.00	1.00	1.00	-	1.00	1.00
MS3*	SS	0.70	0.42	0.60	-	0.70	0.50	0.96	0.97	0.92	-	0.93	0.96
	AB	0.72	0.42	0.66	-	0.74	0.51	0.96	0.97	0.93	-	0.93	0.96
	TB	0.69	0.41	0.56	-	0.67	0.49	0.96	0.97	0.91	-	0.92	0.96
	SL	0.70	0.42	0.59	-	0.69	0.50	0.96	0.97	0.92	-	0.92	0.96
	SpS	0.76	0.43	0.76	-	0.80	0.52	0.97	0.97	0.95	-	0.95	0.97
	RA	-	-	-	-	-	-	-	-	-	-	-	-
	NMA	0.95	0.52	0.97	-	1.00	0.64	1.00	1.00	1.00	-	1.00	1.00
MS4	SS	-	-	-	0.35	-	-	-	-	-	0.83	-	-
	AB	-	-	-	0.48	-	-	-	-	-	0.87	-	-
	TB	-	-	-	0.25	-	-	-	-	-	0.81	-	-
	SL	-	-	-	0.33	-	-	-	-	-	0.83	-	-
	SpS	-	-	-	0.70	-	-	-	-	-	0.92	-	-
	RA	-	-	-	0.86	-	-	-	-	-	0.97	-	-
	NMA	-	-	-	1.00	-	-	-	-	-	1.00	-	-

Legend: SS—supported scaffolds; AR—articulated booms; TB—telescopic booms; SL—scissor lifts; SpS—suspended scaffolds; RA—rope access; NMA—no means of access.

In a first analysis, the results reveal, for all façade claddings, that the alternatives where there is no need for means of access are the most advantageous. Therefore, the results confirm that the planning of accessibility to the different building envelope elements, in the design stage, can have a significant influence on maintenance costs over time. On the other hand, if the use of temporary means of access is indispensable (it should be predicted at design stage also in terms of costs), the results suggest that selecting rope access (for cleaning operations and total replacement of the painted surface, only in maintenance works where technical quality is guaranteed) and/or suspended scaffolds (for other types of interventions) are the most advantageous alternatives.

Concerning the least advantageous alternatives, the results show that MS1 using telescopic booms is the worst option. However, the results reveal that the correct choice of means of access depends on several factors, such as building envelope, safety, interference with users, proximity to road traffic, structural limitations of the building, weather conditions, load capacity, and/or building volume/shape.

On the other hand, the results also show that it is advantageous to opt for more complete maintenance strategies (MS2, MS3, and/or MS4), while MS1 is the least beneficial one. For example, for CTS and RF, in three of the four scenarios analysed, MS3 is the most advantageous alternative. This position only changes to MS2 if more emphasis is placed on maintenance costs (scenario 3). For NSC and ACF, in scenarios 1 and 3, MS2 is the most advantageous alternative, while MS3 is the most beneficial in scenarios 2 and 4. For ETICS, MS3 is the most advantageous alternative for the four scenarios. Finally, for PS, MS4 is the most advantageous alternative for the four scenarios.

Finally, the weights in Tables 3 and 4 are indicative. The weight can be influenced by several factors, such as location, economic context, and owner's requirements, among others. However, subjectivity can be reduced through expert judgements or through other multi-criteria decision methodologies, such as the analytic hierarchy process (AHP), which can help to define priorities through a process of pair-wise comparisons [38,39].

6. Discussion/General Recommendations

From the results, some guidelines can be provided, which can help decision makers to make more rational and informed choices, based on objective criteria. However, the decision process is always influenced by subjective criteria, which depend on the experience and individual perceptions of decision makers. These criteria are practically impossible to model. However, on the other hand, objective criteria can be evaluated depending on the requirements of decision makers. Furthermore, it should be mentioned that all alternatives considered in this study are technically and economically valid. Thus, they can all be successfully applied to building envelope elements.

In a global analysis, the adoption of preventive maintenance activities (such as cleaning operations, local repairs, and local replacements) is beneficial. These actions allow increasing the durability of the façade claddings and reducing their physical degradation. Moreover, the risk to users and owners is reduced and, consequently, users' satisfaction and the aesthetic perception of cities are improved. In Table 5, general recommendations of the most viable maintenance strategies to be implemented in the different façade claddings are presented. There, together with the identification of the most advantageous maintenance strategy and means of access for each façade cladding, the values of the essential parameters to be considered in this analysis are presented. These parameters are:

- The service life: the end of the service life of the façade claddings is reached when the element reaches the degradation condition D;
- The efficiency index: this value ranges between 0 and 1. The higher the EI value, the longer the element remains in the best degradation conditions (A, B, and C) during the time horizon (period of study);
- The annualized maintenance cost: corresponds to the division between the life-cycle cost (includes inspection and maintenance costs) of the maintenance strategy of the element in year 0, by the time horizon;
- The total number of replacements: corresponds to the average number of total replacements to which element is subject during the time horizon.

Table 5. Unit costs, at year 0, for the different maintenance activities without the means of access costs.

Façade Cladding	Maintenance Strategy/Means of Access	Service Life (Years)	Efficiency Index	Annualized Maintenance Costs (€/m^2/Year)	Total Number of Replacements
CTS	MS3* CO: Rope access MI/TR: Suspended scaffolds	108	0.95	0.63	0.9
NSC	MS2 CO: Rope access MI/TR: Suspended scaffolds	148	0.94	0.97	0.5
RF	MS3* CO: Rope access MI/TR: Suspended scaffolds	33	0.96	0.69	3.9
PS	MS4 CO/TR: Rope access	17	0.93	0.54	7.9
ETICS	MS3* CO: Rope access MI/TR: Suspended scaffolds	44	0.92	1.14	2.9
ACF	MS2 CO: Rope access MI/TR: Suspended scaffolds	120	0.94	0.81	0.9

The service life, efficiency index, and total number of total replacements can be consulted in Ferreira et al. [30]. Based on that, it is suggested that MS3 is adopted to ceramic tiling systems (CTS), rendered façades (RF), and ETICS; MS2 is adopted to natural stone claddings (NSC) and architectural concrete façades (ACF); and MS4 is adopted to painted surfaces (PS). Regarding the means of access, the situation where there is no need for means of access is the most advantageous option. This result shows the relevance of having a maintenance plan and considering permanent means of access during the design stage or at least predict temporary means in the maintenance costs. If the building configuration requires the use of means of access, the results suggest opting for rope access in the case of carrying out cleaning operations (when technical viable) and replacing the painted surface and suspended scaffolds for the other types of maintenance activities (minor interventions and total replacement).

7. Conclusions

In this study, a sensitivity analysis of the impact of the means of access costs on the maintenance plans developed for building façade claddings is carried out. Based on this analysis, general recommendations for maintenance strategies are provided. The means of access costs vary between 2.12 €/m^2 (rope access) and 11.84 €/m^2 (telescopic boom). From the analysis, the results and trends on the viability of maintenance strategies, initially identified for supported scaffolds, are maintained for the other types of means of access. In other words, for any means of access, the maintenance costs increase with the complexity of the maintenance strategy. Furthermore, as expected, the lowest maintenance costs occur for the most economical means of access. The situation in which there is no need for means of access (i.e., when the means of access are considered during design stage) is the most economical, leading to an average maintenance cost reduction of 10% to 26% compared to other means of access.

For the situation in which the use of means of access is indispensable, the results suggest opting for rope access in the case of carrying out cleaning operations and replacing the painted surface and suspended scaffolds for the other types of maintenance activities (minor interventions and total replacement). The use of rope access to perform cleaning operations has a considerable impact on the maintenance costs. For a time horizon of 150 years, an average reduction between 15% and 20% is obtained. For the remaining means of access, in comparison with supported scaffolds, suspended scaffolds lead to an average reduction of 5% to 21%; the articulated boom leads to an average reduction of 2% to 6%; the telescopic boom leads to an average increase of 2% to 5%; and the scissor lifts lead to an average increase of 1%.

Based on these results, it is understood that the design of permanent means of access during the design stage may not be economically beneficial for all types of façade claddings. For instance, for a time horizon of 150 years, the average maintenance cost reduction for NSC is 10%. Possibly, the design of the permanent means of access, their construction, and their maintenance over the same horizon may represent a greater burden for the owner/manager than the use of planning temporary means of access when necessary. In addition, façades with nobler coatings may be more difficult to incorporate permanent means of access into the architecture without considerably affecting the aesthetics. However, for façade claddings, such as rendered façades and painted surfaces, the percentage of the means of access in the maintenance costs is already considerable (on average 26%). In these cases, considering permanent means of access can be a more economical solution.

The results of the multi-criteria analysis confirm that there are advantages in selecting more complete maintenance strategies when several criteria are weighed, in addition to costs. When the criteria are assessed individually, the optimal maintenance plan changes according to the assessed criterion. For example, if the main purpose is to increase the service life of the element, more complete maintenance plans considering cleaning operations and minor interventions are preferable. On the other hand, if the purpose is to reduce maintenance costs, simpler maintenance strategies (without preventive maintenance activities) are the most indicated, even though they lead to the elements having higher rates of degradation (or worse degradation conditions) throughout their life cycle and, consequently, higher risks. The multi-criteria analysis helps to evaluate the criteria in a multidimensional way and, therefore, establish a hierarchical scale of preference. The recommendations presented in this paper correspond to the analysis of the results obtained in the different multi-criteria analysis carried out. The results reveal that, when the criteria are evaluated globally, there is a preference for opting for more complete maintenance strategies, while MS1 represents the least beneficial option.

Finally, it should be mentioned that this methodology should be seen as a general guideline for evaluating the impact of the means of access in the maintenance plans of the building envelope elements. Each situation should be evaluated individually, since other criteria, scenarios and weights can be considered, depending on the decision makers' requirements. The choice of materials to be implemented in buildings, maintenance strategies, intervention levels, time intervals between inspections, and means of access always depends on the owners or managers and the results that they want to obtain based on all the available parameters (objective and/or subjective).

Author Contributions: Conceptualization, C.F., I.S.D., A.S., J.d.B. and I.F.-C.; methodology, C.F. and I.S.D.; software, C.F.; validation, C.F.; investigation, C.F., I.S.D., A.S., J.d.B. and I.F.-C.; writing—original draft preparation, C.F.; writing—review and editing, I.S.D., A.S., J.d.B. and I.F.-C.; project administration, A.S. and J.d.B.; funding acquisition, A.S. and J.d.B. All authors have read and agreed to the published version of the manuscript.

Funding: This research was funded by FCT (Portuguese Foundation for Science and Technology) through the project BestMaintenance-LowerRisks (PTDC/ECI-CON/29286/2017).

Data Availability Statement: The data presented in this study are available on request from the corresponding author.

Acknowledgments: The authors gratefully acknowledge the support of CERIS, from Instituto Superior Técnico, Universidade de Lisboa.

Conflicts of Interest: The authors declare no conflict of interest.

References

1. De Silva, N.; Ranasinghe, M.; de Silva, C.R. Risk factors affecting building maintenance under tropical conditions. *J. Financ. Manag. Prop. Constr.* **2012**, *17*, 235–252. [CrossRef]
2. Cardini, E.; Sohn, E.C. Above and beyond: Access techniques for the assessment of buildings and structures. In Proceedings of the AEI 2013: Building Solutions for Architectural Engineering, State College, PA, USA, 3–5 April 2013.
3. Korolchenko, D.A.; Pham, N.T. Loads in rope access system when working at heights. In Proceedings of the 4th World Multidisciplinary Civil Engineering-Architecture-Urban Planning Symposium, Prague, Czech Republic, 18–22 June 2018.
4. Wu, S.; Lee, A.; Tah, J.H.M.; Aouad, G. The use of a multi-attribute tool for evaluating accessibility in buildings: The AHP approach. *Facilities* **2007**, *25*, 375–389. [CrossRef]
5. Kamya, B.M. Industrial rope access approach in the inspection and maintenance management of bridges. In Proceeding of the 5th International Conference on Bridge Management, Surrey, UK, 11–13 April 2005.
6. Marais, K.B.; Saleh, J.H. Beyond its cost, the value of maintenance: An analytical framework for capturing its net present value. *Reliab. Eng. Syst. Saf.* **2009**, *94*, 644–657. [CrossRef]
7. Grosso, R.; Mecca, U.; Moglia, G.; Prizzon, F.; Rebaudengo, M. Collecting built environment information using UAVs: Time and applicability in building inspection activities. *Sustainability* **2017**, *12*, 4731. [CrossRef]
8. Rebaudengo, M.; Piantanida, P. Maintenance of building: Italian examples of deviations between planned and incurred costs. In Proceedings of the 2nd International Conference on Reliability Engineering, Milan, Italy, 20–22 December 2017.
9. Kim, K.; Teizer, J. Automatic design and planning of scaffolding systems using building information modeling. *Adv. Eng. Inform.* **2014**, *28*, 66–80. [CrossRef]
10. El-Haram, M.A.; Horner, M.W. Factors affecting housing maintenance cost. *J. Qual. Maint. Eng.* **2002**, *8*, 115–123. [CrossRef]
11. Ferreira, C.; Silva, A.; de Brito, J.; Dias, I.S.; Flores-Colen, I. Definition of a condition-based model for natural stone claddings. *J. Build. Eng.* **2021**, *33*, 101643. [CrossRef]
12. Rajendran, S.; Gambatese, J. Risk and financial impacts of prevention through design solutions. *Pract. Period. Struct. Des. Constr.* **2013**, *18*, 67–72. [CrossRef]
13. Wang, F.; Tamura, Y.; Yoshida, A. Wind loads on clad scaffolding with different geometries and building opening ratios. *J. Wind Eng. Ind. Aerodyn.* **2013**, *120*, 37–50. [CrossRef]
14. Chew, Y.L.; Tan, P.P. *Staining of Façades*; World Scientific: Singapore, 2003.
15. Chew, Y.L. *Maintainability of Facilities: Green FM for Building Professionals*, 2nd ed.; Scientific Publishing Company: Singapore, 2016.
16. Vossoughi, H.; Siddiqui, R.I. Industrial rope access—An alternate means for inspection, maintenance, and repair of building facades and structures. In *STP1444-EB Building Facade Maintenance, Repair, and Inspection*; ASTM International: West Conshohocken, PA, USA, 2004.
17. McCann, M. Deaths in construction related to personnel lifts, 1992–1999. *J. Saf. Res.* **2003**, *34*, 507–514. [CrossRef]
18. Dogan, E.; Yurdusev, M.A.; Yildizel, S.A.; Calis, G. Investigation of scaffolding accident in a construction site: A case study analysis. *Eng. Fail. Anal.* **2021**, *120*, 105108. [CrossRef]
19. Sevim, B.; Bekiroglu, S.; Arslan, G. Experimental evaluation of tie bar effects on structural behavior of suspended scaffolding systems. *Adv. Steel Constr.* **2017**, *13*, 62–77.
20. Stay, K. Rope access for inspection, testing and maintenance of industrial structures. *J. Prot. Coat. Linings* **2018**, *35*, 44–48.
21. Hold, S. Rope access for inspection and maintenance. *Proc. Inst. Civ. Eng. Munic. Eng.* **1997**, *121*, 206–211. [CrossRef]
22. Pan, C.S.; Hoskin, A.; McCann, M.; Lin, M.L.; Fearn, K.; Keane, P. Aerial lift fall injuries: A surveillance and evaluation approach for targeting prevention activities. *J. Saf. Res.* **2007**, *38*, 617–625.
23. Ferreira, C.; Neves, L.C.; Silva, A.; de Brito, J. Stochastic maintenance models for ceramic claddings. *Struct. Infrastruct. Eng.* **2020**, *16*, 247–265. [CrossRef]
24. Bordalo, R.; de Brito, J.; Gaspar, P.L.; Silva, A. Service life prediction modelling of adhesive ceramic tiling systems. *Build. Res. Inf.* **2011**, *39*, 66–78.
25. Silva, A.; de Brito, J.; Gaspar, P.L. Service life prediction model applied to natural stone wall claddings (directly adhered to the substrate). *Constr. Build. Mater.* **2011**, *25*, 3674–3684.
26. Gaspar, P.L.; de Brito, J. Quantifying environmental effects on cement-rendered facades: A comparison between different degradation indicators. *Build. Environ.* **2008**, *43*, 1818–1828. [CrossRef]
27. Chai, C.; de Brito, J.; Gaspar, P.L.; Silva, A. Predicting the service life of exterior wall painting: Techno-economic analysis of alternative maintenance strategies. *J. Constr. Eng. Manag.* **2014**, *140*, 04013057. [CrossRef]
28. Ximenes, S.; de Brito, J.; Gaspar, P.L.; Silva, A. Modelling the degradation and service life of ETICS in external walls. *Mater. Struct.* **2015**, *48*, 2235–2249. [CrossRef]
29. Serralheiro, M.I.; de Brito, J.; Silva, A. Methodology for service life prediction of architectural concrete facades. *Constr. Build. Mater.* **2017**, *133*, 261–274. [CrossRef]
30. Ferreira, C.; Dias, I.S.; Silva, A.; de Brito, J.; Flores-Colen, I. Criteria for selection of cladding systems based on their maintainability. *J. Build. Eng.* **2021**, *39*, 102260. [CrossRef]
31. ISO 15686-5. *Buildings and Constructed Assets—Service Life Planning—Part 5: Whole Life Costing*; International Organization for Standardization (ISO): Geneva, Switzerland, 2012.

32. Langdon, D. *Life Cycle Costing (LCC) as a Contribution to Sustainable Construction. Guidance on the Use of the LCC Methodology and Its Application in Public Procurement*; Davis Langdon Management Consulting: London, UK, 2007.
33. Sánchez-Silva, M.; Klutke, G.-A. *Reliability and Life-Cycle Analysis of Deteriorating Systems*; Springer International Publishing: Zurich, Switzerland, 2016.
34. CYPE Price Generator. Available online: http://www.geradordeprecos.info/ (accessed on 12 October 2020).
35. Dodgson, J.S.; Spackman, M.; Pearman, A.; Phillips, L.D. *Multi-Criteria Analysis: A Manual*; Department for Communities and Local Government: London, UK, 2009.
36. Cochran, J.K.; Chen, H.N. Fuzzy multi-criteria selection of object-oriented simulation software for production system analysis. *Comput. Oper. Res.* **2005**, *32*, 153–168. [CrossRef]
37. Silva, A.; de Brito, J.; Gaspar, P.L. *Methodologies for Service Life Prediction of Buildings: With a Focus on Façade Claddings*; Springer International Publishing: Zurich, Switzerland, 2016.
38. Arunraj, N.S.; Maiti, J. Risk-based maintenance policy selection using AHP and goal programming. *Saf. Sci.* **2010**, *48*, 238–247. [CrossRef]
39. Bertolini, M.; Bevilacqua, M. A combined goal programming—AHP approach to maintenance selection problem. *Reliab. Eng. Syst. Saf.* **2006**, *91*, 839–848. [CrossRef]

Article

Blockchain-Based Trusted Property Transactions in the Built Environment: Development of an Incubation-Ready Prototype

Srinath Perera *, Amer A. Hijazi, Geeganage Thilini Weerasuriya, Samudaya Nanayakkara and Muhandiramge Nimashi Navodana Rodrigo

Centre for Smart Modern Construction, Western Sydney University, Kingswood, NSW 2747, Australia;
A.Hijazi@westernsydney.edu.au (A.A.H.); T.Weerasuriya@westernsydney.edu.au (G.T.W.);
S.Nanayakkara@westernsydney.edu.au (S.N.); N.Rodrigo@westernsydney.edu.au (M.N.N.R.)
* Correspondence: Srinath.Perera@westernsydney.edu.au

Abstract: Blockchain can be introduced to use cases in the built environment where reliability of transaction records is paramount. Blockchain facilitates decentralised, cryptographically secure, trustworthy, and immutable recordkeeping of transactions. However, more research is urgently required to understand the process and complications in implementing blockchain solutions in the built environment. This paper demonstrates a methodology for developing a blockchain system starting from problem analysis, selection of blockchain platform, system modelling, prototype development, and evaluation. The evolutionary prototyping model was selected as the software development methodology for the use case of property transactions. A systematic process protocol involving the multi-criteria decision-making method, Simple Multi Attribute Rating Technique (SMART), was used to select Hyperledger Fabric as the most suitable blockchain platform for the prototype. The system architecture facilitates a simplified, lean property transaction process implemented through chaincode (smart contract) algorithms and graphical user interfaces. System evaluation through test cases allowed iterative improvements, leading to an incubation-ready software prototype. The contribution to knowledge of this paper is in the demonstration of the process to follow to implement a blockchain solution for a specific domain. The findings provide the foundation for developing proofs of concept for other potential applications of blockchain in the built environment.

Keywords: blockchain; built environment; trust; proof of concept; property transactions; Hyperledger Fabric

Citation: Perera, S.; Hijazi, A.A.;
Weerasuriya, G.T.; Nanayakkara, S.;
Rodrigo, M.N.N. Blockchain-Based
Trusted Property Transactions in the
Built Environment: Development of
an Incubation-Ready Prototype.
Buildings **2021**, *11*, 560. https://
doi.org/10.3390/buildings11110560

Academic Editor: Pierfrancesco De
Paola

Received: 17 October 2021
Accepted: 17 November 2021
Published: 20 November 2021

Publisher's Note: MDPI stays neutral
with regard to jurisdictional claims in
published maps and institutional affiliations.

1. Introduction

In the built environment, with respect to property transactions, where businesses depend on reliability of transaction records, blockchain has been proposed to enable trust and ensure ownership [1–3]. Blockchain technology is suitable for storing and handling thousands of copies of transaction records to enable transaction authentication [1,4–7]. Whilst traditionally the transaction records for built environment are housed in central servers controlled by a single administration point, in blockchain technology, these transaction records are replicated across the network of computers [2,5,8].

Blockchain is considered to be a disruptive technology that has a significant impact in numerous industries in terms of information and trust [2,9–13]. Blockchain is a distributed ledger technology, where data stored in the network is non-centralised, supports peer-to-peer interactions, and creates a cryptographically secured, immutable chain of records [14–17]. As each peer (node) in the blockchain network maintains a copy of the ledger, and the records cannot be modified without detection, blockchain provides a method of ensuring trust in a trustless environment [18], and produces an audit trail of all transactions in the ledger [19].

Maintaining trusted records on a blockchain would enable benefits such as tracking the history of assets, providing trustworthy proof of ownership, reducing transaction times and costs, and minimising fraudulent behaviour [19,20]. However, it is noted that only a few papers have discussed the demonstration of blockchain solutions in the property sector, and more so in the built environment [1,12,21]. The Institution of Civil Engineers (ICE) [14], Centre of Digital Built Britain [22], and ARUP reports regarding the future of smart built environment [23,24] have highlighted that trust is a valuable asset for the construction industry, but that the current system is still facing difficulties to ensure trust. It is recommended to introduce a new system that could aid in reliably sharing information amongst stakeholders [25–28]. The blockchain real estate industry report for the year 2021 by the Foundation for International Blockchain and Real Estate Expertise (FIBREE) pointed out that the property sector is still partly digitalised and mostly paper-based, and the use of blockchain as an innovative solution is still at an early trigger stage [29].

This paper demonstrates a methodology for developing a blockchain system starting from problem analysis, selection of blockchain platform, system modelling, prototype development, and evaluation. It presents the development of an incubation-ready software prototype using blockchain for property transactions as a use case to illustrate the potential of blockchain in the built environment. This is achieved by establishing three objectives. Firstly, to understand the potential applications of blockchain in the built environment, and critically review the need for blockchain in enabling trusted transactions. Secondly, to understand the mechanism of Hyperledger Fabric as a blockchain platform in developing the software prototype. Finally, to design, develop, and evaluate a blockchain-based software prototype for a simplified, lean property transaction process implemented through chaincode (smart contracts) and graphical user interfaces. The findings provide the foundation for developing proofs of concept for other potential applications of blockchain in the built environment.

The rest of this paper is structured as follows: Section 2 presents the literature review including the potential applications of blockchain in the built environment. It also presents a critical review of previous studies on blockchain applications for property transactions, and discusses the suitability of Hyperledger Fabric as a blockchain platform. Section 3 describes the research method to develop the blockchain-based software prototype. Section 4 elaborates the proof of concept, including the business scenario, system overview, architecture, implementation, and system evaluation. Section 5 presents the discussion, and the conclusions are presented in Section 6.

2. Literature Review

In the built environment, with respect to property transactions, where businesses depend on reliability of transaction records, blockchain has been introduced to enable trust and ensure ownership [1–3]. Blockchain technology is suitable for storing and handling thousands of copies of transaction records to enable transaction authentication [1,4–7,30]. Whilst traditionally, the transaction records for built environment are housed in central servers controlled by a single administration point, in blockchain technology, these transaction records are replicated across the network of computers [2,5,8]. The structure of the distributed ledger technology effectively means that all participants in a blockchain network have the same transaction records, and the ability to read and write to the ledger [2,5,14]. Figure 1 shows the differences between centralised databases, traditional decentralised databases, and distributed ledger technology for blockchain.

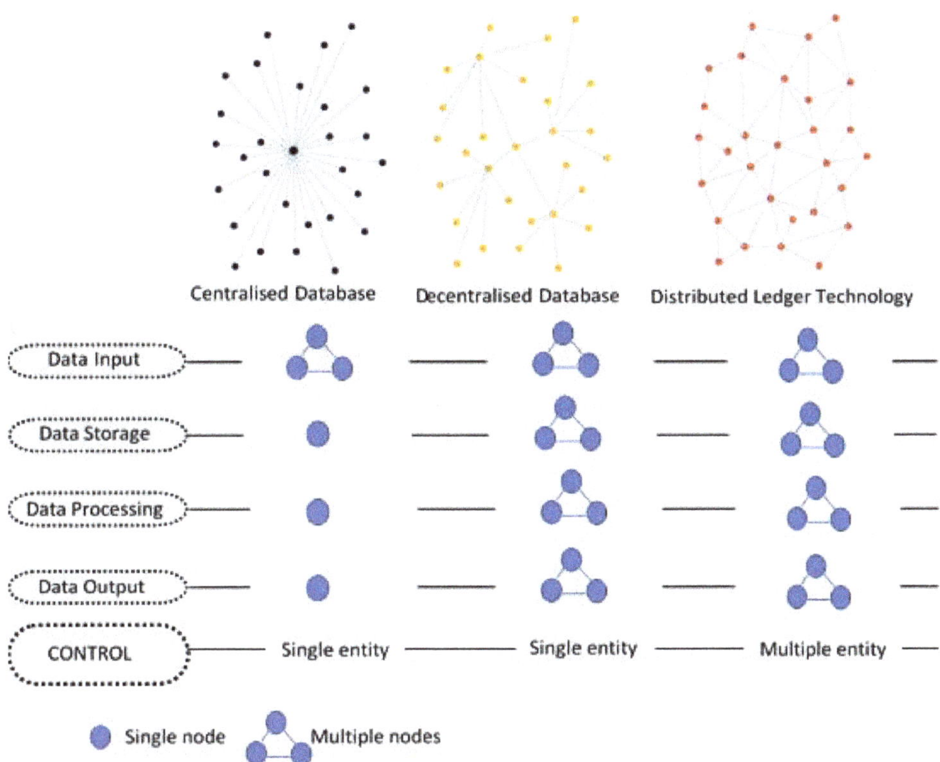

Figure 1. The differences between centralised databases, traditional decentralised databases, and distributed ledger technology (adapted from [2]).

In the built environment and the property sector, still more research is urgently required to exploit the full potential of blockchain as a solution [12,31], such as in title transfer of real estate assets, construction supply chains, and even the integration of Building Information Modelling (BIM) and blockchain.

2.1. The Potential Application of Blockchain in the Built Environment

This section explores some of the main applications of blockchain in the built environment. In theory, one blockchain platform might be implemented to span the entire life cycle of a real estate asset, from materials supply to management of the built asset. It briefly explores real estate title transfer, construction supply chains, and the aspects of BIM and blockchain integration. The applications listed are non-exhaustive but indicate the potential areas involved across the sector.

2.1.1. Title Transfer of Real Estate Assets

Blockchain will ensure the authenticity of land registry title records that could potentially link a real estate asset to the pertinent data relating to a wide range of its stakeholders, thereby streamlining business processes [5]. The data might include information such as digital planning, architecture, certification and verification, specifications, and warranty [3,32] that relate to a given real estate asset. Traditionally, these data have been stored in silos [33,34]. It is still common to have thousands of documents from hundreds of parties [32] where information is disparate, disconnected, and hard to access. Making commitment and collaboration of such information to the blockchain a mechanism for value transfer [11,35] will ensure the integrity of the data for the real estate asset [31]. Blockchain will enable the adoption of a smart contract to optimise contract formulation and negotiation, while transaction ordering through a consensus service will ensure immutability mechanisms for the transfer or creation of value, and transaction validation through a membership service will create trust in trustless environments to enhance auditability and its accountability while automating the execution of the contract [36–39]. This will be discussed in detail in Section 2.2. Given this, a potential adoption of blockchain-based solutions are being conceptualised with an expectation to help smart city developers adopt blockchain as the embodiment of trust to own, trade, and exchange assets, without central servers controlled by a single administration point, enabling a token economy [40–42].

2.1.2. Construction Supply Chains

Traditionally, construction projects are considered unique, but Construction Supply Chain (CSC) processes are often predictable and repeatable. The CSC has an element of value transfer beyond the monetary value within a low-trust environment, where stakeholders have adversarial relationships and interests and mostly work through an intermediary [2,5,7,8,34,43]. This is where the blockchain's key strength could enable a single source of data integrity, by ensuring the immutability of the record and compliance checking for CSC [2,44]. Another pain point in the CSC that blockchain could overcome is traceability as it provides a full audit pathway for the data by creating an immutable record for the CSC activities [44,45]. Blockchain will pave the way to automate the CSC payments through smart contracts, by ensuring that contracted obligations are satisfied along a supply chain by various stakeholders [5,46,47].

2.1.3. Building Information Modelling (BIM)

BIM has played a key role in the digital transformation of the built environment [23,48]. However, the BIM process for supply chain still bears several shortcomings [49] in the absence of a legal context describing the BIM data ownership and no trusted record of the model changes during construction and operation stages [39]. Ready access to the history of project documentation can be an invaluable information source in all phases of the life cycle of the project [50]. There is a value in an immutable record of the BIM model transactions to agree on using it as a single source of truth (SSoT), by enabling BIM data integrity, reliability, and traceability [5,44]. However, the integration is still in its infancy [12,13].

2.2. Need for Blockchain in Enabling Trusted Property Transactions in the Built Environment

A critical review of previous studies on the application of blockchain technology for land registration was conducted to understand the main research focus, issues in maintaining records for property transactions, proposed software prototypes, type of blockchain platform recommended, and challenges and limitations. The Scopus and Web of Science databases were used to search the application of blockchain technology for land registration. Figure 2 shows the search strategy including keywords, inclusion, and exclusion criteria for shortlisting papers.

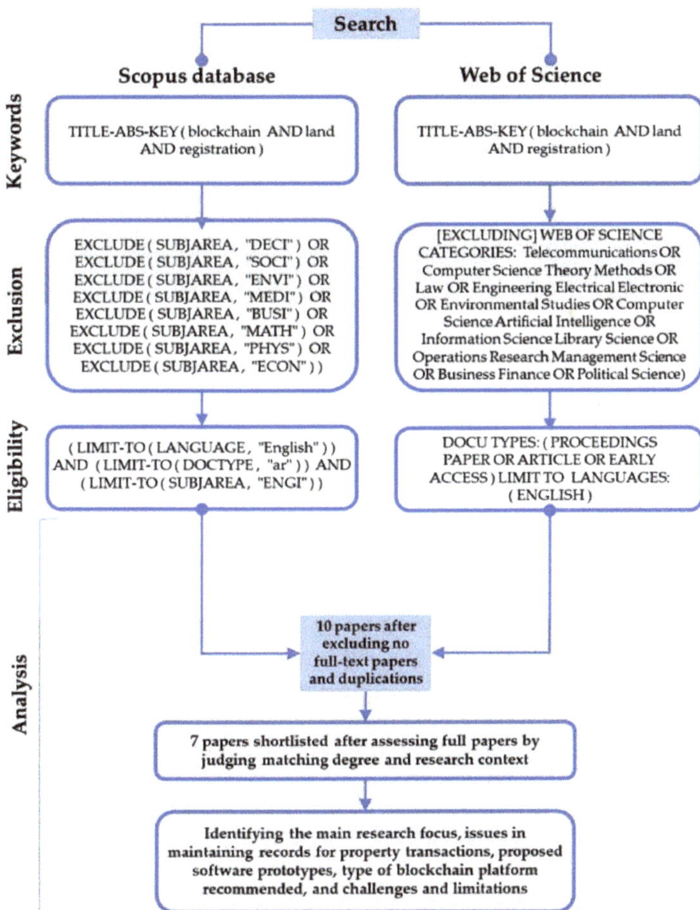

Figure 2. The search strategy.

Most of the previous studies, as illustrated in Table 1, on the application of blockchain for land registration and property transactions are still at a conceptual and/or exploration level, with very few proofs of concept available. This indicates a knowledge gap that needs addressing to examine and validate the usability and challenges in adoption of this technology [1,31,51,52]. However, the previous studies pointed out that the rationale for proposing a blockchain in maintaining records for land registration and property transactions is the capability of this technology to overcome the transparency issue and ensure ownership [1,31,51–55]. In addition, it is argued that adopting blockchain will enable data integrity by assuring the accountability and consistency of data over its entire life cycle, including storing, processing, or retrieving data, while the current system is fragmented with a bureaucratic structure, which leads to ownership conflicts, complexity, and land records documentation issues [31]. There is evidence of corruption in the current system of maintaining records for property transactions in many countries. Adopting blockchain technology will create and maintain a clear audit trail of actions that will help to minimise the possibility of the records being tampered with, by facilitating the litigation proceedings as they arise [1,53,54].

Table 1. A critical review of previous studies on application of blockchain technology for land registration and property transactions.

Main Research Focus	Issues in Maintaining Records for Property Transactions	Proposed Framework/Software Prototype	Type of Blockchain Platform Recommended	Challenges and Limitations	Source
The usage of blockchain for land records management in India. (Conceptual and framework research)	• No data integrity (discrepancies in the records). • Poorly administered. • Lack of transparency. • Current system does not ensure a guarantee of ownership. • Corruption.	• A class diagram and framework of blockchain usage to update land records were proposed. • A software prototype solution was not provided.	Hyperledger Fabric, as it offers a number of Software Development Kits (SDKs) to support various applications.	• Blockchain is still new and very few proofs of concept are available. • Lack of specialised expertise. • Buy-in from participant. • Legal issues (no regulatory standards are available yet). • Instances of security violations. • Cost of implementation.	[1]
To discuss the use of blockchain as a land registration tool in Cyprus. (Horizon scanning research)	• Disputes in land ownership registration and administration. • Lack of transparency. • Lack of accountability.	• Neither a proposed framework nor a software prototype was provided.	Not mentioned.	• Blockchain is still new and more research is urgently required in this area. • Political and legal issues. • Risks of implementation.	[51]
The potential of blockchain application in title registration in Ghana (Exploration and conceptual research)	• Unreliable recordkeeping system. • Land disputes. • Lack of proper boundaries. • Lack of transparency in records verification and transaction.	• Neither a proposed framework nor a software prototype was provided. However, the paper discussed the blockchain-enabled land acquisition and registration model.	Not mentioned.	• Very few proofs of concept are available. • Legal issues (no regulatory standards are available yet).	[52]
An analysis of previous blockchain publications targeted on The Netherlands (Exploration research)	• Lack of transparency. • Lack of accountability. • Fragmented system and no data integrity (data silos).	• Neither a proposed framework nor a software prototype was provided.	Not mentioned.	• Proof of concept study is necessary, limited research has been done into usage of blockchain in the property sector. • Using blockchain to find an innovative configuration for business models for the property sector is not easy.	[31]

Table 1. *Cont.*

Main Research Focus	Issues in Maintaining Records for Property Transactions	Proposed Framework/Software Prototype	Type of Blockchain Platform Recommended	Challenges and Limitations	Source
The potential of blockchain to enable reliable registration of land in real estate (focusing on performance of hash functions in blockchain) (Conceptual and framework research)	• Process in current system is not digitised and leads to records getting tampered with. • Accessing records is time-consuming. • Double registration issue in the establishment of ownership. • Brokerage system (middlemen cost). • Lack of transparency. • Corruption.	• The paper discussed the hashing processing in blockchain and proposed a framework. • A software prototype solution was not provided.	Ethereum as it is the first public blockchain platform supporting smart contracts.	• Not mentioned.	[53]
An analysis of blockchain-based land registration possibilities and challenges (Exploration and conceptual research)	• Complexity of records transfer and registration. • Lack of transparency in records verification and transaction. • Corruption in current system. • No data integrity.	• Neither a proposed framework nor a software prototype was provided. However, the paper discussed some practical applications of blockchain in European countries.	Not mentioned.	• Blockchain in the public type form is not suitable for the specificity of the property sector transfer and land registration. • Defining liability rules is necessary to enable the blockchain in the property sector and land registration. • There is a need to continue the research on optimal legal and technical ways of taking advantage of using blockchain in the property sector.	[54]
The potential of blockchain to enable reliable registration without intermediaries (Horizon scanning research)	• Lack of transparency. • Fragmented system. • Disputes in land ownership registration and administration.	• Neither a proposed framework nor a software prototype was provided.	Ethereum, as it is the first public blockchain platform supporting smart contracts.	• Legal issues (no regulatory standards are available yet). • Buy-in from participant. It is hard to conceive complete protection of consumers in a business-to-consumer relationship in a disintermediated solution such as blockchain.	[55]

Blockchain is still a relatively new technology at an early trigger stage, where the majority of implementations are either in initial or development stage [3,54]. As illustrated in Table 1, most studies use hypothetical cases with a significant gap of a deductive use case, and/or a detailed software prototype to support those hypotheses for land registration and property transactions. Only a few papers have even recommended a specific type of blockchain.

Ethereum was mentioned as the first public blockchain platform supporting smart contracts for mass consumption, such as the financial sector [39,53,55]. However, it is argued that Ethereum is not suitable for the specificity of business, such as the property sector where confidentiality and performance are critical [6,39]. Ethereum as a public (non-permissioned) blockchain platform has a privacy and accountability issue, as it is open to anyone who wishes to participate, and performance and scalability become a challenge as each node has to process each transaction [6,39,54]. Hyperledger Fabric was strongly recommended as a permissioned blockchain platform to satisfy the requirements of privacy, trust, and traceability desirable for a broad range of industry use cases, including property transactions in the built environment [6,39,53]. Hyperledger Fabric offers a number of Software Development Kits (SDKs) based on modular and pluggable components [39] to support various applications, enabling buy-in from participants [53].

Hyperledger Fabric is an open-source blockchain platform hosted by the Linux Foundation that can be used to create permissioned blockchains and develop distributed applications [37]. It is being actively developed by the International Business Machines Corporation (IBM) [56], and has received contributions from Intel and SAP Ariba [37]. The Hyperledger Fabric architecture delivers high degrees of flexibility and confidentiality in its design and implementation, which makes it useful in many of the built environment applications, including property transactions [6,39,56,57]. It helps in achieving privacy, as it requires permission to read and write [6,56]. It also enables auditability by separating transaction processing into three phases; (a) distributed logic processing through Chaincode Services to create a smart contract, which is the business logic code of a transaction in the Hyperledger Fabric platform; (b) transaction ordering through Consensus Services to create blocks of transactions and facilitate trust in the network; and (c) transaction validation through Membership Services to identify the network members and generate a root of trust by satisfying the endorsement policy that defines which peers can run chaincode to execute transaction proposals to be committed to the ledger [36–39]. The use of Hyperledger Fabric as a blockchain platform to enable trusted property transactions is demonstrated in the subsequent sections.

Exploring the use of Hyperledger in the property sector is now increasing: from 6% in 2019 [58] to 14.7% in 2020, according to the FIBREE industry report 2020 for blockchain in the property sector [3]. With only two countries in 2019 using Hyperledger, more than seven countries, including Australia, China, India, the United Kingdom, and the USA, are currently using it. The following section describes the methodology adopted in this paper to provide a use case of blockchain-based trusted property transactions in the built environment.

3. Research Method, Design and Tools

The prototyping model of evolutionary prototyping [59] was selected as the software development methodology for the proposed system. In this method, an initial prototype is constructed, which is then evaluated. Successive prototypes were developed, with additional functionality based on received feedback. The prototyping model was selected due to its advantage of reducing the development time and providing a rapid solution to the identified problem scenario [60]. The steps of the research method followed in this paper are illustrated in Figure 3, and described below.

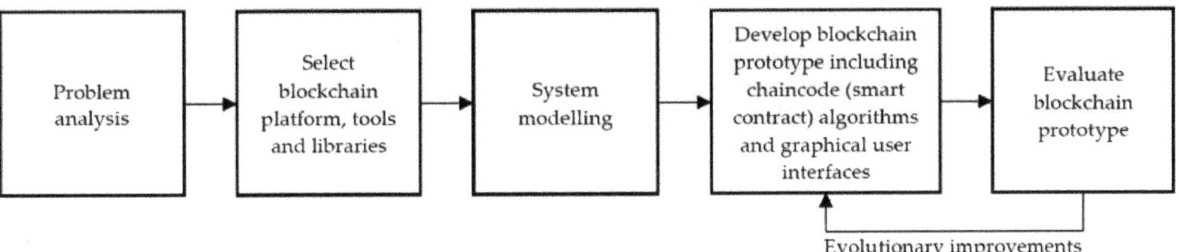

Figure 3. Research method to develop the blockchain-based software prototype.

The scope of the business scenario covers real estate property transactions in New South Wales (NSW), Australia. In a blockchain solution development for an application domain, full modelling of the domain requirements and issues should be undertaken. This may involve data flow modelling, process modelling, and entity-relationship modelling, among others [60]. However, this paper discusses a simplified analysis of the problem domain as presented in Section 4.1, since the focus is on the demonstration of the development of a blockchain system. The system requirements were identified through literature and by reviewing relevant documentation available on NSW government websites. These system requirements are elaborated in Section 4.2.

A systematic process protocol presented in Nanayakkara et al. [61] was followed to identify the most suitable blockchain platform for the prototype. This involved the application of the multi-criteria decision-making method, Simple Multi Attribute Rating Technique (SMART). The SMART method was selected due to the relative simplicity of application to assist accurate decision-making [62]. Blockchain platforms that would suit the system requirements were identified and evaluated based on their salient features. A major initial decision was to choose between permissioned and permissionless blockchain networks. Permissionless networks allow users to freely join the blockchain network and pseudonymously engage in transactions. In contrast, permissioned networks require the participants to be known and authorised to join and transact in the network [6,15]. It was decided that permissioned networks are the most suitable type of blockchain for property transactions, as participants should be known and held accountable for their transactions. Furthermore, permissioned networks will ensure privacy of users' sensitive data. The list of candidate permissioned blockchain platforms included Corda R3, Elements, Hyperledger Fabric, IBM Blockchain, and New Economy Movement (NEM). Subsequent criteria such as cost, level of support, ease of use, performance, and security [61], were considered, and provided weights based on the importance to the system development process. Next, weights for each selection criterion were determined for each of the blockchain platforms based on a two-stage evaluation by the authors, individually and collectively. The finalised platform weights were multiplied by the criterion weights, and finally added together to obtain the total value. Table 2 lists the selection criteria, weights for each criterion, the criteria weights for each platform, calculated values, and total values for the top four blockchain platforms that emerged from the evaluation. The Hyperledger Fabric blockchain platform was selected as the most suitable match for the identified requirements through this method, due to its features and benefits described in Section 2.2.

Next, the tools, libraries and programming languages related to the platform were selected based on the system requirements, following the process protocol of Nanayakkara et al. [61]. The tools, libraries, and programming languages are discussed in Sections 4.3 and 4.4. The system architecture was designed to fulfil the requirements, and is presented in Section 4.3. The blockchain prototype, including chaincode (smart contract) algorithms, was developed, and a graphical user interface was created to facilitate the system operation. Details of the system implementation are provided in Section 4.4. Finally, the functionality of the blockchain prototype was evaluated through test cases. The prototype was iteratively improved based on the evaluation, as discussed in Section 4.5.

Table 2. Top four blockchain platforms suitable for the system requirements based on SMART evaluation (adapted from [61]).

Criteria		A	B	C	D	E	F	G	H	I	Total	Rank
Weight for Criterion		**10**	**8**	**8**	**4**	**10**	**5**	**6**	**7**	**6**		
Hyperledger Fabric	Weight	8	10	8	9	9	9	9	7	9	552	1
	Calculated Value	80	80	64	36	90	45	54	49	54		
Corda	Weight	7	4	7	10	10	6	10	5	10	483	2
	Calculated Value	70	32	56	40	100	30	60	35	60		
NEM	Weight	4	5	9	8	8	3	8	9	7	432	3
	Calculated Value	40	40	72	32	80	15	48	63	42		
IBM Blockchain	Weight	6	3	8	7	8	8	9	6	5	422	4
	Calculated Value	60	24	64	28	80	40	54	42	30		

A—Community availability/learning material/level of support, B—Cost, C—Language and ease of use, D—Performance, E—Permissioned network, F—History and reputation in the industry, G—Security, H—Supports API, I—Updates or release of versions.

4. The Proof of Concept

4.1. The Business Scenario

In NSW, buyers and sellers transfer property ownership digitally by employing the services of real estate agents and registered lawyers or conveyancers [63]. Property titles are stored by the NSW Land Registry Services (NSW LRS) in a centralised land registry. Searches on title and street address, land value, historical data, and so on, are freely available through the NSW LRS online portal. Authorised information brokers provide detailed information related to land and property for a fee [64].

The typical process of property transactions in NSW is detailed as follows, and is illustrated in Figure 4. A person who wishes to sell property ("seller") will contact a real estate agent ("agent"), who will list the property to find prospective buyers [5,65]. The seller will also engage a lawyer to handle the property transaction process. A property buyer will search for properties that match their requirements, and inspect properties on sale. When the buyer identifies a suitable property, the buyer should pay the agent a fee to reserve the property. The buyer will then engage a lawyer to carry out the property transaction. The buyer also has to arrange finances with their bank. The buyer's bank will assign an assessor to verify the market value of the property. A title search will be conducted by the buyer's lawyer through an information broker to ensure that the land being sold is legitimately owned by the seller. Information such as council and water rate adjustments, land type, sewerage line clearance, flood zone, and bushfire prone area, are also queried through the relevant government authorities [64]. Once all the checks have taken place, the buyer pays a deposit to the agent. The seller's lawyer prepares the contract, which is checked by the buyer's lawyer, and the buyer and seller sign the contract. When the mortgage is approved by the buyer's bank, a pre-settlement inspection will be conducted by the buyer or the buyer's lawyer. The final payment is made to the seller's lawyer, who will prepare the documentation to transfer the property title to the buyer, and register the title with the land registry [66,67]. The buyer's lawyer arranges the payment of transfer (stamp) duty to the revenue department, if applicable [68]. The buyer's lawyer will claim all expenses incurred from the buyer. The seller's lawyer will pay the agent, make other deductions, and transfer the payment to the seller.

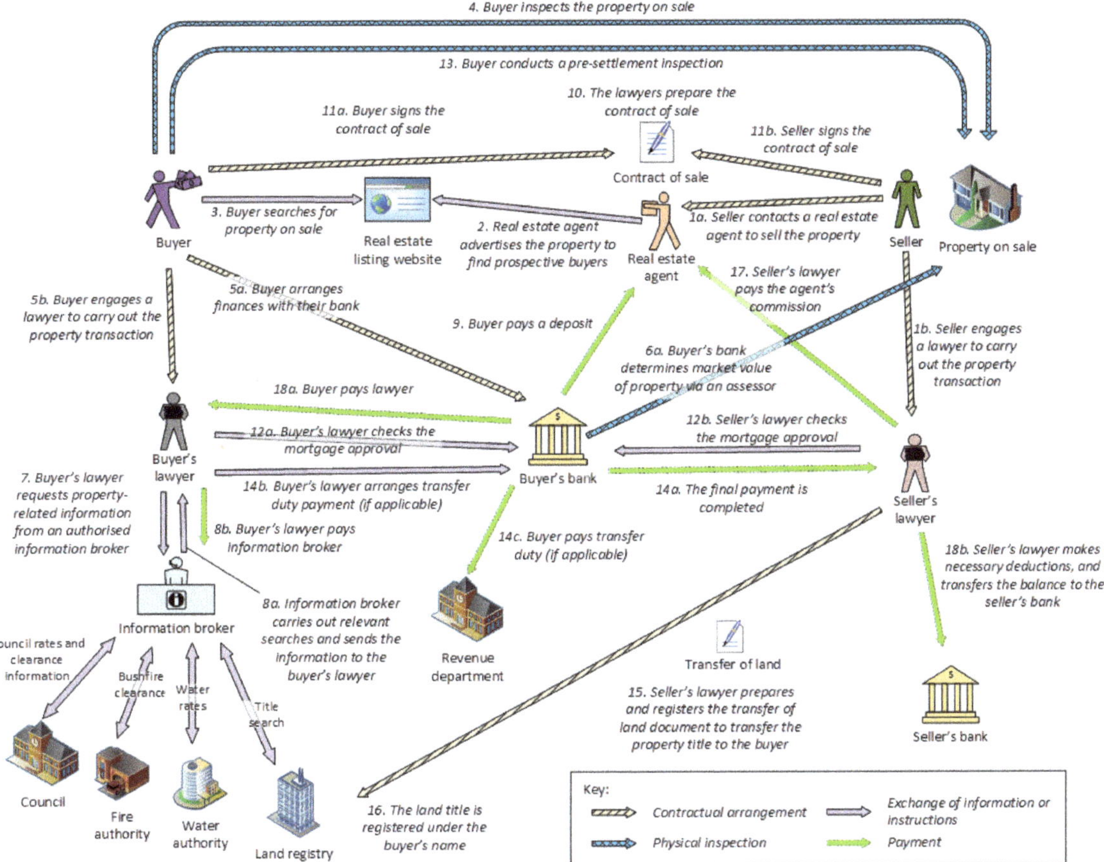

Figure 4. Current process for property transactions.

As seen in Figure 4, this existing process for property transactions is complex, time-consuming, and costly, since many intermediaries are involved. Furthermore, issues such as lack of transparency and data fragmentation are also present. To overcome these issues, a blockchain system architecture that facilitates a simplified, lean property transaction process through Hyperledger Fabric chaincode (smart contracts) and graphical user interfaces is introduced in the following sub-sections.

4.2. System Overview

Registered users of the blockchain system can view details of all properties that they own, advertise their own property to be sold on the land market, search for properties advertised on the market, perform a title search of properties on sale, view the transaction history of their own property or any property on sale, initiate a sale after negotiating with a buyer, and purchase the property once the sale has been initiated by the seller. Clearance information related to the property, such as council and water rates, sewerage line clearance, flood risk and so on, is proposed to be retrieved through chaincode (smart contracts) that will connect to the databases of relevant authorities that provide such information. It is also proposed that property transactions within the system can be performed through exchanging a fiat-collateralised stable cryptocurrency. The detailed operation of the land registry has not been modelled in this prototype, but is indicated here as a requirement to showcase the possibility of the involvement of multiple agencies and stakeholders. The system features are summarised within the use case diagram in Figure 5. When compared to Figure 4, it is apparent that there are significantly less types of participants in the transaction process. The blockchain system will play the role of intermediaries such as real estate agents, lawyers, and information brokers, and automate the transactions as described in the next section.

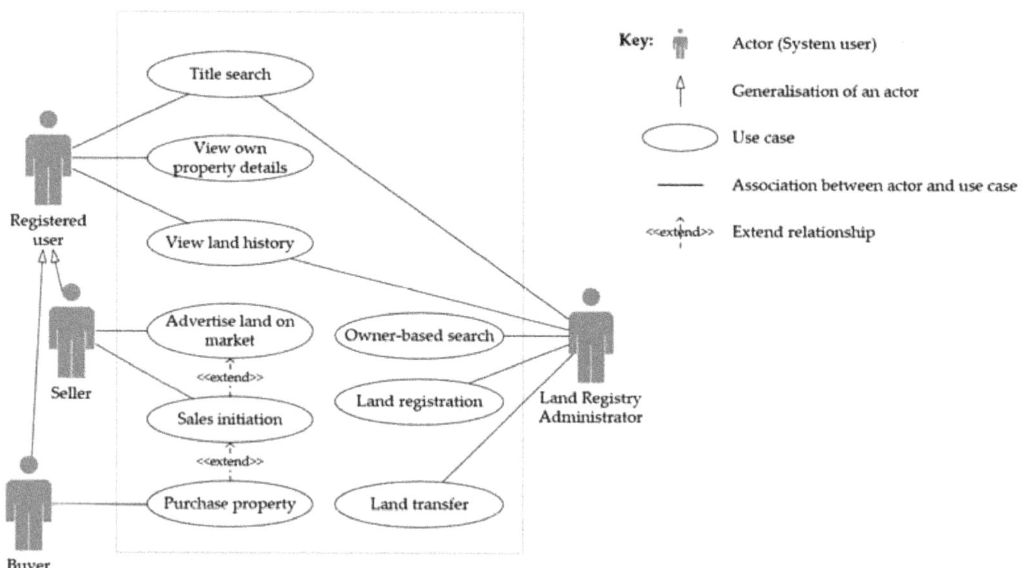

Figure 5. Use case diagram for property transactions.

4.3. System Architecture

This section presents an overview of the system architecture, which is illustrated in Figure 6. The system processes are described using Hyperledger Fabric terminology, and the function of the key terms are summarised in Table 3. The prototype has three main components, namely, the web application for property transactions, the blockchain network, and external chaincodes. The web application (client) provides the graphical user interface for all authorised users to interact with the blockchain network. The blockchain network connects the peers that hold a copy of the ledger and internal chaincode, and the ordering service. The external chaincodes represent processing within external organisations, such as city councils and banks. The client application and the blockchain system are connected through the Hyperledger Fabric Software Development Kit (HLF SDK) Application Programming Interface (API). Authorised users of the system can perform search queries or carry out transactions through the client application, as described in Section 4.2. The client application will connect to the blockchain network and invoke the internal chaincode, which contains the application logic for property transactions, to read or write data from the ledger. The numbers in the following description correspond to the numbering in Figure 6. (1) The command to invoke the chaincode is known as a transaction proposal, which is sent by the client application to the endorsing peers on the blockchain network. (2) The endorsing peers verify the transaction proposal and execute it by invoking the chaincode. (3) This system proposes the use of external chaincodes to obtain relevant data from government authorities, which are called by the internal chaincode when required. (4) The external chaincodes run the application logic and send the responses back to the internal chaincode. (5) The endorsing peers create a proposal response including the transaction results and peer's signature and send it to the client application. (6) The client application verifies the peer signatures and compares the responses. If the transaction proposal was only a query, the process ends by displaying the result to the user. (7) If a ledger update is required, the client application packages the transaction proposal and endorsed responses into a transaction, and broadcasts it to the ordering service. (8) The ordering service receives transactions from the entire network, orders the transactions, and creates a block of transactions. (9) The ordering service transmits the block to the leading peer, (10) which then distributes the block to all the other peers. (11) The peers validate the transactions within the block and tag the transactions as valid or invalid. If the transaction is valid, then the world state is updated, whereas all valid and invalid transactions are added to the blockchain, which ensures auditability. (12) Finally, the peers emit an event to notify the client about the transaction being valid or invalid, and that it has been added to the blockchain, and the result is displayed to the user [69].

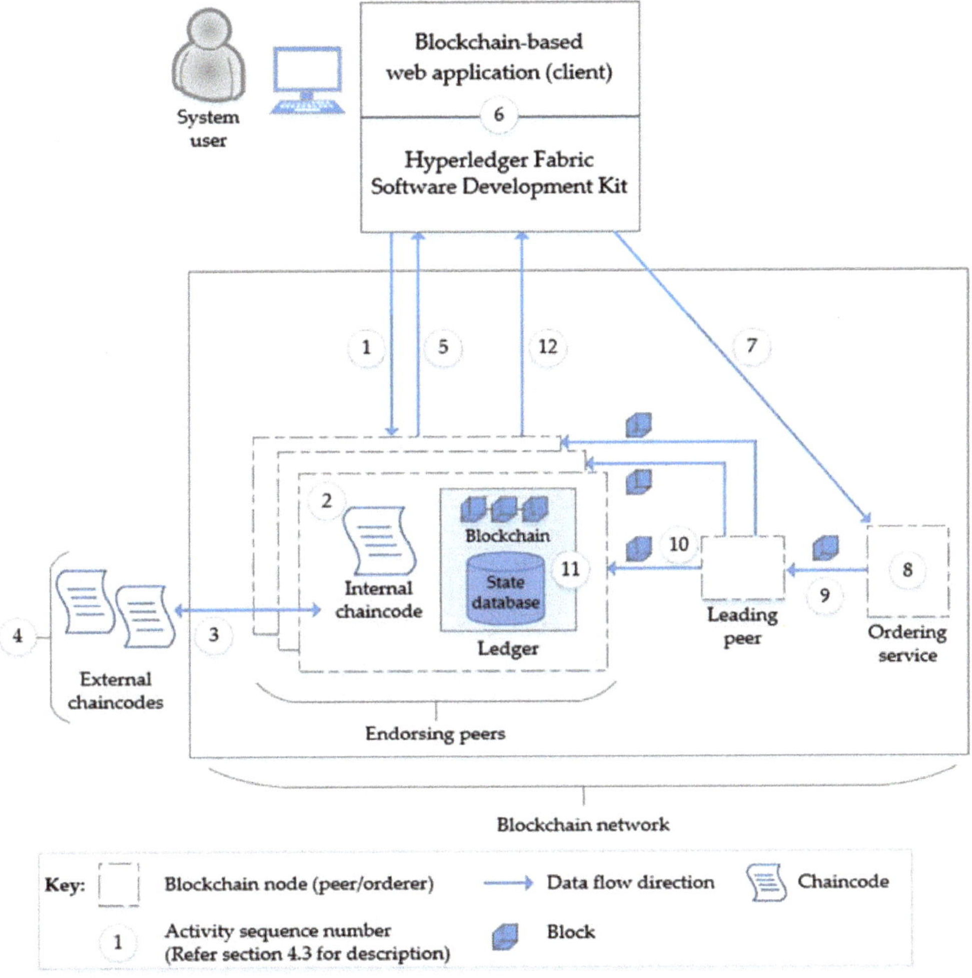

Figure 6. System architecture for the blockchain prototype.

Table 3. The function of Hyperledger Fabric within the prototype.

Term	Description
Ledger	An immutable and sequenced record of all transactions within the prototype. The ledger consists of a blockchain and world state [70].
Blockchain	Consists of blocks, with a sequence of transactions, that are cryptographically linked together [70]. The blockchain will be stored in the file system of each peer connected to the prototype.
World state	Represents the latest values for all keys in the transaction log of the blockchain. A state database is used to store world state data, such as LevelDB and CouchDB. LevelDB is the default state database which stores chaincode data as key-value pairs. Chaincode data is modelled as JSON (JavaScript Object Notation) in CouchDB [70]. This prototype implementation used CouchDB as it allows rich queries of the JSON content.
Chaincode (Smart contract)	Program code installed on peers that implements the application logic, which is invoked by external client applications [70]. The prototype includes internal chaincode for property transactions and external chaincodes to simulate the functionality of connecting to external databases.
Client	An application external to the blockchain network that connects to the network to perform business transactions [70]. All authorised users of the prototype will access the blockchain through the graphical user interfaces in the client application.
Peer	An entity on the blockchain network that maintains the ledger. A subset of peers called *endorsing peers* run chaincode to execute transaction proposals. A *leading peer* communicates with the ordering service and distributes blocks to peers when blocks are received from the ordering service [70]. All peers in the prototype have been defined as endorsing peers.
Ordering service (Orderer)	Performs ordering of transactions into a block and distributes the blocks to connected leading peers. Orderers do not execute or validate transactions [70]. The prototype contains a single ordering service node.

4.4. System Implementation

The process sequence for a successful property transaction within the blockchain prototype, starting from the point a seller advertises a property to the completion of the fund transfer and the consequent update of the blockchain regarding the land title transfer, is illustrated in Figure 7. The interactions between the web application and HLF SDK were omitted from the figure for simplicity; these processes would take place between the internal chaincode and the buyer/seller. Unsuccessful queries and updates are also handled by the prototype, although these processes have not been depicted in the figure for the sake of readability.

The internal chaincode contains all the application logic related to property transactions that were presented in Section 4.2. The external chaincodes would be installed at the government authorities, such as local councils and water authorities, which would be queried through the system to check property title clearance information. Connecting these authorities to the system would reduce the number of intermediaries involved, and decrease the time taken to perform the clearance checks for a given property title, compared to the current process. The study also proposes the use of a fiat-collateralised stable cryptocurrency ("C-AUD") backed by the Reserve Bank of Australia to pay for property transactions. This cryptocurrency will be equal to the Australian Dollar, and there is no price volatility against fiat currency. The bank chaincode contains the logic of checking the cryptocurrency balance of a buyer, and transferring funds from the buyer's account to the seller's account and the transfer duty to the NSW revenue department.

The blockchain prototype uses the Go programming language for the chaincode, Node.js for the web application, and the HLF SDK for Node.js to create the link between the web application and the blockchain network. Go files are written for each chaincode in the system, and contain the instructions to run the blockchain query and update functions.

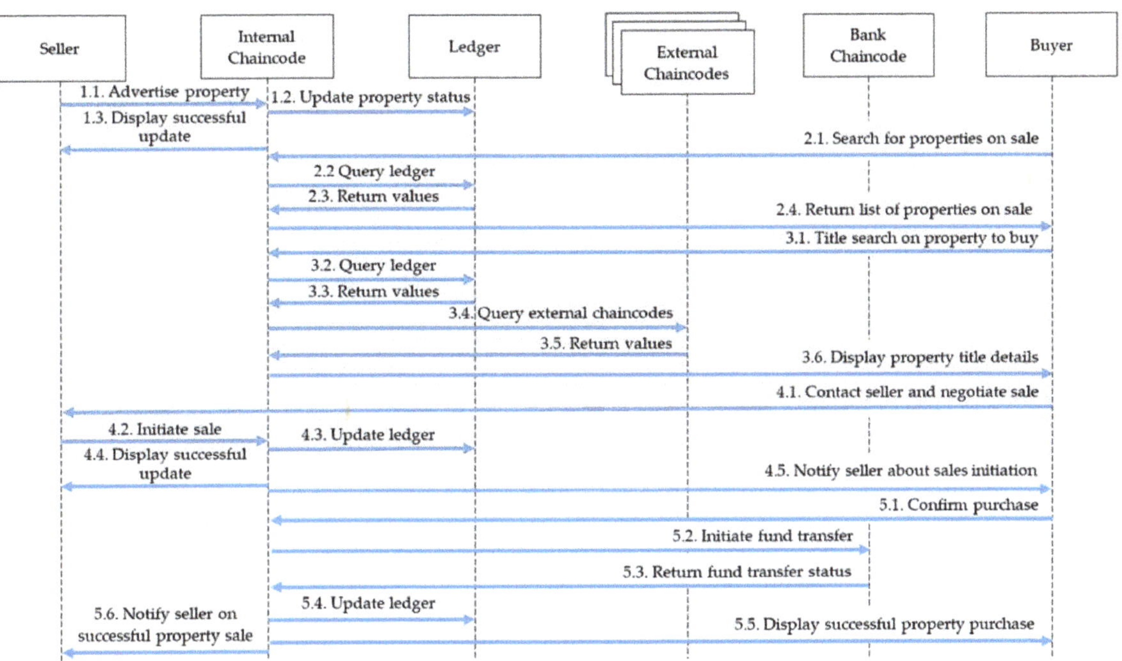

Figure 7. Process sequence for a successful property transaction through the blockchain prototype.

The Hyperledger Explorer tool was configured to be used with the prototype so that the blockchain network administrator can view the blocks, transactions, chaincodes, and other relevant information within the blockchain network. Figure 8 presents a view of the details of a transaction within a particular block using Hyperledger Explorer.

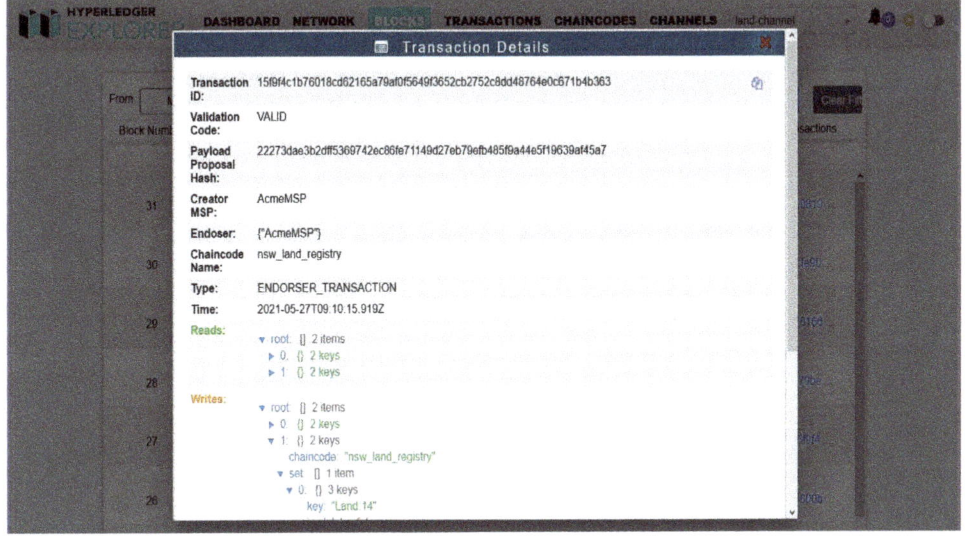

Figure 8. Viewing transaction details in a block through the Hyperledger Explorer tool.

The prototype allows authorised land registry administration to search for land titles, and view land ownership details and transaction history of land titles (Figure 9). Members of the general public will have to register their identity on the blockchain platform using a recognised identification document in order to use the system. Once logged into the system, they can view the details of all properties they own. A prospective seller can advertise their property on the market by listing its selling price, and a prospective buyer can search for properties that are on sale by providing the address, or the name of the street or city, or view all properties on sale (Figure 10). A prospective buyer can perform a title search of the property on sale in order to verify the legitimacy of the sale, and view all clearance information. If satisfied with the land details, the prospective buyer will contact the seller and negotiate the sale of land. When the negotiation is complete, the seller can initiate the sale through the system, and the buyer can purchase the property. The buyers and sellers can view the status of the transactions at any time during the process, which also includes the block details (Figure 11).

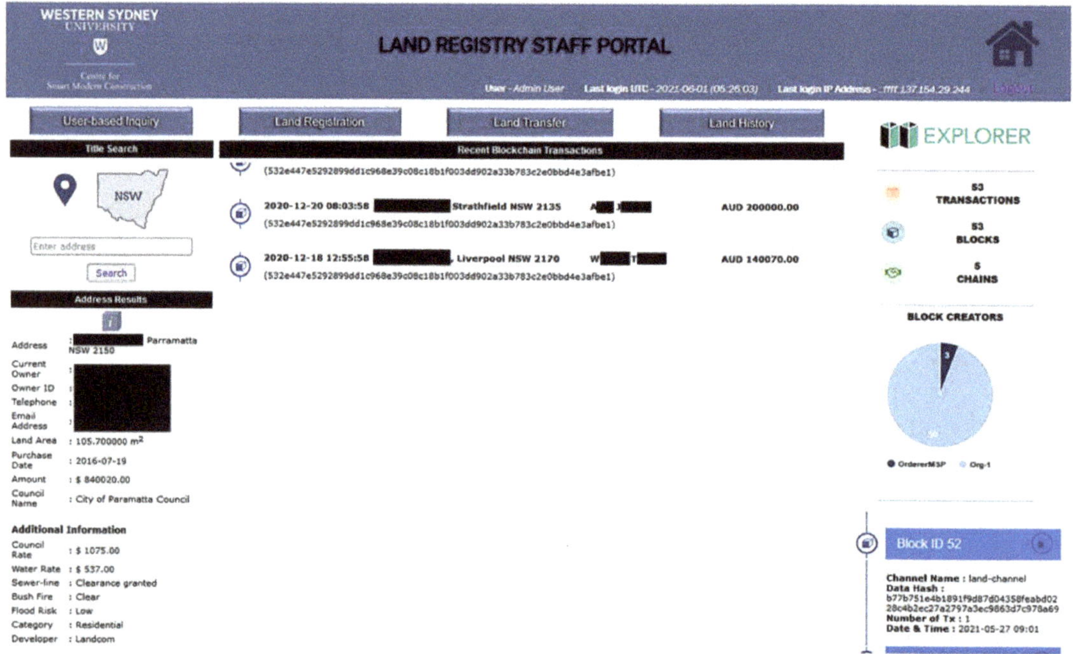

Figure 9. Output of title search blockchain query function available for land registry administration.

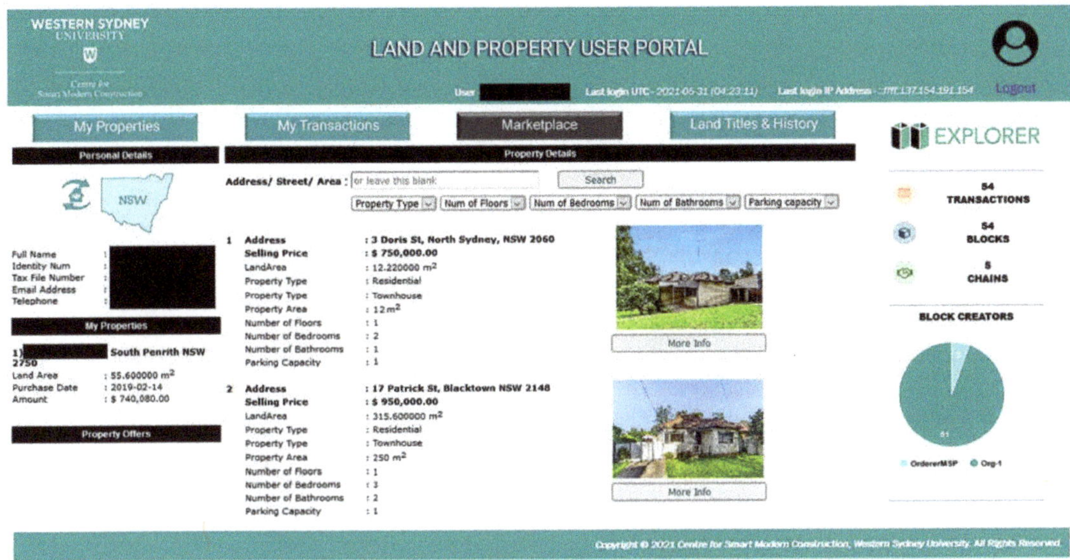

Figure 10. Viewing property on sale in the marketplace.

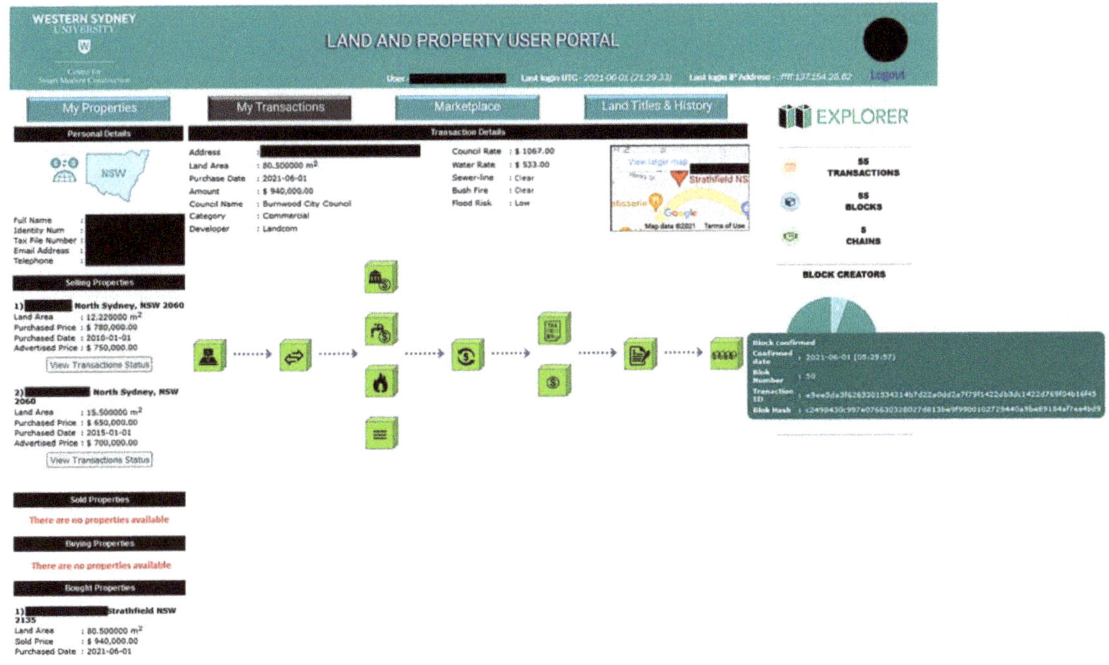

Figure 11. Transaction status view for a property buyer.

4.5. System Evaluation

The blockchain prototype was comprehensively evaluated by the authors to ensure methodological reliability and the accurate execution of smart contracts. The evaluation was conducted using test cases for each of the identified system requirements presented in

Figure 5. A test case scenario for property search in the marketplace using advanced search criteria, such as specifying the property type, number of bedrooms, and so on, is provided in Table 4 as an example. In this scenario, only a blockchain query will be performed, and the blockchain will not be updated. A blockchain update is performed in instances such as confirming the sale of the property. Then, the test cases would also check whether a new block is added, and its details are displayed on the sidebar, and if the number of blocks and transactions are updated on the sidebar. Improvements to the prototype were added iteratively until there were no failures of test cases. The final prototype fulfils all the system requirements, and provides an incubation-ready proof of concept for blockchain-based trusted property transactions. An external validation of the system would be required if the prototype is developed into a minimum viable product that is industry-ready.

Table 4. Example test case scenario.

Test Scenario ID		Property Search–3		Test Case ID	Property Search–3A
Test Case Description		User searches for property on sale with advanced search criteria		Test Priority	High
Pre-Requisite		Successful login to the user portal		Post-Requisite	N/A
Test Execution Steps:					
S. No	Action	Inputs	Expected Output	Actual Output	Test Result
1	Go to Marketplace page	http://landblocks. online/marketplace	Marketplace page	Marketplace page	Pass
2	Input advanced search criteria and click the Search button	City: Penrith Property Type: Townhouse Number of bedrooms: 2 Parking capacity: 2	Display all properties on sale in Penrith with type Townhouse, 2 bedrooms, and parking capacity of 2.	Display all properties on sale in Penrith with type Townhouse, 2 bedrooms, and parking capacity of 2.	Pass

5. Discussion

Blockchain has been proposed as a solution for many transaction-oriented domains. However, its implementation is novel and challenging. Unlike traditional software solutions, it requires a solution based on a full ecosystem. Therefore, development of blockchain solutions for application domains are complicated, and require further research. This research attempted to demonstrate the process of development of a blockchain prototype, taking property transactions as the use case. Maintaining property transaction records on a blockchain will potentially accelerate a shift in the work environment to a transparent and cooperative chain of transactions by assuring the accountability and consistency of data over the entire life cycle, including storing, processing, and retrieving data, leading to data integrity [31,39,71]. The Ethereum platform is still being considered in most use cases in the property sector [3]. However, Hyperledger Fabric was strongly recommended, as it will address the privacy issue of Ethereum, ensure accountability of network participants, and support various applications, as it has modular and pluggable components to satisfy a broad range of industry use cases [6,39,53].

The developed prototype provides a tangible proof of concept for the application of blockchain for property transactions, providing the benefits of blockchain including increased data integrity, transparency of transactions, and increased trust related to property transactions. The proposed connecting of relevant government authorities to the blockchain system through external chaincodes is expected to reduce the number of intermediaries involved, and the time taken for clearance checking. This prototype also differs from other existing and proof of concept property transaction systems through the proposed integration of a fiat-collateralised stable cryptocurrency. This is expected to help ensure trust in the system in its initial stages, rather than by enabling transactions through volatile cryptocurrencies. Future iterations of the system may provide the capability of transacting through alternative payment media mutually agreed upon by buyers and sellers. Currently, these two features require further development due to the limitations discussed below. Furthermore, connecting these external organisations would require integrating disparate systems and ensuring proper data transfer among the systems.

The operation of the blockchain was tested by entering test data for multiple test cases through the graphical user interface. The graphical user interface was created to shield the complexity of the blockchain from a general user of the system; all the blockchain operations occur in the backend of the system, and are not visible to a typical user. However, the Hyperledger Explorer tool within the prototype enables a system administrator to graphically view the operation of the blockchain, and verify blocks, data, and transactions within the blockchain.

The intention of this software prototype is to showcase the potential of a blockchain solution for property transactions in the built environment. Therefore, the prototype has a few limitations, as follows. If users wish to use the blockchain system, they will have to register their identity on the blockchain platform and be provided with login credentials. The user registration process is not covered within the current scope of this prototype. Linking users' bank details to the system would also be part of the user registration process, and is not handled currently. Therefore, user details are represented within the prototype by manually inserting sample data. The prototype demonstrates the capability of connecting the system to external government organisations in order to query required information, by calling external chaincodes through the main chaincode. However, developing the actual data processing code for these external organisations is not within the scope of the prototype, and is only simulated through sample functions within the external chaincode. Furthermore, the property title details would need to be integrated with existing governmental systems in order to ensure proper regulation. The subsequent property transactions could be facilitated by developing the system as an enterprise blockchain system with an appropriate financial model for it to be commercially viable. The proposed system has eliminated most of the intermediaries involved in the property transaction process. However, it is acknowledged that the involvement of lawyers may be required to maintain the legality of transactions.

Apart from the limitations of the proposed software prototype in this research, adoption of blockchain is going beyond just a new technology solution to a digital transformation of the current business model [31], and that will require involvement of all relevant stakeholders as part of this solution, such as legal professionals [52,55], to set a new regulatory standard and define new liability rules, which are not easy to establish. It would also entail transferring the vast amount of existing property title data to the new system, which would involve high costs in terms of time and effort. Given this, the challenges also include buy-in from the participants, as it is hard to conceive complete protection of consumers in a business-to-consumer relationship that bypasses intermediaries [55,72], a lack of specialised expertise required for this adoption [1], and the cost of this adoption [1,51].

6. Conclusions

This paper illustrates the potential of blockchain in the built environment. It uses property transactions as a use case in demonstrating the development of a blockchain system in a step-by-step, systematic approach. Adopting blockchain technology will create and maintain a clear audit trail of actions that will help to minimise the possibility of the records being tampered with by facilitating the litigation proceedings as they arise. Trustworthy and immutable records are essential for establishing ownership of property titles in the built environment. Blockchain will enable the adoption of a smart contract to optimise contract formulation and negotiation, while transaction ordering through a consensus service will ensure immutability mechanisms for the transfer or creation of value, and transaction validation through a membership service will create trust in trustless environments to enhance auditability and its accountability, while automating the execution of the contract.

The prototyping model of evolutionary prototyping was selected as the software development methodology for the proposed system. The SMART method was used due to the relative simplicity of application to assist accurate decision-making to identify the most suitable blockchain platform for the prototype. Hyperledger Fabric, through the development of a prototype, was strongly recommended as a permissioned blockchain platform to satisfy the requirements of privacy of sensitive data, trust, and traceability, which can be desirable for a broad range of industry use cases, including property transactions in the built environment. This study has achieved its aim by understanding the mechanism of the Hyperledger Fabric blockchain platform, and developing a prototype proof of concept that enables trusted property transactions. The blockchain prototype was comprehensively evaluated by the authors to ensure methodological reliability and the accurate execution of smart contracts. The prototype includes functionality for land registry administrators and users of the general public who wish to buy and sell properties. It facilitates execution of smart contracts for property transactions through chaincodes. It also enhances trust in the network by identifying authorised users, and ensuring transactions are validated and propagated through the blockchain network. The proposed system will eliminate the need for many intermediary service providers by connecting users directly with relevant information sources, thereby reducing the time taken for property transactions. This system has been included as a use case for land and property management in the official Hyperledger wiki [73]. Future research development requires a comprehensive analysis of the problem domain that involves the whole property transaction ecosystem. Modelling of the relevant domain is an essential precursor to the development of any blockchain system. The authors envision that the business case for the adoption must remain the focus while technology overcomes the temporary barriers of reliability, traceability, and interoperability. Finally, the findings provide the foundation for developing proofs of concept for other potential applications of blockchain in the built environment.

Author Contributions: Conceptualisation, S.P., A.A.H., G.T.W., S.N. and M.N.N.R.; methodology, S.P., A.A.H., G.T.W., S.N. and M.N.N.R.; software, S.N., G.T.W., M.N.N.R. and A.A.H.; validation, S.N., G.T.W., M.N.N.R. and A.A.H.; writing—original draft preparation, A.A.H. and G.T.W.; writing—review and editing, S.P., S.N. and M.N.N.R.; supervision, S.P. All authors have read and agreed to the published version of the manuscript.

Funding: This research was funded by the Centre for Smart Modern Construction and the School of Engineering, Design and Built Environment, Western Sydney University, Australia.

Acknowledgments: The authors acknowledge the contributors of the Centre for Smart Modern Construction, Western Sydney University, Australia.

Conflicts of Interest: The authors declare no conflict of interest.

References

1. Thakur, V.; Doja, M.; Dwivedi, Y.K.; Ahmad, T.; Khadanga, G. Land records on blockchain for implementation of land titling in India. *Int. J. Inf. Manag.* **2020**, *52*, 101940. [CrossRef]
2. Vadgama, N. *Distributed Ledger Technology in the Supply Chain*; UCL Centre for Blockchain Technologies: London, UK, 2019.
3. FIBREE. *Global Industry Report "Blockchain Real Estate"*; FIBREE: Vienna, Austria, 2020.
4. Niranjanamurthy, M.; Nithya, B.; Jagannatha, S. Analysis of blockchain technology: Pros, cons and SWOT. *Clust. Comput.* **2019**, *22*, 14743–14757. [CrossRef]
5. Davies, E.; Kirby, N.; Bond, J.; Grogan, T.; Moore, A.; Roche, N.; Rose, A.; Tasca, P.; Vadgama, N. *Towards a Distributed Ledger of Residential Title Deeds in the UK*; University College London: London, UK, 2020.
6. Perera, S.; Nanayakkara, S.; Rodrigo, M.; Senaratne, S.; Weinand, R. Blockchain technology: Is it hype or real in the construction industry? *J. Ind. Inf. Integr.* **2020**, *17*, 100125. [CrossRef]
7. Hewavitharana, T.; Nanayakkara, S.; Perera, S. Blockchain as a project management platform. In Proceedings of the 8th World Construction Symposium: Towards a Smart, Sustainable and Resilient Built Environment, Colombo, Sri Lanka, 8–10 November 2019; pp. 137–146.
8. Rodrigo, M.N.N.; Perera, S.; Senaratne, S.; Jin, X. Potential application of blockchain technology for embodied carbon estimating in construction supply chains. *Buildings* **2020**, *10*, 140. [CrossRef]

9. Paulavičius, R.; Grigaitis, S.; Igumenov, A.; Filatovas, E. A decade of blockchain: Review of the current status, challenges, and future directions. *Informatica* **2019**, *30*, 729–748. [CrossRef]
10. Yli-Huumo, J.; Ko, D.; Choi, S.; Park, S.; Smolander, K. Where is current research on blockchain technology?—A systematic review. *PLoS ONE* **2016**, *11*, e0163477.
11. Hijazi, A.A.; Perera, S.; Alashwal, A.; Calheiros, R.N. Blockchain adoption in construction supply chain: A review of studies across multiple sectors. In Proceedings of the International Council for Research and Innovation in Building and Construction (CIB) World Building Congress—Constructing Smart Cities, Hong Kong, China, 17–21 June 2019.
12. Li, J.; Greenwood, D.; Kassem, M. Blockchain in the built environment and construction industry: A systematic review, conceptual models and practical use cases. *Autom. Constr.* **2019**, *102*, 288–307. [CrossRef]
13. Hijazi, A.A.; Perera, S.; Calheiros, R.N.; Alashwal, A. Rationale for the integration of BIM and blockchain for the construction supply chain data delivery: A systematic literature review and validation through focus group. *J. Constr. Eng. Manag.* **2021**, *147*, 03121005. [CrossRef]
14. Institution of Civil Engineers. *Blockchain Technology in the Construction Industry: Digital Transformation for High Productivity*; Institution of Civil Engineers: London, UK, 2018.
15. Belotti, M.; Božić, N.; Pujolle, G.; Secci, S. A vademecum on blockchain technologies: When, which, and how. *IEEE Commun. Surv. Tutor.* **2019**, *21*, 3796–3838. [CrossRef]
16. Singh, S.K.; Sharma, P.K.; Pan, Y.; Park, J.H. BIIoVT: Blockchain-based secure storage architecture for intelligent internet of vehicular things. *IEEE Consum. Electron. Mag.* **2021**. [CrossRef]
17. Bordel, B.; Alcarria, R.; Martin, D.; Sanchez-Picot, A. Trust provision in the internet of things using transversal blockchain networks. *Intell. Autom. Soft Comput.* **2019**, *25*, 155–170. [CrossRef]
18. Ølnes, S.; Ubacht, J.; Janssen, M. Blockchain in government: Benefits and implications of distributed ledger technology for information sharing. *Gov. Inf. Q.* **2017**, *34*, 355–364. [CrossRef]
19. Lemieux, V.L. Evaluating the use of blockchain in land transactions: An archival science perspective. *Eur. Prop. Law J.* **2017**, *6*, 392–440. [CrossRef]
20. Graglia, J.M.; Mellon, C. Blockchain and property in 2018: At the end of the beginning. *Innov. Technol. Gov. Glob.* **2018**, *12*, 90–116. [CrossRef]
21. Lemieux, V.L. A typology of blockchain recordkeeping solutions and some reflections on their implications for the future of archival preservation. In Proceedings of the 2017 IEEE International Conference on Big Data, Boston, MA, USA, 11–14 December 2017; pp. 2271–2278.
22. Centre of Digital Built Britain. *Year One Report: Towards a Digital Built Britain*; Centre of Digital Built Britain: Cambridge, UK, 2018.
23. ARUP. *Blockchain and the Built Environment*; ARUP: London, UK, 2019.
24. Kinnaird, C.; Geipel, M. *Blockchain Technology: How the Inventions Behind Bitcoin are Enabling a Network of Trust for the Built Environment*; ARUP: London, UK, 2017.
25. Love, P.E.; Edwards, D.J. Forensic project management: The underlying causes of rework in construction projects. *Civ. Eng. Environ. Syst.* **2004**, *21*, 207–228. [CrossRef]
26. Bankvall, L.; Bygballe, L.E.; Dubois, A.; Jahre, M. Interdependence in supply chains and projects in construction. *Supply Chain Manag. Int. J.* **2010**, *15*, 385–393. [CrossRef]
27. Thunberg, M. Developing a Framework for Supply Chain Planning in Construction. Ph.D. Thesis, Linköping University, Linköping, Sweden, 2016.
28. Cha, J.; Singh, S.K.; Pan, Y.; Park, J.H. Blockchain-based cyber threat intelligence system architecture for sustainable computing. *Sustainability* **2020**, *12*, 6401. [CrossRef]
29. FIBREE. *Global Industry Report "Blockchain Real Estate"*; FIBREE: Berlin, Germany, 2021.
30. Deng, Z.; Ren, Y.; Liu, Y.; Yin, X.; Shen, Z.; Kim, H.-J. Blockchain-based trusted electronic records preservation in cloud storage. *Comput. Mater. Contin.* **2019**, *58*, 135–151. [CrossRef]
31. Veuger, J. Dutch blockchain, real estate and land registration. *J. Prop. Plan. Environ. Law* **2020**, *12*, 93–108. [CrossRef]
32. McNamara, A.J.; Sepasgozar, S.M. Developing a theoretical framework for intelligent contract acceptance. *Constr. Innov.* **2020**, *20*, 421–445. [CrossRef]
33. Mason, J.; Escott, H. Smart contracts in construction: Views and perceptions of stakeholders. In Proceedings of the FIG Conference, Istanbul, Turkey, 6–11 May 2018.
34. Nanayakkara, S.; Perera, S.; Bandara, H.D.; Weerasuriya, G.T.; Ayoub, J. Blockchain technology and its potential for the construction industry. In Proceedings of the 43rd Australasian Universities Building Education Association (AUBEA) Conference: Built to Thrive: Creating Buildings and Cities that Support Individual Well-Being and Community Prosperity, Noosa, QLD, Australia, 6–8 November 2019; pp. 662–672.
35. Drescher, D. *Blockchain Basics: A Non-Technical Introduction in 25 Steps*, 1st ed.; Apress: Frankfurt am Main, Germany, 2017.
36. Androulaki, E.; Barger, A.; Bortnikov, V.; Cachin, C.; Christidis, K.; De Caro, A.; Enyeart, D.; Ferris, C.; Laventman, G.; Manevich, Y. Hyperledger fabric: A distributed operating system for permissioned blockchains. In Proceedings of the 13th EuroSys Conference, Porto, Portugal, 23–26 April 2018; pp. 1–15.
37. Hyperledger. Hyperledger Fabric. Available online: https://hyperledger-fabric.readthedocs.io/en/release-2.0/whatis.html (accessed on 26 June 2020).

38. Cocco, S.; Singh, G. Top 6 Technical Advantages of Hyperledger Fabric for Blockchain Networks. Available online: https://developer.ibm.com/technologies/blockchain/articles/top-technical-advantages-of-hyperledger-fabric-for-blockchain-networks/ (accessed on 28 July 2020).
39. Nawari, N.O. Blockchain technologies: Hyperledger fabric in BIM work processes. In Proceedings of the International Conference on Computing in Civil and Building Engineering, Melbourne, VIC, Australia, 3–4 February 2020; pp. 813–823.
40. Kundu, D. Blockchain and trust in a smart city. *Environ. Urban. ASIA* **2019**, *10*, 31–43. [CrossRef]
41. Catalini, C.; Gans, J.S. *Initial Coin Offerings and the Value of Crypto Tokens*; National Bureau of Economic Research: Cambridge, MA, USA, 2018.
42. Lemieux, V.L. Blockchain and distributed ledgers as trusted recordkeeping systems. In Proceedings of the Future Technologies Conference (FTC), Vancouver, BC, Canada, 29–30 November 2017.
43. Rodrigo, M.; Perera, S.; Senaratne, S.; Jin, X. Blockchain for construction supply chains: A literature synthesis. In Proceedings of the ICEC-PAQS Conference, Sydney, NSW, Australia, 18–20 November 2018.
44. Hijazi, A.A.; Perera, S.; Alashwal, A.; Calheiros, R.N. Enabling a single source of truth through BIM and blockchain integration. In Proceedings of the International Conference on Innovation, Technology, Enterprise and Entrepreneurship (ICITEE), Manama, Bahrain, 24–25 November 2019; pp. 385–393.
45. Nanayakkara, S.; Perera, S.; Senaratne, S. Stakeholders' perspective on blockchain and smart contracts solutions for construction supply chains. In Proceedings of the CIB World Building Congress, Hong Kong, China, 17–21 June 2019.
46. Nanayakkara, S.; Perera, S.; Senaratne, S.; Weerasuriya, G.T.; Bandara, H.M.N.D. Blockchain and smart contracts: A solution for payment issues in construction supply chains. *Informatics* **2021**, *8*, 36. [CrossRef]
47. Sigalov, K.; Ye, X.; König, M.; Hagedorn, P.; Blum, F.; Severin, B.; Hettmer, M.; Hückinghaus, P.; Wölkerling, J.; Groß, D. Automated Payment and Contract Management in the Construction Industry by Integrating Building Information Modeling and Blockchain-Based Smart Contracts. *Appl. Sci.* **2021**, *11*, 7653. [CrossRef]
48. Lamb, K. *Blockchain and Smart Contracts: What the AEC Sector Needs to Know*; Centre for Digital Built Britain: Cambridge, UK, 2018.
49. Mason, J. The BIM fork—Are smart contracts in construction more likely to prosper with or without BIM? *J. Leg. Aff. Disput. Resolut. Eng. Constr.* **2019**, *11*, 02519002. [CrossRef]
50. Shergold, P.; Weir, B. *Building Confidence: Improving the Effectiveness of Compliance and Enforcement Systems for the Building and Construction Industry across Australia*; Australian Institute of Building Surveyors: Gordon, NSW, Australia, 2018.
51. Yapicioglu, B.; Leshinsky, R. Blockchain as a tool for land rights: Ownership of land in Cyprus. *J. Prop. Plan. Environ. Law* **2020**, *12*, 171–182. [CrossRef]
52. Mintah, K.; Baako, K.T.; Kavaarpuo, G.; Otchere, G.K. Skin lands in Ghana and application of blockchain technology for acquisition and title registration. *J. Prop. Plan. Environ. Law* **2020**, *12*, 147–169. [CrossRef]
53. Ranjith, V.; Sathyajith, R.; Naveen Kumar, U.; Pushpalatha, M. A performance comparison of different security hashing in a blockchain based system that enables a more reliable and SWIFT registration of land. *Int. J. Adv. Sci. Technol.* **2020**, *29*, 1-06.
54. Kaczorowska, M. Blockchain-based land registration: Possibilities and challenges. *Masaryk Univ. J. Law Technol.* **2019**, *13*, 339–360. [CrossRef]
55. Nasarre-Aznar, S. Collaborative housing and blockchain. *Administration* **2018**, *66*, 59–82. [CrossRef]
56. Sajana, P.; Sindhu, M.; Sethumadhavan, M. On blockchain applications: Hyperledger fabric and Ethereum. *Int. J. Pure Appl. Math.* **2018**, *118*, 2965–2970.
57. Cachin, C. Architecture of the hyperledger blockchain fabric. In Proceedings of the Workshop on Distributed Cryptocurrencies and Consensus Ledgers, Chicago, IL, USA, 25 July 2016.
58. FIBREE. *Global Industry Report "Blockchain Real Estate"*; FIBREE: Paris, France, 2019.
59. Sherrell, L. Evolutionary prototyping. In *Encyclopedia of Sciences and Religions*; Runehov, A.L.C., Oviedo, L., Eds.; Springer: Dordrecht, The Netherlands, 2013; p. 803.
60. Rodrigo, M.N.N.; Perera, S.; Senaratne, S.; Jin, X. Systematic development of a data model for the blockchain-based embodied carbon (BEC) estimator for construction. *Eng. Constr. Archit. Manag.* **2021**. [CrossRef]
61. Nanayakkara, S.; Rodrigo, M.N.N.; Perera, S.; Weerasuriya, G.T.; Hijazi, A.A. A methodology for selection of a blockchain platform to develop an enterprise system. *J. Ind. Inf. Integr.* **2021**, *23*, 100215. [CrossRef]
62. Moges Kasie, F. Combining simple multiple attribute rating technique and analytical hierarchy process for designing multi-criteria performance measurement framework. *Glob. J. Res. Eng.* **2013**, *13*, 15–29.
63. NSW LRS. eConveyancing in NSW. Available online: https://www.nswlrs.com.au/eConveyancing (accessed on 24 June 2020).
64. NSW LRS. Find Records. Available online: https://www.nswlrs.com.au/Find-Records (accessed on 24 June 2020).
65. Australia and New Zealand Banking Group Limited (ANZ). Real Estate Agents and Property Advertising. Available online: https://www.anz.com.au/personal/home-loans/tips-and-guides/agents-and-advertising/ (accessed on 24 June 2020).
66. Shuaib, M.; Daud, S.M.; Alam, S.; Khan, W.Z. Blockchain-based framework for secure and reliable land registry system. *Telkomnika* **2020**, *18*, 2560–2571. [CrossRef]
67. Australia and New Zealand Banking Group Limited (ANZ). Property Settlement: What Is It and How Does It Work? Available online: https://www.anz.com.au/personal/home-loans/tips-and-guides/property-settlement-what-is-it-how-does-it-work/ (accessed on 24 June 2020).

68. NSW Government. Exchanging Contracts and Settlement. Available online: https://www.nsw.gov.au/life-events/living-nsw/buying-residential-property-nsw (accessed on 24 June 2020).
69. Hyperledger. Transaction Flow. Available online: https://hyperledger-fabric.readthedocs.io/en/release-2.1/txflow.html (accessed on 15 July 2020).
70. Hyperledger. Glossary. Available online: https://hyperledger-fabric.readthedocs.io/en/release-2.2/glossary.html (accessed on 15 July 2020).
71. Lemieux, V.L. Blockchain recordkeeping: A SWOT analysis. *Inf. Manag.* **2017**, *51*, 20–27.
72. Cheng, M.; Liu, G.; Xu, Y.; Chi, M. When blockchain meets the AEC industry: Present status, benefits, challenges, and future research opportunities. *Buildings* **2021**, *11*, 340. [CrossRef]
73. Hyperledger Wiki. Land and Property Management. Available online: https://wiki.hyperledger.org/display/LMDWG/Land+and+Property+Management (accessed on 20 May 2021).

Article

Cost Modeling from the Contractor Perspective: Application to Residential and Office Buildings

Francisco Pereira Monteiro [1], Vitor Sousa [2], Inês Meireles [3] and Carlos Oliveira Cruz [2,*]

[1] Department of Civil Engineering, Architecture and GeoResources, Instituto Superior Técnico, Universidade de Lisboa, Av. Rovisco Pais, 1049-001 Lisbon, Portugal; franciscomonteiro_@hotmail.com
[2] CERIS, Instituto Superior Técnico, Universidade de Lisboa, Av. Rovisco Pais, 1049-001 Lisbon, Portugal; vitor.sousa@tecnico.ulisboa.pt
[3] RISCO, Department of Civil Engineering, Campus de Santiago, University of Aveiro, 3810-193 Aveiro, Portugal; imeireles@ua.pt
* Correspondence: oliveira.cruz@tecnico.ulisboa.pt

Abstract: For the majority of the contractual arrangements used in construction projects, the owner is not responsible for the cost deviations due to the variability of labor productivity or material price, amongst many other aspects. Consequently, the cost performance of a project may be entirely distinct for the owner and the contractor. Since the majority of quantitative research on cost estimation and deviation found in the literature adopts the owners' perspective, this research provides a contribution towards modeling costs and cost deviation from a contractor's perspective. From an initial sample of 13 residential buildings and 10 office building projects, it was possible to develop models for cost estimation at the early stage of development, including both endogenous and exogenous variables. Although the sample is relatively small, the authors were able to fully analyze all the cost data, using no secondary sources of data (which is very frequent in cost modeling studies). The statistically significant variables in the cost estimation models were the areas above and below ground and the years following the 2008 financial crisis, including the international bailout (2011–2014) period. For estimating the unit cost, a nonlinear model was obtained with the number of underground and total floor, the floor ratio, and the years following the 2008 financial crisis, including the international bailout (2011–2014) period as predictors. For the office buildings, a statistically significant correlation was also found between the cost deviation and number of underground floors.

Keywords: cost estimation; cost deviation; financial crisis; promotor-contractor; statistical modeling

Citation: Monteiro, F.P.; Sousa, V.; Meireles, I.; Oliveira Cruz, C. Cost Modeling from the Contractor Perspective: Application to Residential and Office Buildings. *Buildings* **2021**, *11*, 529. https://doi.org/10.3390/buildings11110529

Academic Editor: David Arditi

Received: 12 October 2021
Accepted: 4 November 2021
Published: 10 November 2021

Publisher's Note: MDPI stays neutral with regard to jurisdictional claims in published maps and institutional affiliations.

1. Introduction

The complexity of construction projects is increasing, both on their "hard" (or tangible) and "soft" (or intangible) dimensions. From new materials to new construction technologies, a multitude of technical solutions have emerged over the last few decades, widening the range of alternative options available for the "hard" dimension of construction projects. Concurrently, the range of aspects to manage in construction project has also increased. The "soft" dimension of construction projects includes the need for satisfying an increasingly broader and stringent social (e.g., health and safety), environmental (e.g., construction and demolition waste management), and economic (e.g., use of life-cycle cost as awarding criteria on public projects in the European Union) requirements. Consequently, construction managers are now facing additional challenges in their projects. To aid them in their tasks, several standards and regulations have been published (e.g., ISO 21,500 family of standards) and new tools are becoming available (e.g., BIM—Building Information Modeling). These provide holistic and consistent guidelines and technological support to tackle the complexity of managing construction projects within this new context.

Despite all these evolutions, the financial control of construction projects is still a dominant dimension in a project's governance. In this regard, cost estimation in the early stages dictates the investment decisions, although, at the early stages, there is a significant risk surrounding the estimation, given the technical uncertainty. Therefore, more accurate cost forecasting in the early stages of the project's development and better quantification/understanding of cost deviations are amongst the key concerns of any construction project manager [1].

Within this research, the contractor perspective is adopted by analyzing the financial performance of 23 building projects of a large industrial group in Portugal (13 residential buildings and 10 office building projects). Among the companies in the group, there is real estate and a contractor that develop, amongst other types of projects, residential and office buildings in collaboration. Although the dataset is relatively small, it is homogenous, in the sense that the contractor was the same company, and the cost analysis used no secondary data. The real estate assumes all the licensing, design, marketing, and commercialization and the contractor executes the projects. The contractor also develops projects for external clients, both private and public, of various types (e.g., commercial, healthcare, and educational buildings; water, transportation, and energy infrastructures).

The paper is organized as follows. After the introduction, Section 2 presents the literature review, Section 3 explains the data used and the methods, Section 4 presents the results, and, finally, Section 5 provides the main conclusions.

2. Literature Review

Historically, there have been several tools for cost estimating at early stages of a project's development. The simplest models are based on parametric estimation of costs, built upon expert judgments (see for instance, [2]). The traditional multiple regression analysis (RA) has been the tool most used by researchers (e.g., [3,4]). Artificial neural networks (ANN) have gained some expression for data modeling in various engineering problems, including cost estimation (e.g., [1,5,6]), and case-based reasoning (CBR) is also being used in various tasks related to construction management (e.g., resource estimation—[7]; duration estimation—[8]). A review on CBR use for construction management can be found in [9] and its use for cost estimation can be found in [10–12]. A comparison between the three methods was done by [13], with the new tools achieving better results than regression models. More recently, [14] developed cost estimation models using support vector machines, along with ANN combined with an unsupervised deep Boltzmann machine, and included exogenous variables (e.g., consumer price index, interest rate for loan, population of the city) in combination with endogenous variables (e.g., total area). Some authors have also developed models to estimate the cost of portions of the projects (e.g., structure—[15,16]).

Table 1 summarizes the main research on the topic, along with the methods and explanatory variables used in each study. It should be noted that some models were developed to estimate the total cost (when the area is included in the model) whereas others were developed to estimate the unit cost (when the area is not included in the model). Some variables listed in Table 1 should be interpreted as a category of variables rather than a single variable, in some cases simply because they are measured differently depending on the author. For instance, the construction area may be gross, usable, or other; the number of stories may also be total, above ground, and underground; or the height may be of the building or of the floor. Others are naturally a category of variables, such as the structural characteristics that may include the type of structure or foundation (e.g., [17]). A few are even impossible to quantify adequately at the early stages of the project development, namely the duration. In fact, it is far more common to use cost as an independent variable to estimate the construction duration (e.g., see [18–20]) for examples of time-relationships), because a cost estimate tends to be done by the designer before the contractor develops the construction schedule.

Table 1. Early-stage cost estimation models for buildings.

Reference	[21]	[22]	[23]	[24]	[25]	[26]	[27]	[28]	[29]	[30]	[31]	[32]	[33]
Method													
Regression Analysis	X		X	X	X	X		X	X	X	X		
Artificial Neural Network		X	X		X								X
Case-Based Reasoning												X	
Other							X						
Variables													
Project related													
Building type		X				X		X		X			
Area	X	X	X					X		X	X	X	X
Number of stories		X	X							X		X	X
Number of households												X	
Height			X	X				X	X				X
Duration		X	X		X				X		X		
Location			X								X		
Above ground external envelope characteristics	X		X								X		
Underground external envelope characteristics	X		X										
Number of lifts			X								X	X	
Number of piloti floors												X	
Structural characteristics			X										
Other			X		X		X			X		X	X
Management related													
Type of contract			X			X	X						
Procurement strategy			X		X	X	X						
Other			X		X								
Other													
Type of client						X	X						
Construction year	X									X			
Designer characteristics							X						
Contractor characteristics								X					
Site characteristics			X										
Sample													
Size	15	30	288	36	50	93	-	30	290	42,340 18,469	75	91	232
Type	R	S	R				-	O		R O	R	R	

R—Residential buildings. O—Office buildings. S—School buildings.

There are also authors attempting to use BIM for conceptual cost estimation (e.g., [34]). However, this approach requires a quantities takeoff, which implies a degree of project development that is incompatible with the early stages of development in this research (definition of general characteristics of the project, such as area and number of floors, and a preliminary sketch). In fact, even some models reported in the literature review presented herein use variables that may be unavailable at this stage of the project development (e.g., proportion of walls and windows in the external envelope). There is a clear trade-off between model adjustment, i.e., estimation accuracy and the availability of information in the early stages of the project. The review presented focused on cost estimation for building projects and is not intended to be exhaustive, but rather illustrates that different tools, sample sizes, and variables have been used. There is also an extensive literature on other types of projects (e.g., transportation infrastructure projects—[35–39]).

The topic of cost deviations is closely related to cost estimation, since a more accurate cost estimation should reduce cost deviations. There is an extensive literature on the magnitude (e.g., [40–46]) and causes (e.g., [47–51]) of cost deviations. The former tends to be quantitative, based on the analysis of the performance of past project, while the latter is mostly qualitative, resorting to questionnaires or interviews with experts.

The research relating the magnitude with the causes of cost deviation is less extensive and the causes are limited to macro variables of the projects, such as: (i) The size of the project ([52,53]); (ii) the nature of ownership/promotor (public or private—e.g., [41,54]); (iii) the type of intervention (new build or refurbishment/rehabilitation—[40]); (iv) the type of project (residential, infrastructure, commercial, and other—e.g., [55]); (v) the procurement model (design-bid-build, design and build, project management—e.g., [52,56]); or (vi) the tender method (open, selection, negotiated tendering—e.g., [57]).

Most research on cost modeling in general (cost estimation and cost deviations) tends to focus on variables endogenous to the projects. Table 1 provides a clear illustration of this claim, with the variables used by the various authors being exclusively related to the project or its management. There is a smaller body of literature on the influence of exogenous variables on the financial performance of construction projects. For instance, Refs. [58–60] demonstrated the relation between political and economic cycles and the cost deviation in public projects.

The quantitative research available in the literature, both in terms of cost estimation and quantitative analysis of cost variations, tends to reflect the construction projects' financial performance from the owners' perspective. The records used by most of the authors were obtained from the owners (or from the contractors) and represent the payments made to the contractors and not the expenses of the contractors. However, the amounts payed by the owners do not match perfectly the amounts spent by the contractors to execute the projects after deducting the profit margin. Regarding the cost estimation, the owners' perspective is affected by the commercial strategy adopted by the contractors in each moment, frequently represented by the margin defined in their bids. In highly competitive contexts, the margins tend to decrease, whereas in low competitive contexts the margins tend to increase. Concerning cost deviations, the variability of materials prices, labor productivity, or site overheads, amongst other potential causes of cost deviation (e.g., accidents, equipment breakdown, or failure) are not measured when analyzing historical construction cost data from the owner's perspective. From the owner's perspective, change orders and errors/omissions (if the design is provided by the owner) are the most relevant causes of cost deviations.

The literature has provided recently an active discussion whether cost deviations are motivated by more technical aspects (e.g., cost escalation, scope changes, unforeseen events/conditions) of the projects ([43,44]) or by estimator bias ([61,62]). However, this discussion is outside of the scope of the present research. This discussion focuses on the cost deviations between the first estimate and final cost, and in the context of major infrastructure projects more applicable to public projects. This includes references to the benefits of the projects for society. Herein, the scope is restricted to private projects and cost deviations between the detailed design and final cost. Furthermore, the cost-benefit ratio is simply the cost of the project versus the income generated by its commercialization. Thus, fundamentally the technical aspects will drive the cost deviations and the potential estimator bias will be more on the expected market valuation of the project.

3. Data and Methods

As referred above, the data used was obtained from a large industrial group in Portugal that include a real estate and a contractor in their portfolio of companies. All projects were developed in collaboration between these two companies of the group and, despite the formal split between them, they end up working as a single entity with complementary expertise.

The 23 building projects were developed mostly in Portugal, with only 2 being abroad (Angola and Mozambique). The projects in Portugal are concentrated in the Lisbon and Porto metropolitan areas (the two major cities in Portugal) and can be classified as premium. The projects in Africa are located in the capital cities of the respective countries (Luanda, in Angola, and Maputo, in Mozambique). Naturally, there are differences between the Portuguese and African contexts at various levels, but the projects are all new developments in consolidated urban areas. Focusing on the Portuguese projects, infrastructures (e.g., roads, water, electricity, communications) and support facilities (e.g., subcontractors, suppliers) are good and can be regarded equivalent in both Lisbon and Porto regions. Furthermore, since the projects are all from the same group, the management approach and skills can be considered identical and, in many cases, the designers were also the same. The projects in Portugal also resorted to the same subcontractor and suppliers in many instances.

Information on the projects includes the: (i) Proportion of the cost by major category of works (structure, architecture, technical installations, and site overheads); (ii) estimated cost; (iii) profit margin; (iv) estimated price; (v) final price; (vi) total area, above ground, and underground gross-built area; and (vii) total floors, above ground, and underground number of floors. There is also information on the start year and duration of the projects. Both the cost and prices of the projects were update to 2019 values using the formulas for price adjustment applicable to public residential and office buildings in Portugal. In Portugal, the reimbursements to contractors in public construction projects are corrected to account for inflation. Since this is mandatory, there are formulas defined by law for estimating the increase (or decrease) in the payments to the contractor for 23 different types of projects (Law-Decree n° 6/2004). These formulas represent the average weight of labor, materials (a selection from 51 different materials), and equipment on the total price of the projects. The price indexes of the labor, materials, and equipment are published monthly by the government based on the official inflation data. The estimated and final unit prices and the cost deviations were calculated from the available data. It was not possible to retrieve all the fields for all the projects, particularly the final price that was available for only 16 projects.

In addition to the endogenous variables, the influence of the 2008 financial crisis and subsequent international bailout that Portugal had between 2011 and 2014 was also included. This exogenous variable was modeled with a categorical predictor assuming the value of 1 between 2008 and 2014 and 0 in the remaining years. A lag of 1 year was also considered at the start and end of the crisis to evaluate if there was a delay between these events and the impact on the cost of the projects.

Due to confidentiality issues regarding some of the data (revealing the cost without the profit margin of the contractor for an external client), indexes were computed dividing the value of each project by the average of all the projects in the sample. This was done particularly for the projects profit margin, total and unit cost, as well as total, unit initial, and final prices. Area and floor ratios were also computed dividing the values above ground by the values underground since there is typically a relation both due to parking requirements.

A statistical approach was used to analyze the data, comprising of two steps: (i) A preliminary data analysis and (ii) data modeling. The preliminary data analysis included calculation of descriptive statistics, assumptions testing, and unidimensional statistical analysis. The normality and homogeneity of variance were tested using the Shapiro–Wilk and Levene's tests, respectively, and the unidimensional analysis was done using either parametric or non-parametric distribution comparison (*t*-test/ANOVA or Mann–Whitney/Kruskal–Wallis), for categorical variables, and correlation (Pearson or

Spearman), for continuous variables. The data modeling was based on the traditional least squares multiple linear regression. Non-linear regression was also used, when necessary, but given the sample size, the use of artificial intelligence tools (e.g., artificial neural networks, support vector machines, random forests) was not considered. Given the small sample size, bootstrapping (1000 simulations with simple sampling and 95% confidence interval based on percentile) was used to strengthen the confidence in the results.

The restriction of the context (projects from a single company), scope (all buildings are classified as premium in terms of quality), and location (the spatial variability of the locations is small) limits the generalization of the results. However, it excludes these variables from the cost estimation and deviations of the projects and enables the possibility of capturing the cost estimation and deviations drivers that are specific to the projects. This is an important difference from most past research effort, which in most cases use data samples with projects that may be very different, developed by distinct contractors, designed by different teams, and, in some cases, promoted by various owners in many locations. This broader scope allows capturing an overall average cost performance of the projects, but it is impossible to assess if it was due to the contractor competence, design quality, owner experience, nature of the project, local factors, or other aspects that are controlled for in the analysis. Consequently, using large mixed samples of data may fail in terms of applicability to a specific project.

4. Results and Discussion

4.1. Preliminary Data Analysis

As defined in Section 3, a preliminary data analysis was carried out comprising an overall statistical characterization of the projects in the sample, followed by the statistical analysis of the distribution of costs by major categories of works (structural works, architectural works, technical installation works, and site overheads). The latter provides information, not only on the typical distribution of costs by category, but assesses if there statistically significant differences depending on the type of building.

The projects totalize a cost of over 155 million euros, with the residential buildings contributing 57% and the office buildings accounting for 43%. The initial price (cost plus typical margin used by the contractor for external clients) of each individual project ranged from 1.5 to 20 million euros. The average initial unit price is 560 €/m^2 for office buildings, and 785 €/m^2 for residential buildings. This difference is, however, strongly influenced by the two residential buildings outside Portugal (one in Angola and another in Mozambique) that had an average initial unit price of 1408 €/m^2. Table 2 presents some descriptive statistics characterizing the dataset.

Figure 1 illustrates the distribution of the cost and price indexes and the weight of each cost category for the residential and office buildings. The number of projects with information regarding the cost and initial price is roughly the same, but there are fewer projects with information regarding the final price. Consequently, analyzing the evolution from cost to final price is not possible (Figure 1 bottom). Considering the substantial price difference of the projects outside Portugal, one of them clearly an outlier identified in Figure 1, they were excluded from the analysis from this point forward.

Comparing the weight of the cost categories between a residential and office building, it is visible a difference in all cost categories except for the site overheads. These differences were found to be statistically significant (Table 3), and the site overheads would also be considered statistically significant for a significance level of 0.10 instead of the typical 0.05. The parametric *t*-test was used since the data was found to be normally distributed for both residential and office buildings subsets according with the Shapiro–Wilk test.

Table 2. Descriptive statistics of some of the main variables in the dataset.

Variable		Sample	Range	Minimum	Maximum	Sum	Mean	Std. Dev.	Skewness	Kurtosis
Floors (-)	Underground	19	4	1	5	64	3.37	1.065	−0.849	1.152
	Above Ground	21	20	3	23	153	7.29	4.880	2.162	4.987
	Total	19	16	5	21	189	9.95	3.837	1.424	2.528
	Ratio	19	6.25	0.75	7.00	42.48	2.24	1.531	1.944	4.278
Area (m^2)	Underground	22	16,893.00	420.00	17,313.00	131,353.75	5970.63	4184.18	0.905	0.926
	Above Ground	22	10,342.00	1557.00	11,899.00	142,095.44	6458.88	2983.97	0.221	−0.740
	Total	23	26,136.00	1977.00	28,113.00	294,621.19	12,809.62	6671.70	0.287	−0.311
	Ratio	22	3.08	0.62	3.71	33.14	1.51	0.833	0.935	0.661
Cost Category Weight (%)	Structure	23	20.00	12.70	32.70	540.30	23.49	5.019	−0.237	−0.398
	Architecture	23	25.10	29.60	54.70	955.70	41.55	7.751	−0.128	−1.420
	Technical Installations	23	24.90	9.50	34.40	532.60	23.16	6.822	−0.425	−0.449
	Site Overheads	23	12.50	7.50	20.00	263.40	11.45	2.988	1.598	2.854
Total Cost Index (-)		21	21	2.11	0.19	2.29	21.00	0.126	0.333	0.501
Margin Index (-)		21	21	1.96	0.36	2.32	21.00	0.134	0.375	0.501
Price (-)	Initial	22	19,367,364.57	1,477,203.03	20,844,567.61	185,850,166.26	8,447,734.83	5,107,220.52	0.873	0.610
	Final	16	19,159,444.23	2,746,435.50	21,905,879.73	155,809,085.39	9,738,067.84	5,404,338.41	0.970	0.421
Unit Price (-)	Initial	22	1401.44	429.25	1830.69	15,022.83	682.86	288.56	3.239	12.577
	Final	16	1441.43	402.96	1844.39	11,576.50	723.53	343.78	2.563	7.779
Cost Deviation (%)		15	15	38.06	−13.41	24.66	57.00	2.153	69.507	0.580
Duration (days)		23	23	240	240	480	7320	14.109	4578.656	0.481

Note: The margin and cost, both total and unit values, were not included due to confidentiality of the data.

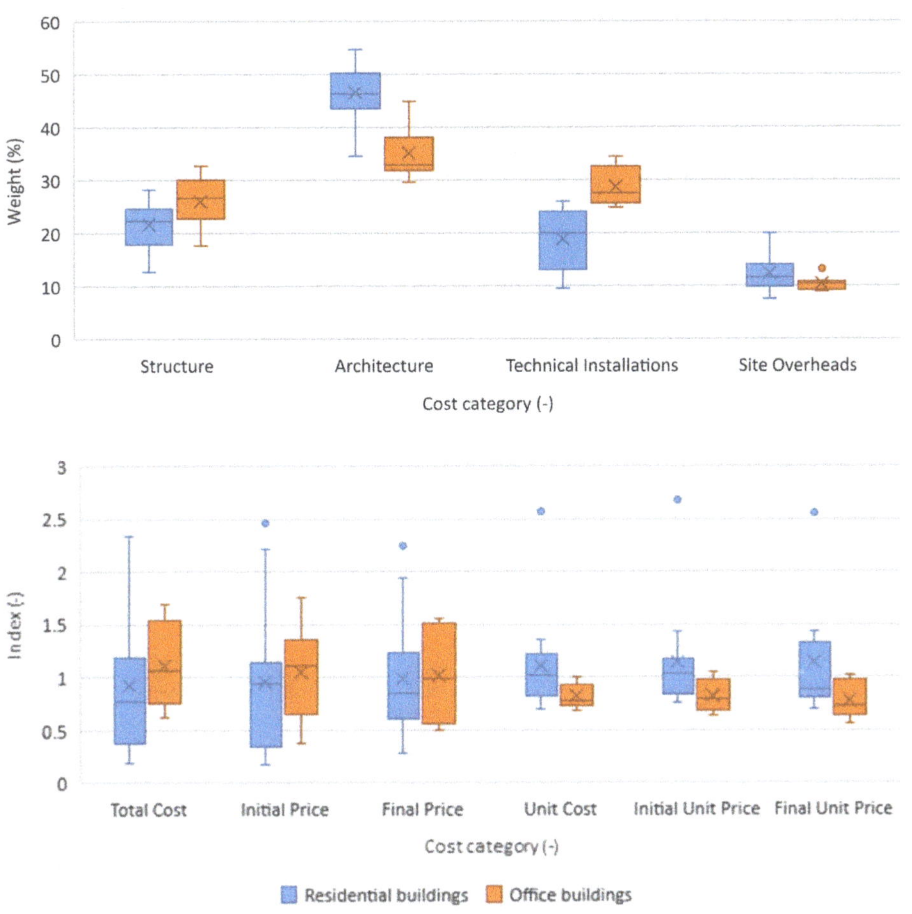

Figure 1. Projects distribution of the weight by cost category (**top**) and cost and prices indexes (**bottom**).

Office buildings present a lower weight of architecture costs, which may be explained by the tendency for open spaces. These savings are partially compensated by more expensive structures and technical installations, since the unit cost difference is only statistically significant for a 10% significance level. Assuming that the open spaces imply wider spans, this may contribute to explain the higher weight of the structures in office buildings. Considering, the demand for heating, ventilation, air conditioning, the requirements regarding electric and telecommunication facilities tend to be higher for office buildings than for residential buildings, which may explain the results. These results were further confirmed by bootstrapping (not presented herein the full table of results), with the significance of the *t*-test result increasing to 0.045, 0.003, and 0.002, for the structure, architecture, and technical installations, respectively.

Table 3. Means comparison between residential and office buildings' cost categories weights and unit cost and prices.

Variables		Levene's Test		t-Test			Difference			
		F	Sig.	t	Df	Sig. (2-Tailed)	Mean	Std. Error	95% Confidence Interval	
									Lower	Upper
Structure	EVA	0.018	0.894	**2.176**	**19**	**0.042**	**4.557**	**2.094**	**0.174**	**8.940**
	EVNA			2.177	18.834	0.042	4.557	2.093	0.173	8.941
Architecture	EVA	0.007	0.935	**−5.043**	**19**	**0.000**	**−11.906**	**2.361**	**−16.848**	**−6.965**
	EVNA			−5.043	18.801	0.000	−11.906	2.361	−16.852	−6.961
Technical Installations	EVA	1.459	0.242	**4.970**	**19**	**0.000**	**9.801**	**1.972**	**5.673**	**13.929**
	EVNA			5.070	17.367	0.000	9.801	1.933	5.729	13.873
Site Overheads	EVA	3.285	0.086	*−1.802*	*19*	*0.087*	*−1.725*	*0.957*	*−3.727*	*0.278*
	EVNA			−1.866	13.814	0.083	−1.725	0.924	−3.710	0.261
Unit Cost	EVA	0.941	0.346	*−2.042*	*17*	*0.057*	*−86.404*	*42.314*	*−175.679*	*2.871*
	EVNA			−2.174	16.949	0.044	−86.404	39.749	−170.286	−2.522
Initial Unit Price	EVA	0.174	0.681	**−2.222**	**18**	**0.039**	**−100.576**	**45.273**	**−195.692**	**−5.460**
	EVNA			−2.222	17.920	0.039	−100.576	45.273	−195.722	−5.430
Final Unit Price	EVA	0.054	0.821	−1.412	12	0.183	−106.575	75.453	−270.974	57.823
	EVNA			−1.443	11.650	0.175	−106.575	73.854	−268.029	54.878

EVA—Equal variances assumed. EVNA—Equal variances not assumed. Italics—result significance at a 0.10 level. Bold—result significance at a 0.05 level.

The unit cost and initial price are also statistically different between residential and office buildings, if a 10% threshold is considered for the unit cost. The same is not verified for the final cost, but this can be attributed to the combination of the cost deviations and, mostly, to the smaller sample of project with final price data available. The bootstrapping results (not presented herein the full table of results) confirms the results obtained for the parameters (unit cost, initial, or final price), with the unit cost difference closer to be statistically significant at a 5% significance level (p-value = 0.055).

It is interesting to notice that the total cost and prices (initial and final) of office buildings are slightly higher than for residential buildings, but the unit cost and prices are slightly lower. This implies that the office buildings in the sample are larger, in average, than the residential buildings, but that the lower expenses on architecture are only partially compensated by the more expensive structure and technical installations.

Table 4 reveals the statistical significance of the influence of the 2008 economic crisis and the international bailout that followed until 2014 on the unit cost and prices of office building projects. Within the residential buildings in Portugal, only 2 were executed between 2008 and 2015 (in 2014 and 2015). As such, it is impossible to assess the influence of this exogenous variable on the financial performance of the residential building projects separately. Considering all projects, the unit cost difference during the crisis is no longer statistically significant and the final cost is only significant for a 10% significance level. However, this may result from the masking effect of mixing residential and office building projects and differences in the sample size for cost and initial and final price. In general, the significance level with bootstrapping decreased for all the projects analyzed together and increased for the office buildings (not presented herein the full table of results). This made the unit cost difference become statistically significant for a 10% significance level (p-value = 0.096). Regarding the office buildings, this made the site overheads and the unit cost difference of office buildings lose their statistical significance.

Table 4. Means comparison between the projects developed during the economic crisis years and during the other years.

Variables		Levene's Test		t-Test			Difference			
		F	Sig.	t	Df	Sig. (2-Tailed)	Mean	Std. Error	95% Confidence Interval Lower	Upper
All buildings										
Structure	EVA	2.616	0.122	−0.653	19	0.521	−1.924	2.944	−8.085	4.238
	EVNA			−0.445	3	0.683	−1.924	4.325	−14.782	10.935
Architecture	EVA	0.026	0.874	1.349	19	0.193	5.919	4.387	−3.262	15.100
	EVNA			1.187	4	0.301	5.919	4.988	−7.919	19.757
Technical Installations	EVA	3.737	0.068	−1.141	19	0.268	−4.199	3.680	−11.901	3.504
	EVNA			−2.048	17	0.056	−4.199	2.050	−8.523	0.126
Site Overheads	EVA	1.252	0.277	−0.203	19	0.841	−0.268	1.316	−3.021	2.486
	EVNA			−0.322	11	0.753	−0.268	0.832	−2.090	1.554
Unit Cost	EVA	0.055	0.818	1.566	17	0.136	83.700	53.461	−29.092	196.492
	EVNA			1.610	5	0.169	83.700	51.981	−50.578	217.979
Initial Unit Price	EVA	0.290	0.597	**2.396**	**18**	**0.028**	**133.254**	**55.626**	**16.388**	**250.119**
	EVNA			2.392	5	0.066	133.254	55.701	−13.522	280.029
Final Unit Price	EVA	0.938	0.352	*1.959*	*12*	*0.074*	*152.192*	*77.701*	*−17.105*	*321.488*
	EVNA			2.284	8	0.052	152.192	66.628	−1.443	305.826
Office Buildings										
Structure	EVA	1.605	0.241	−1.536	8	0.163	−4.719	3.072	−11.802	2.364
	EVNA			−2.222	8	0.058	−4.719	2.124	−9.633	0.195
Architecture	EVA	3.441	0.101	1.216	8	0.259	4.424	3.638	−3.966	12.813
	EVNA			1.703	8	0.127	4.424	2.597	−1.566	10.414
Technical Installations	EVA	3.395	0.103	0.829	8	0.431	2.014	2.431	−3.592	7.620
	EVNA			0.980	6	0.366	2.014	2.055	−3.049	7.078
Site Overheads	EVA	4.993	0.056	**−2.723**	**8**	**0.026**	**−1.719**	**0.631**	**−3.175**	**−0.263**
	EVNA			−2.075	2	0.148	−1.719	0.828	−4.695	1.257
Unit Cost	EVA	6.878	0.039	2.612	6	0.040	98.267	37.628	6.195	190.340
	EVNA			**3.385**	**5**	**0.022**	**98.267**	**29.034**	**21.891**	**174.643**
Initial Unit Price	EVA	10.343	0.012	3.140	8	0.014	150.429	47.907	39.956	260.901
	EVNA			**4.614**	**8**	**0.002**	**150.429**	**32.605**	**74.662**	**226.195**
Final Unit Price	EVA	2.021	0.228	*2.465*	*4*	*0.069*	*181.754*	*73.733*	*−22.961*	*386.469*
	EVNA			2.465	3	0.088	181.754	73.733	−49.712	413.220

EVA—Equal variances assumed. EVNA—Equal variances not assumed. Italics—result significance at a 0.10 level. Bold—result significance at a 0.05 level.

The economic crisis impacted more severely on labor cost (there was a high unemployment and salary cuts) than on materials and equipment (a portion are imported and subject to less devaluation). This is consistent with the statistical significance of the site overheads on the office building projects, considering that a large portion of the cost in this category is due to the management team.

Since the majority of the data was found to be normally distributed based on the Shapiro–Wilk test (the non-normally distributed variables were the site overheads, margin, and the underground and above ground floors), the Pearson correlation was used. The results (Table 5) reveal the expected correlation between the cost and prices with the areas and between the areas and the weight of the structure. Some less obvious results include the negative correlation between the unit cost and prices and the underground area, total area, and area ratio. However, this is logical since the underground areas tend to be for parking spaces, with lower demands for architecture (and technical installations works) that justify a lower unit cost and prices compared to the areas above ground. The negative relation between the unit cost and price and the total may indicate the existence of a scale effect. The bootstrap results confirm the correlations (not presented herein the full table of results). For instance, the 95% confidence interval of the correlation between the total cost and the above ground area is estimated to be between 0.705 and 0.980.

Table 5. Pearson correlation results.

Variables		Structure	Architecture	Technical Installations	Site Overheads	Total Cost	Initial Price	Final Price	Unit Cost	Initial Unit Price	Final Unit Price	Cost Deviation
Structure	Correlation		−0.425 **	0.005	−0.135	**0.340 ***	**0.417 ***	0.376	−0.270	−0.153	−0.221	−0.013
	Sig. (2-tailed)		0.007	0.976	0.397	0.042	0.010	0.062	0.107	0.347	0.273	0.951
	N		21	21	21	19	20	14	19	20	14	13
Architecture	Correlation			−0.543 **	0.053	−0.216	−0.253	−0.143	0.322	0.253	0.319	−0.077
	Sig. (2-tailed)			0.001	0.739	0.196	0.119	0.477	0.054	0.119	0.112	0.714
	N			21	21	19	20	14	19	20	14	13
Technical Installations	Correlation				−0.302	0.205	0.221	0.221	−0.011	−0.216	−0.200	−0.209	−0.128
	Sig. (2-tailed)				0.057	0.221	0.173	0.956	0.196	0.218	0.298	0.542
	N				21	19	20	14	19	20	14	13
Site Overheads	Correlation					**0.413 ***	**0.483 ****	−0.331	0.012	0.005	−0.044	0.297
	Sig. (2-tailed)					0.014	0.003	0.100	0.944	0.974	0.826	0.160
	N					19	20	14	19	20	14	13
Underground Floors	Correlation	0.336	0.062	−0.109	−0.314	0.108	0.280	0.238	−0.088	−0.140	−0.089	0.149
	Sig. (2-tailed)	0.079	0.744	0.568	0.102	0.596	0.142	0.296	0.664	0.463	0.695	0.514
	N	18	18	18	18	16	18	13	16	18	13	13
Above Ground Floors	Correlation	−0.286	**0.380 ***	−0.063	−0.089	−0.090	−0.064	0.082	0.008	−0.049	0.151	0.162
	Sig. (2-tailed)	0.106	0.031	0.719	0.615	0.638	0.726	0.696	0.966	0.785	0.469	0.456
	N	19	19	19	19	17	18	14	17	18	14	13
Total Floors	Correlation	−0.098	0.327	−0.132	−0.183	−0.036	0.021	0.211	−0.072	−0.104	0.053	0.184
	Sig. (2-tailed)	0.587	0.069	0.461	0.313	0.853	0.907	0.325	0.710	0.561	0.806	0.389
	N	18	18	18	18	16	18	13	16	18	13	13
Underground Area	Correlation	**0.558 ****	−0.495 **	0.286	−0.273	**0.579 ****	**0.663 ****	**0.473 ***	−0.520 **	−0.389 *	−0.516 *	−0.051
	Sig. (2-tailed)	0.000	0.002	0.070	0.085	0.001	0.000	0.019	0.002	0.016	0.010	0.807
	N	21	21	21	21	19	20	14	19	20	14	13
Above Ground Area	Correlation	**0.431 ****	−0.100	−0.005	−0.327 *	**0.739 ****	**0.691 ****	**0.758 ****	−0.246	−0.216	−0.231	0.000
	Sig. (2-tailed)	0.007	0.526	0.976	0.040	0.000	0.000	0.000	0.141	0.183	0.250	1.000
	N	21	21	21	21	19	20	14	19	20	14	13
Total Area	Correlation	**0.539 ****	−0.362 *	0.171	−0.350 *	**0.743 ****	**0.800 ****	**0.714 ****	−0.427 *	−0.358 *	−0.407 *	0.000
	Sig. (2-tailed)	0.001	0.022	0.277	0.027	0.000	0.000	0.000	0.011	0.027	0.043	1.000
	N	21	21	21	21	19	20	14	19	20	14	13
Area Ratio	Correlation	**0.539 ****	−0.362 *	0.171	−0.350 *	**0.743 ****	**0.800 ****	**0.714 ****	−0.427 *	−0.358 *	−0.407 *	0.000
	Sig. (2-tailed)	0.001	0.022	0.277	0.027	0.000	0.000	0.000	0.011	0.027	0.043	1.000
	N	21	21	21	21	19	20	14	19	20	14	13

Bold—statistical significant result. **—Correlation is significant at the 0.01 level (2-tailed). *—Correlation is significant at the 0.05 level (2-tailed).

For the variables that are not normally distributed, the non-parametric Spearman correlation was also used (not presented herein), leading to similar results. The exception was a positive statistically significant correlation between the number of floors above ground and the weight of the architecture costs.

4.2. Data Modeling

The previous unidimensional statistical analysis provides some insight on the data, but fails to account for the potential interaction between the variables. In fact, a comparison of mean assumes that all the projects in each category are identical regarding all other variables and the same applies for the correlation between two variables. Since all projects are distinct amongst them, modeling the data with multiple linear regression allows identifying the independent variables that are statistically significant to explain the dependent variable, while controlling for the influence of the other independent variables variability. This approach has its own limitations, namely the fact that a linear relation and specific relation (sum) of the variables is assumed.

The cost and prices, both total and unit, were selected as independent variables, along with the cost deviation. All other variables were considered as potential predictors. A hybrid approach was used to select the predictors to include in the models, combining expert judgment and the best subsets tool with the Akaike Information Criterion. The option for this hybrid approach resulted from an experimental stage using only statistical tools to select the predictors (stepwise and best subsets using the Akaike Information Criterion, Ajusted R2 and Overfit Prevention Criterion) that produced models with a very high fit, but were not robust from an engineering point of view. Furthermore, the models for predicting total cost and price were developed without intercept to ensure that the value tends to zero when the project size decreases. There were no signs of heteroscedasticity (White and Breusch–Pagan tests), non-normal distribution of the residuals (Shapiro–Wilk test), or influential observations (Cook's distance) in all hybrid models. Still, robust standard errors were used in all models. There is also no evidence of specification problems (linktest), and the functional forms seem appropriate (Ramsey test).

The regression models for the initial and final price model are presented in Table 6. The R2 of the models is 0.92. Given the high R2 obtained, the models with the predictors selected with statistical tools alone produced similar results in terms of fit to the data. For instance, using the best subsets with the adjusted R2 as the criterion to select variables, it was possible to obtain a model for the initial price with an R2 of 0.95 using the following variables: (i) Area above ground; (ii) area x type; (iii) floors above ground; (iv) total floors; and (v) area ratio. However, this comes with a cost in terms of outliers (3 cases were identified as outliers using the Cook's distance) and represents a potential overfit (a model with 5 variables for a dataset with 18 cases). Due to the reduced size of the sample available (8 residential and 6 office buildings) for developing the final price model, the result should be looked with due care.

Due to confidentiality, the model for the total cost cannot be disclosed. The variables in the models were the same of the initial price models, which is logical since the difference between both is the margin set by the contractor. However, the results of the model are depicted in Figure 2, corresponding to an R2 of 0.94.

Both total and unit cost or prices are connected, but the high correlation between the total cost or price and the construction area may mask the influence of other variables. Considering the confidentiality issues and the limitations of sample size, only the initial unit price was modeled. The first model obtained attained an R2 of 0.505 using as predictors the variables: (i) Floors above ground; (ii) total floors; (iii) floor ratio; and (iv) economic crisis.

However, since a clear non-linear pattern was visible when plotting observed versus predicted initial unit prices, a non-linear multiple regression model was developed. The non-linearity was accounted for by including power coefficients in the scale predictors. The best model resulted in a power of 1.011 for the floors above ground and 1.608 for the total floors, increasing the R2 to 0.720 (Table 7).

Table 6. Regression models for the initial and final price.

Parameter	B	Robust Std. Error [a]	t	Sig.	95% Confidence Interval	
					Lower Bound	Upper Bound
Initial Price						
Above Ground Area (AGA)	735.860	138.565	5.311	0.000	443.512	1028.207
Underground Area (UGA)	462.428	121.467	3.807	0.001	206.155	718.701
Area X Crisis	−102.426	36.276	−2.824	0.012	−178.961	−25.890
Final Price						
Above Ground Area	1393.707	399.891	3.485	0.005	513.554	2273.860
Underground Area	232.331	127.608	1.821	0.096	−48.531	513.194
Area X Type	−181.507	118.842	−1.527	0.155	−443.077	80.062

[a]—HC3 method.

Table 7. Regression models for the initial unit price.

Parameter	B	Robust Std. Error [a]	t	Sig.	95% Confidence Interval	
					Lower Bound	Upper Bound
Intercept	503.309	36.238	13.889	0.000	425.022	581.596
Above Ground Floors [1.011]	−160.284	30.403	−5.272	0.000	−225.966	−94.602
Total Floors [1.608]	17.286	3.129	5.524	0.000	10.525	24.046
Floor Ratio	117.935	25.915	4.551	0.001	61.949	173.920
Economic Crisis = 0	211.752	36.914	5.736	0.000	132.005	291.499
Economic Crisis = 1	0.000					

[a]—HC3 method.

There is the influence of the economic crisis, but the proportion of underground and above ground floors became statistically significant with the removal of the area from the model. The difference between the linear and non-linear models can be observed in Figure 3, evidencing the fit increase in the latter.

The apparently lower fit of the models for the unit price is misleading. In fact, multiplying the area by the initial unit prices estimated with the non-linear model to determine that the total initial price achieves an R2 of 0.97 (Figure 4). This fit difference between the models for the total and unit prices results from the correlation between the total area and the number of floors. This correlation produces multicollinearity between the variables, resulting in the exclusion of the number of floors from any model in which the area is also used. Removing the influence of the area by modeling the unit price allows for the influence of the number of floors to be accounted for, which explains the accuracy increase.

Bootstrapping was also used in the development of the regression models and confirm the statistical significance of the regression coefficients for a 95% confidence interval. Generally, the significance of the regression coefficients decreased, but the p-value remained lower than 0.05 in all cases except for the final price model. For this model, the regression coefficients of the Underground Area and Area X Type already exceeded the 5% significance threshold even without bootstrapping, which can be attributed to the small number of projects for which the final price was available.

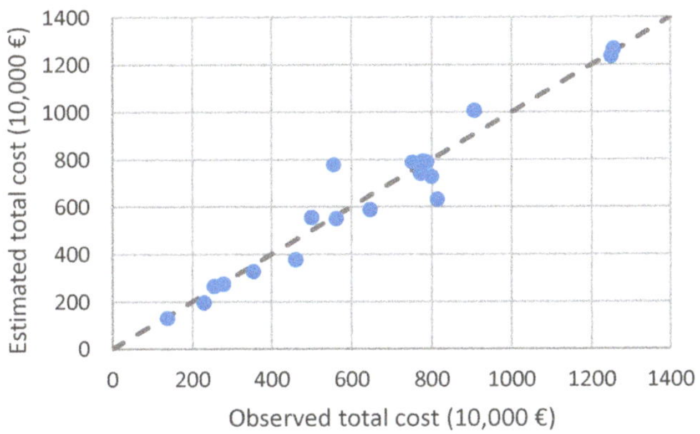

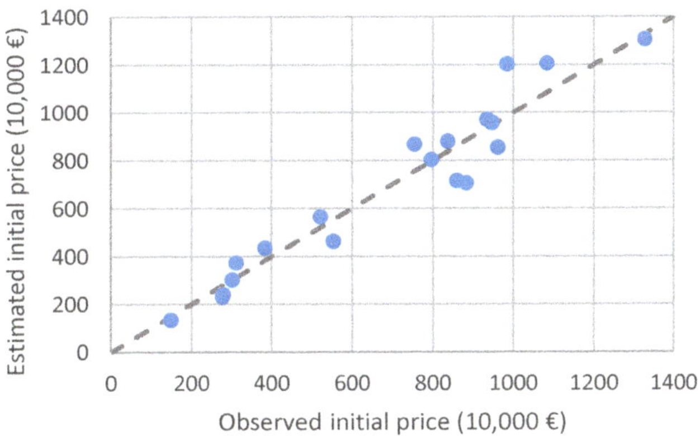

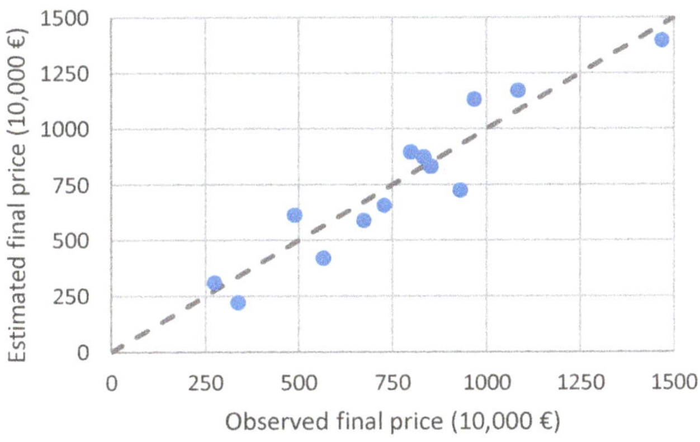

Figure 2. Observed versus estimated total cost and initial and final price.

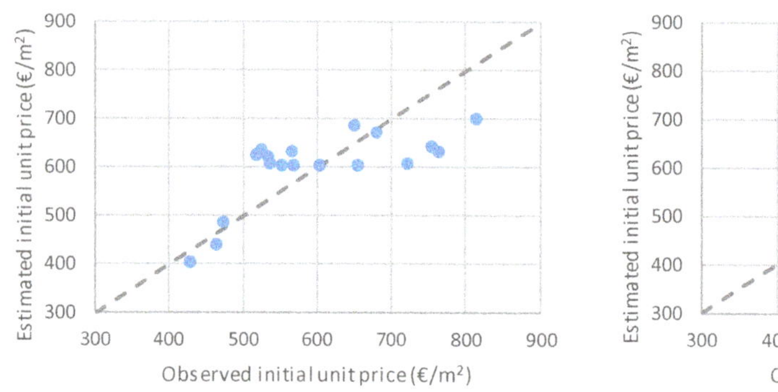
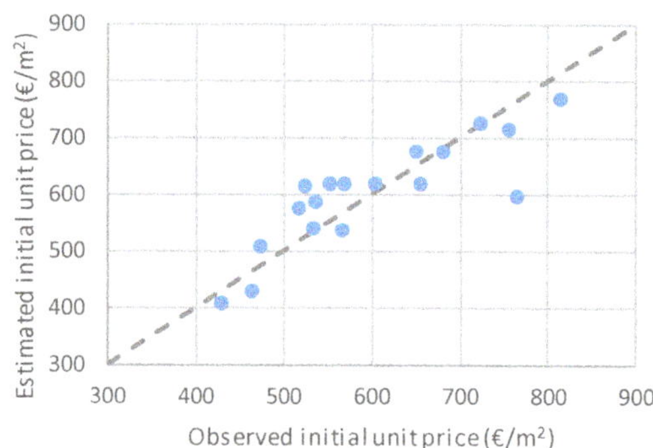

Figure 3. Observed versus estimated initial unit price (**left**: Linear model; **right**: Non-linear model).

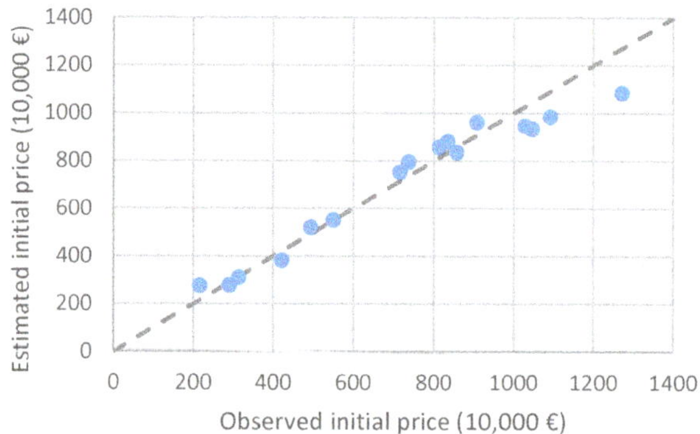

Figure 4. Observed versus estimated initial price using the non-linear initial unit price model.

The three cost estimation models developed for which the mathematical formulation can be disclosed as follows:

$$\text{Initial Price } (\text{€}) = 735.86 \cdot \text{AGA} + 462.428 \cdot \text{UGA} - 102.426 \cdot \text{TA} \cdot \text{C}$$

$$\text{Final Price } (\text{€}) = 1393.707 \cdot \text{AGA} + 232.331 \cdot \text{UGA} - 181.507 \cdot \text{TA} \cdot \text{T}$$

$$\text{Initial unit price } \left(\frac{\text{€}}{\text{m}^2} \right) = 503.309 - 160.284 \cdot \text{AGF}^{1.011} + 17.286 \cdot \text{TF}^{1.608} + 117.935 \cdot \frac{\text{AGF}}{\text{UGF}} + 211.752 \cdot (1 - \text{C})$$

where AGA is the above ground area (m^2); UGA is the underground area (m^2); A is the total area (m^2); C represents the economic crisis (takes the value of 1 if in crisis and 0 otherwise); T represents the type of building (takes the value of 1 if residential and 2 if office); AGF is the number of floors above ground; UGF is the number of underground floors; and TF is the total number of floors.

With the purpose of testing and validating the models developed in this research, the model for the initial price was applied to a project currently under development by the organization. Considering that the project used for validation was estimated in over 45 million euros, significantly higher than the projects in the dataset, and that the difference to the price estimated by the organization was less than 5%, there was positive feedback from the organization regarding the accuracy and extrapolation capability of the model.

In the sample of 13 projects (6 office and 7 residential) for which initial and final prices were available, an average cost deviation of 3.5% was obtained. Only 3 projects had a final price lower than the initial estimate (average of −6.5%). The projects with positive cost deviations were, on average, 6.5% costlier and there was no project without a cost deviation. Comparing with the literature available, which generally adopts the owner perspective, the magnitude of the cost deviation is clearly smaller than usually reported and it becomes evident that the contractor always experiences some cost deviation, even if that is not reflected on the bill of the owner.

Either due to the limitations of the dataset, the fact that the projects are limited in type, the spatial context and stakeholders involved, or a combination of these and other factors, the cost deviation depends on specific aspects of each project that are not captured by the general information used herein and it was not possible to model them. The only statistically significant result obtained was the high Person correlation (0.814) between the number of underground floors and the cost deviation of office buildings. The corresponding regression model indicates that the average cost deviation in office buildings increase 0.65% per the underground floor, but this was obtained from a sample of only 6 projects and its validity is questionable.

5. Conclusions

This research revisits the topic of cost estimation and deviation of construction projects, but adopts the innovative perspective of a contractor, which seems uncommon in the literature review carried out. Furthermore, to the best of our knowledge, this is one of the few efforts linking endogenous and exogenous variables in cost estimation functions.

Contrarily to most research available, only similar projects (premium residential and office buildings) from a single promotor-contractor are used. This compromises the size of the database available, but eliminates the variability of cost estimates and deviations due to: (i) Factors related to the contractor or the designer (e.g., experience; competencies; dimension; management models); (ii) characteristics of the projects (e.g., premium buildings, social buildings, public buildings); (iii) relation between owner and contractor (e.g., type of owner—public, private; type of contract—design-bid-build, design-build; payment method—lump sum, unit prices); and (iv) aspects associated with the location (e.g., weather conditions; laws and regulations). Since the projects are promoted by the real estate company of the same group, the commercial strategy issues related to the degree of competition of the market has less effect on the cost of the projects. The contract does not have to adjust its margin to win the contract and so the influence of the level of competition in the market is only limited to the portion of the project that is executed by subcontractors. By doing so, the results presented herein grasp the "real" cost estimation and deviations driven by project-related factors. The high accuracy of the cost estimation may be partially due to the reduced sample size, but it must be taken into account that the variables that have been reported to influence cost performance are strongly restricted. The results obtained with these restrictions support the importance of the technical expertise of the involved parties in the cost estimation and deviations reported in the literature. Comparing the average and range of the cost deviations in this study with other authors, it is licit to assume that, at least, a portion of the difference is due to the experience of the teams involved and not only due to project (e.g., construction technology) or context (e.g., weather conditions) specificities. Another factor possibly underlying the differences in terms of the magnitude of the cost deviations is the collaborative effort of promoter and contractor in this case, reducing the conflicts that are not rare in the traditional design-bid-build contracts where the promoter has limited expertise/resources regarding the execution stage of the construction project.

Despite the reduced sample size when compared to other studies, it is noticeable that the cost deviations in this context are smaller than what is typically reported when adopting the owner, either public or private, perspective. The generalization of the results may be limited, but they do provide a source for other contractors benchmark their performance and the methodology proposed sets a basis for developing similar studies both in research or practical contexts. In fact, the linear and non-linear regression models developed are of easy interpretation and assessment from an expert, which was done with good results, whereas artificial intelligence models are black-boxes impossible or very difficult to be validated by experts. The practical expert validation carried out, along with the bootstrapping results, reinforce the applicability of the models for the specific context in which it was developed and corroborates the applicability of the methodology in other contexts.

The models developed for estimating costs have a very high fit to the data and highlight the influence of the economic crisis and international bailout on the construction costs. In Portugal, the price of construction projects in open competition also suffered a strong reduction during this period due to the lack of both private and public construction projects. However, since the price is driven not only by the cost but also by the market conditions (e.g., relation between demand and supply), the variation is not necessarily identical, and this research is able to capture the pattern of the cost.

The cost deviations seem to depend more in particular aspects of each project than overall characteristics, despite the positive statistically significant relation between the number of underground floors and the cost deviations in office buildings found.

Author Contributions: Conceptualization, V.S., I.M. and C.O.C.; methodology, V.S.; software, F.P.M.; validation, V.S., I.M. and C.O.C.; formal analysis, F.P.M.; investigation, F.P.M.; resources, V.S.; data curation, F.P.M. and V.S.; writing—original draft preparation, V.S. and F.P.M.; writing—review and editing, I.M. and C.O.C.; visualization, F.P.M.; supervision, V.S.; All authors have read and agreed to the published version of the manuscript.

Funding: This research received no external funding.

Institutional Review Board Statement: Not applicable.

Informed Consent Statement: Not applicable.

Data Availability Statement: Some or all data, models, or code generated or used during the study are proprietary or confidential in nature and may only be provided with restrictions. The estimated cost and initial and final values of each individual project can only be disclosed normalized format (actual value divided by the sample average). The model for the estimated cost cannot be disclosed.

Conflicts of Interest: The authors declare no conflict of interest.

References

1. Hegazy, T.; Ayed, A. Neural Network Model for Parametric Cost Estimation of Highway Projects. *J. Constr. Eng. Manag.* **1998**, *124*, 210–218. [CrossRef]
2. Cruz, C.O.; Branco, F. Reconstruction Cost Model for Housing Insurance. *J. Leg. Aff. Disput. Resolut. Eng. Constr.* **2020**, *12*, 05020007. [CrossRef]
3. Lowe, D.J.; Emsley, M.W.; Harding, A. Predicting Construction Cost Using Multiple Regression Techniques. *J. Constr. Eng. Manag.* **2006**, *132*, 750–758. [CrossRef]
4. Trost, S.M.; Oberlender, G.D. Predicting Accuracy of Early Cost Estimates Using Factor Analysis and Multivariate Regression. *J. Constr. Eng. Manag.* **2003**, *129*, 198–204. [CrossRef]
5. Kim, G.H.; Seo, D.S.; Kang, K.I. Hybrid Models of Neural Networks and Genetic Algorithms for Predicting Preliminary Cost Estimates. *J. Comput. Civ. Eng.* **2005**, *19*, 208–211. [CrossRef]
6. Sonmez, R. Conceptual cost estimation of building projects with regression analysis and neural networks. *Can. J. Civ. Eng.* **2004**, *31*, 677–683. [CrossRef]
7. Soto, B.G.; Streule, T.; Klippel, M.; Bartlomé, O.; Adey, B.T. Improving the planning and design phases of construction projects by using a Case-Based Digital Building System. *Int. J. Constr. Manag.* **2018**, *20*, 1–12. [CrossRef]
8. Jin, R.; Han, S.; Hyun, C.; Cha, Y. Application of Case-Based Reasoning for Estimating Preliminary Duration of Building Projects. *J. Constr. Eng. Manag.* **2016**, *142*, 04015082. [CrossRef]
9. Hu, X.; Xia, B.; Skitmore, M.; Chen, Q. The application of case-based reasoning in construction management research: An overview. *Autom. Constr.* **2016**, *72*, 65–74. [CrossRef]
10. Kim, K.J.; Kim, K. Preliminary Cost Estimation Model Using Case-Based Reasoning and Genetic Algorithms. *J. Comput. Civ. Eng.* **2010**, *24*, 499–505. [CrossRef]
11. Ji, C.; Hong, T.; Hyun, C. CBR Revision Model for Improving Cost Prediction Accuracy in Multifamily Housing Projects. *J. Manag. Eng.* **2010**, *26*, 229–236. [CrossRef]
12. Ahn, J.; Ji, S.-H.; Ahn, S.J.; Park, M.; Lee, H.-S.; Kwon, N.; Lee, E.-B.; Kim, Y. Performance evaluation of normalization-based CBR models for improving construction cost estimation. *Autom. Constr.* **2020**, *119*, 103329. [CrossRef]

13. Kim, G.-H.; An, S.-H.; Kang, K.-I. Comparison of construction cost estimating models based on regression analysis, neural networks, and case-based reasoning. *Build. Environ.* **2004**, *39*, 1235–1242. [CrossRef]
14. Rafiei, M.H.; Adeli, H. Novel Machine-Learning Model for Estimating Construction Costs Considering Economic Variables and Indexes. *J. Constr. Eng. Manag.* **2018**, *144*, 04018106. [CrossRef]
15. Günaydın, H.M.; Doğan, S.Z. A neural network approach for early cost estimation of structural systems of buildings. *Int. J. Proj. Manag.* **2004**, *22*, 595–602. [CrossRef]
16. Doğan, S.Z.; Arditi, D.; Günaydın, H.M. Determining Attribute Weights in a CBR Model for Early Cost Prediction of Structural Systems. *J. Constr. Eng. Manag.* **2006**, *132*, 1092–1098. [CrossRef]
17. Jin, R.; Cho, K.; Hyun, C.; Son, M. MRA-based revised CBR model for cost prediction in the early stage of construction projects. *Expert Syst. Appl.* **2012**, *39*, 5214–5222. [CrossRef]
18. Sousa, V.; Almeida, N.M.; Dias, L.A. Role of Statistics and Engineering Judgment in Developing Optimized Time-Cost Relationship Models. *J. Constr. Eng. Manag.* **2014**, *140*, 04014034. [CrossRef]
19. Sousa, V.; Almeida, N.M.; Dias, L.A.; Branco, F.A. Risk-Informed Time-Cost Relationship Models for Sanitation Projects. *J. Constr. Eng. Manag.* **2014**, *140*, 06014002. [CrossRef]
20. Sousa, V.; Meireles, I. The Influence of the Construction Technology in Time-Cost Relationships of Sewerage Projects. *Water Resour. Manag.* **2018**, *32*, 2753–2766. [CrossRef]
21. Thalmann, P. A low-cost construction price index based on building functions. In Proceedings of the 15th International Cost Engineering Congress, ICEC, Rotterdam, The Netherlands, 1 March 1998.
22. Elhag, T.M.S.; Boussabaine, A.H. An artificial neural system for cost estimation of construction projects. In Proceedings of the 14th Annual ARCOM Conference, Reading, UK, 9–11 September 1998; pp. 219–226.
23. Emsley, M.W.; Lowe, D.J.; Duff, A.R.; Harding, A.; Hickson, A. Data modelling and the application of a neural network approach to the prediction of total construction costs. *Constr. Manag. Econ.* **2002**, *20*, 465–472. [CrossRef]
24. Picken, D.H.; Ilozor, B.D. Height and construction costs of buildings in Hong Kong. *Constr. Manag. Econ.* **2003**, *21*, 107–111. [CrossRef]
25. Attalla, M.; Hegazy, T. Predicting Cost Deviation in Reconstruction Projects: Artificial Neural Networks versus Regression. *J. Constr. Eng. Manag.* **2003**, *129*, 405–411. [CrossRef]
26. Skitmore, R.M.; Ng, S.T. Forecast models for actual construction time and cost. *Build. Environ.* **2003**, *38*, 1075–1083. [CrossRef]
27. Elhag, T.; Boussabaine, A.; Ballal, T. Critical determinants of construction tendering costs: Quantity surveyors' standpoint. *Int. J. Proj. Manag.* **2005**, *23*, 538–545. [CrossRef]
28. Li, H.; Shen, Q.; Love, P.E. Cost modelling of office buildings in Hong Kong: An exploratory study. *Facilities* **2005**, *23*, 438–452. [CrossRef]
29. Stoy, C.; Schalcher, H.-R. Residential Building Projects: Building Cost Indicators and Drivers. *J. Constr. Eng. Manag.* **2007**, *133*, 139–145. [CrossRef]
30. Wheaton, W.C.; Simonton, W.E. The secular and cyclic behavior of 'true' construction costs. *J. Real Estate Res.* **2007**, *29*, 1–26. Available online: https://www.jstor.org/stable/24888194 (accessed on 1 October 2021). [CrossRef]
31. Stoy, C.; Pollalis, S.; Schalcher, H.-R. Drivers for Cost Estimating in Early Design: Case Study of Residential Construction. *J. Constr. Eng. Manag.* **2008**, *134*, 32–39. [CrossRef]
32. Jin, R.; Han, S.; Hyun, C.; Kim, J. Improving Accuracy of Early Stage Cost Estimation by Revising Categorical Variables in a Case-Based Reasoning Model. *J. Constr. Eng. Manag.* **2014**, *140*, 04014025. [CrossRef]
33. Bayram, S.; Ocal, M.E.; Oral, E.L.; Atis, C. Comparison of multi layer perceptron (MLP) and radial basis function (RBF) for construction cost estimation: The case of Turkey. *J. Civ. Eng. Manag.* **2015**, *22*, 480–490. [CrossRef]
34. Muratova, A.; Ptukhina, I. *BIM as an Instrument of a Conceptual Project Cost Estimation*; Springer: Cham, Switzerland, 2020; Volume 70. [CrossRef]
35. Karaca, I.; Gransberg, D.D.; Jeong, H.D. Improving the Accuracy of Early Cost Estimates on Transportation Infrastructure Projects. *J. Manag. Eng.* **2020**, *36*, 04020063. [CrossRef]
36. Swei, O.; Gregory, J.; Kirchain, R. Construction cost estimation: A parametric approach for better estimates of expected cost and variation. *Transport. Res. B-Meth.* **2017**, *101*, 295–305. [CrossRef]
37. Flyvbjerg, B.; Hon, C.; Fok, W.H. Reference class forecasting for Hong Kong's major roadworks projects. *Proc. Inst. Civ. Eng.-Civ. Eng.* **2016**, *169*, 17–24. [CrossRef]
38. Gunduz, M.; Ugur, L.O.; Ozturk, E. Parametric cost estimation system for light rail transit and metro trackworks. *Expert Syst. Appl.* **2011**, *38*, 2873–2877. [CrossRef]
39. Al-Tabtabai, H.; Alex, A.P.; Tantash, M. Preliminary cost estimation of highway construction using neural networks. *Cost Eng.* **1999**, *41*, 19–24.
40. Shehu, Z.; Endut, I.R.; Akintoye, A.; Holt, G.D. Cost overrun in the Malaysian construction industry projects: A deeper insight. *Int. J. Proj. Manag.* **2014**, *32*, 1471–1480. [CrossRef]
41. Sweis, G.J.; Sweis, R.; Rumman, M.A.; Hussein, R.A.; Dahiya, S.E. Cost overruns in public construction projects: The case of Jordan. *J. Am. Sci.* **2013**, *9*, 134–141. [CrossRef]
42. Love, P.E.D.; Wang, X.; Sing, C.-P.; Tiong, R.L.K. Determining the probability of cost overruns in Australian construction and engineering projects. *J. Constr. Eng. Manag.* **2013**, *139*, 321–330. [CrossRef]

43. Love, P.; Ahiaga-Dagbui, D. Debunking fake news in a post-truth era: The plausible untruths of cost underestimation in transport infrastructure projects. *Transp. Res. Part A Policy Pr.* **2018**, *113*, 357–368. [CrossRef]

44. Love, P.E.; Ika, L.A.; Ahiaga-Dagbui, D.D. On de-bunking 'fake news' in a post truth era: Why does the Planning Fallacy explanation for cost overruns fall short? *Transp. Res. Part A Policy Pr.* **2019**, *126*, 397–408. [CrossRef]

45. Love, P.; Sing, C.P.; Carey, B.; Kim, J.T. Estimating Construction Contingency: Accommodating the Potential for Cost Overruns in Road Construction Projects. *J. Infrastruct. Syst.* **2015**, *21*, 04014035. [CrossRef]

46. Love, P.E.D.; Sing, X.C.P.; Tiong, R.L.K. Determining the Probability of Project Cost Overruns. *J. Constr. Eng. Manag.* **2020**, *146*, 04020060. [CrossRef]

47. Kaming, P.F.; Olomolaiye, P.O.; Holt, G.; Harris, F.C. Factors influencing construction time and cost overruns on high-rise projects in Indonesia. *Constr. Manag. Econ.* **1997**, *15*, 83–94. [CrossRef]

48. AbuSafiya, H.A.M.; Suliman, S.M.A. Causes and Effects of Cost Overrun on Construction Project in Bahrain: Part I (Ranking of Cost Overrun Factors and Risk Mapping). *Mod. Appl. Sci.* **2017**, *11*, 20. [CrossRef]

49. Derakhshanalavijeh, R.; Teixeira, J.M.C. Cost overrun in construction projects in developing countries, Gas-Oil industry of Iran as a case study. *J. Civ. Eng. Manag.* **2016**, *23*, 125–136. [CrossRef]

50. Annamalaisami, C.D.; Kuppuswamy, A. Reckoning construction cost overruns in building projects through methodological consequences. *Int. J. Constr. Manag.* **2019**, *1*, 1–11. [CrossRef]

51. Balali, A.; Moehler, R.C.; Valipour, A. Ranking cost overrun factors in the mega hospital construction projects using Delphi-SWARA method: An Iranian case study. *Int. J. Constr. Manag.* **2020**, *1*, 1–9. [CrossRef]

52. Shrestha, P.P.; Fathi, M. Impacts of Change Orders on Cost and Schedule Performance and the Correlation with Project Size of DB Building Projects. *J. Leg. Aff. Disput. Resolut. Eng. Constr.* **2019**, *11*, 04519010. [CrossRef]

53. Flyvbjerg, B.; Holm, M.K.S.; Buhl, S.L. What Causes Cost Overrun in Transport Infrastructure Projects? *Transp. Rev.* **2004**, *24*, 3–18. [CrossRef]

54. Flyvbjerg, B.; Holm, M.S.; Buhl, S. Underestimating Costs in Public Works Projects:Error or Lie? *J. Am. Plan. Assoc.* **2002**, *68*, 279–295. [CrossRef]

55. Pearl, R.G.; Akintoye, A.; Bowen, P.A.; Hardcastle, C. Analysis of tender sum forecasting by quantity surveyors and contractors in South Africa. *J. Phys. Dev. Sci. Acta Stuctilla* **2003**, *10*, 5–35.

56. Bucciol, A.; Chillemi, O.; Palazzi, G. Cost overrun and auction format in small size public works. *Eur. J. Political Econ.* **2013**, *30*, 35–42. [CrossRef]

57. Reyers, J.; Mansfield, J. The assessment of risk in conservation refurbishment projects. *Struct. Surv.* **2001**, *19*, 238–244. [CrossRef]

58. Catalão, F.P.; Cruz, C.O.; Sarmento, J.M. Exogenous determinants of cost deviations and overruns in local infrastructure projects. *Constr. Manag. Econ.* **2019**, *37*, 697–711. [CrossRef]

59. Catalão, F.P.; Cruz, C.O.; Sarmento, J.M. The determinants of cost deviations and overruns in transport projects, an endogenous models approach. *Transp. Policy* **2019**, *74*, 224–238. [CrossRef]

60. Catalão, F.P.; Cruz, C.O.; Sarmento, J.M. Public management and cost overruns in public projects. *Int. Public Manag. J.* **2020**, 1–27. [CrossRef]

61. Flyvbjerg, B.; Ansar, A.; Budzier, A.; Buhl, S.; Cantarelli, C.; Garbuio, M.; Glenting, C.; Holm, M.S.; Lovallo, D.; Molin, E.; et al. On de-bunking "Fake News" in the post-truth era: How to reduce statistical error in research. *Transp. Res. Part A Policy Pr.* **2019**, *126*, 409–411. [CrossRef]

62. Flyvbjerg, B.; Ansar, A.; Budzier, A.; Buhl, S.; Cantarelli, C.; Garbuio, M.; Glenting, C.; Holm, M.S.; Lovallo, D.; Molin, E.; et al. Five things you should know about cost overrun. *Transp. Res. Part A Policy Pr.* **2018**, *118*, 174–190. [CrossRef]

Article

Critical Review of the Evolution of Project Delivery Methods in the Construction Industry

Salma Ahmed and Sameh El-Sayegh *

Civil Engineering Department, American University of Sharjah, Sharjah 26666, UAE; g00043157@alumni.aus.edu
* Correspondence: selsayegh@aus.edu

Abstract: Selecting the appropriate project delivery method (PDM) is a very significant managerial decision that impacts the success of construction projects. This paper provides a critical review of related literature on the evolution of project delivery methods, selection methods and selection criteria over the years and their suitability in the construction industry of today's world. The literature review analysis has concluded that project delivery methods evolve at a slower rate compared to the evolution of the construction industry. The paper also suggests features of an evolved project delivery method that is digitally integrated, people-centered, and sustainability-focused. Moreover, the paper highlights the latest selection criteria such as risk, health and wellbeing, sustainability goals and technological innovations. Furthermore, the paper concluded that advanced artificial intelligence techniques are yet to be exploited to develop a smart decision support model that will assist clients in selecting the most appropriate delivery method for successful project completion. Additionally, the paper presents a framework that illustrates the relationship between the different PDM variables needed to harmonize with the construction industry. Last, but not least, the paper fills a gap in the literature as it covers a different perspective in the field of project delivery methods. The paper also provides recommendations and future research ideas.

Keywords: project delivery methods; construction; PDM selection criteria; PDM selection methods

Citation: Ahmed, S.; El-Sayegh, S. Critical Review of the Evolution of Project Delivery Methods in the Construction Industry. *Buildings* **2021**, *11*, 11. https://dx.doi.org/10.3390/buildings11010011

Received: 12 November 2020
Accepted: 23 December 2020
Published: 26 December 2020

1. Introduction

The construction industry is a major contributor to any country's economy. The impact of this contribution largely depends on the successful and efficient delivery of construction projects. One of the critical success factors in any construction project is the managerial decision of the project delivery method [1]. This is due to the fact that it has a direct effect on key performance indicators such as cost, schedule, quality, project execution and safety [2].

The term delivery method refers to the assignment of responsibilities to the different parties involved in a project in order to establish a framework of the entire design, procurement and construction process [1]. There are various delivery methods available in the construction industry, from the traditional design-bid-build (DBB) to other alternative methods such as design-build (DB) and construction manager at risk (CMR). Using DBB, the owner issues two separate contracts, one with the consultant for the design phase of the project and the other with a construction professional for project execution [3]. On the other hand, in DB, a single legal entity is given the sole responsibility to hire both the consultant and the contractor under one contract representing a single commitment [4]. Furthermore, CMR is a delivery method in which the construction manager is recruited during the design phase of the project, giving him the responsibilities of both a project coordinator and a general contractor [5]. Additionally, collaborative delivery methods such as integrated project delivery, alliancing and partnering represent emerging forms of delivery methods that emphasize features such as collaboration, trust, commitment, as well as co-learning [6].

When it comes to choosing the project delivery method, many owners rely on a list of predefined selection criteria and selection methods to assist them in the decision process. These methods and criteria are not comprehensive and may not be applicable enough in today's modern construction industry as conventional project management practices are not being updated at an appropriate rate to embrace changes that have already transformed the construction industry, such as technological advancements and greening practices [7,8]. Referred to as "construction 4.0" is a term that was conceived from the concept of industry 4.0, which is viewed as the fourth industrial revolution that originated from Germany [9]. Construction 4.0 is a digital transformation of the industry through the use of sophisticated gadgets such as laser scanning, drones, and 3D printing in order to enhance the management of construction projects throughout the different phases, which will enable the establishment of smarter and sustainable buildings [10].

Apart from the digital transformation in construction project management, there are other changes that further differentiate the construction industry today from the past. The construction context is very different today with the introduction of sustainable and green construction. As the industry changes and with the increasing global awareness about the negative impacts brought upon the environment through construction activities, project managers are under extreme pressure to steer their construction projects towards sustainable development by implementing green measures [11]. Additionally, the construction environment itself is not the same; it was some sixty years ago with growing populations and changing lifestyles worldwide. This will ultimately have an impact on altering customer expectations. Since customers are often regarded as the ultimate stakeholder, it is essential that project managers always update themselves in terms of customer expectations [12]. Consequently, with this in mind, the evolution of the project delivery methods, selection criteria and models have become more critical to be able to satisfy the demands of the modern construction industry.

Therefore, the aim of this paper is to conduct a systematic literature review on the project delivery methods available in the construction industry, the selection criteria that are identified in the literature, as well as the selection models and decision support tools used by owners to choose the appropriate project delivery method. This research answers critical empirical questions by highlighting the new selection criteria for project delivery methods in today's construction industry. Additionally, this research classifies the project delivery selection models according to the progression of rigor by academics. Moreover, the results of this literature review will contribute to the body of research knowledge as it will provide a detailed review of the evolution of project delivery over the past sixty years. Furthermore, new selection criteria will be highlighted, and new features of project delivery methods will be identified. The study addresses the following three research questions:

1. What research has been carried out on delivery methods, selection criteria and selection methods of delivery methods?
2. What are the new selection criteria for project delivery methods highlighted post literature analysis?
3. What are the features of the project delivery method that future research should focus on to fill the gaps in the literature?

2. Theoretical Background

2.1. Project Delivery Methods

Selecting the appropriate project delivery method is one of the most important managerial decisions as it has a direct impact on the success of the project since it affects key performance indicators such as cost, quality, schedule and safety. Indeed, project delivery methods have evolved over the years, and there have been many variations and alternatives introduced in the construction industry to meet various consumer demands.

To begin with, up until 1990, the traditional delivery system, design-bid-build (DBB), was considered the dominant method where professionals were endorsing and standardizing its features throughout almost all construction projects [13]. DBB, also known as the conventional method, where the owner issues two separate contracts, one with the consultant for the design phase and the second contract is with a construction professional for the execution of the project. This disconnection, however, has led to several disputes and resulted in an increase in the number of claims and change orders, which ultimately lead to cost and time overruns [14]. In fact, this delivery method is usually associated with the single fixed-price or the lump sum contract strategy where the contractor performs a specified for a specific amount of money. Such a contract removes the risk of any changes to the final cost for the owner [15].

As the demand for heavy engineering projects increased, it became difficult to precisely quantify the required work, making the lump sum contract incapable of achieving the project's objective. Therefore, a unit price contract strategy was developed, where the owner divides the work into bid items with an estimated quantity of work for each item. After this, the contractor bids the direct cost of each item and must account for overhead, profit and other project expenses [16]. Moreover, as the 20th-century progressed, and with the increase in the complexity of buildings, the need for more coordination between stakeholders emerged, which urged the need for alternative delivery systems. This is when the design build (DB) started gaining popularity in the construction industry, in which the project delivery culture was significantly transformed as the project owner's contracts for both design and construction from a single entity called DB. Indeed, the shift was challenging, and owners were reluctant in the beginning as they feared that they would no longer have contractual advocacy and the quality of construction projects would be compromised [17]. However, as the process evolved, these fears vanished as DB has proved to provide benefits such as collaborative construction effort since the designer and contractor work as one entity. Moreover, DB also allows fast track alternative where some portion of construction can be started while the design is still ongoing; therefore, this can result in cost and time savings [18].

Over the years, there have been other variations to design build, including bridging, novation DB, package deals, direct DB, develop and construct, turnkey method and build operate transfer. Each one of these variations is designed to meet diverse scenarios of construction settings [19,20]. Another delivery system that emerged around the same time as DB was construction management (CM), where the owner hires both a design firm and a construction project firm early in the preconstruction phase of the project. The construction manager would then advise the owner in matters regarding design and managing construction activities. Although it is true that this method leads to a high level of collaboration between project participants, it also requires high owner involvement, which dictates the need for a sophisticated owner [21]. A derivative of construction management is the construction management at risk (CMR) approach. This is where the role of the construction manager shifts from being an advisor to a vendor, where they will act as both a project coordinator and general contractor to execute the construction activities. This method is associated with a guaranteed maximum price contract, which is an advantage to the owners [5]. It also leads to decreased change order and increased cost certainty as well as superiority in product and service quality levels when compared to the traditional DBB delivery method [22,23].

Nonetheless, it can be seen that these delivery methods were developed to target specific objectives with a restricted focus, which leads to fragmented approaches as the improvement of the overall delivery system is yet to be achieved in the construction industry [24]. Researchers argue that the recent development of integrated project delivery (IPD) systems is the solution to this problem [25]. IPD is defined as a "method that integrates people, systems, business structures and practices into a process that collaboratively harnesses the talents and insights of all participants to reduce waste and optimize efficiency through all phases of design, fabrication and construction" [26]. Moreover, Azhar et al. [27] listed six features that characterize IPD. These include early involvement of key participants, shared risk and reward, multiparty contract, collaborative decision-making and control, liability waivers among participants as well as jointly developed project goals. Furthermore, a need for more integration in delivering construction projects is critical to cover the limitations of the traditional DBB method, which leads to the development of various cultures that results in severe inefficiency and high costs of inadequate interoperability as well as high levels of data and team fragmentation that even CM and CMR methods were not able to overcome [28–31]. Indeed, these traditional delivery approaches have historically resulted in a profound number of claims, high risks, delayed schedules and over-priced projects [32]. It is about time that integration is taken to another level in these delivery methods where project members are engaged in a much faster way that allows for real-time monitoring through intuitive interfaces with the help of the technological innovations that transformed the construction industry [33–36]. In fact, Demetracopoulou et al. [37] confirmed that there is a strong positive correlation between the characteristics that lead to innovation opportunities and the level of integration between designer and contractor.

Among other efforts to increase collaboration in project delivery methods is the introduction of lean delivery methods. Lean delivery consists of four phases. The first phase is the project definition phase, which deals with determining the needs and value of the client. In contrast, the second phase is the lean design phase, where decisions regarding product and process are made simultaneously to create a conceptual design. Furthermore, the third phase is the lean supply phase, which consists of transforming the conceptual design to detailed engineering documents such as components fabrication and logistics of deliveries. The last phase is the lean assembly phase, which begins with the delivery of materials, tools, labor or other components to the project is finished and handed over to the client [38]. Under the umbrella of collaborative delivery methods also comes alliances and partnering in which project alliancing is a delivery method that allows the owner and other participants to work together as an integrated and collaborative team with faith and trust to manage risks jointly and share the project outcome in the end. While partnering is a method used by two organizations who share mutual goals to reach specific business objectives. It constitutes an agreed-upon method to solve conflicts with the aim of continuous improvement [39].

2.2. Selection Criteria for Project Delivery Methods

Owners are presented with various options for their project delivery process from traditional DBB to DB or CMR. Ideally, project delivery selection would be based on which success factors offer the greatest likelihood of achieving the desired success criteria of a project. Over the years, there have been many changes in the construction industry that have caused frequent updates to the list of success factors either by adding more factors or prioritizing some factors over the others.

To begin with, up until the 1970s and 1980s, the delivery method was selected mainly on a cost-oriented basis. However, beyond the 1980s, the customers' demands have evolved where they were looking for more integration and mutual cooperation between project members [40]. As the interaction increased, the owners realized that this decreased disputes and change orders, which ultimately reduced delay in schedules and a rise in costs. Hence this caused factors such as communication, cost and schedule growth to be included in the selection criteria list as they lead to the more efficient selection of project delivery [41]. Furthermore, around this period, the construction industry witnessed the age of information technology, which brought advances in engineering software. For instance, the application of Building Information Modelling (BIM) technology in each of the different delivery systems to integrate various disciplines during the design and construction phases [42]. This technological boom that the construction industry-endorsed has further emphasized the significance of the communication selection criteria.

Moreover, around the year 1987, the concept of sustainability invaded the construction industry. Although the literature does represent some papers that discuss the effect of sustainability on project delivery, such as Korkmaz et al. [43], who presented evaluation metrics for sustainable project delivery, the research in this field still does not suffice. Indeed, this area of study is still in its embryonic stage, and more digging is required about the inclusion of sustainability goals in the selection criteria list for the various project delivery methods selection. Unfortunately, this shows that even though the construction industry witnessed the move towards sustainability a long time ago, project management is still struggling to incorporate sustainability in the selection criteria list for project delivery selection. This proves that there is a lag between the rate of evolution of the construction industry and the rate at which the selection criteria list is being updated, indicating that there is still much room for improvement.

At the start of the 21st-century, more criteria were included in the selection set. Among those criteria was quality as customers have started paying more attention to the quality of the delivered project rather than just economic and transaction-specific measures [44,45]. Moreover, analysis of literature has revealed that more papers in the 21st-century were directed towards studying risk as a selection factor upon which the project delivery method would be selected [46–49]. Furthermore, Gransberg et al. [50] claimed that even though all of these selection criteria are relevant, the owner's characteristics and his experience on how to handle disputes as well as his willingness to take risk affects all other factors and, therefore, should play a major role in selecting the most appropriate project delivery method. Additionally, the health and wellbeing of the workers in the construction industry is another selection criterion that has been the center of attention in recent papers [51,52]. Not only this but, around the year 2011, there has been a huge digital transformation in the construction industry where drones, laser technologies and artificial intelligence started being used in the construction process [9]. However, there is very little research on the contribution of these technological advances to the list of criteria used to select the most appropriate delivery method, which creates a gap that needs to be bridged in future research.

2.3. Selection Methods of Project Delivery Methods

Selecting the most suitable project delivery method is a complex and lengthy process that demands a comprehensive analysis of various success factors and criteria, and it does not follow a one size fits all approach [53]. Traditionally, project managers relied on their gut feelings and the delivery methods they are most familiar with to help them choose. However, with the increasing complexity and evolution of the construction projects, project managers realized that there was a need for a structured mechanism or tool to assist them in choosing the most suitable delivery method for a specific construction project [54].

It began with a simplified version of a scoring and decision chart where each project delivery method was assigned a score using a numerical scale that measured its ability to fulfill a specific criterion. After this, the evaluation criteria were weighed to identify the relative significance of each of the selected criteria. The overall score of each project delivery was then calculated by adding up all the scores from each criterion, and then finally, the project delivery with the highest score was identified as the most appropriate alternative [55]. However, Like the age of information technology arrived by the year 1975, the decision-making tools grew more sophisticated with the introduction of multi-attribute utility theory (MAUT) and analytical hierarchy process (AHP) to help improve the objectivity of the selection process and make it less subjective.

In MAUT, the project manager initially identifies a utility function for each criterion. These functions are later used to compute the utility score of each project delivery method with regard to different criteria. Similar to the weighted sum approach, weights are assigned to each criterion individually to indicate their relative significance. After this, the utility scores for all the various criteria are weighted and summed to calculate a global utility score for a specific delivery method. Finally, the project delivery method with the highest global utility score is selected [56]. While in AHP, the first step in the process is identifying the different project delivery methods and developing a hierarchy of the selected criteria. The main step in the process is the conduction of the pair-wise comparison of project delivery methods where project managers are to compare all methods with reference to the evaluation criteria, respectively. Ratio scales are then used to measure the manager's comparative preferences and integrated to compute an overall weight for each project delivery method [57].

After introducing AHP in the early 1980s, Saaty [58] introduced the analytic network process (ANP) around 1996, which was considered as the general form of AHP. It was used in order to overcome the limitations underlying the assumption of independence between criteria in which the ANP model allowed for complicated interrelations between various criteria elements. Furthermore, with the introduction of selection criteria such as quality, flexibility and speed by the beginning of the 21st-century, it was challenging to measure them using numerical values. This was when the method referred to as the fuzzy logic approach was introduced in the construction industry to select the project delivery method. Ng et al. [59] explained the fuzzy approach where the integral function in this method was the membership function. These functions were used to assign a criterion in a fuzzy set to either 0 or 1, where 1 indicated a member and 0 indicated otherwise. This helped in the conversion of linguistic terms such as low, medium or high into numerical values. However, there is no evidence in the literature that the current methods are fit to quantify other selection criteria that have been added due to the evolution of the construction industry, such as the parameters of sustainability, for instance. This, in turn, creates a gap that the selection methods that have not evolved or matured enough to catch up with the pace of the construction industry's evolution.

In addition to new embellishments in the criteria elements list, the digital transformation that invaded the construction industry has also brought along with it some changes in the selection methods used to choose the project delivery method. For instance, the development of the Monte Carlo simulation algorithm, which is a technique used to randomly generate input variables from statistical distributions to model a stochastic process [60]. The outputs of the simulation then result from conducting a large number of iterations to account for risk and uncertainty. Some project managers also opt to use a mix of methods to help them in the decision-making process of selecting the most appropriate delivery method, such as combining both ANP and Monte Carlo simulation to reach optimum results. Furthermore, over the years, there have been several advances in decision-making tools such as tools that formally separate project characteristics from project goals to assist decision-makers in selecting an optimum delivery method based on their institutional needs and requirements [61]. Although there has been much sophistication in the selection methods over the years, there are still some limitations that need to be covered. For instance, the development of selection models that take into account the interdependencies between different projects basically defines the construction industry of today, where all projects are interconnected in one way or the other. Another limitation that needs to be fulfilled is the development of an optimization model that considers different scenarios of time and cost tradeoffs in order to satisfy the new selection criteria presented in the previous section [62].

3. Research Methodology

3.1. Research Design

This paper follows a systematic literature review that was conducted as per the guidelines of preferred reporting items for systematic reviews and meta-analyses (PRISMA), which is an evidence-based set of 4 stages to report a wide array of systematic reviews as illustrated in Figure 1. The first stage is the identification of the review characteristics, which includes scope definition, databases as well as search criteria. The second stage is a screening of the relevant scientific contributions. While the third stage is eligibility evaluation, and the last stage is data analysis and synthesis.

1. Identification of review characteristics: The scope of the review focuses on the evolution of project delivery methods, selection criteria and selection models over the years. The database used to conduct this search was mainly Scopus, as it incorporates relevant sources of peer-reviewed studies.

2. Screening: The research included only journal articles and books (conference papers were excluded) that were published in the English language with no specific time period to provide a comprehensive overview of the evolution of the construction industry and project delivery methods over the years. The search string used was "TITLE-ABS-KEY" using the keywords "project delivery methods" or "project delivery systems" and "construction".

3. Eligibility analysis: The first step is abstract analysis to evaluate if the paper fits the scope of the research, and if it does not fit, then it automatically gets excluded. After this, full-text analysis is done to select eligible documents.

4. Data analysis and synthesis: The selected papers were first classified according to the publication date in order to determine whether they belong to the past or present or future stages of project delivery methods evolution. After this, the papers were categorized, whether they are empirical or conceptual studies. The selected studies were further analyzed to develop a list of 3 research targets: evolution of project delivery methods, evolution of project delivery selection criteria, evolution of project delivery selection models/methods.

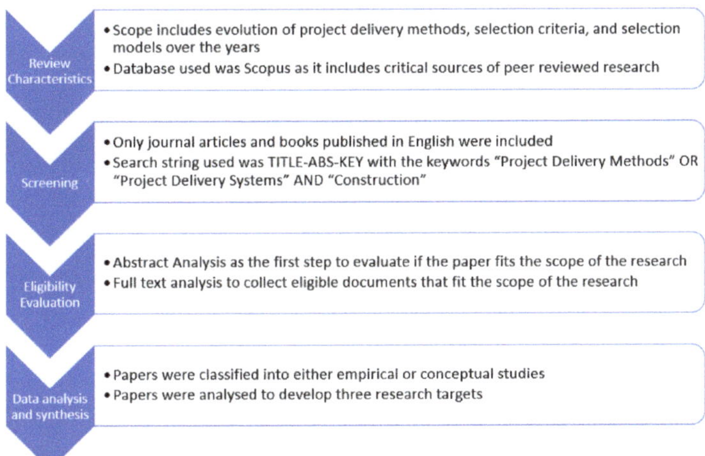

Figure 1. Preferred reporting items for systematic reviews and meta-analyses (PRISMA) checklist.

3.2. Data Collection

Using the keywords "project delivery methods" or "project delivery systems" and "construction" on Scopus with the limitation of only English language and the exclusion of conference papers, a total of 328 papers were collected. These selected papers were further filtered manually to eliminate the studies that fall outside the scope of the research. As a result of this filtration process, a total of 103 studies were eliminated, and only 225 references were included in the end. Simple statistical analysis was done on these 225 references to show the number of publications per year (Figure 2) and the number of publications per country as well (Figure 3). The results show that there is no clear trend for the number of publications per year, but rather it presents a cyclic timeline with peaks at certain time periods. While Figure 3 illustrates that the United States is the leading country in this field of research with the highest number of publications in this area of study.

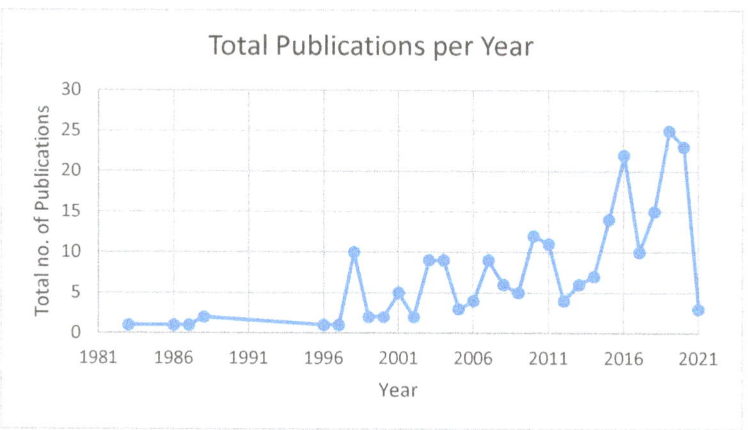

Figure 2. Number of publications per year.

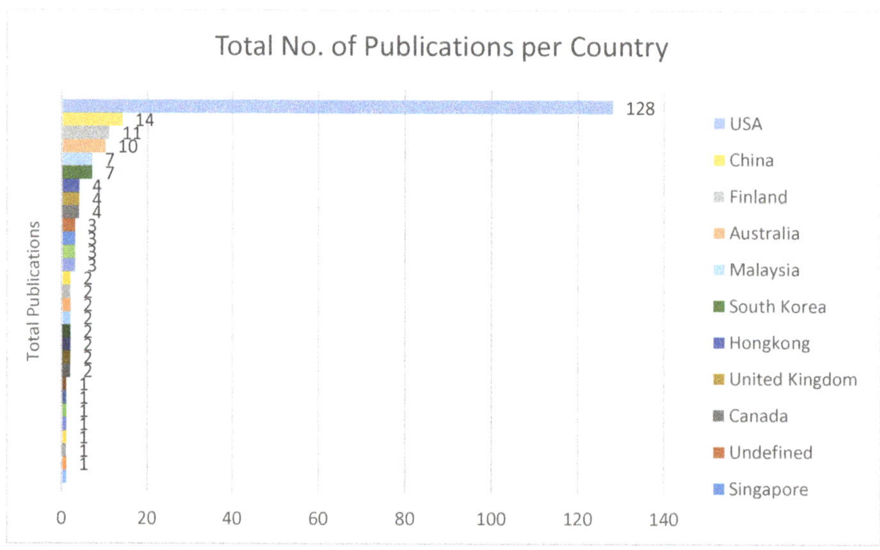

Figure 3. Number of publications per country.

3.3. Analysis

The project delivery methods were divided into four categories based on the major changes in the contractual relationships among the key contracting parties and the rough timeframe for the emergence of these delivery methods as reported in the literature. The first category, referred to in this paper as PDM 1.0, refers to the pre-1850s era and includes the master-builder method. During that era, construction was mainly labor-intensive, and arrangements such as master builder were the most dominant ones [63]. The second category, referred to in this paper as PDM 2.0, includes the design–bid–build (DBB) method, which emerged in the 1850s in response to the emergence of specialized disciplines and the separation of design and construction as professional disciplines. The contractual relationships have changed, and clients now have two contracts: one with the designer and one with the contractor. Most literature sources refer to this method (DBB) as the traditional project delivery method. PDM 3.0 represents the emergence of alternative delivery methods, such as design-build and construction management. A review of related literature showed that as time passed, the construction industry became more complex, and clients became aware of the many problems associated with the traditional DBB method. Literature analysis has shown that the contractual arrangements have changed, and clients looked for arrangements such as construction management to act as their representative and coordinate/manage the construction project. In addition, some clients looked for arrangements that integrate design and construction, such as using design-build. Later, the literature showed that clients started to use CMR, where one entity will handle construction management and general contracting services. According to reviewed literature, this era started in the late 1950s and early 1960s. The methods included in PDM 3.0 are often referred to in the literature as alternative project delivery methods as in alternatives to the traditional DBB method. PDM 4.0 represents the collaborative delivery methods, which include IPD, alliancing, partnering and relationship-based contracting, which have only started gaining significant attention in the literature in the past 10–12 years. This category includes methods that promote collaboration and a team atmosphere as a solution to the many problems in the construction industry. In other words, PDM 4.0 represents features of a project delivery method that answer the demands of the modern construction industry. This version constitutes of digitally integrated, people-centered innovation and sustainability-focused delivery methods [64,65].

Similarly, the evolution of selection criteria for project delivery methods was divided into four stages: selection criteria 1.0, selection criteria 2.0, selection criteria 3.0 and selection criteria 4.0. This division was based on the changes in clients' expectations, as reported in the literature, and the evolution of project delivery methods that required different selection criteria. Mostly, this categorization goes along with the PDM categorization. As time passes and the expectations of customers in the construction industry change, the selection criteria list gets updated accordingly to match these demands. From an observational point of view, clients historically relied only on their gut feelings to select the project delivery method with no specific criteria. This is referred to as selection criteria 1.0. Furthermore, literature analysis shows that earlier studies, conducted before the 2000s, emphasized the importance of cost and economic measures to achieve customer satisfaction [66]. Based on this, the paper categorized selection criteria 2.0 as the time when the cost was the most dominant criterion in the selection of project delivery methods. However, the onset of the 21st century shifted the expectations of stakeholders, where they demanded other criteria besides transaction-focused ones such as quality, cooperation, the interaction between the different project parties, shared risks [67,68]. Moreover, literature analysis has shown that almost all of the studies done in the 2000s related to the field of project delivery methods included a multi-attribute selection criteria list that includes quality, time, cost, cost growth, schedule growth, risk, communication, owner characteristics, project type and complexity, market competitiveness and contractor's abilities. Therefore, the paper categorized this stage as selection criteria 3.0. Additionally, as time passed, clients became more aware of sustainability issues and technological advancements in construction and started demanding new dimensions such as management of environmental and related know-how on site, management of work safety, cleanliness and order on-site, as well as an innovation [69]. In fact, analysis of literature also has shown that new selection criteria such as sustainability and technological innovations have been only getting more attention in research since 2006 onwards, where only 12 papers were reported from the literature regarding these selection criteria items in this study. That is why this paper categorized this phase as selection criteria 4.0 to highlight the new selection criteria that need to be investigated further in literature and added to the multi-attribute criteria list of selection 3.0, such as sustainability, health and wellbeing as well as advanced technological innovations.

The evolution of the selection methods of the project delivery methods was analyzed in a similar way, and four categories were identified: selection methods 1.0, selection methods 2.0, selection methods 3.0 and selection methods 4.0. This division was based on the fact that as time passes and more technological advancements invade the construction industry, more advancement and sophistication is also witnessed in the field of selection methods development. In this paper, selection methods 1.0 represents the period when there was no structured decision-making tool, and it was solely based on gut feelings. While selection methods 2.0 is used to represent the emergence of simple scoring charts and basic weighted sum approaches to choose the project delivery method. Over time, more complicated and sophisticated selection methods were introduced into literature such as AHP, ANP, MUAT and web-based methods, as well as knowledge-based and risk-based approaches. This paper uses the term selection methods 3.0 to define this stage. Moreover, the increase in digital transformation in the construction industry has led to exploring more artificial intelligence techniques to develop selection models such as Analytical Neural Network (ANN) and fuzzy logic approaches [9]. This paper categorizes this stage as selection methods 4.0.

4. Results

4.1. PDM 4.0

Table 1 below shows the synthesized literature collected for the two stages of PDM 3.0 and PDM 4.0. Where PDM 3.0 consists of DB, CM and CMR, PDM 4.0 includes integrated project delivery (IPD), alliances, partnerships and lean project delivery.

Table 1. Overview of people-centered innovations and mass production (PDM 3.0) and PDM 4.0.

Stage	PDM	Research Type	Sources
PDM 3.0	Design build	Conceptual empirical	[13,70–95] [3,34,96–144]
	CMR	Conceptual empirical	[70–72,74,76,79,93,145–147] [22,23,30,34,97,109,110,113,117–119,122,127,148–152]
	CM	Conceptual empirical	[28,73,77,111,153,154] [48,96,99,101,108,111,112,115,119,123–125,133,138,139,155–158]
PDM 4.0	IPD	Conceptual empirical	[6,39,70,146,159–161] [27,30,31,34,35,102,111,119,150,162–173]
	Alliancing	Conceptual empirical	[6,39,174,175] [176–181]
	Partnerships	Conceptual empirical	[6,39,175,182–185] [107,117,186–189]
	Lean project delivery	Conceptual empirical	[190–194] [38,157,162,195–200]

Figures 4 and 5 below illustrate the evolution of project delivery methods and features of PDM 4.0, respectively. PDM 1.0 represents the period pre-1850s, where the master builder was the most dominant delivery method as there were no specialized disciplines [63]. Moreover, the PDM 2.0 stage is highlighted mainly by design–bid–build. Furthermore, PDM 3.0 represents alternative delivery methods, such as DB and CM. The last phase, PDM 4.0, represents collaborative project delivery methods. Last, but not least, the main features of PDM 4.0 include mass-production, digital integration, collaboration and integrated delivery methods, as well as a focus on sustainability.

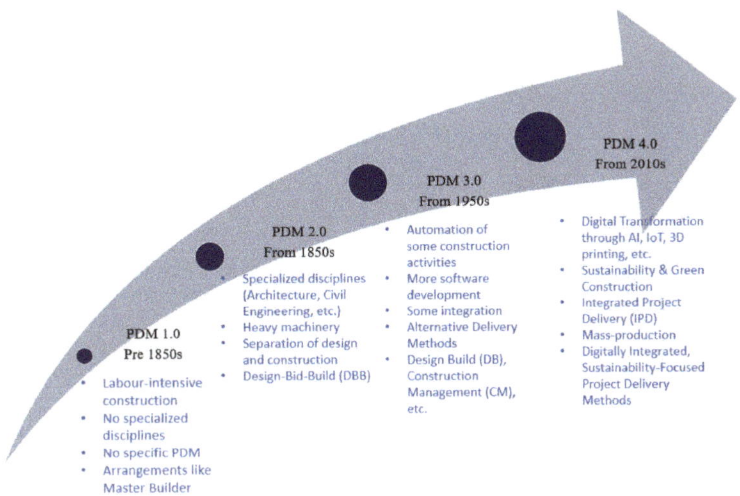

Figure 4. Evolution of project delivery methods.

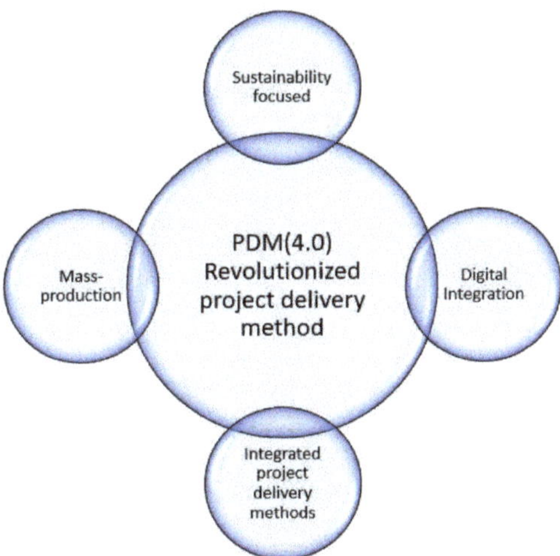

Figure 5. Features of PDM 4.0.

4.2. Selection Criteria 4.0

Post completion of the critical review analysis of the evolution of project delivery methods' selection criteria, they were categorized into 4 phases. The first phase is referred to as selection criteria 1.0, where managers selected the delivery method based on their gut feelings with no specified factors. The second stage is called selection criteria 2.0, where cost was the most dominant success factor. Followed by selection criteria 3.0, where a multi-attribute criteria list was developed that included quality, time, cost, cost growth, schedule growth, risk, communication, owner characteristics, project type and complexity, market competitiveness and contractor's abilities. The last phase, selection criteria 4.0, includes the multi-attribute criteria list from selection criteria 3.0 with the addition of new selection criteria such as sustainability, advanced technological innovations as well as health and wellbeing. The evolution of selection criteria is illustrated in Figure 6 below. Table 2 shows an overview of a selection of criteria for project delivery methods synthesized from literature analysis.

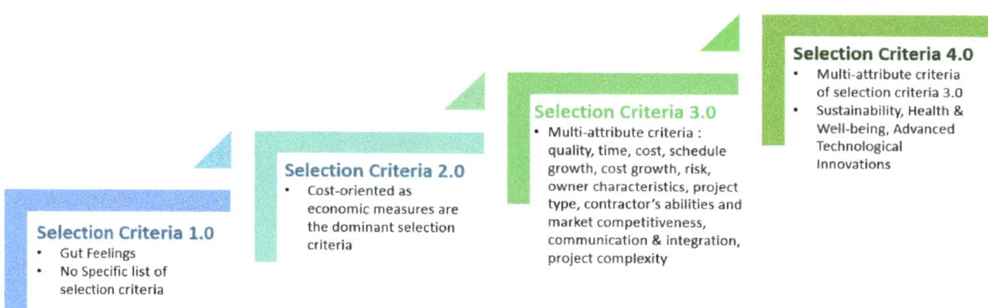

Figure 6. Evolution of selection criteria.

Table 2. Overview of selection criteria.

Criteria	Sources	# of Citations
Quality	[41,46,55,143,178,201–207]	12
Owner involvement	[46,57,157,202,206,208–210]	8
Time/delivery speed	[40,55,207,210,211]	5
Project cost	[55,203,206,210,212–214]	7
Cost growth	[1,46,57,203,206,215,216]	7
Project type	[40,41,80,89,124,207,210]	7
Project manager's characteristics	[41,46,59,124,125,217–221]	10
Schedule growth	[1,40,46,54,57,59,124,157,203,206,207,216]	12
Market competitiveness	[59,204,205,209,222–224]	7
Contractor's abilities	[46,204,206,225–227]	6
Sustainability goals	[70,167,202,206,228–230]	7
Technological innovations	[223,224,231–233]	5
Risk	[1,46,57,112,202–204,206,210,234–238]	14
Complexity	[46,57,73,202,204,206,207,209,224]	9
Communication	[163,239]	2

4.3. Selection Methods 4.0

Post completion of the literature review analysis, the evolution of the selection methods can be categorized into four stages. The first stage, referred to as selection methods 1.0, represents no structured method where the delivery method was selected based on gut feelings. While selection methods 2.0 include simple scoring charts and a basic weighted sum approach. Moreover, selection methods 3.0 represent multi-attribute approaches such as AHP, ANP, MAUT and knowledge as well as risk-based approaches. The last stage, which is selection methods 4.0, represents AI approaches such as ANN, fuzzy logic and smart decision models. Figure 7 illustrates the evolution of project delivery selection methods. Table 3 represents an overview of project delivery selection methods.

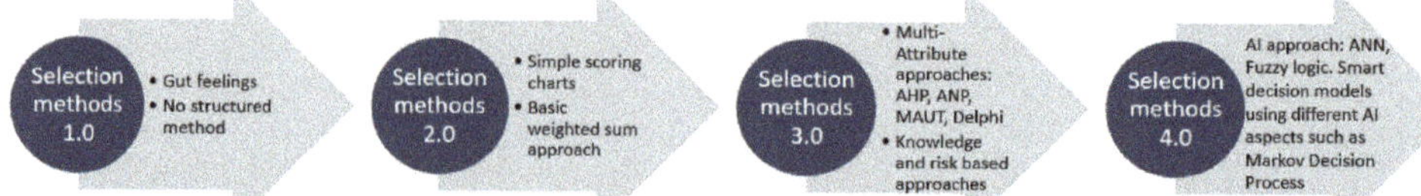

Figure 7. Evolution of project delivery selection methods.

Table 3. Overview of project delivery selection methods.

No	Method	Source	Total
1	Weighted sum approach	[55,240,241]	3
2	AHP	[41,46,54,57,197,201,217]	7
3	ANP	[60,157,242,243]	4
4	Multi-attribute decision models	[1,54,56,205,219,240,244,245]	8
5	Fuzzy approach	[197,232,233,246–253]	11
6	Simulation decision models	[224,254–256]	4
7	ANN	[257–260]	4
8	Web-based approach	[61,202,261]	3
9	Case-based reasoning	[218,231,232,262–264]	6
10	Risk-based approach	[265,266]	2

5. Discussion

The construction industry has witnessed many changes over the past years that have led to the formation of the modern construction industry. The main features of this modern form include the digital transformation where the use of drones, laser technologies, 3D printing and artificial intelligence have overwhelmed the construction processes [9,267]. Furthermore, the use of the Internet of things (IoT) and radio frequency identification (RFID) has created a smart construction site where effective tracking of equipment and tools has been enabled through automation of the construction process [268]. Additionally, the simulation of the complex nature of construction project works has been made possible through BIM along with virtual reality and 3D printing [269]. Moreover, prefabrication is another process change that has had a huge impact on the transformation of the construction industry and led to an efficient implementation of waste reduction management strategies [28,270]. Not only this but, apart from digital transformation, sustainability also has been another major change that transformed the construction industry. With the use of green building technologies and green procurement to integrate environmental aspects into the whole building supply chain, the enhanced environmental performance of the building industry has been made possible [271]. Yet, with all this sophistication in the construction industry, professionals are still not utilizing these capabilities to their full potential. This could be attributed to the fact that most clients are still using traditional methods, and construction professionals are not efficiently updating the conventional project management practices at an appropriate rate in order to embrace the changes that these technological advancements and greening practices have brought into the sector [7,8].

This paper analyzed the evolution of project delivery methods and divided them into four stages: PDM 1.0, which is the phase of master builder with no specialized designs, PDM 2.0, which is DBB, PDM 3.0, which represents the alternative delivery methods such as DB and CMR and PDM 4.0 which represents collaborative delivery methods such as alliances, partnerships, lean and IPD. Although the delivery methods have evolved over the years to keep up with the changes in the construction industry, there is still a lag between the rate at which the construction industry is changing, and the rate at which project management practices are being updated as features such as sustainability, digital integration and mass production that have already changed the construction industry are yet to be incorporated in project delivery methods. Furthermore, the paper listed the features of PDM 4.0 that would match the demands of the modern construction industry. These features include mass production, digital integration, people-centered innovation and integrated project delivery methods with a focus on sustainability.

Similarly, the paper analyzed the evolution of the selection criteria of the project delivery methods in relation to how the customer expectations and demands in the construction industry have changed over time. The results presented four stages of selection criteria. Selection criteria 1.0 represent the stage where there were no specific criteria, and delivery methods were chosen based on gut feelings. The second stage, which is selection 2.0, represents the stage where customer's demands were mainly transaction-focused [66]. While the third stage, which is selection criteria 3.0, represents the addition of multi-attribute criteria to the original list that only contained economic measures as customers started demanding more criteria such as quality, cooperation, interaction and shared risks [67,68]. The last stage, which is selection criteria 4.0, includes both the multi-attribute criteria from selection criteria 3.0 and the addition of other criteria such as sustainability, health and wellbeing as well as technological innovations in order to match the demands of the customers in the modern construction industry of today [69]

Last, but not least, the paper analyzed the evolution of selection methods and presented selection methods 4.0, which deals with more exploration of AI techniques. In fact, a potential smart decision model that may deem feasible is the use of the Markov decision process (MDP). MDP is an optimization decision-making tool where the output depends on the input provided by the user. This decision method has been applied to construction site management in Cameroon and has proven to be very successful [272]. It, therefore, has the feasibility potential to be applied as a decision support tool for project delivery selection that may enable time cost tradeoffs or account for project interdependencies.

All in all, a framework was developed in order to illustrate the relationship between the PDM variables. The framework shows that selection methods 4.0 that represent Artificial Intelligence (AI) approaches and smart decision models will incorporate the selection criteria 4.0, which includes the multi-attribute criteria from selection criteria 3.0 and the new selection criteria such as sustainability, health and wellbeing as well as advanced technological innovations. These will then be used to choose an optimal delivery method that will consist of features such as sustainability focus, digital integration, people-centered innovations and mass production (PDM 4.0). Figure 8 below illustrates the framework.

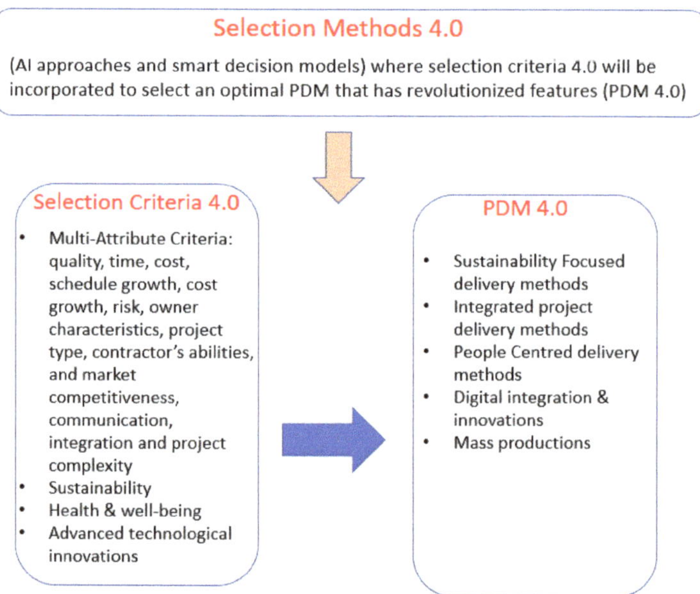

Figure 8. Framework of the relationship between PDM variables.

6. Concluding Remarks and Recommendations

6.1. Concluding Remarks

Research in the area of construction project delivery methods is very rich, as shown in the high number of references cited below. This paper represents a comprehensive literature review related to the evolution of project delivery methods, selection criteria and selection models in the construction industry. The paper discussed and evaluated the different project delivery methods available in construction. Furthermore, the paper also highlighted new selection criteria for the selection of project delivery methods. This covers an important literature gap and offers new directions of research that focuses on the transition required in traditional project delivery methods, selection criteria and selection models to meet the demands of the modern construction industry. Based on the reviewed literature, the main conclusion are as follows:

- Despite the major changes in the selection criteria and models of project delivery methods over the years, there is still a profound lag between the rate of the evolution of the construction industry and the rate at which project delivery methods, selection criteria and selection models are being updated which creates a critical gap that needs to be bridged;
- PDM 4.0 represents features of a project delivery method which is characterized by digitally integrated and sustainably focused project delivery methods to meet the demands of the construction industry;
- Selection criteria 4.0 consists of new success factors such as sustainability goals, advanced technological innovations, health and wellbeing to be added to the success factors list in order to satisfy the needs of the construction industry;
- Selection methods 4.0 features smart decision models that exploit different and advanced aspects of artificial intelligence to fulfill the requirements of the digitally transformed construction industry and meet limitations such as projects interdependencies and time–cost tradeoffs.

The construction industry is definitely approaching an evolutionary era where traditional project delivery methods, selection criteria and methods will no longer be able to compete in the modern industry of today. Several changes need to be updated in these management practices to guarantee the success of future construction projects. Indeed, with the use of PDM 4.0, selection criteria 4.0 and selection methods 4.0, the delivery of construction projects is bound to improve and will harmonize with the characteristics of the construction industry.

6.2. Recommendations and Future Research

The effort to update project delivery management practices to deal with the ever-changing construction industry is growing. However, the rate at which this is happening is very slow compared to the rate at which the industry is evolving. The biggest changes that the construction industry has been facing are by far related to sustainability and digital transformation. Most of the research done in these two areas regarding project delivery methods is still in its infancy stage and still has a long way to reach its mature stage. To overcome some of the challenges brought upon by the evolution of the construction industry, more research is needed to measure the effectiveness of different delivery methods in achieving sustainability goals. Another direction is to investigate the role of technological innovations in developing more sophisticated delivery methods, which are digitally integrated and sustainability-focused.

Some of the project success challenges in the construction industry could be overcome by frequently updating and revising the list of selection criteria used to choose the most appropriate delivery method. For example, including sustainability goals, health and wellbeing as well as advanced technological innovations. In fact, Governmental entities and professional organizations should establish codes and regulations to ensure that project delivery methods are selected based on the new selection criteria added to the traditional list. Furthermore, construction professionals play a crucial role in the implementation of safety management protocols as well as sustainable measures when selecting their project delivery method.

Apart from the selection criteria list, there is a need for construction and project management innovations to update the decision support models that owners use to select the delivery method. A potential research idea would be the exploitation of different and advanced artificial intelligence techniques to establish smart decision models that will assist project managers in choosing the most appropriate delivery method. Indeed, major stakeholders need to work together to study the challenges, integration aspects and training skills required to be able to utilize such technology. If deemed feasible, this can open the gate to a major new level of effectiveness in the project delivery selection process.

Author Contributions: Conceptualization, S.A. and S.E.-S.; methodology, S.A. and S.E.-S.; formal analysis, S.A. and S.E.-S.; investigation, S.A. and S.E.-S.; writing—original draft preparation, S.A.; writing—review and editing, S.E.-S.; visualization, S.A.; supervision, S.E.-S.; project administration, S.E.-S.; funding acquisition, S.E.-S. All authors have read and agreed to the published version of the manuscript.

Funding: The work in this paper was supported, in part, by faculty research grant number [EFRG18-SCR-CEN-42] from the American University of Sharjah. The APC is funded by grant number [EFRG18-SCR-CEN-42]. This paper represents the opinions of the authors and does not mean to represent the position or opinions of the American University of Sharjah.

Data Availability Statement: Not Applicable.

Conflicts of Interest: The authors declare no conflict of interest. The funders had no role in the design of the study; in the collection, analyses, or interpretation of data; in the writing of the manuscript, or in the decision to publish the results.

References

1. Oyetunji, A.; Anderson, D. Relative Effectiveness of Project Delivery and Contract Strategies. *J. Constr. Eng. Manag.* **2006**, *132*, 3–13. [CrossRef]
2. El-Sayegh, S. Evaluating the effectiveness of project delivery methods. *J. Constr. Manag. Econ.* **2008**, *23*, 457–465.
3. Hale, R.; Shrestha, P.; Gibson, G.; Migliaccio, C. Empirical Comparison of Design/Build and Design/Bid/Build Project Delivery Methods. *J. Constr. Eng. Manag.* **2009**, *135*, 579–587. [CrossRef]
4. Tenah, A. Project delivery systems for construction: An overview. *Cost Eng* **2001**, *43*, 30.
5. Akpan, E.O.P.; Amade, B.; Okangba, B.; Ekweozor, C.O. Constructability practice and project delivery processes in the Nigerian construction industry. *J. Build. Perform.* **2014**, *5*, 10–21.
6. Engebø, A.; Lædre, O.; Young, B.; Larssen, P.F.; Lohne, J.; Klakegg, O.J. Collaborative project delivery methods: A scoping review. *J. Civ. Eng. Manag.* **2020**, *26*, 278–303. [CrossRef]
7. Pishdad, P.B.; Beliveau, Y.J. Analysis of existing project delivery and contracting strategy (PDCS) selection tools with a look towards emerging technology. In Proceedings of the 46th Annual International Associated school of Construction (AsC), Boston, MA, USA, 4–7 April 2010.
8. Robichaud, B.; Anantatmula, S. Greening Project Management Practices for Sustainable Construction. *J. Manag. Eng.* **2011**, *27*, 48–57. [CrossRef]
9. Osunsanmi, T.; Aigbavboa, C.; Oke, A. Construction 4.0: The future of the construction industry in South Africa. *World Acad. Sci. Eng. Technol. Int. J. Civ. Env. Eng.* **2018**, *12*, 206–212.
10. Craveiroa, F.; Duartec, J.P.; Bartoloa, H.; Bartolod, P.J. Additive manufacturing as an enabling technology for digital construction: A perspective on Construction 4.0. *Sustain. Dev.* **2019**, *12*, 150–156. [CrossRef]
11. Hwang, B.-G.; Ng, W.J. Project management knowledge and skills for green construction: Overcoming challenges. *Int. J. Proj. Manag.* **2013**, *31*, 272–284. [CrossRef]
12. Sims, B.L.; Anderson, W. Meeting Customer Expectations in the Construction Industry. *Lead. Manag. Eng.* **2001**, *1*, 29–32. [CrossRef]
13. Friedlander, C. FEATURE: Design/Build Solutions. *J. Manag. Eng.* **1998**, *14*, 59–64. [CrossRef]
14. Azhar, N.; Kang, Y.; Ahmad, I.U. Factors Influencing Integrated Project Delivery in Publicly Owned Construction Projects: An Information Modelling Perspective. *Procedia Eng.* **2014**, *77*, 213–221. [CrossRef]
15. Griffiths, F. Project contract strategy for 1992 and beyond. *Int. J. Proj. Manag* **1989**, *7*, 69–83. [CrossRef]
16. Teicholz, P.M.; Ashley, D.B. Optimal bid prices for unit price contract. *J. Constr. Div.* **1978**, *104*, 57–67.
17. Gransberg, D.D.; Koch, J.A.; Molenaar, K.R. *Preparing for Design-Build Projects: A Primer for Owners, Engineers, and Contractors*; ASCE: Reston, VA, USA, 2006.
18. Okere, G. Comparison of DB to DBB on highway projects in Washington State, USA. *J. Constr. Supply Chain Manag.* **2018**, *8*, 73–86. [CrossRef]
19. Xia, B.; Chan, P.A.; Yeung, F.J. Developing a Fuzzy Multicriteria Decision-Making Model for Selecting Design-Build Operational Variations. *J. Constr. Eng. Manag.* **2011**, *137*, 1176–1184. [CrossRef]
20. Algarni, M.; Arditi, D.; Polat, G. Build-Operate-Transfer in Infrastructure Projects in the United States. *J. Constr. Eng. Manag.* **2007**, *133*, 728–735. [CrossRef]

21. Gould, F.E. *Managing the Construction Process: Estimating, Scheduling, and Project Control*; Prentice Hall: Boston, MA, USA, 2012. (In English)
22. Rojas, M.; Kell, I. Comparative Analysis of Project Delivery Systems Cost Performance in Pacific Northwest Public Schools. *J. Constr. Eng. Manag.* **2008**, *134*, 387–397. [CrossRef]
23. Carpenter, N.; Bausman, C. Project Delivery Method Performance for Public School Construction: Design-Bid-Build versus CM at Risk. *J. Constr. Eng. Manag.* **2016**, *142*, 05016009. [CrossRef]
24. Azari-Najafabadi, R.; Ballard, G.; Cho, S.; Kim, Y.W. A Dream of Ideal Project Delivery System. In Proceedings of the AEI 2011: Building Integration Solutions, Oakland, CA, USA, 30 March–2 April 2012; pp. 427–436. [CrossRef]
25. Al Mousli, M.H.; El-Sayegh, S. Assessment of the design–construction interface problems in the UAE. *Arch. Eng. Des. Manag.* **2016**, *12*, 353–366. [CrossRef]
26. Eckblad, S.; Ashcraft, H.; Audsley, P.; Blieman, D.; Bedrick, J.; Brewis, C.; Stephens, N.D. *Integrated Project Delivery—A Working Definition*; AIA California Council: Sacramento, CA, USA, 2007.
27. Azhar, N.; Kang, Y.; Ahmad, I. Critical Look into the Relationship between Information and Communication Technology and Integrated Project Delivery in Public Sector Construction. *J. Manag. Eng.* **2015**, *31*, 04014091. [CrossRef]
28. Tatum, C.B. Issues in professional construction management. *J. Constr. Eng. Manag.* **1983**, *109*, 112–119. [CrossRef]
29. Gallaher, M.; O'Connor, A.; Dettbarn, J.; Gilday, L. *Cost Analysis of Inadequate Interoperability in the US Capital Facilities Industry (NIST GCR 04-867)*; National Institute of Standards and Technology: Gaithersburg, MD, USA, 2004.
30. Choi, J.; Yun, S.; Leite, F.; Mulva, S. Team Integration and Owner Satisfaction: Comparing Integrated Project Delivery with Construction Management at Risk in Health Care Projects. *J. Manag. Eng.* **2019**, *35*, 05018014. [CrossRef]
31. David, C.; Becerik-Gerber, G. Understanding Construction Industry Experience and Attitudes toward Integrated Project Delivery. *J. Constr. Eng. Manag.* **2010**, *136*, 815–825. [CrossRef]
32. AbouDargham, S.; Bou Hatoum, M.; Tohme, M.; Hamzeh, F. Implementation of Integrated Project Delivery in Lebanon: Overcoming the Challenges. In Proceedings of the 27th Annual Conference of the International. Group for Lean Construction, Dublin, Ireland, 1–9 July 2019; pp. 917–928. [CrossRef]
33. Levitt, R. Towards project management 2.0. *Eng. Proj. Organ. J.* **2011**, *1*, 197–210. [CrossRef]
34. El Asmar, M.; Hanna, A.; Loh, W.-Y. Quantifying Performance for the Integrated Project Delivery System as Compared to Established Delivery Systems. *J. Constr. Eng. Manag.* **2013**, *139*, 04013012. [CrossRef]
35. Hanna, A. Benchmark Performance Metrics for Integrated Project Delivery. *J. Constr. Eng. Manag.* **2016**, *142*, 04016040. [CrossRef]
36. Sepasgozar, S.; Karimi, R.; Shirowzhan, S.; Mojtahedi, M.; Ebrahimzadeh, S.; McCarthy, D. Delay causes and emerging digital tools: A novel model of delay analysis, including integrated project delivery and PMBOK. *Buildings* **2019**, *9*, 191. [CrossRef]
37. Demetracopoulou, V.; O'Brien, W.; Khwaja, N. Lessons Learned from Selection of Project Delivery Methods in Highway Projects: The Texas Experience. *J. Leg. Aff. Disput. Resolut. Eng. Constr.* **2020**, *12*, 04519040. [CrossRef]
38. Ghosh, S.; Robson, K.F. Analyzing the Empire State Building Project from the Perspective of Lean Delivery System—A Descriptive Case Study. *Int. J. Constr. Educ. Res* **2015**, *11*, 257–267. [CrossRef]
39. Lahdenperä, P. Making sense of the multi-party contractual arrangements of project partnering, project alliancing and integrated project delivery. *Constr. Manag. Econ.* **2012**, *30*, 57–79. [CrossRef]
40. Konchar, M.; Sanvido, V. Comparison of U.S. Project Delivery Systems. *J. Constr. Eng. Manag.* **1998**, *124*, 435–444. [CrossRef]
41. Alhazmi, T.; McCaffer, R. Project procurement system selection model. *J. Constr. Eng. Manag* **2000**, *126*, 176–184. [CrossRef]
42. Kunz, A.; Ballard, H. Global Project Delivery Systems Using BIM. In Proceedings of the AEI 2011: Building Integration Solutions, Oakland, CA, USA, 30 March–2 April 2012; pp. 472–479. [CrossRef]
43. Korkmaz, S.; Riley, D.; Horman, M. Piloting Evaluation Metrics for Sustainable High-Performance Building Project Delivery. *J. Constr. Eng. Manag* **2010**, *136*, 877–885. [CrossRef]
44. Benson, L.; Bodniewicz, B.; Vittands, J.; Carr, J.; Watson, K. Innovative design-build procurement approach for large wastewater facility. *Proc. Water Env. Fed.* **2013**, *2013*, 7253–7269. [CrossRef]
45. Giachino, J.; Cecil, M.; Husselbee, B.; Matthews, C. Alternative project delivery: Construction management at risk, design-build and public-private partnerships. *Proc. Water Environ. Fed.* **2015**, *2015*, 1–11. [CrossRef]
46. Mahdi, I.; Alreshaid, K. Decision support system for selecting the proper project delivery method using analytical hierarchy process (AHP). *Int. J. Proj. Manag* **2005**, *23*, 564–572. [CrossRef]
47. Tran, D.; Molenaar, K. Critical Risk Factors in Project Delivery Method Selection for Highway Projects. *Constr. Res. Congr.* **2012**, 331–340. [CrossRef]
48. Farnsworth, C.; Warr, R.; Weidman, J.; Mark, D. Effects of CM/GC Project Delivery on Managing Process Risk in Transportation Construction. *J. Constr. Eng. Manag.* **2016**, *142*, 04015091. [CrossRef]
49. Al-Sobiei, O.S.; Arditi, D.; Polat, G. Predicting the risk of contractor default in Saudi Arabia utilizing artificial neural network (ANN) and genetic algorithm (GA) techniques. *Constr. Manag. Econ.* **2005**, *23*, 423–430. [CrossRef]
50. Gransberg, D.; Dillon, W.; Reynolds, L.; Boyd, J. Quantitative analysis of partnered project performance. *J. Constr. Eng. Manag.* **1999**, *125*, 161–166. [CrossRef]
51. Loudoun, R.; Townsend, K. Implementing health promotion programs in the Australian construction industry. *Eng. Constr. Arch. Manag.* **2017**, *24*, 260–274. [CrossRef]

52. Hanna, E.; Markham, S. Constructing better health and wellbeing? Understanding structural constraints on promoting health and wellbeing in the UK construction industry. *Int. J. Workplace Health Manag.* **2019**, *12*, 146–159. [CrossRef]
53. Pooyan, M.-R. A Model for Selecting Project Delivery Systems in Post-Conflict Construction Projects. Master's Thesis, Concordia University, Montreal, QC, Canada, 2012.
54. Cheung, S.-O.; Lam, T.-I.; Wan, Y.-W.; Lam, K.-C. Improving Objectivity in Procurement Selection. *J. Manag. Eng.* **2001**, *17*, 132–139. [CrossRef]
55. Al-Sinan, F.M.; Hancher, D.E. Facility project delivery selection model. *J. Manag. Eng.* **1998**, *4*, 244–259. [CrossRef]
56. Skitmore, M.; Marsden, D. Which procurement system? Towards a universal procurement selection technique. *Constr. Manag. Econ.* **1988**, *6*, 71–89.
57. Al Khalil, M. Selecting the appropriate project delivery method using AHP. *Int. J. Proj. Manag.* **2002**, *20*, 469–474. [CrossRef]
58. Saaty, T.L. *Decision Making with Dependence and Feedback: The Analytic Network Process*; RWS Publ.: Pittsburgh, PA, USA, 1996.
59. Ng, S.; Luu, D.; Chen, S.; Lam, K. Fuzzy membership functions of procurement selection criteria. *Constr. Manag. Econ.* **2002**, *20*, 285–296. [CrossRef]
60. El-Abbasy, M.; Zayed, T.; Ahmed, M.; Alzraiee, H.; Abouhamad, M. Contractor Selection Model for Highway Projects Using Integrated Simulation and Analytic Network Process. *J. Constr. Eng. Manag.* **2013**, *139*, 755–767. [CrossRef]
61. Khwaja, N.; O'Brien, W.; Martinez, M.; Sankaran, B.; O'Connor, J.; Hale, W. Innovations in Project Delivery Method Selection Approach in the Texas Department of Transportation. *J. Manag. Eng.* **2018**, *34*, 05018010. [CrossRef]
62. Hartmann, A.; Love, P.E.D.; Ibbs, W.; Chih, Y. Alternative methods for choosing an appropriate project delivery system (PDS). *Facilities* **2011**, *29*, 527–541. [CrossRef]
63. Yates, J.; Battersby, L. Master Builder Project Delivery System and Designer Construction Knowledge. *J. Constr. Eng. Manag.* **2003**, *129*, 635–644. [CrossRef]
64. Whyte, J. How digital information transforms project delivery models. *Proj. Manag. J.* **2019**, *50*, 177–194. [CrossRef]
65. Paolillo, W.; Olson, B.; Straub, E. People centered innovation: Enabling lean integrated project delivery and disrupting the construction industry for a more sustainable future. *J. Constr. Eng.* **2016**, *2016*, 1–7. [CrossRef]
66. Anderson, E.W.; Fornell, C.; Lehmann, D.R. Customer satisfaction, market share, and profitability: Findings from Sweden. *J. Mark.* **1994**, *58*, 53–66. [CrossRef]
67. Homburg, C.; Rudolph, B. Customer satisfaction in industrial markets: Dimensional and multiple role issues. *J. Bus. Res.* **2001**, *52*, 15–33. [CrossRef]
68. Torbica, M.; Stroh, C. Customer Satisfaction in Home Building. *J. Constr. Eng. Manag.* **2001**, *127*, 82–86. [CrossRef]
69. Kärnä, S.; Junnonen, J.-M.; Kankainen, J. Customer satisfaction in construction. In Proceedings of the 12th Annual Conference on Lean Construction, Helsingøv, Denmark, 3–5 August 2004; pp. 476–488.
70. Raouf, A.; Al-Ghamdi, S. Effectiveness of Project Delivery Systems in Executing Green Buildings. *J. Constr. Eng. Manag.* **2019**, *145*, 03119005. [CrossRef]
71. Sullivan, J.; Asmar, E.; Chalhoub, J.; Obeid, H. Two Decades of Performance Comparisons for Design-Build, Construction Manager at Risk, and Design-Bid-Build: Quantitative Analysis of the State of Knowledge on Project Cost, Schedule, and Quality. *J. Constr. Eng. Manag.* **2017**, *143*, 04017009. [CrossRef]
72. Zuber, S.; Nawi, M.; Abdul Nifa, F.; Bahaudin, A. An overview of project delivery methods in construction industry. *Int. J. Supply Chain Manag.* **2018**, *7*, 177–182.
73. Uhlik, F.T.; Eller, M.D. Alternative delivery approaches for military medical construction projects. *J. Arch. Eng.* **1999**, *5*, 149–155. [CrossRef]
74. Steiman, H.; Hickey, T.; Callahan, N. Use and benefits of alternative capital project delivery strategies: Design-build and construction management at risk. *J. N. Engl. Water Work. Assoc.* **2010**, *124*, 7–18.
75. Park, M.; Ji, S.H.; Lee, H.S.; Kim, W. Strategies for design-build in Korea using system dynamics modeling. *J. Constr. Eng. Manag.* **2009**, *135*, 1125–1137. [CrossRef]
76. Molenaar, K.R.; Yakowenko, G. *Alternative Project Delivery, Procurement, and Contracting Methods for Highways*; American Society of Civil Engineers: Reston, VA, USA, 2007; pp. 1–155.
77. Mulvey, D.L. Project delivery trends: A contractor's assessment. *J. Manag. Eng.* **1998**, *14*, 51–54. [CrossRef]
78. Retherford, N. FEATURE: Project Delivery and the US Department of State. *J. Manag. Eng.* **1998**, *14*, 55–58. [CrossRef]
79. Kalach, M.; Abdul-Malak, M.A.; Srour, I. Architect and Engineer's Spectrum of Engagement under Alternative Delivery Methods: Agreement Negotiation and Formation Implications. *J. Leg. Aff. Disput. Resolut. Eng. Constr.* **2020**, *12*. [CrossRef]
80. Brahim, J.; Latiffi, A.; Fathi, M.S. Application of building information modelling (bim) in design and build (D&B) projects in Malaysia. *Malays. Constr. Res. J.* **2018**, *25*, 29–41.
81. Gad, G.M.; Adamtey, S.A.; Gransberg, D. Trends in quality management approaches to design-build transportation projects. *Transp. Res. Rec.* **2015**, *2504*, 87–92. [CrossRef]
82. Gransberg, D. Design/build in transportation from the research perspective. *Lead. Manag. Eng.* **2003**, *3*, 133–136. [CrossRef]
83. Connor, M.C. Legal exposure in design/build contracts. *Ashrae J.* **2003**, *45*, 49.
84. Lam, E.; Chan, A.; Chan, D. Potential problems of running design-build projects in construction. *Hkie Trans. Hong Kong Inst. Eng.* **2003**, *10*, 8–14. [CrossRef]

85. Lahdenperä, P. *Design-Build Procedures Introduction, Illustration and Comparison of U.S. Modes*; VTT Publications: Espoo, Finland, 2001; pp. 2–155.

86. Sweeney, N.J. Who pays for defective design? *J. Manag. Eng.* **1998**, *14*, 65–68. [CrossRef]

87. Friedlander, M.C.; Roberts, K.M. Single entity option. *Indep. Energy* **1997**, *27*, 28–30.

88. Kirschenman, M.D. Total project delivery systems. *J. Manag. Eng.* **1986**, *2*, 222–230. [CrossRef]

89. Kanoglu, A. An integrated system for design/build firms to solve cost estimation problems in the design phase. *Arch. Sci. Rev.* **2003**, *46*, 37–47. [CrossRef]

90. Gibson, G.E.; O'connor, J.T.; Migliaccio, G.; Walewski, J. Key implementation issues and lessons learned with design-build projects. In *Alternative Project Delivery, Procurement, and Contracting Methods for Highways*; American Society of Civil Engineers: Reston, VA, USA, 2007; pp. 1–19.

91. Thomas, L.M.; Lester, H.D. Project delivery systems: Architecture/engineering/construction industry trends and their ramifications. In *The Routledge Companion for Architecture Design and Practice: Established and Emerging Trends*; Routledge: Abingdon, UK, 2016; pp. 429–436.

92. Gard, P.T. Fast and innovative delivery of high performance building: Design-build delivers with less owner liability. *Strat. Plan. Energy Environ.* **2004**, *23*, 7–22. [CrossRef]

93. Diekmann, J.E. Past perfect: Historical antecedents of modern construction practices. *J. Constr. Eng. Manag.* **2007**, *133*, 652–660. [CrossRef]

94. Gad, G.M.; Gransberg, D.; Loulakis, M. Policies and procedures for successful implementation of alternative technical concepts. *Transp. Res. Rec.* **2015**, *2504*, 78–86. [CrossRef]

95. Kennedy, R.; Sidwell, A.C. Re-engineering the construction delivery process: The Museum of Tropical Queensland, Townsville—A Case Study. *Constr. Innov.* **2001**, *1*, 77–89. [CrossRef]

96. Papajohn, D.; El Asmar, M. Impact of Alternative Delivery on the Response Time of Requests for Information for Highway Projects. *J. Manag. Eng.* **2020**, *37*. [CrossRef]

97. Franz, B.; Molenaar, K.R.; Roberts, B.A.M. Revisiting project delivery system performance from 1998 to 2018. *J. Constr. Eng. Manag.* **2020**, *146*. [CrossRef]

98. Moon, H.; Kim, K.; Lee, H.S.; Park, M.; Williams, T.P.; Son, B.; Chun, J.Y. Cost Performance Comparison of Design-Build and Design-Bid-Build for Building and Civil Projects Using Mediation Analysis. *J. Constr. Eng. Manag.* **2020**, *146*. [CrossRef]

99. Noorzai, E. Performance Analysis of Alternative Contracting Methods for Highway Construction Projects: Case Study for Iran. *J. Infrastruct. Syst.* **2020**, *26*. [CrossRef]

100. Abou Chakra, H.; Ashi, A. Comparative analysis of design/build and design/bid/build project delivery systems in Lebanon. *J. Ind. Eng. Int.* **2019**, *15*, 147–152. [CrossRef]

101. Arthur, A.; Alleman, D.; Molenaar, K. Examination of Project Duration, Project Intensity, and Timing of Cost Certainty in Highway Project Delivery Methods. *J. Manag. Eng.* **2019**, *35*, 04018049. [CrossRef]

102. Adamtey, S.A. A Case Study Performance Analysis of Design-Build and Integrated Project Delivery Methods. *Int. J. Constr. Educ. Res.* **2019**, 1–17. [CrossRef]

103. Tran, D.; Diraviam, G.; Minchin, R.E. Performance of Highway Design-Bid-Build and Design-Build Projects by Work Types. *J. Constr. Eng. Manag.* **2018**, *144*, 04017112. [CrossRef]

104. Shrestha, P.; Fernane, J. Performance of Design-Build and Design-Bid-Build Projects for Public Universities. *J. Constr. Eng. Manag.* **2017**, *143*, 04016101. [CrossRef]

105. Park, H.-S.; Lee, D.; Kim, S.; Kim, J.-L. Comparing project performance of design-build and design-bid-build methods for large-sized public apartment housing projects in Korea. *J. Asian Arch. Build. Eng.* **2015**, *14*, 323–330. [CrossRef]

106. Shrestha, P.; O'Connor, J.; Gibson, G.E. Performance Comparison of Large Design-Build and Design-Bid-Build Highway Projects. *J. Constr. Eng. Manag.* **2012**, *138*, 1–13. [CrossRef]

107. Chasey, A.D.; Maddex, W.E.; Bansal, A. Comparison of public-private partnerships and traditional procurement methods in North American highway construction. *Transp. Res. Rec.* **2012**, *2268*, 26–32. [CrossRef]

108. Gransberg, D.; Molenaar, K. Critical Comparison of Progressive Design-Build and Construction Manager/General Contractor Project Delivery Methods. *Transp. Res. Rec.* **2019**, *2673*, 261–268. [CrossRef]

109. Feghaly, J.; El Asmar, M.; Ariaratnam, S.T. State of Professional Practice for Water Infrastructure Project Delivery. *Pr. Period. Struct. Des. Constr.* **2020**, *25*. [CrossRef]

110. Mehany, M.; Bashettiyavar, G.; Esmaeili, B.; Gad, G. Claims and Project Performance between Traditional and Alternative Project Delivery Methods. *J. Leg. Aff. Disput. Resolut. Eng. Constr.* **2018**, *10*. [CrossRef]

111. Ibrahim, M.W.; Hanna, A.; Kievet, D. Quantitative Comparison of Project Performance between Project Delivery Systems. *J. Manag. Eng.* **2020**, *36*. [CrossRef]

112. Bypaneni, S.P.K.; Tran, D.Q. Empirical Identification and Evaluation of Risk in Highway Project Delivery Methods. *J. Manag. Eng.* **2018**, *34*. [CrossRef]

113. Hasanzadeh, S.; Esmaeili, B.; Nasrollahi, S.; Gad, G.M.; Gransberg, D. Impact of Owners' Early Decisions on Project Performance and Dispute Occurrence in Public Highway Projects. *J. Leg. Aff. Disput. Resolut. Eng. Constr.* **2018**, *10*. [CrossRef]

114. McWhirt, D.; Ahn, J.; Shane, J.S.; Strong, K.C. Military construction projects: Comparison of project delivery methods. *J. Facil. Manag.* **2011**, *9*, 157–169. [CrossRef]

115. Koppinen, T.; Lahdenperä, P. Realized Economic Efficiency of Road Project Delivery Systems. *J. Infrastruct. Syst.* **2007**, *13*, 321–329. [CrossRef]
116. Lahdenperä, P.; Koppinen, T. Financial analysis of road project delivery systems. *J. Financ. Manag. Prop. Constr.* **2009**, *14*, 61–78. [CrossRef]
117. Ghavamifar, K.; Touran, A. Alternative Project Delivery Systems: Applications and Legal Limits in Transportation Projects. *J. Prof. Issues Eng. Educ. Pr.* **2008**, *134*, 106–111. [CrossRef]
118. Bingham, E.; Gibson, G.; El Asmar, M. Identifying Team Selection and Alignment Factors by Delivery Method for Transportation Projects. *J. Constr. Eng. Manag.* **2019**, 145. [CrossRef]
119. Francom, T.; Ariaratnam, S.T.; El Asmar, M. Industry perceptions of alternative project delivery methods applied to trenchless pipeline projects. *J. Pipeline Syst. Eng. Pr.* **2016**, 7. [CrossRef]
120. Chini, A.; Ptschelinzew, L.; Minchin, R.E.; Zhang, Y.; Shah, D. Industry Attitudes toward Alternative Contracting for Highway Construction in Florida. *J. Manag. Eng.* **2018**, 34. [CrossRef]
121. Ernzen, J.; Schexnayder, C.; Flora, G. Design-Build Effects on a Construction Company: A Case Study. *Transp. Res. Rec.* **1999**, *1654*, 181–187. [CrossRef]
122. Shrestha, P.; Maharajan, R.; Batista, J.R.; Shakya, B. Comparison of Utility Managers' and Project Managers' Satisfaction Rating of Alternative Project Delivery Methods Used in Water and Wastewater Infrastructures. *Public Work. Manag. Policy* **2015**, *21*, 263–279. [CrossRef]
123. Koppinen, T.; Lahdenperä, P. *Road Sector Experiences on Project Delivery Methods*; VTT Tiedotteita—Valtion Teknillinen Tutkimuskeskus; VTT: Espoo, Finland, 2004; pp. 3–216.
124. Koppinen, T.; Lahdenperä, P. *The Current and Future Performance of Road Project Delivery Methods*; VTT Publications: Espoo, Finland, 2004.
125. Bilec, M.; Ries, R. Preliminary Study of Green Design and Project Delivery Methods in the Public Sector. *J. Green Build.* **2007**, *2*, 151–160. [CrossRef]
126. Molenaar, K.R.; Bogus, S.M.; Priestley, J.M. Design/build for water/wastewater facilities: State of the industry survey and three case studies. *J. Manag. Eng.* **2004**, *20*, 16–24. [CrossRef]
127. Sindhu, J.; Choi, K.; Lavy, S.; Rybkowski, Z.K.; Bigelow, B.F.; Li, W. Effects of Front-End Planning under Fast-Tracked Project Delivery Systems for Industrial Projects. *Int. J. Constr. Educ. Res.* **2018**, *14*, 163–178. [CrossRef]
128. Signore, A. Design/build project delivery method: Strategic opportunities for pharmaceutical facilities. *Pharm. Eng.* **1998**, *18*, 84–90.
129. Feghaly, J.; El Asmar, M.; Ariaratnam, S.; Bearup, W. Design-Build Project Administration Practices for the Water Industry. *J. Pipeline Syst. Eng. Pr.* **2020**, 12. [CrossRef]
130. Tran, D.Q.; Nguyen, L.D.; Faught, A. Examination of communication processes in design-build project delivery in building construction. *Eng. Constr. Arch. Manag.* **2017**, *24*, 1319–1336. [CrossRef]
131. Vashani, H.; Sullivan, J.; El Asmar, M. DB 2020: Analyzing and forecasting design-build market trends. *J. Constr. Eng. Manag.* **2016**, 142. [CrossRef]
132. Ramsey, D.; El Asmar, M.; Gibson, G.E. Quantitative performance assessment of single-step versus two-step design-build procurement. *J. Constr. Eng. Manag.* **2016**, 142. [CrossRef]
133. Antoine, A.L.C.; Molenaar, K.R. Empirical study of the state of the practice in alternative technical concepts in highway construction projects. *Transp. Res. Rec.* **2016**, *2573*, 143–148. [CrossRef]
134. Chang, A.S.; Shen, F.Y.; Ibbs, W. Design and construction coordination problems and planning for design-build project new users. *Can. J. Civ. Eng.* **2010**, *37*, 1525–1534. [CrossRef]
135. El Asmar, M.; Lotfallah, W.; Whited, G.; Hanna, A.S. Quantitative methods for design-build team selection. *J. Constr. Eng. Manag.* **2010**, *136*, 904–912. [CrossRef]
136. Gransberg, D.; Senadheera, S.P. Design-build contract award methods for transportation projects. *J. Transp. Eng.* **1999**, *125*, 565–567. [CrossRef]
137. Flora, G.; Ernzen, J.J.; Schexnayder, C. Field-level management's perspective of design/build. *Pr. Period. Struct. Des. Constr.* **1998**, *3*, 180–185. [CrossRef]
138. Tran, D.Q.; Harper, C.M.; Smadi, A.M.; Mohamed, M. Staffing needs and utilization for alternative contracting methods in highway design and construction. *Eng. Constr. Arch. Manag.* **2020**, *27*, 2163–2178. [CrossRef]
139. Tran, D.Q.; Molenaar, K.R.; Alarcön, L.F. A hybrid cross-impact approach to predicting cost variance of project delivery decisions for highways. *J. Infrastruct. Syst.* **2016**, 22. [CrossRef]
140. Lam, E.W.M.; Chan, A.P.C.; Chan, D.W.M. Development of the design-build procurement system in Hong Kong. *Arch. Sci. Rev.* **2004**, *47*, 387–397. [CrossRef]
141. Arditi, D.; Lee, D.E. Assessing the corporate service quality performance of design-build contractors using quality function deployment. *Constr. Manag. Econ.* **2003**, *21*, 175–185. [CrossRef]
142. Ling, F.Y.Y.; Chan, S.L.; Chong, E.; Ee, L.P. Predicting performance of design-build and design-bid-build projects. *J. Constr. Eng. Manag.* **2004**, *130*, 75–83. [CrossRef]
143. Lee, D.E.; Arditi, D. Total quality performance of design/build firms using quality function deployment. *J. Constr. Eng. Manag.* **2006**, *132*, 49–57. [CrossRef]

144. Liang, L.Y. Grouping decomposition under constraints for design/build life cycle in project delivery system. *Int. J. Technol. Manag.* **2009**, *48*, 168–187. [CrossRef]
145. Kaplin, J.; Conley, J. Construction management-at-risk as a delivery method for water projects. *J. N. Engl. Water Work. Assoc.* **2010**, *124*, 219–226.
146. Kantola, M.; Saari, A. Project delivery systems for nZEB projects. *Facility* **2016**, *34*, 85–100. [CrossRef]
147. Shane, J.S.; Gransberg, D. Coordination of design contract with construction manager-at-risk preconstruction service contract. *Transp. Res. Rec.* **2010**, *2151*, 55–59. [CrossRef]
148. Francom, T.; El Asmar, M.; Ariaratnam, S.T. Longitudinal study of construction manager at risk for pipeline rehabilitation. *J. Pipeline Syst. Eng. Pr.* **2017**, *8*. [CrossRef]
149. Francom, T.; El Asmar, M.; Ariaratnam, S.T. Performance Analysis of Construction Manager at Risk on Pipeline Engineering and Construction Projects. *J. Manag. Eng.* **2016**, *32*. [CrossRef]
150. Bilbo, D.; Bigelow, B.; Escamilla, E.; Lockwood, C. Comparison of Construction Manager at Risk and Integrated Project Delivery Performance on Healthcare Projects: A Comparative Case Study. *Int. J. Constr. Educ. Res.* **2015**, *11*, 40–53. [CrossRef]
151. Shrestha, P.; Davis, B.; Gad, G.M. Investigation of Legal Issues in Construction-Manager-at-Risk Projects: Case Study of Airport Projects. *J. Leg. Aff. Disput. Resolut. Eng. Constr.* **2020**, *12*. [CrossRef]
152. Minchin, R.E.; Thakkar, K.; Ellis, R.D. Miami intermodal center-introducing "CM-at-Risk" to transportation construction. In *Alternative Project Delivery, Procurement, and Contracting Methods for Highways*; American Society of Civil Engineers: Reston, VA, USA, 2007; pp. 46–59.
153. West, N.; Gransberg, D.; McMinimee, J. Effective tools for projects delivered by construction manager-general contractor method. *Transp. Res. Rec.* **2012**, *2268*, 33–39. [CrossRef]
154. Kluenker, C.H. Construction manager as project integrator. *J. Manag. Eng.* **1996**, *12*, 17–20. [CrossRef]
155. Diab, M.; Gebken, R.J.; Mehany, M.S.H. Strategies to Leverage Contractor Experience. *J. Leg. Aff. Disput. Resolut. Eng. Constr.* **2020**, *12*. [CrossRef]
156. Gransberg, D. Comparing construction manager-general contractor and federal early contractor involvement project delivery methods. *Transp. Res. Rec.* **2016**, *2573*, 18–25. [CrossRef]
157. Khalafallah, A.; Fahim, A. Project delivery systems for healthcare projects: To lean or not to lean. *Lean Constr. J.* **2018**, *2018*, 47–62.
158. Kartam, N.A.; Al-Daihani, T.G.; Al-Bahar, J.F. Professional project management practices in Kuwait: Issues, difficulties and recommendations. *Int. J. Proj. Manag.* **2000**, *18*, 281–296. [CrossRef]
159. Zuber, S.Z.S.; Nawi, N.M.; Nifa, F.A.A. Construction procurement practice: A review study of integrated project delivery (IPD) in the Malaysian construction projects. *Int. J. Suppl. Chain Manag.* **2019**, *8*, 777–783.
160. Boon, L.H.; Saar, C.C.; Lau, S.E.N.; Aminudin, E.; Zakaria, R.; Hamid, A.R.A.; Sarbini, N.N.; Zin, R.M. Building information modelling integrated project delivery system in Malaysia. *Malays. Constr. Res. J.* **2019**, *6*, 144–152.
161. Widjaja, K. Sustainable design in project delivery: A discussion on current and future trends. *J. Green Build.* **2016**, *11*, 39–56. [CrossRef]
162. Mesa, H.A.; Molenaar, K.R.; Alarcón, L.F. Comparative analysis between integrated project delivery and lean project delivery. *Int. J. Proj. Manag.* **2019**, *37*, 395–409. [CrossRef]
163. Mesa, H.A.; Molenaar, K.R.; Alarcón, L.F. Exploring performance of the integrated project delivery process on complex building projects. *Int. J. Proj. Manag.* **2016**, *34*, 1089–1101. [CrossRef]
164. Ling, F.Y.Y.; Teo, P.X.; Li, S.; Zhang, Z.; Ma, Q. Adoption of Integrated Project Delivery Practices for Superior Project Performance. *J. Leg. Aff. Disput. Resolut. Eng. Constr.* **2020**, *12*. [CrossRef]
165. Engebø, A.; Klakegg, O.J.; Lohne, J.; Lædre, O. A collaborative project delivery method for design of a high-performance building. *Int. J. Manag. Proj. Bus.* **2020**, *13*, 1141–1165. [CrossRef]
166. Singleton, M.S.; Hamzeh, F.R. Implementing integrated project delivery on department of the navy construction projects. *Lean Constr. J.* **2011**, *2011*, 17–31.
167. Rosayuru, H.; Waidyasekara, K.; Wijewickrama, M. Sustainable BIM based integrated project delivery system for construction industry in Sri Lanka. *Int. J. Constr. Manag.* **2019**, 1–15. [CrossRef]
168. El Asmar, M.; Hanna, A.S.; Loh, W.Y. Evaluating integrated project delivery using the project quarterback rating. *J. Constr. Eng. Manag.* **2016**, *142*. [CrossRef]
169. Piroozfar, P.; Farr, E.R.P.; Zadeh, A.H.M.; Timoteo, I.; Kilgallon, S.; Jin, R. Facilitating Building Information Modelling (BIM) using Integrated Project Delivery (IPD): A UK perspective. *J. Build. Eng.* **2019**, *26*, 100907. [CrossRef]
170. Cho, S.; Ballard, G. Last planner and integrated project delivery. *Lean Constr. J.* **2011**, *2011*, 67–73.
171. Teng, Y.; Li, X.; Wu, P.; Wang, X. Using cooperative game theory to determine profit distribution in IPD projects. *Int. J. Constr. Manag.* **2019**, *19*, 32–45. [CrossRef]
172. Laurent, J.; Leicht, R.M. Practices for Designing Cross-Functional Teams for Integrated Project Delivery. *J. Constr. Eng. Manag.* **2019**, *145*. [CrossRef]
173. Osman, W.N.; Nawi, M.N.M.; Zulhumadi, F.; Shafie, M.W.M.; Ibrahim, F.A. Individual readiness of construction stakeholders to implement integrated project delivery (IPD). *J. Eng. Sci. Technol.* **2017**, *12*, 229–238.
174. Johnson, T.R.; Feng, P.; Sitzabee, W.; Jernigan, M. Federal acquisition regulation applied to alliancing contract practices. *J. Constr. Eng. Manag.* **2013**, *139*, 480–487. [CrossRef]

175. Walker, D.; Hampson, K. Procurement Choices. In *Procurement Strategies: A Relationship-Based Approach*; Blackwell Science Ltd.: Hpboken, NJ, USA, 2008; pp. 11–29.
176. Fernandes, D.A.; Costa, A.A.; Lahdenperä, P. Key features of a project alliance and their impact on the success of an apartment renovation: A case study. *Int. J. Constr. Manag.* **2018**, *18*, 482–496. [CrossRef]
177. Hauck, A.J.; Walker, D.H.T.; Hampson, K.D.; Peters, R.J. Project alliancing at national museum of Australia—Collaborative process. *J. Constr. Eng. Manag.* **2004**, *130*, 143–152. [CrossRef]
178. Heidemann, A.; Gehbauer, F. The way towards cooperative project delivery. *J. Financ. Manag. Prop. Constr.* **2011**, *16*, 19–30. [CrossRef]
179. Rahmani, F. Challenges and opportunities in adopting early contractor involvement (ECI): Client's perception. *Arch. Eng. Des. Manag.* **2020**, 1–20. [CrossRef]
180. Sagvekar, S.; Wayal, A.S. Early contractor involvement (ECI): Indian scenario of construction project delivery. *Int. J. Sci. Technol. Res.* **2019**, *8*, 807–811.
181. Scheepbouwer, E.; Humphries, A.B. Transition in adopting project delivery method with early contractor involvement. *Transp. Res. Rec.* **2011**, *2228*, 44–50. [CrossRef]
182. Bolton, B.; Heller, J. Effective airport project delivery, leadership and culture. *J. Airpt. Manag.* **2019**, *13*, 6–16.
183. Vilasini, N.; Neitzert, T.R.; Rotimi, J.O. Correlation between construction procurement methods and lean principles. *Int. J. Constr. Manag.* **2011**, *11*, 65–78. [CrossRef]
184. Balzani, M.; Raco, F.; Zaffagnini, T. Learning for placement fostering innovation in the construction sector through public-private partnership in the Emilia-Romagna region. *Turk. Online J. Educ. Technol.* **2017**, *2017*, 404–410.
185. Ptschelinzew, L.; Minchin, R.E.; Chini, A.; Zhang, Y. Relationship Management Strategies for Identifying Party Discord and Misperceptions. *J. Leg. Aff. Disput. Resolut. Eng. Constr.* **2020**, *12*. [CrossRef]
186. Ngoma, S.; Mundia, M.; Kaliba, C. Benefits, constraints and risks in infrastructure development via public-private partnerships in Zambia. *J. Constr. Dev. Ctries.* **2014**, *19*, 15–33.
187. Gajurel, A. *Performance-Based Contracts for Road Projects: Comparative Analysis of Different Types*; Springer: Berlin/Heidelberg, Germany, 2014; pp. 1–159.
188. Kantola, M.; Saari, A. Ensuring functionality of a nearly zero-energy building with procurement methods. *Facility* **2014**, *32*, 312–323. [CrossRef]
189. Barlow, J.; Köberle-Gaiser, M. The private finance initiative, project form and design innovation. The UK's hospitals programme. *Res. Policy* **2008**, *37*, 1392–1402. [CrossRef]
190. Brioso, X.; Aguilar, R.; Calderón-Hernandez, C. *Synergies Between Lean Construction and Management of Heritage Structures and Conservation Strategies—A General Overview*; RILEM Bookseries; Springer: Berlin/Heidelberg, Germany, 2019; Volume 18, pp. 2142–2149.
191. Ballard, G. The lean project delivery system: An update. *Lean Constr. J.* **2008**, *2008*, 1–19.
192. Moaveni, S.; Banihashemi, S.Y.; Mojtahedi, M. A conceptual model for a safety-based theory of lean construction. *Building* **2019**, *9*, 23. [CrossRef]
193. Darrington, J. Using a design-build contract for lean integrated project delivery. *Lean Constr. J.* **2011**, *2011*, 85–91.
194. Koskela, L.; Howell, G.; Ballard, G.; Tommelein, I. The foundations of lean construction. *Des. Constr.* **2007**, *291*, 211–226.
195. Forbes, L.H.; Ahmed, S.M. *Modern Construction: Lean Project Delivery and Integrated Practices*; CRC Press: Boca Raton, FL, USA, 2010; pp. 1–491.
196. Brioso, X.; Calderón, C.; Aguilar, R.; Pando, M.A. *Preliminary Methodology for the Integration of Lean Construction, BIM and Virtual Reality in the Planning Phase of Structural Intervention in Heritage Structures*; RILEM Bookseries; Springer: Berlin/Heidelberg, Germany, 2019; Volume 18, pp. 484–492.
197. Schultz, R.; Sarfaraz, A.; Jenab, K. Analysis of risk and reliability in project delivery methods. In *Transportation Systems and Engineering: Concepts, Methodologies, Tools, and Applications*; IGI Global: Hershey, PA, USA, 2015; Volume 2–3, pp. 612–622.
198. Brioso, X.; Humero, A.; Murguia, D.; Corrales, J.; Aranda, J. Using post-occupancy evaluation of housing projects to generate value for municipal governments. *Alex. Eng. J.* **2018**, *57*, 885–896. [CrossRef]
199. Lapinski, A.R.; Horman, M.J.; Riley, D.R. Lean processes for sustainable project delivery. *J. Constr. Eng. Manag.* **2006**, *132*, 1083–1091. [CrossRef]
200. Klotz, L.; Horman, M.; Bodenschatz, M. A lean modeling protocol for evaluating green project delivery. *Lean Constr. J.* **2007**, *3*, 1–18.
201. Hwang, B.-G.; Lim, E.S.J. Critical Success Factors for Key Project Players and Objectives: Case Study of Singapore. *J. Constr. Eng. Manag.* **2013**, *139*, 204–215. [CrossRef]
202. Feghaly, J.; El Asmar, M.; Ariaratnam, S.; Bearup, W. Selecting project delivery methods for water treatment plants. *Eng. Constr. Arch. Manag.* **2019**, *27*, 936–951. [CrossRef]
203. Bingham, E.; Gibson, G.E.; Asmar, M.E. Measuring User Perceptions of Popular Transportation Project Delivery Methods Using Least Significant Difference Intervals and Multiple Range Tests. *J. Constr. Eng. Manag.* **2018**, *144*. [CrossRef]
204. Qiang, M.; Wen, Q.; Jiang, H.; Yuan, S. Factors governing construction project delivery selection: A content analysis. *Int. J. Proj. Manag.* **2015**, *33*, 1780–1794. [CrossRef]
205. Li, H.; Wang, Z.; Liu, H. Design framework for construction project delivery systems. *Tech. Technol. Educ. Manag.* **2010**, *5*, 847–852.

206. Chen, Y.Q.; Lu, H.; Lu, W.; Zhang, N. Analysis of project delivery systems in Chinese construction industry with data envelopment analysis (DEA). *Engineering, Construction and Architectural Management* **2010**, *17*, 598–614. [CrossRef]
207. Liu, B.; Huo, T.; Liang, Y.; Sun, Y.; Hu, X. Key Factors of Project Characteristics Affecting Project Delivery System Decision Making in the Chinese Construction Industry: Case Study Using Chinese Data Based on Rough Set Theory. *J. Prof. Issues Eng. Educ. Pr.* **2016**, 142. [CrossRef]
208. Liu, B.; Huo, T.; Shen, Q.; Yang, Z.; Meng, J.; Xue, B. Which owner characteristics are key factors affecting project delivery system decision making? Empirical analysis based on the rough set theory. *J. Manag. Eng.* **2015**, *31*, 05014018. [CrossRef]
209. Ding, J.; Wang, N.; Hu, L. Framework for Designing Project Delivery and Contract Strategy in Chinese Construction Industry Based on Value-Added Analysis. *Adv. Civ. Eng.* **2018**, *2018*. [CrossRef]
210. Martin, H.; Lewis, T.M.; Petersen, A. Factors affecting the choice of construction project delivery in developing oil and gas economies. *Arch. Eng. Des. Manag.* **2016**, *12*, 170–188. [CrossRef]
211. Touran, A.; Gransberg, D.; Molenaar, K.; Ghavamifar, K. Selection of project delivery method in transit: Drivers and objectives. *J. Manag. Eng.* **2011**, *27*, 21–27. [CrossRef]
212. Alleman, D.; Antoine, A.; Stanford, M.S.; Molenaar, K. Project Delivery Methods' Change-Order Types and Magnitudes Experienced in Highway Construction. *J. Leg. Aff. Disput. Resolut. Eng. Constr.* **2020**, *12*, 04520006. [CrossRef]
213. Li, H.; Arditi, D.; Wang, Z. Factors that affect transaction costs in construction projects. *J. Constr. Eng. Manag.* **2013**, *139*, 60–68. [CrossRef]
214. Sirbovan, B.; DiProspero, D.; Larson, B. Primer of design and construction delivery methods for today's modern pharmaceutical and biotech facilities. *Pharm. Eng.* **2006**, *26*, 8–18.
215. Creedy, G.D.; Skitmore, M.; Wong, J.K.W. Evaluation of risk factors leading to cost overrun in delivery of highway construction projects. *J. Constr. Eng. Manag.* **2010**, *136*, 528–537. [CrossRef]
216. Franz, B.; Leicht, R.; Molenaar, K.; Messner, J. Impact of Team Integration and Group Cohesion on Project Delivery Performance. *J. Constr. Eng. Manag.* **2017**, *143*, 04016088. [CrossRef]
217. Mafakheri, F.; Dai, L.; Slezak, D.; Nasiri, F. Project delivery system selection under uncertainty: Multicriteria multilevel decision aid model. *J. Manag. Eng.* **2007**, *23*, 200–206. [CrossRef]
218. Kumaraswamy, M.M.; Dissanayaka, S.M. Developing a decision support system for building project procurement. *Build. Env.* **2001**, *36*, 337–349. [CrossRef]
219. Molenaar, K.R.; Songer, A.D. Model for public sector design-build project selection. *J. Constr. Eng. Manag.* **1998**, *124*, 467–479. [CrossRef]
220. Aldossari, K.M.; Lines, B.C.; Smithwick, J.B.; Hurtado, K.C.; Sullivan, K.T. Best practices of organizational change for adopting alternative project delivery methods in the AEC industry. *Eng. Constr. Archit. Manag.* **2020**. [CrossRef]
221. Moradi, S.; Kähkönen, K.; Aaltonen, K. Project Managers' Competencies in Collaborative Construction Projects. *Buildings* **2020**, *10*, 50. [CrossRef]
222. Moon, H.; Cho, K.; Hong, T.; Hyun, C. Selection Model for Delivery Methods for Multifamily-Housing Construction Projects. *J. Manag. Eng.* **2011**, *27*, 106–115. [CrossRef]
223. Liu, B.; Xue, B.; Huo, T.; Shen, G.; Fu, M. Project external environmental factors affecting project delivery systems selection. *J. Civ. Eng. Manag.* **2019**, *25*, 276–286. [CrossRef]
224. Ding, X.; Sheng, Z.; Du, J.; Li, Q. Computational experiment study on selection mechanism of project delivery method based on complex factors. *Math. Probl. Eng.* **2014**, *2014*. [CrossRef]
225. Ibbs, C.W.; Kwak, Y.; Ng, T.; Odabasi, A.M. Project Delivery Systems and Project Change: Quantitative Analysis. *J. Constr. Eng. Manag.* **2003**, *129*, 382–387. [CrossRef]
226. Ahn, Y.H.; Pearce, A.R.; Holley, P.W. Project delivery system for foreign manufacturers in the United States. *Int. J. Constr. Educ. Res.* **2009**, *5*, 149–166. [CrossRef]
227. Lines, B.C.; Ravi, K. Developing More Competitive Proposals: Relationship between Contractor Qualifications-Based Proposal Content and Owner Evaluation Scores. *J. Constr. Eng. Manag.* **2018**, 144. [CrossRef]
228. Mollaoglu-Korkmaz, S.; Swarup, L.; Riley, D. Delivering Sustainable, High-Performance Buildings: Influence of Project Delivery Methods on Integration and Project Outcomes. *J. Manag. Eng.* **2013**, *29*, 71–78. [CrossRef]
229. Montalbán-Domingo, L.; García-Segura, T.; Amalia, S.; Pellicer, E. Social Sustainability in Delivery and Procurement of Public Construction Contracts. *J. Manag. Eng.* **2019**, *35*, 04018065. [CrossRef]
230. Yun, S.; Jung, W. Benchmarking sustainability practices use throughout industrial construction project delivery. *Sustainability* **2017**, *9*, 7. [CrossRef]
231. Luu, D.T.; Ng, S.T.; Chen, S.E. Formulating procurement selection criteria through case-based reasoning approach. *J. Comput. Civ. Eng.* **2005**, *19*, 269–276. [CrossRef]
232. Luu, D.T.; Ng, S.T.; Chen, S.E. A case-based procurement advisory system for construction. *Adv. Eng. Softw.* **2003**, *34*, 429–438. [CrossRef]
233. Mostafavi, A.; Karamouz, M. Selecting Appropriate Project Delivery System: Fuzzy Approach with Risk Analysis. *J. Constr. Eng. Manag.* **2010**, *136*, 923–930. [CrossRef]
234. Lee, Z.P.; Rahman, R.A.; Doh, S.I. Key drivers for adopting design build: A comparative study between project stakeholders. *Phys. Chem. Earth Parts A/B/C* **2020**, *120*, 102945. [CrossRef]

235. El-Said, M.; El-Dokhmaesy, A.; Younis, M.E. Integrated project delivery and associated risk reduction in construction projects in Egypt. *J. Eng. Appl. Sci.* **2019**, *66*, 837–859.
236. Tran, D.Q.; Molenaar, K.R. Exploring critical delivery selection risk factors for transportation design and construction projects. *Eng. Constr. Arch. Manag.* **2014**, *21*, 631–647. [CrossRef]
237. Tran, D.Q.; Molenaar, K.R. Impact of risk on design-build selection for highway design and construction projects. *J. Manag. Eng.* **2014**, *30*, 153–162. [CrossRef]
238. Osipova, E.; Eriksson, P.E. How procurement options influence risk management in construction projects. *Constr. Manag. Econ.* **2011**, *29*, 1149–1158. [CrossRef]
239. Jefferies, M.; Brewer, G.J.; Gajendran, T. Using a case study approach to identify critical success factors for alliance contracting. *Eng. Constr. Arch. Manag.* **2014**, *21*, 465–480. [CrossRef]
240. Gordon, C. Choosing Appropriate Construction Contracting Method. *J. Constr. Eng. Manag.* **1994**, *120*, 196–210. [CrossRef]
241. Meshref, A.N.; Elkasaby, E.A.; Wageh, O. Innovative reliable approach for optimal selection for construction infrastructures projects delivery systems. *Innov. Infrastruct. Solut.* **2020**, *5*. [CrossRef]
242. Popic, Z.; Moselhi, O. Project Delivery Systems Selection for Capital Projects Using the Analytical Hierarchy Process and the Analytical Network Process. *Constr. Res. Congr* **2014**, 1339–1348. [CrossRef]
243. Saaty, T.L. The analytic hierarchy and analytic network processes for the measurement of intangible criteria and for decision-making. In *International Series in Operations Research and Management Science*; Springer: New York, NY, USA, 2016; Volume 233, pp. 363–419.
244. Marzouk, M.; Elmesteckawi, L. Analyzing procurement route selection for electric power plants projects using SMART. *J. Civ. Eng. Manag.* **2015**, *21*, 912–922. [CrossRef]
245. Li, H.; Qin, K.; Li, P. Selection of project delivery approach with unascertained model. *Kybernetes* **2015**, *44*, 238–252. [CrossRef]
246. Khanzadi, M.; Nasirzadeh, F.; Hassani, S.M.H.; Mohtashemi, N.N. An integrated fuzzy multi-criteria group decision making approach for project delivery system selection. *Sci. Iran.* **2016**, *23*, 802–814. [CrossRef]
247. Cao, Y.; Li, H.; Su, L. Decision-making for project delivery system with related-indicators based on pythagorean fuzzy weighted muirhead mean operator. *Information* **2020**, *11*, 451. [CrossRef]
248. An, X.; Wang, Z.; Li, H.; Ding, J. Project Delivery System Selection with Interval-Valued Intuitionistic Fuzzy Set Group Decision-Making Method. *Group Decis. Negot.* **2018**, *27*, 689–707. [CrossRef]
249. Su, L.; Li, H.; Cao, Y.; Lv, L. Project delivery system decision making using pythagorean fuzzy TOPSIS. *Eng. Econ.* **2019**, *30*, 461–471. [CrossRef]
250. Liu, X.; Qian, F.; Lin, L.; Zhang, K.; Zhu, L. Intuitionistic fuzzy entropy for group decision making of water engineering project delivery system selection. *Entropy* **2019**, *21*, 101. [CrossRef]
251. Martin, H.; Lewis, T.; Petersen, A.; Peters, E. Cloudy with a Chance of Fuzzy: Building a Multicriteria Uncertainty Model for Construction Project Delivery Selection. *J. Comput. Civ. Eng.* **2017**, *31*, 04016046. [CrossRef]
252. Nguyen, P.; Tran, D.Q.; Lines, B.C. Empirical Inference System for Highway Project Delivery Selection Using Fuzzy Pattern Recognition. *J. Constr. Eng. Manag.* **2020**, *146*, 12. [CrossRef]
253. Tsai, T.C.; Yang, M.L. Risk assessment of design-bid-build and design-build building projects. *J. Oper. Res. Soc. Jpn.* **2010**, *53*, 20–39. [CrossRef]
254. Zhu, J.W.; Zhou, L.N.; Li, L.; Ali, W. Decision simulation of construction project delivery system under the sustainable construction project management. *Sustainability* **2020**, *12*, 2202. [CrossRef]
255. Francis, A. Simulating Uncertainties in Construction Projects with Chronographical Scheduling Logic. *J. Constr. Eng. Manag.* **2017**, *143*. [CrossRef]
256. Gil, N.; Tommelein, I.D.; Ballard, G. Theoretical comparison of alternative delivery systems for projects in unpredictable environments. *Constr. Manag. Econ.* **2004**, *22*, 495–508. [CrossRef]
257. Chen, Y.Q.; Liu, J.Y.; Li, B.; Lin, B. Project delivery system selection of construction projects in China. *Expert Syst. Appl.* **2011**, *38*, 5456–5462. [CrossRef]
258. Kumaraswamy, M.M. Industry development through creative project packaging and integrated management. *Eng. Constr. Arch. Manag.* **1998**, *5*, 229–237.
259. Ling, F.Y.Y.; Liu, M. Using neural network to predict performance of design-build projects in Singapore. *Build. Env.* **2004**, *39*, 1263–1274. [CrossRef]
260. Flood, I. Towards the next generation of artificial neural networks for civil engineering. *Adv. Eng. Inf.* **2008**, *22*, 4–14. [CrossRef]
261. Molenaar, K.R.; Songer, A.D. Web-based decision support systems: Case study in project delivery. *J. Comput. Civ. Eng.* **2001**, *15*, 259–267. [CrossRef]
262. Luu, D.T.; Ng, S.T.; Chen, S.E. Parameters governing the selection of procurement system–an empirical survey. *Eng. Constr. Arch. Manag.* **2003**, *10*, 209–218.
263. Zhu, X.; Meng, X.; Chen, Y. A novel decision-making model for selecting a construction project delivery system. *J. Civ. Eng. Manag.* **2020**, *26*, 635–650. [CrossRef]
264. Yoon, Y.; Jung, J.; Hyun, C. Decision-making support systems using case-based reasoning for construction project delivery method selection: Focused on the road construction projects in Korea. *Open Civ. Eng. J.* **2016**, *10*, 500–512. [CrossRef]

265. Tran, D.; Molenaar, K. Risk-Based Project Delivery Selection Model for Highway Design and Construction. *J. Constr. Eng. Manag.* **2015**, *141*, 04015041. [CrossRef]
266. Molenaar, K.R. Programmatic cost risk analysis for highway megaprojects. *J. Constr. Eng. Manag.* **2005**, *131*, 343–353. [CrossRef]
267. El-Sayegh, S.; Romdhane, L.; Manjikian, S. A critical review of 3D printing in construction: Benefits, challenges, and risks. *Arch. Civ. Mech. Eng.* **2020**, *20*, 1–25. [CrossRef]
268. Costin, A.M.; Teizer, J. Fusing passive RFID and BIM for increased accuracy in indoor localization. *Vis. Eng.* **2015**, *3*, 1–20. [CrossRef]
269. Mak, S. A model of information management for construction using information technology. *Autom. Constr.* **2001**, *10*, 257–263. [CrossRef]
270. Zhao, L.; Liu, Z.; Mbachu, J. Optimization of the Supplier Selection Process in Prefabrication Using BIM. *Buildings* **2019**, *9*, 222. [CrossRef]
271. Zhang, Z.; Hu, J.; Shen, L. Green Procurement Management in Building Industry: An Alternative Environmental Strategy. In Proceedings of the 20th International Symposium on Advancement of Construction Management and Real Estate, Hangzhou, China, 23–25 October 2017; Springer: Singapore, 2017; pp. 1217–1228.
272. Manjia, M.B.; Abanda, F.H.; Pettang, C. *Using Markov Decision Process for Construction Site Management in Cameroon*; University of Yaoundé: Yaoundé, Cameroon, 2014.

 *buildings* 

Article

Construction Disputes in the UAE: Causes and Resolution Methods

Sameh El-Sayegh, Irtishad Ahmad *, Malak Aljanabi, Rawan Herzallah, Samuel Metry and Omar El-Ashwal

Civil Engineering Department, American University of Sharjah, Sharjah P.O. Box 26666, UAE; selsayegh@aus.edu (S.E.-S.); g00061357@aus.edu (M.A.); g00058017@aus.edu (R.H.); b00056992@aus.edu (S.M.); B00054449@aus.edu (O.E.-A.)
* Correspondence: irahmad@aus.edu

Received: 13 August 2020; Accepted: 23 September 2020; Published: 25 September 2020

Abstract: Claims and disputes occur frequently in the construction industry between different contracting parties, mainly the owner, the designer and the contractor. Consequently, valuable time and a significant amount of money are lost. The United Arab Emirates (UAE) construction industry, one of the most vibrant sectors globally, is experiencing a high level of construction disputes and claims. This paper aims to identify and assess the major causes of disputes in the UAE and weigh the effectiveness of the methods used for their avoidance and resolution. The sources of disputes, and their avoidance/resolution methods, were identified through a comprehensive literature review. A survey was then developed and sent to 150 construction professionals. Fifty-four responses were received and analyzed. The results show that the top five sources of disputes in the UAE are variations initiated by the owner, obtaining permit/approval from the municipality and other governmental authorities, material change and approval during the construction phase, the slowness of the owner in decision-making, and the short time available during the design phase. As for the avoidance and the resolution method, the most effective method was found to be negotiation.

Keywords: disputes; claims; dispute resolution methods; construction industry; UAE

1. Introduction

Disputes are common in the construction industry. Disputes arise due to disagreements between any of the contracting parties. Disputes have a devastating effect on construction projects, as they may result in cost overruns, delays, and loss of productivity. It is vital to understand the causes of disputes to complete a construction project within cost and time [1]. Construction disputes impact project objectives and strain relationships between contracting parties [2]. A dispute in construction projects is considered to be an impediment to the path of successful project completion [3]. Disputes are resource consuming, unpleasant, and expensive [4]. Conflicts disrupt the flow of work, resulting in additional costs, delays, and other negative impacts [5]. These problems might lead to construction claims and disputes. Arditi and Pattanakitchamroon [6] discussed various reasons for such problems, including incorrect design/specifications, unusually severe weather condition, change orders and extra work. Disputes mainly arise from claims resulting from differing site conditions, delays, design errors or changes, acceleration or suspension of work, construction failures, and additional or deleted work [7]. Mitropoulos and Howell [8] suggested that the main factors that influence disputes are project uncertainty, opportunistic behavior, and contractual problems.

Disputes, if not addressed and resolved properly, can create significant losses in the project and for the company [9]. Artan and Bakioglu [10] argued that risk is a major factor leading to disputes. Cheung and Pang [11] divided construction disputes into two types based on the sources: contractual and speculative. The main causes of both types of construction disputes are ambiguity, deficiency, inconsistency, and defectiveness, which are grouped under the "contract incompleteness" [11]. They further identified factors such as risk, uncertainty and conflicts as other causes of contractual disputes. On the other hand, human perception factors such as ambiguity of contractual clauses that may cause interpretational difficulties, opportunistic behavior and conflicts could be other causes of disputes [11].

The problem of disputes and claims plagues the construction industry. The leading causes need to be identified in specific regions, since each region has a unique setting due to their differing legal, political and cultural aspects. The identification of causes will lead to appropriate resolution methods being chosen depending on the specific case. In this paper, the country investigated is the United Arab Emirates, which has a very vibrant economy due to growing tourism and business activities. As a consequence, there is a continuous growth in the construction industry in the United Arab Emirates (UAE), which attracts international contractors and investors to its large and unique projects. Construction disputes have increased in recent years due to the increase in the size and complexity of construction projects. Awwad et al. [5] found that, in the Middle East, the UAE is the second highest country in terms of the value of construction disputes after the Kingdom of Saudi Arabia (KSA). Moreover, the Middle East has the longest average duration of dispute settlement, which is 14.6 months [5]. The UAE construction industry is experiencing fast growth with many ongoing projects, investment into green open spaces and strong government support. The construction industry contributes more than 10% to the United Arab Emirates' (UAE) gross domestic production (GDP). The UAE construction industry is growing consistently and expected to grow more in the near future to accommodate the UAE's strategic goals, which include significant spending in infrastructure construction. The construction activity is projected to continue rising as a percentage of real GDP in the UAE; from 10.3% in 2011 to 11.5% in 2021 [12]. The growth of the construction industry in the UAE is accompanied by an increase in claims and disputes, which, as a result, leads to delays, and additional cost to the project.

The two most common ways to mitigate disputes are avoidance and resolution. Resolution methods are further divided into early and late categories. Negotiation, risk allocation, early non-binding neutral evaluation, and partnering are included under avoidance methods. Early resolution methods included negotiation, conciliation, and mini-trial/executive tribunal. Late resolution methods included, negotiation, arbitration, mediation, adjudication, dispute review boards, and litigation.

Studies show that the most frequently used methods to resolve disputes are negotiation, arbitration, and litigation. Alternative Dispute Resolution (ADR) methods proved to be more efficient in settling disputes, with less cost and time compared to those most used in the Middle East [5]. The Hong Kong International Arbitration Center (HKIAC) stated that increasingly complicated construction projects over the last two decades have resulted in a growing number of projects adopting Alternative Dispute Resolution (ADR) [9]. The construction industry has called for alternatives owing to the high cost, in terms of time and money, of taking things to a court or organizing for arbitration [9]. Martin and Thompson [13] discussed the various methods of dispute resolution and provided tools that parties can use to manage the selected process of dispute resolution. Martin and Thompson [13] identified five basic forms of construction dispute resolutions, which are collaborating, Dispute Review Board (DRB), mediation, arbitration, and litigation. According to Yates [7], there are different traditional resolution techniques such as arbitration, litigation, and negotiation. The problem with litigation and arbitration is that both of them are adversarial processes, in other words, one side wins and the other side lose [7]. Using adversarial methods could weaken relationships between the two sides involved [7]. Cheung et al. [14] found that negotiation between the disputants takes first place in resolving any problem. Cheung et al. [14] stated that the styles of negotiation depend on the personality of the disputants, situation volume and type. Negotiation helps to prevent members of different construction firms from using undesirable methods such as arbitration or litigation; in addition, it saves time, money and the reputation of the firm [7]. Mitropoulos and Howell [4] showed that the prevention of high-cost, complicated disputes mainly depends on the problem-solving ability and planning of the project organization. The main methods used to resolve the claims in the UAE are negotiation, meditation, arbitration, and litigation [15]. Ho and Liu [16] stated that claims and disputes are integrated and interrelated. Ho and Liu [16] developed a Model based on the "Game Theory" or "Nash-Equilibrium" that helps the owner to avoid the presence of construction claims due to opportunistic bidding. Haugen and Singh [9] identified three relevant impacting factors in deciding the ADR method: The market position of the individual parties, the relationship between parties, and the source of dispute.

The main objectives of the paper are twofold. First, to identify and present the major causes and sources of disputes and claims in the UAE construction industry. Second, to outline the effective methods of avoidance and resolution. The major causes and leading methods of mitigation are identified from the available literature and publications. These findings provided the basis of a survey instrument conducted among the construction industry professionals in UAE. The method used is described in the next section.

2. Materials and Methods

In order to accomplish the objectives of this research study, the first step undertaken was a thorough review of the existing literature on the subject. This helped researchers to identify the main sources of construction disputes and their avoidance and resolution methods. Twenty-seven causes of disputes were identified. They are grouped in five categories: Design-related, owner-related, contractor-related, contractual, and 'other' disputes. The literature review was the basis of a survey instrument that was developed to be distributed to the UAE construction professionals.

The survey included three sections. The first section obtained basic information about the profile of the respondents. This included questions about the company type, the company's years of experience, industry sector (type), the company type, and the size of the projects undertaken by the company. The second section identified the major causes of disputes based on their frequency of occurrence. Respondents were asked to use the Likert Scale with the ratings: very high, high, moderate, low, and very low. The third section measured the effectiveness of dispute resolution and avoidance methods as practiced in the UAE by using the ratings extremely effective, effective, neutral, not effective, extremely not effective and not applicable. The survey was sent to 150 construction professionals. Fifty-four responses were received, corresponding to a 36% response rate. Out of the 54 responses, 28 were local companies and 26 were international companies. Table 1 shows the respondents' profile.

Table 1. Respondents' Profile.

Category		Respondents (54)	
		Number	**%**
Years of experience	>20 years	9	16.7%
	11–20 years	16	29.6%
	5–10 years	20	37.0%
	<5 years	9	16.7%
Role	Owner	4	7.4%
	Consultant	16	29.6%
	Contractor	29	53.7%
	Management Consultant	5	09.3%
Average Project Size [1]	<50 (Million AED)	8	14.8%
	50–200 (Million AED)	15	27.8%
	200–500 (Million AED)	8	14.8%
	>500 (Million AED)	23	42.6%

[1] 1 USD = 3.67 AED (2020 Currency).

The survey results are analyzed to identify the frequently occurring disputes in the UAE. The weighted average is calculated for each of the disputes and resolution/avoidance method.

The weight is assigned for each dispute source, based on frequency of occurrence, as 5, 4, 3, 2, and 1 for very high, high, moderate, low and very low, respectively. Likewise, the weight assigned for the effectiveness of each resolution/avoidance method is 5, 4, 3, 2, 1, and 0 for extremely effective, effective, neutral, not effective, extremely not effective and not applicable, respectively. The Weighted Average (WA) is calculated using Equation (1)

$$\text{Weighted Average, } WA = \frac{\sum_{i=1}^{5} W_i \times X_i}{\sum_{i=1}^{5} X_i} \tag{1}$$

where:

W_i = Weight assigned to ith response; W_i = 1, 2, 3, 4 and 5 for i = 1, 2, 3, 4 and 5, respectively;

X_i = Frequency of the ith response;

i = Response category index = 1, 2, 3, 4 and 5 for very low, low, moderate, high and very high, respectively.

Equation (1) is also used for the effectiveness of the dispute resolution methods, but the response category index (i) was assigned 5, 4, 3, 2, 1, and 0 for extremely effective, effective, neutral, not effective, extremely not effective and not applicable, respectively. The causes of disputes are ranked based on their weighted average scores. The cause with the highest average is ranked 1, and so on. The same process is used to rank the dispute resolution methods; however, the ranking is within each group only. For comparison purposes and to study the strength of relationship between two sets of ranking, the Spearman rank correlation coefficient (RHO) was determined using the IBM SPSS 26 software (IBM corporation, Armonk, NY, USA). A higher value of RHO (approaching 1) indicates a strong correlation.

3. Sources of Construction Disputes

Twenty-seven sources of disputes were identified through the literature review and divided into five groups: designer, owner, contractor, contractual and other. Table 2 lists the sources of disputes along with the references.

Table 2. Sources of Disputes suggested by the literature.

	Sources of disputes	Literature Source
Designer-related	Time limitation in the design phase	[17–22]
	Poor design	[23]
	Inadequate or incomplete technical plans/specification	[5]
	Poor preparation and approval of drawings	[15,17,24–32]
	Material change and approval during the construction phase	[15,17–19,22,24,33,34]
Owner-related	Slowness of the owner's decision-making process	[17,24]
	Inadequate early planning of the project	[17,24]
	Failure to make interim awards on extensions of time and compensating by the owner	[5]
	Variations initiated by the owner (additive/deductive)	[5]
	Poor Financing by the owner	[15,24,29,32]
Contractor-related	low Financing by the contractor during construction	[15,24,29,32]
	Shortage and unproductive workers	[24]
	Inadequate site investigation	[15,25–27,32]
	Poorly defined scope of work	[17,21,22,24,34–36]
	Poor supervision and site management	[17,24]
	Unsuitable leadership style of construction/project manager	[17–19,24,37–39]
	Underestimation and incompetence of contractors	[5]
Contractual	Poorly written contracts	[9,15,17–22,24,25,29,31,34,37,40]
	Differing Site Conditions	[23]
	Contract Amendments	[5,41]
	Contradictory and inaccurate information in the contract documents	[5]
Other	Obtaining Permit/Approval from the municipality/different government authority	[15,17–19,24,37,40,42,43]
	Modifying legislation and regulations	[5]
	Inappropriate weather conditions	[44]
	Impact on locality in terms of noise, traffic, and pollution/contamination	[45]
	Lack of communication and coordination between parties during construction	[17–19,21,24,33,34,40]
	impact of local cultures and social values in the settlement of conflicts	[5]

3.1. Designer-Related Disputes

Time limitation in the design phase occurs as clients typically allow a very limited time to complete and submit the designs. If the design time is very limited, the design may lack specific details and accuracy. Poor design occurs when the design is nonfunctional, has missing elements and does not meet the owner's requirements. Inadequate or incomplete technical plans and specifications can cause delays during construction. Changes in material specifications and the subsequent approval process may take time, and cause disputes regarding the additional costs of the new materials and delays due to shipping and the difficulty procuring the material. All these cost the owner and designer, which, in turn, causes disputes [46].

3.2. Owner-Related Disputes

One of the most common causes of disputes is the slowness of the owner's decision-making process, in which the owner takes a long time making decisions, which delays the construction process. Inadequate early planning of the project creates disputes and causes additional cost to the owner [32]. Therefore, detailed early planning is required to avoid conflicts and extra costs. Failure to make interim awards on extensions of time and compensation by the owner is a common practice amongst owners. This may cause even more difficult-to-resolve disputes towards the end of the project. Waiting until the end of the project to deal with disputes makes them harder and costlier to resolve [47]. Owners may request variations such as adding or deducting from the previously agreed scope, which cause disputes since some variations may result in additional time and cost. If the owner cannot finance the project on time, construction will be delayed and the project might stop for a certain period until the owner is ready to finance the project, which creates disputes [8].

3.3. Contractor-Related Disputes

Inadequate financing by the contractor during construction leads to delays, work interruption, and poor quality of subcontractors' work. If the contractor has a shortage of workers, a delay in the building process occurs and this can create disputes. Disputes can be created due to low productivity [10]. Inadequate site investigation can create multiple disputes. A poorly defined scope of work is a type of contractor dispute. Poor supervision and site management by the contractor can cause accidents and delay the construction process. Site management is a major factor of construction disputes led by the contractor. Likewise, an unsuitable leadership style from the construction/project manager takes place when there is an incompetent person lacking appropriate qualifications in a position of construction/project manager, which results in them making the wrong decisions during construction [13]. Contractors can be asked to stop the construction process if the contracting company is not capable to continue to make progress in the project [48].

3.4. Contractual Disputes

A poorly written contract leads to different interpretations of the same issue, and that leads to an argument, which later on develops into a dispute. In addition, differing site condition is also considered a contractual dispute. This is when the contractor encounters unknown physical conditions of an unusual nature that differ materially from those that are ordinarily encountered and generally recognized as inherent in the work at the project's location [8]. Differing site conditions is the top reason for claims [6]. Contract amendments are used when the parties want to modify the terms of an existing legal agreement. Contradictory and inaccurate information in the contract document is when the contract needs to be very accurate and needs to be revised before an agreement between the parties is made. The statements in the contract should not contradict the scope of work and should be clarified properly between the parties.

3.5. Other Disputes

A source of other disputes is obtaining permits and/or approvals from the municipality/different government authority. Delays in getting approvals and permissions from official governmental offices might lead to disputes. Modifying legislation and regulations can cause disputes. Each country has its own law and rules that are subject to change. These changes might cause disputes during the construction process. Similarly, inappropriate weather conditions can cause delays and cost overruns, which consequently result in disputes. Some construction projects might result in traffic jams due to road closure or noise, especially in areas that have schools or hospitals, for example, neighbors' complaints [5]. The lack of communication and coordination between parties during construction causes confusion and misunderstanding of the scope of work, and this causes social disputes between the parties.

4. Dispute Avoidance and Resolution Methods in Construction

Table 3 shows the dispute avoidance and resolution methods along with their literature sources.

Table 3. Summary of dispute avoidance and resolution methods.

S/N	Method	Literature Source
	Dispute Avoidance Methods	
1	Negotiation	[5,9,23,45,49,50]
2	Risk Allocation	[5]
3	Early Non-Binding Neutral Evaluation	[5,51]
4	Partnering	[5,9]
	Early Resolution Methods	
5	Negotiation	[5,9,23,45,49,50]
6	Conciliation	[9,23,49,50,52]
7	Mini-Trial/Executive Tribunal	[23,52]
	Late Resolution Methods	
8	Negotiation	[5,9,23,45,49,50]
9	Arbitration	[23]; [49–51,53]
10	Mediation	[5,9,13,23,49–51,54]
11	Adjudication	[23,49,50,52]
12	Dispute Review Boards	[5,23,51,52]
13	Litigation	[5,9,13,51,52]

4.1. Dispute Avoidance Methods

Dispute avoidance methods are used to prevent disputes from occurring. At the beginning of any claim, usually, all parties would choose the resolution methods that eliminate the dispute at its root [55]. Dispute avoidance methods include negotiation, risk allocation, early non-binding neutral evaluation and partnering. Negotiation is always the best resolution method to prevent disputes from happening, as it is requiring less time and saves cost down the road. Risk allocation promotes balanced risk distribution among the contracting parties. Steen [47] stated that allocating fair contract risk is one of the main ways to prevent litigation and solve construction disputes. An unfair shifting of risk later causes the parties to spend more time and effort finding ways to stay afloat in business [47]. Early non-binding neutral evaluation could be an alternative. The neutral entity is chosen by the parties to resolve the dispute with no intention to be biased [16]. This requires preselecting an interdependent "neutral" entity to serve the parties as an observer, fact-finder and dispute-resolver for as long as the construction is in process [47]. Some clients choose partnering to be their resolution method. Steen [47] recommended building teams as a key to prevent disputes. Building teams helps improve cooperation and coordination among different parties and helps to establish a better understanding between the parties [47].

4.2. Early Resolution Methods

Early resolution methods attempt to reach a satisfactory and acceptable solution to both parties, in which they seek to minimize the disputed amount (usually $ volume) or prevent moving to a more expensive and time-consuming method [15]. Once a dispute occurs, companies have various choices in picking an early resolution method according to their preference. Usually, disputants settle their issues in the early stages by conciliation. This arises out of a clause in a construction contract whereby the parties agree to attempt to resolve their disputes through pacification and appeasement. The clause requires a conciliator, appointed by an agreement between the parties or by a specific institution. In addition, mini-trail or executive tribunal could be an alternative [9]. This process involves a panel, which consists of senior management representatives from each party as well as a neutral third party, or mediator who has been selected by the members of the organizations involved in the dispute.

4.3. Late Resolution Methods

Late resolution methods are used in the last stages of the dispute occurrence. These approaches are used in case of the failure of both avoidance and early resolution methods. While these methods can be very effective and efficient in settling the disputes, they are very expensive and consume considerable time and effort. In large and international companies, usually, the disputants use dispute review boards (DRB) instead of arbitration to provide an efficient and cost-effective means of dispute resolution [23]. On the other hand, local and small companies usually refer to mediation as an alternative solution. This method is a voluntary non-binding process in which a mediator assists the parties to retain full control over resolving the dispute [23]. Adjudication is another voluntary nonbinding process in which a mediator assists the parties in achieving a negotiated settlement [23]. Arbitration and adjudication can take place in all companies as the contract can identify everyone's rights without referring to the court. However, if all these methods are not applicable, the parties have no choice but litigation in the court of law. Usually, disputants try to avoid litigation, as this stage of dispute settlement is very costly and time-consuming to all parties involved.

5. Results

5.1. Assessment of the Sources of Construction Disputes in the UAE

Table 4 shows the sources of construction disputes, ranked in terms of their weighted average score based on the survey results.

Table 4. Main construction disputes in the United Arab Emirates (UAE).

Sources of Disputes	Weight	Rank
Variations initiated by the owner (additive/deductive)	4.06	1
Obtaining permit/approval from the municipality/different government authority	3.87	2
Material change and approval during the construction phase	3.83	3
Slowness of the owner's decision-making process	3.81	4
Time limitation in the design phase	3.72	5
Lack of communication and coordination between parties during construction	3.7	6
Poor financing by the owner	3.69	7
Inadequate early planning of the project	3.67	8
Poor preparation and approval of drawings	3.65	9
Underestimation and incompetence of contractors	3.63	10
Low financing by the contractor during construction	3.59	11
Unsuitable leadership style of construction/project manager	3.54	12
Failure to make interim awards on extensions of time and compensating by the owner	3.52	13
Shortage and unproductive manpower	3.48	14
Modifying legislation and regulations	3.46	15
Poor supervision and site management	3.44	16
Inadequate or incomplete technical plans/specification	3.43	17
Poorly written contracts	3.35	18
Contradictory and inaccurate information in the contract documents	3.26	19
Poor design	3.17	20
Contract amendments	3.17	21
Differing site conditions	3.13	22
Inadequate site investigation	3.11	23
Poorly defined scope of work	3.11	24
Inappropriate weather conditions	2.8	25
Impact on locality in terms of noise, traffic, and pollution/contamination	2.76	26
Impact of local cultures and social values in the settlement of conflicts	2.72	27

The highest ranked dispute is "variations initiated by the owner", with a weighted average of 4.06. Most owners have limited or very little knowledge about planning, engineering and management of construction projects. This lack of knowledge frequently leads to multiple variations throughout the construction process. Another reason could be the typical rush through design and construction to complete the project. The second ranked dispute is "obtaining permit/approval from the municipality/different government authority", with a weighted average of 3.87. Obtaining approvals consume a lot of time and leads to delays in project completion. Due to the construction volume in the UAE, obtaining approval may take longer than expected. These delays usually result in disputes. The third ranked dispute is "material change and approval during the construction phase", with a weighted average of 3.83. The owners and the consultants typically initiate changes in material specifications during construction, causing delays in additional negotiation and approval. These kinds of delays give rise to disputes.

The fourth highest ranked dispute is "slowness of the owner's decision-making process", with a weighted average of 3.81. Such actions halt the contractor's production process and result in disputes. The fifth highest ranked dispute is "time limitation in the design phase", with a weighted average of 3.72. If the designer is limited by time, the quality of design will be adversely affected. This lack of quality contributes to disputes. The weighted average of the top five disputes range from 4.06 to 3.72, which is not very wide, indicating a consensus among the respondents. In addition, the range indicates how these causes of disputes occur in the UAE with a high frequency.

Figure 1 shows the weighted average of each of the five dispute categories. The numbers represent the average of all causes of disputes within the category. The highest is the 'owner', followed by the designer, contractor, contractual and other categories.

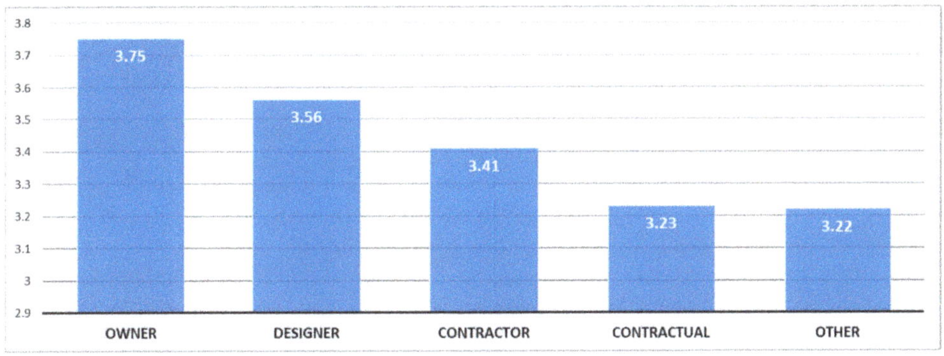

Figure 1. Weighted average of dispute categories in the UAE construction industry.

5.2. Assessment of the Dispute Avoidance and Resolution Methods

Based on the survey results, the effectiveness of each dispute avoidance and resolution method is determined as a weighted average of the responses. The results are summarized in Table 5.

Negotiation is found to be the most effective method in dispute avoidance, followed by risk allocation, early non-binding neutral evaluation and, finally, partnering. Negotiation is an effective avoidance method because of its ease of application. Construction organizations usually meet before the occurrence of potential disputes or issues to negotiate possible actions that can be taken before they happen. In addition, the negotiation method did not receive 'not applicable' response in the survey, meaning that it should be considered as an applicable method according to the respondents. Risk Allocation, Early Non-Binding Neutral Evaluation and Partnering methods received 'not applicable' responses by from some respondents. Partnering is the least effective method, due to the fact that not all parties are prepared to collaborate to resolve a dispute that has not occurred.

Table 5. Assessment of dispute avoidance and resolution methods.

Method	Weighted Average
Dispute Avoidance Methods	
Negotiation	4.15
Risk Allocation	3.5
Early Non-Binding Neutral Evaluation	3.3
Partnering	3.09
Early Resolution Methods	
Negotiation	4
Conciliation	3.59
Mini-Trial/Executive Tribunal	2.94
Late Resolution Methods	
Negotiation	3.72
Arbitration	3.31
Mediation	3.28
Litigation	3.28
Adjudication	2.87
Dispute Review Board	2.70

Negotiation was also found to be the most effective method as an early resolution method due to its ease and flexibility in reaching a resolution to the dispute at hand. (Normally, these kinds of negotiations are formal. Informal negotiations usually precede formal ones, to facilitate them.) Conciliation came in second place. Consulting with a neutral third party is favored because a neutral opinion on the matter of dispute is valued by the disputants. Finally, the 'Mini-trial' came in last because of reluctance to involve the court system and lawyers to solve the disputes. The involvement of lawyers and the court is always the last resort and the least preferred option to solve a dispute. However, if necessary, it is used.

As for the late resolution methods, the following methods are ranked from the most effective to the least effective, respectively, as Negotiation, Arbitration, Litigation Mediation, Adjudication, and Dispute Review Boards. Negotiation was shown to be an extremely effective method to avoid and resolve disputes, due to its ease of use, lack of complications, lack of involvement of other parties, and quickness. However, sometimes it is inapplicable depending on the severity of the issues causing the disputes. The second most effective method found is Arbitration, a method to solve disputes without the involvement of the court by consulting one or more arbitrators to determine what to do with the dispute at hand. This method helps to reach a solution faster than involving the court. The third most effective method is Litigation, in which is the court is involved to resolve the disputes according to the laws and regulations of the country. This method can be costly and time-consuming.

5.3. Comparative Analysis

The data were analyzed based on the different respondents' perspectives. The Spearman Rank Correlation Coefficient (RHO) was used to compare the resulting rankings. Table 6 shows the comparative results of the different categories. All the results are significant at the 0.01 level (two-tailed), with the exception of the contractor vs. consultant (in terms of resolution methods), which was significant at the 0.05 level (2-tailed). The Spearman Rank Correlation Coefficient (RHO) values were all positive and show strong agreement on the rankings of the causes of disputes and the dispute resolution methods.

Table 6. Comparative analysis.

Categories	Spearman Rank		
	RHO	**Pval**	**Significance**
Causes of Disputes			
Contractors vs. Consultants	0.723	0.000	Significant at the 0.01 level (2-tailed)
Years of Experience (0–10 vs. >10 years)	0.757	0.000	Significant at the 0.01 level (2-tailed)
Size (0–200M vs. >200M AED)	0.72	0.000	Significant at the 0.01 level (2-tailed)
Local vs. International	0.754	0.000	Significant at the 0.01 level (2-tailed)
Resolution Methods			
Contractors vs. Consultants	0.681	0.011	Significant at the 0.05 level (2-tailed)
Years of Experience (0–10 vs. >10 years)	0.871	0.000	Significant at the 0.01 level (2-tailed)
Size (0–200M vs. >200M AED)	0.83	0.000	Significant at the 0.01 level (2-tailed)
Local vs. International	0.691	0.009	Significant at the 0.01 level (2-tailed)

Table 7 summarizes the results from the contractor and consultant perspectives.

Table 7. Sources of disputes—comparative results (contractors vs. consultants).

Sources of Disputes	Contractor		Consultant	
	Average	**Rank**	**Average**	**Rank**
Time limitation in the design phase	3.72	6	3.72	8
Poor design	3.21	20	3.12	23
Inadequate or incomplete technical plans/specification	3.34	17	3.52	13
Poor preparation and approval of drawings	3.83	4	3.44	14
Material change and approval during the construction phase	3.9	1	3.76	6
Slowness of the owner's decision-making process	3.79	5	3.84	2
Inadequate early planning of the project	3.66	8	3.68	10
Failure to make interim awards on extensions of time and compensating by the owner	3.62	10	3.4	15
Variations initiated by the owner (additive/deductive)	3.9	2	4.24	1
Poor financing by the owner	3.55	13	3.84	3
low financing by the contractor during construction	3.48	14	3.72	9
Shortage and unproductive workers	3.59	12	3.36	16
Inadequate site investigation	3.07	23	3.16	21
Poorly defined scope of work	2.97	24	3.28	17
Poor supervision and site management	3.31	18	3.6	11
Unsuitable leadership style of construction/project manager	3.48	15	3.6	12
Underestimation and incompetence of contractors	3.48	16	3.8	5
Poorly written contracts	3.62	11	3.04	24
Differing site conditions	3.1	22	3.16	22
Contract amendments	3.14	21	3.2	19
Contradictory and inaccurate information in the contract documents	3.28	19	3.24	18
Obtaining permit/approval from the municipality/different government authority	3.9	3	3.84	4
Modifying legislation and regulations	3.69	7	3.2	20
Inappropriate weather conditions	2.97	25	2.6	27
Impact on locality in terms of noise, traffic, and pollution/contamination	2.83	26	2.68	25
Lack of communication and coordination between parties during construction	3.66	9	3.76	7
Impact of local cultures and social values in the settlement of conflicts	2.76	27	2.67	26

6. Discussion

Construction is a complex process. There are several factors contributing to this complexity. One of the major factors is the interaction between different entities, such as the owner (client), the designers (architect/engineer) and the constructors (contractor/subcontractors) with conflicting objectives. Identification of the causes of disputes and mitigation (avoidance and resolutions) methods in the context of UAE is the subject of this paper. It is not too difficult to appreciate the fact that disputes in construction, where money is involved in great amounts, is inevitable. It is also not difficult to come up with a list of the main causes of disputes. In this paper, however, the causes of disputes and methods of mitigation were identified and ranked in order of their frequency, as perceived by the construction professionals in UAE. Professionals were drawn from the main entities, owners, designers and constructors, and a survey was conducted. The results of the survey, as presented and discussed here, are the main contributions of this paper and are expected to enrich the body of knowledge on this subject.

As for the owners-related disputes, the highest weighted average is 4.06 for "variations initiated by the owner". As for designer-related disputes, the highest weighted average was 3.83 for "Material change and approval during the construction phase". The change in material occurs frequently, which can be initiated by any of the three main parties. Contractors can sometimes change materials in order to reduce cost; the owner, on the other hand, can do the same for decorative or aesthetic reasons or cost reduction. Sometimes, the specified material may not be available, or the market price of a specific material may become inhibitive and a replacement may become necessary. As for contractor-related disputes, the highest weighted average is 3.63 for "underestimation and incompetence of contractors". This cause of dispute is a consequence of contractors underestimating the scope and requirement of the job. It may even be due to the inability and lack of competency of the contractor to do the job. On the other hand, the lowest weighted average obtained was 3.11 for both "inadequate site investigation" and "poorly defined scope of work".

As for contractual-related disputes, the highest ranked source of dispute is "poorly written contract" by a weighted average of 3.35 and the "differing site condition" is ranked lowest, with 3.13. A contract is written between the parties, and details the rules and conditions for the construction project. A poorly written contract causes disputes between the parties, as the project is based on a weak agreement, leading to miscommunication between the parties. As for 'other' disputes, the highest rank is 3.87, for "obtaining permit/approval from municipality". The government regulations in obtaining approval vary in complexity and time requirement depending on jurisdictions. Often, the government regulations follow bureaucratic procedures for authorization of the approval required to proceed, thus causing unpredictable delays.

According to the survey results, the negotiation method is most recommended, whether it is during the avoidance, the early resolution or the late resolution phases. This is consistent with other studies [7]. A total of 90 to 95% of the construction claims are solved by the method of negotiation in the construction industry [7]. Negotiation is the number one process that is usually used when contractors try to deal with construction claims. If negotiation were not applicable, then it would be preferable to go with the risk allocation method during the avoidance phase, and the Conciliation method would be recommended during the early resolution phase. Regarding the late resolution phase, either the arbitration or mediation method, as these two methods both obtained the second highest responses.

The findings of this paper are expected to increase awareness in the UAE construction industry about the roles each major entity, the owners, the designers and the contractors, play in a construction project in relation to disputes and the methods of their mitigation. Their improved understanding of the causes of disputes is expected to have a positive impact by reducing the effects of disputes in construction.

7. Conclusions

At present, the UAE is one of the most vibrant countries as far as the construction sector is concerned. As a result, it is confronted with multiple issues including a significant number of construction disputes. This paper identifies the main causes of disputes in the UAE in terms of their frequency of occurrence. In addition, the paper identifies the main dispute avoidance and resolution methods and presents a comparative analysis of their effectiveness. The disputes are categorized in five different groups based on their sources. These are design, owner, contractor, contractual, and other. Owner-related causes are found to be the most predominant, as changes and modifications in the scope of the project and time taken by them to make decisions usually become grounds for dispute.

Mitigation techniques are investigated in three different phases of a construction project based on their appropriateness and effectiveness. They are avoidance, early resolution and late resolution. Negotiation is found to be the most effective at all three stages. As far as the resolution methods are considered, conciliation, arbitration, mini-trial, and the use of Dispute Review Boards are found to be applied at various degrees. Litigation or settlement in a court is found to be the least desired by all entities.

Author Contributions: Conceptualization, S.E.-S. and I.A.; methodology, S.E.-S. and I.A.; validation, M.A., R.H., S.M. and O.E.-A.; formal analysis, M.A., R.H., S.M. and O.E.-A.; investigation, M.A., R.H., S.M. and O.E.-A.; resources, M.A., R.H., S.M. and O.E.-A.; data curation, M.A., R.H., S.M. and O.E.-A.; writing—original draft preparation, M.A., R.H., S.M. and O.E.-A.; writing—review and editing, S.E.-S. and I.A.; visualization, S.E.-S.; supervision, S.E.-S. and I.A.; project administration, S.E.-S. and I.A.; funding acquisition, S.E.-S. and I.A. All authors have read and agreed to the published version of the manuscript.

Funding: The work in this paper was supported, in part, by the Open Access Program from the American University of Sharjah. The APC was funded by grant number [OAP-CEN-091].

Conflicts of Interest: The authors declare no conflict of interest. The funders had no role in the design of the study; in the collection, analyses, or interpretation of data; in the writing of the manuscript, or in the decision to publish the results. This paper represents the opinions of the authors and does not mean to represent the position or opinions of the American University of Sharjah.

References

1. Viswanathan, A.; Panwar, A.; Kar, S.; Lavingiya, R.; Jha, K. Causal Modeling of Disputes in Construction Projects. *J. Leg. Aff. Disput. Resolut. Eng. Constr.* **2020**, *12*, 04520035. [CrossRef]
2. Jagannathan, M.; Delhi, V. Litigation proneness of dispute resolution clauses in construction contracts. *J. Leg. Aff. Disput. Resolut. Eng. Constr.* **2019**, *11*, 04519011. [CrossRef]
3. Patil, S.; Iyer, K.; Chaphalkar, N. Influence of extrinsic factors on construction arbitrators' decision making. *J. Leg. Aff. Disput. Resolut. Eng. Constr.* **2019**, *11*, 04519021. [CrossRef]
4. Kisi, K.; Lee, N.; Kayastha, R.; Kovel, J. Alternative dispute resolution practices in international road construction contracts. *J. Leg. Aff. Disput. Resolut. Eng. Constr.* **2020**, *12*, 04520001. [CrossRef]
5. Awwad, R.; Barakat, B.; Menassa, C. Understanding dispute resolution in the Middle East region from perspectives of different stakeholders. *J. Manag. Eng.* **2016**, *32*, 05016019. [CrossRef]
6. Arditi, D.; Pattanakitchamroon, T. Analysis methods in time-based claims. *J. Constr. Eng. Manag.* **2008**, *134*, 242–252. [CrossRef]
7. Yates, J. The art of negotiation in construction contract disputes. *J. Leg. Aff. Disput. Resolut. Eng. Constr.* **2011**, *3*, 94–96. [CrossRef]
8. Mitropoulos, P.; Howell, G. Model for understanding, preventing, and resolving project disputes. *J. Constr. Eng. Manag.* **2001**, *127*, 223–231. [CrossRef]
9. Haugen, T.; Singh, A. Dispute resolution strategy selection. *J. Leg. Aff. Disput. Resolut. Eng. Constr.* **2015**, *7*, 05014004. [CrossRef]
10. Artan, D.; Bakioglu, G. Modeling the relationship between risk and dispute in subcontractor contracts. *J. Leg. Aff. Disput. Resolut. Eng. Constr.* **2018**, *10*, 04517022. [CrossRef]
11. Cheung, S.; Pang, K. Anatomy of construction disputes. *J. Constr. Eng. Manag.* **2013**, *139*, 15–23. [CrossRef]
12. Dubai Chamber of Commerce Study, Dubai, December 2012. Available online: https://www.dubaichamber.com/whats-happening/chamber_news/construction-sector-to-contribute-11-1-to-uae%E2%80%99s-gdp-in-2015 (accessed on 24 September 2020).
13. Martin, G.; Thompson, A. Effective management of construction dispute resolution. *J. Leg. Aff. Disput. Resolut. Eng. Constr.* **2011**, *3*, 67–70. [CrossRef]
14. Cheung, S.; Yiu, T.; Yeung, S. A study of styles and outcomes in construction dispute negotiation. *J. Constr. Eng. Manag.* **2006**, *132*, 805–814. [CrossRef]
15. Zaneldin, E. Construction claims in United Arab Emirates: Types, causes, and frequency. *Int. J. Proj. Manag.* **2006**, *24*, 453–459. [CrossRef]
16. Ho, S.; Liu, L. Analytical model for analyzing construction claims and opportunistic bidding. *J. Construct. Eng. Manag.* **2004**, *130*, 94–104. [CrossRef]
17. Al Mousli, M.; El-Sayegh, S. Assessment of the design–construction interface problems in the UAE. *Archit. Eng. Des. Manag.* **2016**, *12*, 353–366. [CrossRef]
18. Arain, F.; Pheng, L.; Assaf, S. Contractors' views of the potential causes of inconsistencies between design and construction in Saudi Arabia. *J. Perform. Constr. Facil.* **2006**, *20*, 74–83. [CrossRef]

19. Arain, F.; Sui, P.; Assaf, S. Consultant's prospects of the sources of design and construction interface problems in large building projects in Saudi Arabia. *Environ. Des. Sci.* **2007**, *5*, 15–37. [CrossRef]
20. Lopez, R.; Love, P.E. Design error costs in construction projects. *J. Constr. Eng. Manag.* **2012**, *138*, 585–593. [CrossRef]
21. Love, P.E.; Cheung, S.O.; Irani, Z.; Davis, P.R. Causal discovery and inference of project disputes. *IEEE Trans. Eng. Manag.* **2011**, *58*, 400–411. [CrossRef]
22. Love, P.E.; Lopez, R.; Kim, J.; Kim, M. Influence of organizational and project practices on design error costs. *J. Perform. Constr. Facil.* **2014**, *28*, 303–310. [CrossRef]
23. Gad, G.; Kalidindi, S.; Shane, J.; Strong, K. Analytical framework for the choice of dispute resolution methods in international construction projects based on risk factors. *J. Leg. Aff. Disput. Resolut. Eng. Constr.* **2011**, *3*, 79–85. [CrossRef]
24. Faridi, A.; El-Sayegh, S. Significant factors causing delay in the UAE construction industry. *Constr. Manag. Econ.* **2006**, *24*, 1167–1176. [CrossRef]
25. Semple, C.; Hartman, F.; Jergeas, G. Construction claims and disputes: Causes and cost/time overruns. *J. Constr. Eng. Manag.* **1994**, *120*, 785–795. [CrossRef]
26. Mehany, M.; Grigg, N. Causes of road and bridge construction claims: Analysis of Colorado Department of Transportation projects. *J. Leg. Aff. Disput. Resolut. Eng. Constr.* **2014**, *7*, 04514006. [CrossRef]
27. Bu-Bshait, K.; Manzanera, I. Claim management. *Int. J. Proj. Manag.* **1990**, *8*, 222–228. [CrossRef]
28. Tanaka, T. Analysis of Claims in U.S. Construction Projects. Master's Thesis, Department of Civil Engineering, Massachusetts Institute of Technology, Cambridge, MA, USA, 1988.
29. Farooqui, R.; Azhar, S.; Umer, M. Key causes of disputes in the Pakistani construction industry—Assessment of trends from the viewpoint of contractors. In Proceedings of the 50th ASC International Conference Proceedings, Washington, DC, USA, 26–28 March 2014.
30. Jergeas, G. Claims and disputes in the construction industry. *AACE Int. Trans.* **2001**, C.D. 31.
31. Rhys, J. How constructive is construction law? *Constr. Law J.* **1994**, *10*, 28–38.
32. Mishmish, M.; El-Sayegh, S. Causes of claims in road construction projects in the UAE. *Int. J. Constr. Manag.* **2016**, *18*, 26–33. [CrossRef]
33. Al-Dubaisi, A. Change Orders in Construction Projects in Saudi Arabia. Master's Thesis, King Fahd University of Petroleum and Minerals, Dhahran, Saudi Arabia, 2000.
34. Weshah, N.; Ghandour, W.; Jergeas, G.; Falls, L. Factor analysis of the interface management (IM) problems for construction projects in Alberta. *Can. J. Civ. Eng.* **2013**, *40*, 848–860. [CrossRef]
35. Dumont, P.; Gibson, G.; Fish, J. Scope management using project definition rating index. *J. Manag. Eng.* **1997**, *13*, 54–60. [CrossRef]
36. Ling, F.; Poh, B. Problems encountered by owners of design–build projects in Singapore. *Int. J. Proj. Manag.* **2008**, *26*, 164–173. [CrossRef]
37. El-Sayegh, S. Risk assessment and allocation in the UAE construction industry. *Int. J. Proj. Manag.* **2008**, *26*, 431–438. [CrossRef]
38. Al-Hammad, A.; Assaf, S. Design–construction interface problems in Saudi Arabia. *Build. Res. Inf.* **1992**, *20*, 60–63. [CrossRef]
39. Elmualim, A.; Gilder, J. BIM: Innovation in design management, influence and challenges of implementation. *Archit. Eng. Des. Manag.* **2014**, *10*, 183–199. [CrossRef]
40. Al-Hammad, A. Common interface problems among various construction parties. *J. Perform. Constr. Facil.* **2000**, *14*, 71–74. [CrossRef]
41. Bristow, D.; Vasilopoulos, R. The new CCDC 2: Facilitating dispute resolution of construction projects. *Constr. Law J.* **1995**, *11*, 95–117.
42. Chow, L.; Ng, S. Delineating the performance standards of engineering consultants at design stage. *Constr. Manag. Econ.* **2010**, *28*, 3–11. [CrossRef]
43. Assaf, S.; Al-Hammad, A. The effect of economic changes on construction cost. *Am. Assoc. Cost Eng. Trans.* **1988**, 63–67.
44. Scott, S.; Harris, R. United Kingdom construction claims: Views of professionals. *J. Constr. Eng. Manag.* **2004**, *130*, 734–741. [CrossRef]
45. Wong, P.; Maric, D. Causes of disputes in construction planning permit applications. *J. Leg. Aff. Disput. Resolut. Eng. Constr.* **2016**, *8*, 04516006. [CrossRef]

46. Chester, M.; Hendrickson, C. Cost impacts, scheduling impacts, and the claims process during construction. *J. Constr. Eng. Manag.* **2005**, *131*, 102–107. [CrossRef]

47. Steen, R. Five steps to resolving construction disputes—Without litigation. *J. Manag. Eng.* **1994**, *10*, 19–21. [CrossRef]

48. Cheung, S.; Suen, H.; Ng, S.; Leung, M. Convergent views of neutrals and users about alternative dispute resolution. *J. Manag. Eng.* **2004**, *20*, 88–96. [CrossRef]

49. Tanielian, A. Arbitration still best road to binding dispute resolution. *J. Leg. Aff. Disput. Resolut. Eng. Constr.* **2013**, *5*, 90–96. [CrossRef]

50. Musonda, H.; Muya, M. Construction dispute management and resolution in Zambia. *J. Leg. Aff. Disput. Resolut. Eng. Constr.* **2011**, *3*, 160–169. [CrossRef]

51. Treacy, T. Use of alternative dispute resolution in the construction industry. *J. Manag. Eng.* **1995**, *11*, 58–63. [CrossRef]

52. Fenn, P.; Lowe, D.; Speck, C. Conflict and dispute in construction. *Constr. Manag. Econ.* **1997**, *15*, 513–518. [CrossRef]

53. El-adaway, I.; Kandil, A. Multi-agent system for construction dispute resolution (MAS-COR). *J. Constr. Eng. Manag.* **2010**, *136*, 303–315. [CrossRef]

54. Kelleher, T.; Walters, G. *Smith, Currie and Hancock's Common-Sense Construction Law: A Practical Guide for the Construction Professional*, 4th ed.; Wiley: Hoboken, NJ, USA, 2009.

55. Thomas, H.; Heuer, D.; Filippelli, R. Settlement of construction jurisdictional disputes. *J. Constr. Eng. Manag.* **1984**, *110*, 165–177. [CrossRef]

MDPI
St. Alban-Anlage 66
4052 Basel
Switzerland
www.mdpi.com

Buildings Editorial Office
E-mail: buildings@mdpi.com
www.mdpi.com/journal/buildings